On Growth and Form

WILEY SERIES IN MATHEMATICAL AND COMPUTATIONAL BIOLOGY

CHAPLAIN/SINGH/MCLACHLAN—On Growth and Form: Spatio-temporal Pattern Formation in Biology

Reflecting the rapidly growing interest and research in the field of mathematical biology, this outstanding new book series examines the integration of mathematical and computational methods into biological work. It also encourages the advancement of theoretical and quantitative approaches to biology, and the development of biological organisation and function.

The scope of the series is broad, ranging from molecular structure and processes to the dynamics of ecosystems and the biosphere, but unified through evolutionary and physical principles, and the interplay of processes across scales of biological organisation.

Topics to be covered in the series include:

- Cell and molecular biology
- Functional morphology and physiology
- Neurobiology and higher function
- Immunology
- Epidemiology
- Ecological and evolutionary dynamics of interacting populations

A fundamental research tool, the *Wiley Series in Mathematical and Computational Biology* provides essential and invaluable reading for biomathematicians and development biologists, as well as graduate students and researchers in mathematical biology and epidemiology.

On Growth and Form

Spatio-temporal Pattern Formation in Biology

Edited by

M. A. J. Chaplain

University of Dundee, UK

G. D. Singh

University of Dundee, UK

J. C. McLachlan

St. Andrews University, UK

JOHN WILEY & SON, LTD

Chichester • New York • Weinheim • Brisbane • Singapore •Toronto

Baffins Lane, Chichester,
West Sussex, PO19 1UD, England

National 01243 779777
International (+44) 1243 779777

e-mail (for orders and customer service enquiries): cs-books@wiley.co.uk

Visit our Home Page on http://www.wiley.co.uk or http://www.wiley.com

Other Wiley Editorial Offices

John Wiley & Sons, Inc., 605 Third Avenue,
New York, NY 10158-0012, USA

Weinheim • Brisbane • Singapore • Toronto

Library of Congress Cataloging-in-Publication Data

On growth and form : spatiotemporal pattern formation in biology / edited by M.A.J. Chaplain, G.D. Singh, J.C. McLachlan.
p. cm. — (Wiley series in mathematical and computational biology)
Includes bibliographical references and index.
ISBN 0-471-98451-5 (alk. paper)
1. Pattern formation (Biology). 2. Growth. I. Series. II. Chaplain, M. A. J. III. McLachlan, J. C. IV. Gurdev Singh, 1944-
QH491 O5 1999 99-035919
571.3 21—dc21

British Library Cataloguing in Publication Data

A catalogue record for this book is available from the British Library

ISBN 0 471 98451 5

Produced from PostScript files supplied by the Editor
Printed and bound in Great Britain by Antony Rowe Ltd, Chippenham
This book is printed on acid-free paper responsibly manufactured from sustainable forestry in which at least two trees are planted for each one used for paper production.

To Fiona
with love and thanks

ἀγεωμέτρητος μηδεὶς εἰσίτω

Contents

Foreword

Much has been written about D'Arcy Wentworth Thompson and his famous book, including an excellent biography by his daughter (Ruth D'Arcy Thompson, 1958, *D'Arcy Wentworth Thompson*, Oxford University Press). *On Growth and Form* was first published in 1917 and now this volume, stemming from a conference on current work, honors that book 82 years later. It is really quite a remarkable record of publishing endurance, and while it cannot compete with Shakespeare or the Bible, in all fairness, they had a head start.

I have asked before, and I ask again, how can one explain this extraordinary longevity of his great work. The answer is not to be found solely in the increased interest today among biologists in the use of mathematics for understanding living phenomena, although that is certainly an important factor. But there are other reasons as well. In 1961 Cambridge University Press sent me a thick package of reviews, and while the majority were from newspapers and the usual biological publications, there were reviews in specialized journals for psychiatrists, veterinarians, and orthodontists! But most interesting to me has been the fascination in the book by the art world: painters, sculptors, crafts workers of all kinds, and architects. It was even listed and highly recommended in the Whole Earth catalogue, that symbol of the revolutionary days of the 1960s and 1970s. There is something about form in nature that has a universal appeal.

I suspect that one of the most important reasons for its continued popularity is the beautiful prose. It is a sheer pleasure to read. His style has few equals; there is poetry in his sentences. Even when he says things that are no longer true, the reading of them still stands.

It is hard to imagine, however, that this is the reason biologists still hold him in such high esteem. That I think comes more from his vision of how mathematics and physics can contribute to the form of living creatures. Nowadays, with the great flowering of molecular developmental biology, there is an increasing awareness that even though genes and the proteins they produce are enormously important, they alone cannot account for the growth and development of animals and plants. One must also consider the physical forces – which can be expressed in simple mathematical terms – for no matter how clever are the activities of genes, they do not work in a vacuum. The substances which they create, and the environments in which they are created, are subject to physical laws and constraints that tell us much about the forms that are produced. Only part of the show is in the genes; the rest has to do with the properties of matter. Even though D'Arcy Thompson dismissed genetics, he more than anyone

On Growth and Form: Spatio-temporal Pattern Formation in Biology.
Editor Mark A.J. Chaplain, G.D. Singh, J.C. McLachlan

began our thinking about the other part of the equation, the appreciation of which is almost more relevant today than it was in 1917. So his book and his ideas survive with vigor because molecular developmental genetics has exposed what else is needed.

I became intimately familiar with the book when I rashly agreed to do an abridged edition in 1960. It was an enormously difficult job, largely because I knew I was tampering with an acknowledged classic. I was determined not to alter the prose, but somehow I had to retain the book's continuity while removing large sections. Once I started I soon saw my difficulties: it was like being asked to reduce *Hamlet* to a third of its length without rewriting a word. My problem was helped a bit, adding commentary here and there, but nevertheless I knew I was playing with fire. As one might expect, the reaction was mixed. Many people, especially those to whom the book was new, actually read the abridged version whereas they might have been overawed by the huge 1942 edition. At the same time many scholars, especially mathematicians, were furious at my presumption. A distinguished colleague in the Mathematics Department at Princeton, whom I thought might be sympathetic, told me angrily that I had committed an unpardonable sin by removing most of the equations. Since it was published in 1961 I have on two occasions given a lecture at St Andrews University on my experimental work with cellular slime molds, and both times I was introduced in a funereal tone as the man who abridged D'Arcy Thompson's *On Growth and Form*. Their great professor could manage quite well on his own – my help was not needed.

Clearly they have it right. The mere fact that an international conference took place in his honor is evidence that his influence continues today. Biologists and mathematicians interested in living forms remain inspired by his message. Here is the essence of that message in his own words: "Cell and tissue, shell and bone, leaf and flower, are so many portions of matter, and it is in obedience to the laws of physics that their particles have been moved, moulded and conformed. ... Their problems of form are in the first instance mathematical problems, their problems of growth are essentially physical problems, and the morphologist is, ipso facto, a student of physical science."

John Tyler Bonner

Series Preface

Theoretical biology is an old subject, tracing back centuries. At times, theoretical developments have represented little more than mathematical exercises, making scant contact with reality. At the other extreme have been those works, such as the writings of Charles Darwin, or the models of Watson and Crick, in which theory and fact are intertwined, mutually nourishing one another in inseparable symbiosis. Indeed, one of the most exciting developments in biology within the last quarter-century has been the integration of mathematical and theoretical reasoning into all branches of biology, from the molecule to the ecosystem. It is such a unified theoretical biology, blending theory and empiricism seamlessly, that has inspired the development of this series.

This series seeks to encourage the advancement of theoretical and quantitative approaches to biology, and to the development of unifying principles of biological organization and function, through the publication of significant monographs, textbooks and synthetic compendia in mathematical and computational biology. The scope of the series is broad, ranging from molecular structure and processes to the dynamics of ecosystems and the biosphere, but unified through evolutionary and physical principles, and the interplay of processes across scales of biological organization.

The principal criteria for publication beyond the intrinsic quality of the work are substantive biological content and import, and innovative development or application of mathematical or computational methods. Topics will include but not be limited to cell and molecular biology, functional morphology and physiology, neurobiology and higher function, immunology and epidemiology, and the ecological and evolutionary dynamics of interacting populations. The most successful contributions, however, will not be so easily categorized, crossing boundaries and providing integrative perspectives that unify diverse approaches; the study of infectious diseases, for example, ranges from the molecule to the ecosystem, involving mechanistic investigations at the level of the cell and the immune system, evolutionary perspectives as viewed through sequence analysis and population genetics, and demographic and epidemiological aspects at the level of the ecological community.

The objective of the series is the integration of mathematical and computational methods into biological work; the volumes published hence should be of interest both to fundamental biologists and to computational and mathematical scientists, as well as to the broad spectrum of interdisciplinary researchers that comprise the continuum connecting these diverse disciplines.

Simon Levin, 1999

On Growth and Form: Spatio-temporal Pattern Formation in Biology.
Editor Mark A.J. Chaplain, G.D. Singh, J.C. McLachlan

Editors' Preface

Formam virumque cano, Edinae qui primus ab arce
Dundoniam fato profugus Taiaque venit
Litora, ...

Our book is concerned with growth and form, and with the underlying associated spatio-temporal processes; but it is also concerned with the contribution to this area of one man, D'Arcy Wentworth Thompson (1860-1948). Having left Edinburgh Academy with prizes in classics, modern languages and mathematics, he started a medicine course at Edinburgh University and then changed to study science at Cambridge. He graduated with a B.A. in Zoology in 1883 and the following year in 1884 D'Arcy Thompson was appointed Professor of Biology at University College Dundee. For thirty-three years he remained there, completing his magnum opus *On Growth and Form* in 1917. That same year he was appointed to the Chair of Natural History in St Andrews where he spent the remaining thirty years of his life.

The contents of this volume have their origins in the plenary talks given at the D'Arcy Thompson Meeting, "On Growth and Form: Spatio-temporal Pattern Formation in Biology", which took place at the West Park Conference Centre, University of Dundee, 20-24 September 1998. Before discussing the book and its contents we must mention the meeting out of which this book arose.

Around 100 participants took part in the meeting from a number of scientific disciplines and backgrounds – theoretical and experimental; biologists, mathematicians and statisticians. It turned out to be a unique and "timely" event, given that part of the reason for holding the meeting was to mark the 50th anniversary of D'Arcy's death. There was an atmosphere of genuine open discussion and debate and in addition to the plenary talks around 50 poster presentations were made on the themes of the meeting: On the Growth of Form; On the Mathematics of Form; On the Dynamics of Form; On the Description of Form; On the Form of Form. Concerning the organisation and running of the meeting thanks are due to British Society of Developmental Biology, the Wellcome Trust, the London Mathematical Society, the Royal Bank of Scotland, SmithKline Beecham (in particular Dr Frank Tobin), the Society for Mathematical Biology, the European Society for Mathematical and Theoretical Biology and the Edinburgh Mathematical Society for generous financial sponsorship at the meeting. Thanks also go to Drs Fordyce Davidson, Sandy Anderson and Danny Feltham for their help and support in the day-to-day running and administration of the meeting.

Each and every participant must be thanked for fully contributing to and taking part

On Growth and Form: Spatio-temporal Pattern Formation in Biology.
Editor Mark A.J. Chaplain, G.D. Singh, J.C. McLachlan

in all the discussion sessions which made the meeting so successful. The wholehearted participation generated a warm and collegial atmosphere, showing that *ἀνθρωπος φύσει πολιτικὸν ζῷον* – man is by nature a social being.

Thanks are also due to all the plenary speakers firstly for giving their talks and of course secondly for their collaboration and participation in submitting their chapters for this book. Each chapter was prepared with great care to a very high standard, leaving a minimal amount of work for the task of editing.

Although this book developed out of a meeting, it should in no way be viewed as simply another "conference proceedings". The scope of the book, like D'Arcy, is broad, ranging from genetics and developmental biology, through the dynamics of spatio-temporal systems to physiology and morphometrics, unified through evolutionary and physical principles, and the interplay of processes across scales of biological organization from the cell and molecular biology up to functional morphology. The book contains a substantive biological content and an overview of the state-of-the-art in the mathematical modelling of spatio-temporal systems. The book as a whole is thus not so easily categorized, crossing boundaries and providing integrative perspectives that unify diverse approaches. Having read the book, perhaps we may come to conclusion of D'Arcy himself:

"... something of the use and beauty of mathematics I think I am able to understand. I know that in the study of material things number, order, and position are the threefold clue to exact knowledge: and that these three, in the mathematician's hands, furnish the first outlines for a sketch of the Universe."

And so coming full circle, it is fitting to end with a few words in Latin to the work and memory of D'Arcy Thompson, growth and form, biology and mathematics, and his continuing influence on mathematical biology to this day. Thanks are due to Dr Philip Burton, School of Greek, Latin and Ancient History, University of St Andrews who kindly provided the following lines:

Crescentes rerum cecinit qui pectore pleno
formas, naturae felixque recondita norat;
Scotia quem genuit; misere ex arce superba
Castra Puellarum, Taias qui venit ad oras
Dundoniae, posthac urbis decus Andreanae;
–Hunc canere incipio, cuius non gloria cedet
aequora dum pisces, volucres dumque aera tranant.

Mark A.J. Chaplain
September 1999

G.D. Singh
J.C. McLachlan

1

Problems of Development: The Microcosm and the Macrocosm

Jonathan M.W. Slack

Developmental Biology Programme, Department of Biology and Biochemistry, University of Bath, Bath BA2 7AY, UK

1.1 Introduction: The Problem of Regional Specification Is Largely Solved

As late as 1982, the mechanisms of embryonic development were almost as mysterious as they had been at the turn of the century. Massive progress in understanding the mechanism of heredity had not yet fed into the understanding of development, except to the extent that it was generally understood that different cell types would express different subsets of genes. No problem seemed more mysterious than that of regional specification, how one end of an apparently uniform field of cells could produce a head and the other end a tail. Today, 16 years later in 1998, we feel we more or less understand it. At least we feel that we are most of the way there with *Drosophila* and perhaps halfway there with the vertebrates. Although there are doubtless many more genes and signalling molecules to be discovered, and many more papers in *Cell and Development* to be published on this topic, it now seems unlikely that any more wholly new principles will emerge in this area.

As an example of the level of current knowledge, consider the anteroposterior patterning of *Xenopus* (Isaacs *et al.*, 1995; Pownall *et al.*, 1996; Isaacs *et al.*, 1998). This is drawn from our own work but it is typical of work on other problems in many labs around the world. It has been known for some time that the anteroposterior (AP) pattern in the central nervous system depends on a signal from the posterior end of the forming body. We now know that, during gastrulation, fibroblast growth factors

On Growth and Form: Spatio-temporal Pattern Formation in Biology.
Editor Mark A.J. Chaplain, G.D. Singh, J.C. McLachlan

(FGFs) are expressed around the closing blastopore, which becomes the posterior end as it closes. The FGFs are secreted from these cells, diffuse a few cell diameters, and stimulate the cells they contact to activate a group of transcription factors coded by *Cdx* genes. The Cdx factors in turn activate posterior *Hox* genes, that is *Hox* genes of paralogue groups 6 and above. These are normally expressed in the trunk–tail region, and are required for the regional patterning this part of the body. The Cdx proteins also repress genes such as *otx*, responsible for formation of the fore- and midbrain. It is possible to alter the specificity of transcription factors by swapping the normal activation domain for a repressor domain, or vice versa. If such modified genes or gene products are introduced into the embryos then the results in terms of deletions of heads or tails are exactly as predicted. So this is a mechanism that is quite well understood, and it is typical of much work in developmental biology in recent years.

Of course, some people will never consider that a problem is really understood and will always want more and more detail. But there comes a point beyond which more detail is just boring. There will always be details that we do not know, but if we understand a system well enough to design completely new genes, introduce them into the embryo, and get a predicted alteration in the course of development, then we understand it pretty well. This level of understanding exists for regional patterning of the body of *Drosophila* and *Xenopus*, and for the development of various organs such as the chick limb and the *C. elegans* vulva (Wolpert, 1998). Since there are now many thousands of people engaged in working out the remaining problems, they are bound to be solved to a similar precision within a few years.

I have attempted to generalise this level of understanding by asking whether it is possible to write a genetic program to build a whole animal. The answer is yes, at least superficially, and an exercise of this sort is described in the appendix of my book *From Egg to Embryo* (Slack, 1991). The animal was called the Crendonian snapper, as I was living in the Buckinghamshire village of Long Crendon at the time. It has an unusual anatomy, but is complex enough to be regarded as a real animal and can be built using just 14 genes.

So if we understand regional specification, at least in principle, do we understand development as a whole? I think not, because there are many other problems, lurking to a large extent in the shadows of the media spotlights, which need to be solved, and some of which still seem deeply puzzling at the moment. A closer look shows there are in fact major gaps in the snapper program that are not easily apparent when it is presented on paper, but would be very obvious if the snapper were to be constructed in reality. I shall try to draw attention to some of these problem areas today. For each of them it turns out that we do know a certain amount about the molecular biology of the situation (the microcosm), but we do not understand how the molecules bring about the macroscopic behaviours that are of interest to us (the macrocosm). Theoretical biology is at its best when it can provide the first possible solution to a mysterious problem. So these are all areas where an input from theoretical biology may prove worthwhile.

1.2 Morphogenetic Movements

Regional specification subdivides the embryo into differently determined regions that can behave in different ways. Among these behaviours, the movements of cells and tissues are of pre-eminent importance. They are often described by the term "morphogenesis", used in its narrower sense rather than the wider sense which embraces all of development. Morphogenetic movements are of several types which recur repeatedly in many different circumstances during development. For example:

- Cellular aggregation to form clumps, e.g. in cartilage formation.
- Formation and maintenance of boundaries by differential adhesion, e.g. for the different germ layers.
- Mesenchymal to epithelial transition to form cell sheets, e.g. in formation of coelomic epithelium.
- Transformation between cell sheets, tubes and vesicles, e.g. in kidney development.
- Epithelial to mesenchymal transition, e.g. in neural crest formation.
- Active elongation (convergent extension), driven by cellular intercalation, e.g. in anteroposterior axis elongation of CNS and notochord.
- Chemokinesis or chemotaxis of individual cells, e.g. in wound healing.

In this area the microcosm poses little problem. We think we know the molecular identity of most of the key players in the cytoskeleton, and have a qualitative view of how microfilaments grow and decay, and how motor proteins can operate. But molecular biologists tend to think only in terms of which genes are expressed in which cell or which proteins bind to each other. A proper understanding of morphogenetic movements obviously involves an explanation at the level of physical forces and their effects on cells and tissues. This is a difficult area as it involves both viscosity and elasticity, but it should provide rich pickings for mathematically minded biologists because there are clearly critical thresholds at which the morphogenetic behaviour jumps from one mode to another. A beginning was made some years ago by authors such as Odell *et al.* based on a stretch-activated contraction of cytoskeletal elements (Odell *et al.*, 1981), but more recently interest in this type of work seems to have faded and I am not aware of any current activity.

1.3 The French Flag Problem

Wolpert's famous French flag problem received a lot of attention from theoreticians in the 1970s (Wolpert, 1968). The problem was in two parts: firstly how can a field of cells become determined to differentiate into one- third red, one- third white and one-third blue cells? Secondly, if part of the pattern is removed, how can the 1:1:1 proportion of red white and blue be restored? Nobody seems very interested in the French flag problem today because it is perceived as having been solved. But actually only the first part relating to regional specification has been solved, the answer being that red, white and blue are defined by combinations of transcription factors that are activated by a gradient of a diffusible morphogen of a growth factor. The second part relating to proportion regulation has still not been solved.

This is partly because in the context of early development it is by no means clear that it is a real problem. The theoreticians of the 1970s made strenuous attempts to devise mechanisms that would be completely scale invariant, but actual data on phenomena such as the twinning of vertebrate embryos suggests that proportions do change when the overall size of the embryo is altered. In the preimplantation mouse embryo the change is considerable, and there is little or no proportion regulation

(Tarkowski and Wroblewska, 1967). In the chick there is so much extraembryonic tissue that a change of proportion on twinning would be hard to detect. In *Xenopus*, twins produced by separation of the first two blastomeres do show a demonstrable shift of their proportions away from the normal and towards the dorsoanterior (Cooke and Webber, 1985).

Although the French flag problem posed the questions of pattern and proportion as being intimately linked, they are in fact logically quite distinct. This is because there remains a problem of proportion control even when the arrangement or shape of the tissue masses is quite irregular and there is effectively no pattern. I wish to argue that the control of proportion is a real problem, not so much during early development, but rather during the late postembryonic stages, and that it is intimately connected to the problem of growth control.

Wolpert originally proposed two possible models for the French flag. One involved a gradient, not a simple gradient of the sort that is now familiar, but a gradient in which new sources and sinks could arise at newly created termini. This has been developed by Gierer and Meinhardt as their first class of model and can provide a successful explanation of proportion control in organisms such as *Hydra* and slime moulds (Meinhardt, 1982). The other involved the titration of a diffusible inhibitor. Roughly if the red cells secrete an inhibitor of red which diffuses across the whole flag, then its concentration depends on the proportion of the flag that is red. The red zone expands until the concentration of inhibitor is just sufficient to stop further expansion. This type of model is an extension of a concept proposed initially by Weiss (Weiss and Kavanau, 1957; Rose, 1957). It does not necessarily have any connection to spatial patterning, because it could operate in a situation where the red, white and blue zones were irregular or broken up. It does require that there is communication between the zones and it may have a relevance to postembryonic porportion control, considered further below.

1.4 Regeneration

The French flag problem represents only one of the issues raised by the biological phenomena of regeneration. There is a long-term confusion between regeneration, meaning the regrowth of a part that has been removed; and regulation, meaning the readjustment of a fate map across a domain of undifferentiated tissue. The extreme cases are clear: the formation of two approximately normal twins after separation of the first two blastomeres of an embryo is regulation whereas the regrowth of an amputated amphibian limb is regeneration. But there can be situations in which the distinction is less clear. For example, it is not entirely clear which process is more nearly approximated by *Drosophila* imaginal discs regenerating in the abdomen of an adult fly. A comparable distinction is often made between two typs of regeneration: morphallaxis and epimorphosis. Morphallaxis has never been satisfactorily defined, but carries the idea that the whole animal rearranges itself with extensive cell movement and long-range signalling. Epimorphosis by contrast is easier to recognise as it involves three distinct processes: the dedifferentiation of tissue near the wound, the formation of a blastema, and the regeneration of levels of pattern in conjunction with growth of the blastema.

Regeneration involves quite a few unsolved problems:

- determination of polarity
- mechanism of dedifferentiation
- restrictions on cell type transformation
- identification of permissive factors
- patterning of the regenerate
- control of size and proportion

Of this set the main theoretical problems concern the control of polarity, cell type transformation and control of size. The other topics, although once mysterious, are now perhaps more squarely within the range of experimental molecular biology. First, let us consider polarity. Nobody studies regeneration for long without encountering the famous "law of distal transformation" and marvelling at the fact that an amphibian limb will regenerate the same set of distal structures from either a distal-facing or a proximal-facing cut surface (Butler, 1955). However, this is the easy part! It is even more difficult to imagine how the cells of a hydroid or a worm can regenerate a head, and the same cells regenerate a tail when the cut surface faces the other way (Slack, 1980). In fact it is commonly found that small slices of the body will regenerate the same extremum at both ends. This type of heteromorphosis is found in hydroids, planaria and annelids, and shows no less than three properties of the system. Firstly that there is a mechanism for polarity control, secondly that it is susceptible to error, and thirdly that the determination of polarity must precede the regeneration of pattern. The mechanism must involve some intercellular communication because the cells at the cut surface need to know what structures are in their vicinity in order to know whether to become a head or a tail.

There are two main reasons why there has been little progress in the study of regeneration in the last 20 years. One is that it is unfashionable, and all the effort has gone into early development. The other is that regeneration does not occur in any of the six standard organisms for the study of animal development (mouse, chick, *Xenopus*, zebrafish, *Drosophila*, *C. elegans*). Unless you work on one of these organisms, you are deprived of a whole mass of technical facilities, including methods of keeping the animals, procuring the embryos, microinjecting them or culturing them. Furthermore, if you want any genes or probes, you have to clone them yourself and this represents an enormous investment of work and expense. Things will become worse and not better when genome maps are finished for mouse, *Drosophila* and *C. elegans*, because this will widen even further the gap between what can be done with the model organisms and what can be done with an outsider such as a starfish or a planarian worm.

1.5 Size and Proportion

The concentration on regional specification over the last 20 years has diverted attention away from growth, which has been considered as "merely" an increase in size. But there are some very interesting questions associated with growth:

1. What controls the absolute size of the whole, or why are we bigger than mice?
2. Within a whole, what maintains the constancy of proportions of individual parts?
3. How is a possible change of relative proportions (allometry) produced?

Consider the first question. Some sort of limits to growth must exist for all organisms, but for indeterminate ones the size of the overall colony may depend only on nutrition. So a plant may grow to the size capable of being supported by its root system and no further. If cuttings are taken, they will grow equally large in their turn. If a graft is made to a different variety of rootstock, then the rootstock determines the size of the whole. Fungal mycelia, which have more or less unlimited access to surrounding nutrients, can grow to enormous and indeterminate sizes. Some colonial animals such as hydroids, bryozoa or colonial ascidians are also indeterminate, but most animals have an obvious boundary and a characteristic final shape and size. Even in indeterminate organisms, the individual parts such as leaves, flowers or polyps do have a characteristic size, and the problem this represents is similar to those relating to the overall size of determinate animals.

Growth involves increases in cell number, cell size and the quantity of extracellular material, but attention is normally focused on cell division. A great deal is now known about the microcosm, or the molecular biology of the cell cycle, but virtually nothing is known about how cell division relates to the size of the organism.

One possibility is that growth to final size is completely autonomous to the cells of an early organ rudiment. They require no communication with each other or with the surroundings, they just grow until it is time to stop. If cells know when to stop, then question 1 would truly belong in the microcosm of intracellular molecular biology, and questions 2 and 3 may not exist at all, as everything is simply programmed to go on until it is time to stop.

How could cells know when to stop? Could they measure elapsed time? This seems very unlikely as cold-blooded organisms grow to a final size at very different rates depending on food supply and temperature. At best they would be measuring Bergsonian rather than Newtonian time, i.e. not absolute time but time in the sense of progression of chemical and physical events. Could cells count cell divisions? Maybe they can, at least approximately. An endogenous inhibitor of cyclin dependent kinase (cdk) and thus an inhibitor of cell division, p16 (Ink4), accumulates gradually during fibroblast culture (Hara *et al.*, 1996). Cells from p16 (Ink4) knockout mice do not senesce in culture (Loh and Nakayama, 1996) although it must be admitted that the mice themselves are of normal size, so cell division does not proceed without limit in vivo.

The idea that the growth of parts is purely autonomous, with no long-range interactions, is compatible with findings from interspecies grafts of organ rudiments between amphibian species. Complete eye or limb rudiments grafted from one species to another do, in general, show autonomous growth to their own donor-characteristic size (Twitty, 1930). There are few reported experiments looking at grafts between animals of the same species but different size. Although the limited data are reported as showing "size regulation" in fact they suggest autonomy, as each part seems to grow as it would in its original location (Twitty, 1934).

Despite these results, it seems more likely that control of growth resides at some level in the macrocosm, the whole organism. This is partly because animal embryos may increase in volume by 109-fold between the stage of determination of organ rudiments and the adult size. A very small error in counting of cell divisions, or whatever underlies autonomous growth, could make a big difference to the eventual outcome. There is also a certain amount of experimental evidence pointing to at least some systemic controls over some aspects of growth.

1. Removal of part of the liver is followed by rapid regeneration to the original volume. The systemic signal can be transmitted from one animal to another. The effective growth factor is hepatocyte growth factor (HGF), but the "size signal" from the partial liver is still not known.

2. Removal of one kidney causes the other to hypertrophy. This is reversed if the other kidney is replaced.

3. A small liver or kidney transplanted into a large animal will grow faster than usual and catch up with its host (Kam *et al.*, 1987).

4. If cell size is varied by altering the ploidy (in amphibians), then the overall size is unaltered and the cell number per tissue structure is changed in inverse proportion to cell size (Fankhauser, 1945).

5. The anterior pituitary produces growth hormone (GH) under the control of releasing factors from the hypothalamus. This is required for postnatal growth, particularly of the cartilage growth plates in the long bones. Much of its effect works through local release of insulin-like growth factors (IGFs). Mutants unable to produce GH remain small.

The whole organism could potentially know how large it is by titrating two variables that depend on different dimensionality, e.g. volume against surface area. A growth inhibitor that was produced all over the volume of the organism and removed only at the surface would have a higher steady-state concentration with increasing overall volume. Alternatively, each tissue or part could titrate its own volume against the surface of the whole organism. Known growth inhibitors are extracellular (TGF beta) and intracellular (cdk inhibitors) although TGF beta is not normally active in the circulation as it is sequestered into an inactive complex. Conversely, there might be a growth activator produced at the surface and removed all over the volume. Numerous extracellular mitogens are known, e.g. EGFs, FGFs, PDGFs, HGF, IGFs. What is the surface of an animal? Presumably it is the organ across which materials are exchanged with the exterior, i.e. the lungs, gills or alimentary canal. It seems unlikely that O_2/CO_2 would be a critical variable for assessing overall size and there is no evidence that more or less of the common respiratory gases can affect growth rate. However, the exchange/absorptive organs could produce the relevant factor wherever they were in contact with the external environment, and the concentration of this would provide a measure of surface area.

In the simple case of a solid organism whose surface is the exchange organ, the V/S ratio will be proportional to linear dimension e.g. for a sphere $V/S = r/3$. But exchange organs may be highly branched and dendritic and they may have a fractal dimension rather than a simple dimension; this would complicate calculation.

It is difficult to think of any alternative to dimensional comparison for assessing overall size. One possible method would be for the organism to carry within it an absolute size standard. This could certainly work at the single cell level. Cell size probably does depend on a titration of a soluble molecule with the DNA, for which the quantity is fixed at $1n$ or $2n$. For example, in the cleavage of a zygote, the cells halve volume in each cycle. But after a certain number of divisions, when a certain cell size is reached, the cleavages stop (Newport and Kirschner, 1982). The number of divisions can be experimentally varied by changing the ratio of nucleus to cytoplasm, showing that some titration is going on. In amphibians it is possible to vary the ploidy by cold shock shortly after fertilisation, and there is then a rough proportionality throughout life between the cell size and the ploidy (Fankhauser, 1945). Cell size is an important contributory factor in growth, but there is a great deal of cell division as well, so it would be necessary for absolute size perception to operate above the single cell level.

Assuming that there is some systemic mechanism for the determination of final size, there must also be one or more mechanisms for ensuring the correct proportions of parts. This is because the comparison of dimension involved in the control of final size does not give any information about relative proportions within the organism. Because of the 109-fold increase in size during development, a very small error in relative growth rates would result in severe disproportion if there were not some mechanism for comparing the relative size of parts within the framework of the absolute size of the organism. Much has been written about allometry, or the change in relative proportions occuring with age or between different species. But we do not even understand how constant proportions are maintained, let alone the programmed deviations from constant proportions represented by allometry.

Assuming that there is a systemic mechanism for the control of proportion, this might involve a comparison of the volume of each part, or organ, with the volume of the whole, or it might involve comparisons with one or more other parts. The idea of systemic tissue or organ-specific inhibitors was first postulated by Weiss and developed in the chalone theory of Bullough (Bullough, 1975). Although this fell out of fashion in the 1970s, there probably do exist such inhibitors. For example, GDF-9 (myostatin) seems to be an inhibitor of muscle growth, and knockout mice have disproportionatly large muscles (McPherron *et al.*, 1997). The idea is that the organ will produce its inhibitor, which spreads out over the whole organism. If the organ is too big, the inhibitor will be in excess and growth will slow down. If the organ is too small, the inhibitor will be in deficit and growth will accelerate. Attractive though this idea is, it does not fit well with the limited data on grafting between animals of the same species but different age. Here the general rule seems to be that a small organ in a large host will grow fast, to catch up, regardless of whether there is already a full complement of the particular organ present. This would fit better with a model in which each organ secreted an activating signal that worked not on itself, but on other organs. Of course the function of such interactions would need to be heavily damped otherwise there would be an autocatalytic explosion.

The problem of allometry cannot even be considered until we have some idea of how constant proportions are maintained. It is well known that the allometry formula of Huxley does not have a sound theoretical basis (Reeve and Huxley, 1945) and it is to be hoped that a new theory of relative growth can be worked out once the fundamentals of size and proportion have been understood.

1.6 Stochastic Character of Cell Differentiation

Most tissues in the adult body are composed of several cell types. To some extent this is because tissues have more than one embryological origin and cells are different because their precursor territories were specified as different at an earlier stage. Most tissues are formed from an epithelial component and a mesenchymal component, with the epithelium forming the functional part and the mesenchyme the connective tissue. In the case of the gut, for example, the epithelium comes from the endoderm and the mesenchyme from the splanchnic mesoderm of the phylotypic stage embryo. In addition there are various migratory cell populations in the embryo; for example, histiocytes and other macrophage-like cells come from the bone marrow, and pigment cells come from the neural crest.

But in most tissues there is, in addition to a diversity of cellular origins, also a formation of multiple cell types from a single cell population. Sometimes this continues over the entire life of the animal, as in the blood or the intestine. In other cases most of the cell production occurs during growth and the tissue is mitotically quiescent thereafter, as in most connective tissues and glandular organs. In either case the progenitor cells are often called stem cells (Morrison *et al.*, 1997). There is no precise definition of a stem cell but it is normally thought to be able both to reproduce itself and to produce differentiated progeny, and these differentiated progeny will usually include more than one cell type. Stem cells are normally conceived of as long-lived terminal cells, i.e. they do not autonomously change into something else in the course of time. This excludes early embryonic cells from the definition, although they may also have the ability to generate multiple cell types and hence present the same problem. The best evidence for the existence of stem cells fitting this definition comes from the haemopoietic system where they have been identified as cells with a particular combination of surface antigens (Morrison *et al.*, 1995). There is also good evidences for the existence of stem cells in the intestine and in the skin (Potten and Loeffler, 1990), and it is quite likely that all tissues have their own stem cells that function at least during the phase of growth.

Studies utilising monoclonal culture in vitro have shown the capacity of single cells to give rise to clones of mixed character in the case of the haemopoietic and central nervous systems (Temple, 1989; Suda *et al.*, 1983). A similar result has been obtained for embryonic ectoderm cells (Godsave and Slack, 1989), and it can also be shown that imposition of inducing factors can alter the proportions of what you get but without suppressing the ability to produce more than one cell type.

On the face of it this seems surprising. We have become accustomed to thinking of differentiated cell types as very stable and only changing their differentiated states under the most exceptional circumstances. Although stem cells are not differentiated at the light microscope level, they are long-lived cell types presumably characterised by a particular combination of gene activities, so they should also have some degree of stability. If we think in the language of dynamical system theory, rather than molecular biology, there is the same problem. Any stable cell type, including stem cells, would be conceived of as sitting on an attractor in a multidimensional state space of gene activities. Attractors are necessarily isolated and a transition from one attractor to another could only be achieved by forcing a change in at least one system variable by external intervention. There has been a school of thought maintaining that external signals are always necessary to achieve any change in cell state and that any stem cell population will on close inspection prove to be partitioned into more than one microenvironment to direct formation of the various cell types. There is doubtless an element of truth in this, but the ability of monoclonal cultures to generate multiple differentiated cell types argues that diversity can also arise from mechanisms intrinsic to the cell. This is also suggested by recent studies of cell lineage in vivo using retroviral lineage tracers (Cepko *et al.*, 1995). These have shown that formation of multiple cell types is the norm even for very late divisions and that there are rather few restrictions on what can be formed so long as it conforms with the normal cell type inventory of the tissue.

Such examples drive us to admit that there must be mechanisms intrinsic to stem cells that can generate more than one kind of progeny. Doubtless the proportions can be influenced by external signals as in the example of embryonic ectoderm given above, but there must also be an essential stochasticity in the behaviour of the cells. How can it be explained when all the chemical processes of the cell seem to be deterministic? There are perhaps two different types of possibility. Firstly it might be that there is a special subset of reactions (= gene control circuits) in stem cells whose behaviour is chaotic. This would make effectively non-deterministic the choice of which daughter cell(s) were to mature and in which direction they would differentiate. If this is the case then there is an obvious potential interest in identifying the genes responsible and understanding how the dynamics of the reactions generates chaos. The other possibility is that all the chemical processes of the cell are indeed deterministic but that there is a "graininess" superimposed on this by the existence of a limited number of molecules involved in some rate-limiting steps. The most obvious molecules of which there are just a few copies are the genes themselves, so this may be a good place to start looking for sources of heterogeneity.

Experimental evidence on this question mainly derives from a series of experiments by Ko (Ko, 1992). He studied a cell line carrying a *lacZ* reporter gene coupled to a glucocorticoid response element. When exposed to hydrocortisone the gene was activated and the cells turned blue. A particular cell was either blue or white, and the proportion of blue cells increased smoothly with dose of the hormone. Ko showed that a process that appeared smooth and deterministic when viewed at the population level, for example by measuring lacZ activity produced by different hydrocortisone doses, was actually grainy at the cellular level. His interpretation was that the residence time of the receptor on the response element was quite long. Suppose that the overall degree of saturation of the response element was 50%, at a particular time a given cell might have 0, 1 or 2 sites occupied and the corresponding number of genes active.

If this is really so then the control of differentiation could be absurdly simple. Two cell types could be regulated by one gene whose effector was at some concentration sufficient to give an intermediate level of expression between fully on and fully off and whose product was very unstable, thus giving an instantaneous readout of gene occupancy. The decision on cell type could made by reading the product concentration in a critical time window after division. If more than two cell types are produced from one sort of stem cell, this could easily be arranged by increasing the number of regulatory genes with unstable products and borderline effectors, using a combinatorial code. So two genes could code for 4 cell types, three genes for 8 and so on. Such a system can easily be made to adapt to external signals as they could bias the expression of the regulatory genes. In an extreme case, say a large excess of one particular factor, the gene combination may be jammed into one particular state and only a single cell type produced. More typically it is likely that external factors and their signal transduction pathways would bias the responses and increase or decrease the proportions of particular gene combinations and hence cell types. The essential ingredients of this model are

- very short-lived gene products or metabolic consequences thereof
- low affinity genetic regulatory factors

Detection of both of these might provide some problems. In particular, the existing methods for the study of transcription factors (e.g. bandshift, footprints) all assume tight binding. The components of the system could perhaps be identified genetically in a suitable organism such as *Drosophila*, but an awareness of these problems would be required during the molecular biology follow-up. Once again, components which are thought to be well understood at the level of the microcosm turn out to generate behaviour which is not at all understood at the level of the macrocosm.

REFERENCES

Bullough, W.S. (1975) Mitotic control in adult mammalian tissues. *Biol. Rev.* 50: 99–127.

Butler, E.G. (1955) Regeneration of the urodele forelimb after reversal of its proximodistal axis. *J. Morphol.* 96: 265–281.

Cepko, C., Ryder, E.F. *et al.* (1995) Lineage analysis using retroviral vectors. *Meth. Enzymol.* 254: 387–419.

Cooke, J. and Webber, J.A. (1985) Dynamics of the control of body pattern in the development of *Xenopus laevis*. II Timing and pattern in the development of single blastomeres (presumptive lateral halves) isolated at the 2 cell stage. *J. Embryol. Exp. Morph.* 88: 113–133.

Fankhauser, G. (1945) The effects of changes in chromosome number on amphibian development. *Quart. Rev. Biol.* 20: 20–78.

Godsave, S.F. and Slack, J.M.W. (1989) Clonal analysis of mesoderm induction in *Xenopus laevis*. *Dev. Biol.* 134: 486–490.

Hara, E., Smith, R. *et al.* (1996) Regulation of p16(CDKN2) expression and its implications for cell immortalization and senescence. *Molec. Cell Biol.* 16: 859–867.

Isaacs, H.V., Pownall, M.E. *et al.* (1995) eFGF is expressed in the dorsal midline of *Xenopus laevis*. *Int. J. Dev. Biol.* 39: 575–579.

Isaacs, H.V., Pownall, M.E. *et al.* (1998) Regulation of Hox gene expression and posterior development by the *Xenopus* caudal homologue Xcad3. *EMBO Journal* 17: 3413–3427.

Kam, I. *et al.* (1987) Evidence that host size determines liver size: studies in dogs receiving orthotopic liver transplants. *Hepatology* 7: 362–366.

Ko, M.S.H. (1992) Induction mechanism of a single gene molecule: stochastic or deterministic. *Bioessays* 14: 341–346.

Loh, D.Y. and Nakayama, K. (1996) Mice lacking p27(Kip1) display increased body size, multiple organ hyperplasia, retinal dysplasia and pituitary tumours. *Cell* 85: 707–720.

McPherron, A. C., Lawler, A.M. *et al.* (1997) Regulation of skeletal muscle mass in mice by a new TGFbeta superfamily member. *Nature* 387: 83–90.

Meinhardt, H. (1982) *Models of Biological Pattern Formation*. London: Academic Press.

Morrison, S.J., Uchida, N. *et al.* (1995) The biology of haematopoietic stem cells. *Ann. Rev. Cell Dev. Biol.* 11: 35–71.

Morrison, S.J., Shah, N.M. *et al.* (1997) Regulatory mechanisms in stem cell biology. *Cell* 88: 287–298.

Newport, J. and Kirschner, M. (1982) A major developmental transition in early *Xenopus* embryos: 1. Characterisation and timing of cellular changes at the midblastula stage. *Cell* 30: 675–290.

Odell, G.M., Oster, G. *et al.* (1981) The mechanical basis of morphogenesis. I. Epithelial folding and invagination. *Dev. Biol.* 85: 446–462.

Potten, C. and Loeffler, M. (1990) Stem cells: attributes, cycles, spirals, pitfalls and uncertainties. Lessons from the crypt. *Development* 110: 1001–1020.

Pownall, M.E., Tucker, A.S. *et al.* (1996) eFGF, Xcad3 and Hox genes form a molecular pathway that establishes the anteroposterior axis in *Xenopus*. *Development* 122: 3881–3892.

Reeve, E.C.R. and Huxley, J.S. (1945) Some problems in the study of allometric growth. In *Essays on Growth and Form*, pp. 121–156. P.B. Medawar and W.E.L. Clark, eds. Oxford: Oxford University Press.

Rose, S.M. (1957) Cellular interactions during differentiation. *Biol. Rev.* 32: 351–382.

Slack, J.M.W. (1980) A serial threshold theory of regeneration. *J.theor. Biol.* 82: 105–140.

Slack, J.M.W. (1991) *From Egg to Embryo. Regional Specification in Early Development.* 2nd edn. Cambridge: Cambridge University Press.

Suda, T., Suda, J. *et al.* (1983) Single cell origin of mouse haemopoietic colonies expressing multiple lineages in variable combinations. *Proc. Natl. Acad. Sci. USA* 80: 6689–6693.

Tarkowski, A.K. and Wroblewska, J. (1967) Development of blastomeres of mouse eggs isolated at the 4 and 8 cell stage. *J. Embryol. Exp. Morph.* 18: 155–180.

Temple, S. (1989) Division and differentiation of isolated CNS blast cells in microculture. *Nature* 340: 471–473.

Twitty, V.C. (1930) Regulation in the growth of transplanted eyes. *J. Exp. Zool.* 55: 43–52.

Twitty, V.C. (1934) The relative growth of the amphibian eye studied by means of transplantation. *J. Exp. Zool.* 68: 247–291.

Weiss, P. and Kavanau, J.L. (1957) A model of growth and growth control in mathematical terms. *J. Gen. Physiol.* 41: 1–47.

Wolpert, L. (1968) The French Flag problem. A contribution to the discussion on pattern development and regulation. *Towards a Theoretical Biology*, vol 1 Prologemena. Edinburgh: Edinburgh University Press.

Wolpert, L. (1998) *Principles of Development.* Oxford: Oxford University Press.

2

Development of the Vertebrate Limb: A Model for Growth and Patterning

Cheryll Tickle

Department of Anatomy and Physiology, The Wellcome Trust Building, University of Dundee, Dow Street, Dundee DD1 5EH, UK

2.1 Introduction

The development of the vertebrate limb is an excellent example of how growth and form is generated in embryos and is also a very good model for pattern formation. The limb develops from a small bud of undifferentiated mesenchyme cells encased in ectoderm. Not only does shaping of the bud take place - for example, bud elongation and sculpting of digits - but also complex arrangements of tissues arise within the limb, including the different elements of the skeleton, over 50 discrete muscles, tendons etc. These different tissues originate from the mesenchyme cells of the bud while the ectoderm forms the epidermis of the skin.

The cell interactions in the developing limb that control the processes of growth and patterning were defined some time ago by classical embryological experiments (Saunders, 1977; Tickle and Eichele, 1994). A separate set of interactions governs patterning along each of the three limb axes; proximodistal, anteroposterior and dorsoventral. The characterisation of these interactions by embryological manipulations then led to the formulation of various models to account for how these processes occur. More recently, some of the molecules that mediate these interactions have been discovered and this means that models should now be testable at the biochemical level.

On Growth and Form: Spatio-temporal Pattern Formation in Biology.
Editor Mark A.J. Chaplain, G.D. Singh, J.C. McLachlan

2.2 Chick Wing Development

Experimental analysis of vertebrate limb development has focused on chick embryos. This is because one can make a hole in the eggshell and carry out manipulations through the hole while the embryo remains within the egg. The consequences for limb development can then be examined some time later. The effects on pattern are usually monitored in terms of the skeleton. The skeleton is first laid down as cartilage and the cartilage matrix can be stained in whole mounts with alcian green or alcian blue. The cartilage skeleton of the chick wing follows the general plan for vertebrate limbs (Figure 2.1C) and is laid down in a proximodistal sequence. First the humerus forms at the proximal end of the wing (nearest to the body wall); this is then followed by radius and ulna, and finally, distally, at the tip of the wing, by the digits. The chicken wing has three digits, a small anterior digit 2, a middle digit with two phalanges, digit 3, and a posterior digit with one phalange, digit 4. The numbers denote the accepted homology with the pentadactyl limb plan. The cartilage elements lie in the core of the limb and are flanked dorsally by extensor muscles and ventrally by flexors.

2.3 Growth and Shaping of the Limb Bud

The first signs of limb development are the small buds that arise at appropriate levels along the head to tail axis (Figure 2.1A). Rather surprisingly, these early buds do not form because of an increase in cell proliferation at limb-forming levels, but because cell proliferation falls in the interlimb (flank) region (Searls and Janners, 1971). This appears to be a general feature of budding; facial primordia also form in the same way with local budding being due to maintenance of high levels of proliferation (Minkoff and Kuntz, 1978).

The early limb bud is an almost symmetrical bulge. As the bud first elongates, it sticks out perpendicularly to the body wall but very soon it begins to extend more posteriorly so that it becomes slanting (Figure 2.1B). Further changes in bud shape are mediated by local regions of programmed cell death (Saunders and Fallon, 1967). For example, a necrotic zone develops at the anterior of the bud that generates a sharp angle halfway along the anterior bud margin marking the position in the wing where the elbow will form (and in the leg, the knee). Later on, as digits begin to form, sculpting out of tissue between each individual digit takes place in zones of interdigital cell death.

Recently I have collaborated with Jon Clarke to make detailed fate maps of the early chick limb buds (Vargesson *et al.*, 1997; Figure 2.2). This was done by labelling, in a systematic fashion, small populations of mesenchyme cells with lipophilic dyes (DiI or DiA). These fate maps reveal the structures that mesenchyme cells in particular regions of the bud will form. The maps also provide information about the way in which the bud grows and changes shape. The fate maps made with Di labelling broadly confirm and extend earlier fate maps made by using carbon particle marks or by transplanting labelled tissue (Saunders, 1948; Stark and Searls 1973; Bowen *et al.*, 1989).

One very clear result which emerges from the fate maps of the early limb bud is that cell populations at the tip expand considerably while cell populations in the proximal part of the bud expand rather little. The distal cell populations are pulled out into discrete elongated streams of cells as the bud grows out (Figure 2.2); in

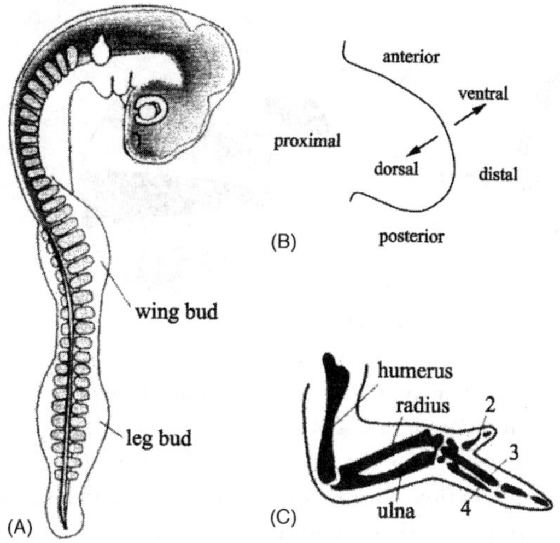

Figure 2.1 Stages in chick limb development. A. Chick embryo at around 3 days of development showing the small buds that will give rise to the wings and legs. B. Wing bud about 24 hours later. Considerable outgrowth has occurred, particularly in the posterior part of the bud, giving rise to elongated slanting shape. The various axes are indicated: proximal–distal; anterior–posterior; dorsal–ventral. C. Chick wing at 10 days of development showing skeleton laid down as cartilage. Note three digits: anterior digit 2, middle digit 3 and posterior digit 4.

contrast, proximal cell populations remain more rounded in shape and there can also be considerable intermingling between neighbouring cell populations. The fate maps of early buds also show that digits come from the posterior part of the bud (Bowen *et al.*, 1989; Vargesson *et al.*, 1997). The early wing bud forms from the region of lateral plate mesoderm that lies opposite somite 15 to 20. However, distal limb mesenchyme cells opposite somite 17/18 form streams of cells that contribute to the most anterior digit, digit 2; while cells opposite 18/19 give rise to digit 3, and digit 4 arises from cells opposite somite 19/20. Thus all three digits arise from a relatively restricted region at the posterior of the early bud. Furthermore, it is clear when one compares the expansion of labelled distal cell populations that cells in the posterior third of the bud give rise to a much larger proportion of the bud compared with cells in the anterior or middle third. Note also that streams of cells which originate from posterior labelling run straight along the posterior margin of the elongated bud, while more medial streams of cells bend anteriorly and fan out into the anterior part of the bud tip (Figure 2.2). It has been suggested that evolution of the hand plate in higher vertebrates is due to bending of the metapterygial axis (Shubin and Alberch, 1986; Coates, 1995). However, the bending does not appear to originate at the posterior margin of the limb, as would be predicted.

Several different mechanisms contribute to growth and shaping of the early bud. The increase in size of DiI and DiA labelled cell populations must reflect the local rate of cell proliferation. Hornbruch and Wolpert (1970) showed that the mitotic index is highest at the distal tip of the wing bud, which correlates with the considerable

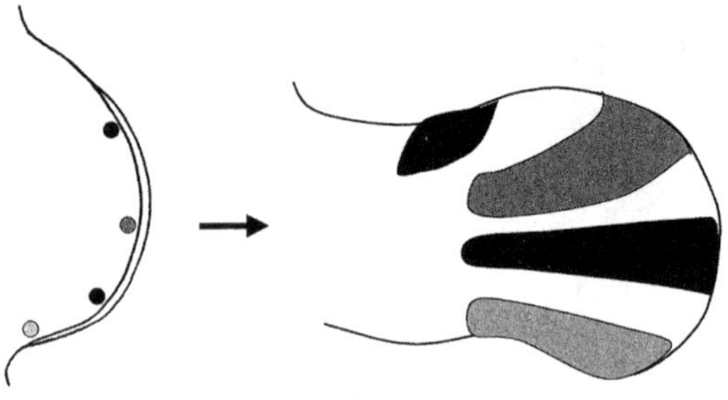

Figure 2.2 Behaviour of mesenchyme cells as limb bud develops. (a) Cells just beneath the apical ectodermal ridge in stage 20 chick wing buds are labelled with DiI or DiA at four different positions along the anteroposterior axis. (b) Distribution of the labelled cells from each location 48 hours later. Different shades of grey have been used to make it easier to match the population of labelled cells at 48 hours with the original labelled cell population. Expansion of the anterior cell population is much less than expansion of the posterior cell populations.

expansion seen here and work by Cooke and Summerbell (1980) looking at S-phase labelling indicated that cell proliferation is higher posteriorly. It is striking, however, that the labelled cell populations at the tip of the bud do not enlarge uniformly in all directions but preferentially along the proximodistal axis to form streams. The formation of streams could occur if cell divisions were oriented with respect to the long axis of the bud, and Trelstad (1977) using the Golgi apparatus as an indicator of cell polarity showed that mesenchyme cells were orientated in the limb bud. Another possibility is that new cells generated by proliferation take up positions preferentially along the long axis. Both of these mechanisms would contribute to bud elongation. Yet another idea is that external constraints on the expanding cell population could dictate its shape. There are precedents in other budding systems for such a mechanism. The best example is epithelial budding of the quail submaxillary gland. Proliferation appears to be uniform throughout the gland rather than just at the tip. Nevertheless, the gland develops into an elongated tongue rather than a round ball. In this case the basement membrane appears to be important in shaping the expanding epithelium and preventing it from blowing up like a balloon (Nogawa, 1981). The view that the ectoderm of the limb bud could shape the expanding mesenchyme was championed a long time ago by Amprino (1965). However, this raises the question of how the shape and size of the ectodermal jacket itself are controlled.

The signals that control limb bud outgrowth are produced by the thickened epithelium at the tip of the bud, the apical ectodermal ridge. When the apical ridge is surgically removed from developing wing buds, the resulting wings are truncated (Saunders, 1948; Summerbell, 1974). The outgrowth-promoting property of the ridge can also be demonstrated by grafting an additional apical ridge to the dorsal surface of a bud. The grafted ridge will induce a secondary outgrowth from the dorsal surface of the host bud (Saunders *et al.*, 1976).

A considerable number of genes have now been identified that are expressed in the apical ridge (Ferretti and Tickle, 1997). The apical ridge expresses three members of the FGF (fibroblast growth factor) family, FGF8, FGF2 and FGF4 (Martin, 1998). Beads soaked in FGF4, FGF2 or FGF8 can rescue limb bud outgrowth after apical ridge removal and allow the formation of distal structures including digits (Niswander *et al.*, 1993; Fallon *et al.*, 1994; Crossley *et al.*, 1996). *Fgf4* is expressed only in the posterior part of the apical ridge while *Fgf8* and FGF2 appear to be expressed in cells throughout the ridge. It is possible that the posterior of the bud grows more simply because more FGF is produced there. Another possibility is that preferential expansion of the posterior part of the bud is due specifically to Fgf4 signalling in this region. (The subtle effect of this particular FGF could be masked in bead experiments). There are indeed indications that *Fgf4* may fulfil other special functions (see later).

The importance of FGF signalling in limb bud outgrowth has been confirmed by recent mouse knockout studies. Mice which are homozygous null for either *Fgf4* and *Fgf8* die before limbs develop and so do embryos in which the fibroblast growth factor receptor-1 gene is mutated but inactivation of *Fgf8* specifically in the apical ridge at limb bud stages leads to limb truncations; see the results of M. Lewandoski *et al.* cited in Martin (1998). Furthermore, in the chicken limbless mutant in which limb buds form but do not grow out, *Fgf8* is not expressed (Grieshammer *et al.*, 1996; Noramly *et al.*, 1996; Ros *et al.*, 1996).

It has been suggested that FGF signalling may not only promote proliferation but also help to position mesenchyme cells in the developing limb bud. Based on the behaviour of labelled cell populations in intact limb buds in which FGF beads have been implanted, it has been suggested that FGF is chemotactic (Li *et al.*, 1996). This could lead to displacement of new cells generated by proliferation towards the source of FGF and thus contribute to stream formation. In outgrowths induced by FGF, in the absence of the bud tip and apical ridge, streams of DiI labelled cells can still be seen (Kostakopoulou *et al.*, 1997). The idea of chemotaxis might also explain why cell streams from the posterior fan out into the anterior. As the bud grows out, fate mapping studies show that the apical ridge, including the region that expresses *Fgf4*, is displaced anteriorly (Vargesson *et al.*, 1997).

Other growth factors are expressed in the ridge including three members of the BMP (bone morphogenetic protein) family. It has been suggested that BMP signalling may anatagonise FGF signalling and thus modulate outgrowth (Niswander and Martin, 1993). Another possibility is that BMP signalling may control the number of signalling cells by controlling cell death in the ridge (Ferrari *et al.*, 1998).

The apical ridge may play a mechanical role in shaping the bud addition to providing signal(s) that control mesenchymal cell proliferation. One possibility is that the ridge may act like a stiffening in a petticoat. The idea is that the mechanical properties of the ridge may stiffen the bud rim along the anteroposterior axis and thus at later stages keep the tip of the limb bud paddle-shaped. When the apical ridge is removed and replaced by FGFs, outgrowth continues but the bud appears to be bulbous rather than paddle-shaped. The end result is that the digits that develop are bunched together instead of lying side by side in the handplate. One interpretation of this result is that FGFs can replace the signalling role of the ridge in bud outgrowth but not its mechanical role (Niswander *et al.*, 1994a).

The dynamics of the apical ridge cell population during limb bud outgrowth are not well understood. When a quail ridge is grafted in place of the ridge in a chick embryo, the composition of the ridge does not change during subsequent development

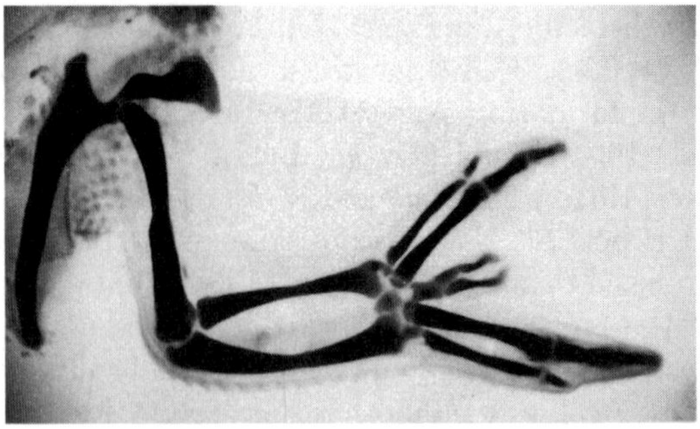

Figure 2.3 Skeleton of a chick wing that developed following graft of a polarising region to the anterior margin of the bud at stage 20. A mirror-image pattern of digits has been produced.

(Saunders *et al.*, 1976). Thus the ridge does not appear to recruit non-ridge ectodermal cells as the bud develops. However, our recent cell lineage analysis of the ridge shows that cells leave the ridge at the anterior end and appear to become non-ridge cells (Vargesson *et al.*, 1997). This suggests that the ridge must be maintained by cell proliferation of pre-existing ridge cells.

2.4 Tissue Patterning

There are two basic ways in which patterns of differentiated cells and tissues could be generated. Cells could be told how to differentiate and then move to the correct positions; alternatively cells could be told where they are and then differentiate accordingly. In the developing limb, connective tissue patterning appears to depend generally on cells differentiating according to their position.

2.4.1 Anteroposterior patterning

The best example of positional signalling is the polarising region that controls the anteroposterior pattern of tissues. The polarising region is a small group of mesenchyme cells at the posterior margin of the limb bud and was discovered by Saunders and Gasseling (1968). When the polarizing region is grafted to the anterior margin in a second wing bud, there is a dramatic change in digit pattern. The normal wing has three digits in the pattern 234 (reading from anterior to posterior); when an additional polarising region is grafted, six digits develop instead of three and they are arranged in a mirror-image symmetrical pattern, 432234 (Figure 2.3). However when a polarising region is grafted posteriorly there is no change in digit pattern.

There have been several different models to account for polarising region signalling. One class consists of gradient models in which the polarising region produces a signal which falls off with increasing distance away from the polarising region (Wolpert, 1969; Figure 2.4). Thus cells at different distances from the polarising region will be exposed to different concentrations of the signal and this will determine whether they will form posterior or anterior structures. When an additional polarising region is grafted to a wing bud, this will then act as a second source of signal and a U-shaped distribution

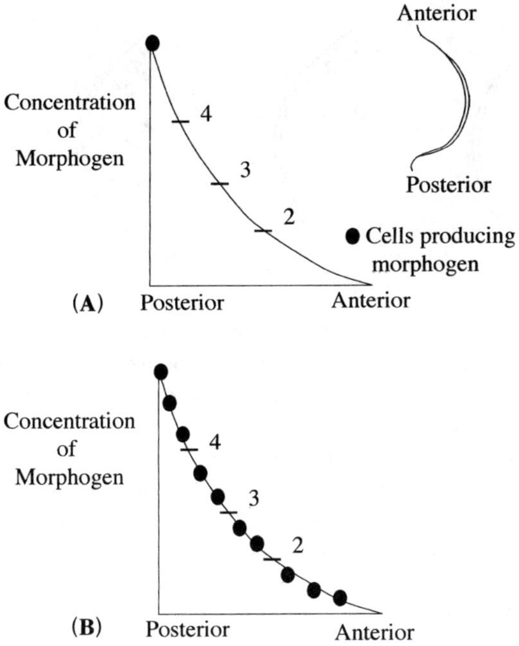

Figure 2.4 Sketch of wing bud indicating anteroposterior axis. A. Cells at the posterior margin of the limb bud produce a long-range diffusible signal and a concentration gradient across the wing bud is produced. Cells at different positions across the anteroposterior axis are exposed to different concentrations of morphogen. Threshold concentrations to specify some of the wing digits are indicated. B. Cells at the posterior margin of the wing bud produce a short-range signal. This induces nearby cells to produce the same signal but at an attenuated level. This sequential signalling process is then repeated across the wing bud.

of morphogen will result (Tickle *et al.*, 1975).

A second class of model that could explain the formation of additional digits following grafts of the polarising region is one that involves intercalation. According to this type of model (French *et al.*, 1976), extra digits would be generated by apposition of posterior and anterior cells. Intercalation depends on short-range interactions while a gradient could be established either by long-range signalling or by sequential short-range signalling accompanied by attenuation (Figure 2.4). It is also possible that position could be specified by a series of different short-range signals.

Several key results from the numerous embryological manipulations that have been carried out appear to rule out an intercalation model. When two polarising regions are placed close together anywhere around the wing bud rim, anterior digits do not form between them (Hornbruch and Wolpert, 1981). Furthermore, signalling by the polarising region is dose-dependent. For example, when signalling by polarising region grafts is attenuated by irradiation (Smith *et al.*, 1978) or when small numbers of signalling cells are implanted (Tickle, 1981), only anterior digits are specified. Polarising region grafts also appear to be able to signal at a distance. Thus when a barrier of leg tissue of around 200μm thickness is placed between the polarising

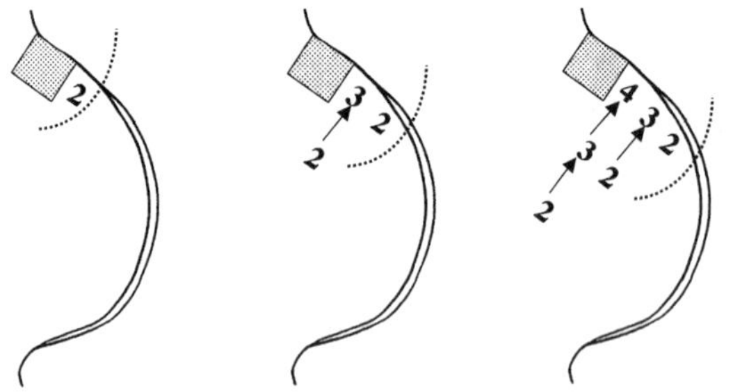

Figure 2.5 A model for concentration-dependent specification of additional digits with time. A source at the anterior margin begins to produce a polarising signal and cells near the source are exposed to a low concentration of morphogen and are specified as digit 2. As the source continues to produce morphogen, the concentration increases near the source and this new level is sufficent to specify digit 3. At the same time, the morphogen is spreading further into the limb and thus cells further away from the source are now exposed to the level of morphogen that specifies digit 2. This process is repeated again to specify a stable pattern of additional digits, 432.

region and responding wing tissue, additional wing digits can still be specified (Honig, 1981). This observation is compatible both with a long-range signalling mechanism and with a propagated series of short-range signals.

The time course of polarising region signalling has been determined by removing the polarising region at different times after grafting (Smith, 1980). This showed that additional digits are specified in an anterior to posterior sequence. Thus when the polarising region was removed after 16 hours only an additional digit 2 is specified. It is difficult to accommodate these data in a model of sequential signalling (Tickle, 1995). However, this sequence of digit specification is consistent with a model in which there is gradual build-up and spread of a long-range signal. According to this model, cells would first form anterior digits and then become promoted so that more posterior digits would be formed (Tickle, 1995; Figure 2.5)

A number of different signalling molecules - retinoic acid, sonic hedgehog (Shh) and BMPs - appear to be associated with the polarising region (Johnson and Tabin, 1997). Now that these molecules have been identified, it should be possible to work out the biochemistry of polarising region signalling. Retinoic acid and Shh can both produce digit duplications and mimic the effect of polarising region grafts but BMPs only have weak polarising activity. Retinoic acid can signal in a dose-dependent fashion and is freely diffusible in the wing bud. Furthermore, chick wings contain endogenous retinoic acid and also express retinoic acid receptors that are necessary to transduce retinoid signals (Hofmann and Eichele, 1994). However, since application of retinoic acid to the anterior margin of the bud induces expression of *Shh*, Shh may mediate retinoic acid signalling and therefore the focus has shifted to the mechanism of Shh signalling in the limb bud. Several different laboratories have shown that retinoids are necessary for limb initiation in chick embryos (Helms *et al.* 1996; Stratford *et al.*, 1996) suggesting that retinoids play a role in early limb outgrowth before Shh acts. This is consistent with the morphology of *Shh* knockout mice in which proximal limb structures are still present although distal structures are missing (Chiang *et al.*, 1996). However, it should be borne in mind that there could be parallel and interacting pathways that operate throughout bud development.

Recent experiments show that Shh can signal in a dose-dependent fashion and that alterations in cell fate occur at a distance from a source of Shh in chick wing buds (Yang *et al.*, 1997). When beads soaked in high concentrations of Shh are implanted at the anterior of chick wing buds, an additional digit 4 can be induced, while beads soaked in low concentrations give only an extra digit 2. DiI labelling studies also indicate that cells that would normally not form digits contribute to the extra digits after Shh bead implantation. Even more strikingly, it can be demonstrated directly that cells are promoted from forming anterior digits to forming posterior digits. Thus when the Shh bead is removed at 16 hours, cells near the bead contribute to the formation of the extra digit 2 that forms; while cells in the same position contribute to the anterior digit 3 in the 4334 digit pattern that results when the Shh bead is left in place. Although all these data are consistent with the idea that the concentration of Shh could determine digit identity, it is not clear how far Shh produced by the polarising region could diffuse within the chick limb bud. Antibody localisation studies indicate that the protein distribution closely matches that of *Shh* transcripts (Marti *et al.*, 1995) and it has emerged that Shh produced by cells is modified autocatalytically and tagged with cholesterol which would tether the protein to the cell plasma membrane (Tabin and McMahon, 1997). Furthermore, Cos cells expressing Shh anchored to a membrane protein can induce a pattern of additional digits when implanted at the anterior of a chick wing bud. All these considerations make it difficult to see how Shh could exert long-range effects. Shh does not induce *Shh* expression in neighbouring cells and this therefore rules out the possibility that Shh can act at a distance by sequential short-range signals. Therefore it seems likely that Shh signalling is mediated by a second signal.

BMPs are good candidates for mediating long-range effects of Shh. Expression of both *Bmp-2* and *Bmp-7* is activated in anterior mesenchyme cells in response to Shh (Laufer *et al.*, 1994). This would provide an interesting parallel with *Drosophila*. BMP2 is encoded by the vertebrate gene homologue of the *Drosophila* gene *dpp* and, in *Drosophila*, hh signalling is mediated by dpp. BMPs could perhaps extend the signalling range although it is not clear how far BMPs could diffuse in limb tissue. Application of either BMP2 or BMP4 to the anterior margin of a chick limb bud, however, does not induce additional digits (Francis *et al.*, 1994). Therefore it appears that if BMPs are the positional signals, they can only operate in cells which have already been exposed to Shh. It is also possible that BMPs are more effective in producing patterning signals when in heterodimeric forms. It will be important to investigate all these various possibilities in order to identify the signal that directly governs behaviour of cells in different positions across the anteroposterior axis of the limb.

2.4.2 Proximodistal patterning

Patterning along the proximodistal axis of the limb appears to be controlled by a timing mechanism rather than a morphogen gradient. Ridge removal experiments show that the structures of the limb are laid down in sequence from proximal to distal under the influence of apical ridge signalling. When the apical ridge is removed early the truncations are very severe, perhaps just a humerus forms; when the ridge is removed later only distal structures are lost. But when the apical ridge is exchanged between limb buds at different stages of bud development, the limbs are always normal. This suggests that the signal from the apical ridge does not itself specify which structure in the proximodistal pattern will be formed at any particular time. Furthermore, in experiments where the tips of limb buds are exchanged between different stages, the outcome is always a normal limb pattern, and this shows there is no signalling from proximal structures to specify which structure should be produced next. Based on these data indicating an autonomous change in the specification of cells at the limb bud tip, it has been suggested that a timing mechanism determines whether cells will form proximal or distal structures. The region of undifferentiated mesenchyme cells at the tip of the limb bud where this timing mechanism operates has been called the progress zone. Cells which leave the progress zone early form proximal structures whereas distal structures are formed from cells that leave later (Summerbell *et al.*, 1973). When proximal cells are grafted to the tip of a limb bud, they form distal structures. In a classical experiment by Saunders and Gasseling (1959), presumptive thigh cells were grafted to the tip of a chick wing. Here the cells were reprogrammed to form distal structures; but because they came from the leg, they formed toes rather than fingers. This change in cell fate can be understood in terms of restarting the progress zone clock. The effects of killing mesenchyme cells in early limb buds can also be interpreted in terms of the progress zone model. Under these circumstances, in which the progress zone is depleted of cells, proliferation of remaining cells will first have to replenish the progress zone, so when cells leave the progress zone they will form distal rather than proximal structures (Wolpert *et al.*, 1979).

At present, it is not clear how cells measure time in the progress zone and how this is related to the production of FGFs by the apical ridge. One possibility is that FGF is required just to keep the cells in a proliferative undifferentiated state. Another possibility is that the amount of FGF received by a cell is integrated over time and that the total amount of FGF then determines whether the cells form distal rather than proximal structures (Figure 2.6). A number of genes are known to be expressed at the tip of the limb bud under the influence of FGF. One of these is *Msx-1*, a homeobox-containing gene and it has been suggested that *Msx-1* could be responsible for maintaining the progress zone. Attention has focused on this particular gene because its expression prevents differentiation of a muscle cell line in culture. When proximal cells that do not express *Msx-1* are grafted at the tip of a bud, *Msx-1* expression is induced (Davidson *et al.*, 1991). Furthermore, expression of *Msx-1* can be induced in proximal cells by application of FGF. *Msx-2*, another homeobox-containing gene has an overlapping expression pattern and other work suggests that *Msx-2* plays a role in cell death.

2.4.3 Dorsoventral patterning

The pattern of structures across the dorsoventral axis of the limb is controlled by ectodermal signalling. When the ectodermal jacket from a right limb bud is recombined with the ectoderm from a left bud so that dorsal ectoderm now lies over ventral mesoderm, the polarity of newly generated distal structures is dictated by the polarity of the ectoderm (MacCabe *et al.*, 1974). Furthermore, when an apical

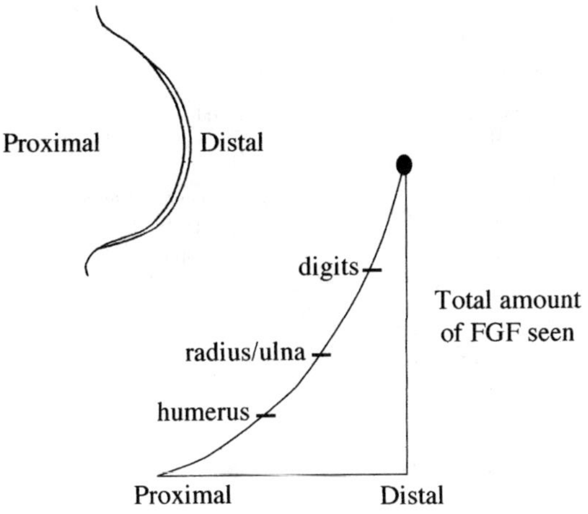

Figure 2.6 A model for patterning along the proximodistal axis indicated on sketch of early wing bud. The apical ridge at the distal tip of the wing bud produces FGF. Cells integrate the amount of FGF seen with time. Different total amounts of FGF specify humerus, radius/ulna, digits as indicated on the diagram.

ridge is grafted to the dorsal surface of a wing bud, the outgrowth that is induced is covered on both sides by dorsal ectoderm and has a double dorsal pattern in terms of muscle and tendons, shape of skeletal elements and differentiation of skin appendages. For example, feather buds develop earlier on the dorsal side of the wing than on the ventral side.

A number of different gradient models have been suggested to account for dorsoventral patterning (Parr and McMahon, 1995; Akita, 1996; Tickle 1995). In all of these, long-range signals emanating from dorsal and ventral ectoderm and/or from the apical ridge have been suggested to account for the arrangement of cells and tissues (Figure 2.7A). A series of short-range signals propagated from either dorsal or ventral ectoderm is also possible. All these ideas are rather similar to those that have been proposed for anteroposterior patterning. However, it is not at all clear that dorsal and ventral mesenchyme cells differentiate according to their distance from the ectoderm and that dose-dependent signalling is important. Instead a rather different mechanism may be involved in which cells are told how to differentiate and then become displaced to the correct position in the limb (Figure 2.7B). For example, early muscle and tendon primordia appear to form immediately beneath the ectoderm (Murray and Wilson, 1997; Ros *et al.*, 1995; Patel *et al.*, 1996) and then later are found deeper within the limb near the skeletal elements.

Recently some of the molecules involved in dorsoventral patterning have been discovered. The dorsal ectoderm of mouse and chick limb buds expresses *Wnt7a* and this provides a dorsalising signal; furthermore, when *Wnt7a* has been functionally inactivated, the distal paw has a double ventral pattern (Parr and McMahon, 1995). A transcription factor, *engrailed*, is expressed in ventral ectoderm and, when *engrailed* is functionally inactivated, Wnt7a is expressed in both dorsal and ventral ectoderm and the paws are double ventral (Loomis *et al.*, 1996). It is not clear whether a Wnt signal produced by the ectoderm could travel very far into the limb mesenchyme. The

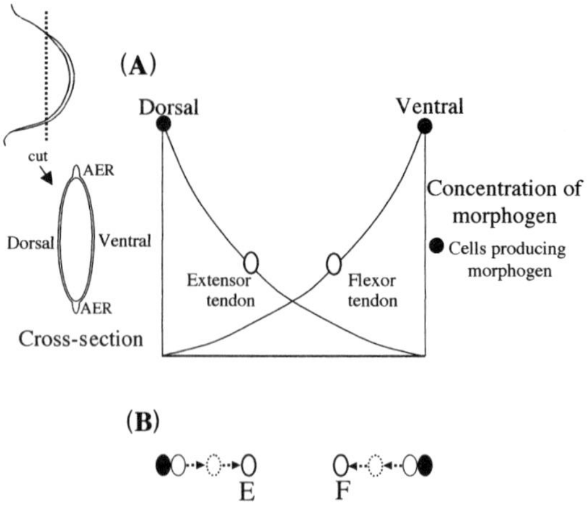

Figure 2.7 Models for patterning across the dorsoventral axis. Dorsoventral axis shown in cross-section of early chick wing bud. A. Cells in dorsal ectoderm and cells in ventral ectoderm produce a long-range diffusible signal. This sets up concentration gradients across the bud in both directions, and structures across the dorsoventral axis develop at particular threshold concentrations. B. Cells in dorsal and ventral ectoderm produce short-range signals that specify adjacent mesenchyme cells to be tendon; these mesenchyme cells then translocate to the appropriate position in the limb to act as extensors dorsally and flexors ventrally.

identity of the signal(s) produced by ventral ectoderm that controls ventral pattern is not known.

2.5 Coordination of Signalling and Responses

Signalling interactions are coordinated during limb bud outgrowth. *Shh* transcription in posterior mesenchyme depends on *Fgf* expression in the posterior apical ridge, and in turn the expression of *Fgf4* in this region appears to depend on *Shh* expression in nearby mesenchyme (Laufer *et al.*, 1994; Niswander *et al.*, 1994b). Furthermore *Wnt7a* expression in dorsal ectoderm helps to maintain *Shh* expression in posterior mesenchyme (Yang and Niswander, 1995). As a result of this coordination of signal production, each new "segment" of the limb produced along the proximodistal axis will be correctly patterned with respect to both anteroposterior and dorsoventral axes. Within each signalling region, coordination of signal production by individual cells may be coordinated by cell–cell signalling via gap junctions (Green *et al.*, 1994) and the number of signalling cells may be regulated by feedback mechanisms involving apoptosis.

As a result of signalling in the bud, position-dependent patterns of expression of the regulatory genes are established. For example, regulatory gene *Lmx1* is expressed in dorsal mesenchyme under the influence of Wnt7a signalling (Vogel *et al.*, 1995). Furthermore, as a result of coordinated signalling by polarising region and apical ridge signals, expression of genes in the 5′ regions of two homeobox clusters, *Hoxd 9–13* and *Hoxa 9–13* is activated in overlapping domains in limb bud mesenchyme such that cells at the distal and posterior regions of the bud express the most posterior and most 5′ genes (Izpisua-Belmonte and Duboule, 1992). It is interesting that the timing and sequence of activation of expression of these genes correlates very well with the way in which pattern seems to be established. Thus the embryological manipulations reviewed above show that pattern is specified in the proximal to distal direction and the anterior to posterior direction. Anterior structures can be promoted to posterior, and proximal structures to distal. This fits well with the characteristics of gene activation in the *Hox* complexes, in which activation of genes in the cluster is unidirectional (from 3′ to 5′) and irreversible. The modelling of these patterns of regulatory gene expression has therefore to consider how signals cooperate. Again, it is critical to discover whether FGF is merely keeping cells in a condition which allows them to respond to polarising signals or whether the responses depend more precisely on the relative concentrations of both signals (Papageorgiou, 1998).

2.5.1 Other signalling interactions

It is interesting that many of the same molecules that operate in the limb are also responsible for patterning other parts of the vertebrate embryos. For example, periodic feather patterns emerge through signalling by FGFs, Shh and BMPs. In this case it has been proposed that a reaction-diffusion mechanism operates in which BMP acts as an inhibitor and Shh and FGF act as activators (Jung *et al.*, 1998). At present, it is not clear how this fits with the postulated roles of the same two molecules in limb patterning.

2.6 Tissue Arrangements Including Musculature

One of the most difficult remaining problems in limb development is how one goes from position-dependent patterns of gene expression to anatomy. Each of the three digits of the chick wing contains the same cell types which express the same characteristic set of differentiated cell products, such as cartilage cells expressing type II collagen. The development of the precise pattern of individual muscles presents some additional problems. This is because some of the cells in the apparently homogeneous population of mesenchyme cells in the early wing bud originate in the somites and these cells are already committed to become the myogenic cells of the limb muscles. These presumptive myogenic cells migrate into the limb-forming region of the lateral plate mesoderm at an early stage (Christ *et al.*, 1977) and then, as the limb develops, become localised in the premuscle masses (Amthor *et al.*, 1998). These masses then split in a stereotypical fashion to give the individual muscles (Schroeter and Tosney 1991). The presumptive myogenic cells in the early limb bud have no knowledge of which muscle they will form; this information resides in the connective tissues which are derived from the lateral plate mesoderm. This can be shown by somite grafting experiments. When somites from leg level are transplanted to a wing-forming level, the muscle cells that migrate from the leg-level somites will nevertheless participate in forming a normal pattern of wing muscles (Butler *et al.*, 1988). When the skeletal pattern of a chick wing is duplicated by grafting an additional polarising region or adding a bead soaked in retinoic acid, the muscle pattern is also duplicated (Shellswell and Wolpert, 1977; Robson *et al.*, 1994). Thus the appropriate arrangements of the different tissues appear to depend on common connective tissue patterning which seems likely to involve the expression of regulatory genes such as *Hox* genes (Yamamoto *et al.*, 1998)

Acknowledgments
My research is supported by the MRC, BBSRC, the Wellcome Trust, the Leverhulme Trust, Human Frontiers Science Programme, the Anatomical Society and the Royal Society. I thank J.J. Sanz-Ezquerro and M. Davey for help with the figures.

REFERENCES

Akita, K. (1996) The effect of the ectoderm on the dorsoventral pattern of epidermis, muscles and joints in the developing chick leg: a new model. *Anat. Embryol.* 193: 377–386.

Amprino, R. (1965) Aspects of limb morphogenesis in the chicken. In: *Organogesis.* (R.L. DeHaan and H. Ursprung, eds.) pp. 255–281. Holt, Rinehart and Winston, New York.

Amthor, H., Christ, B., Weil, M. and Patel, K. (1998) The importance of timing differentiation during limb muscle development. *Curr. Biol.* 8: 642–652.

Bowen, J., Hinchliffe, J.R., Horder, T.J. and Reeve, A.M.F. (1989) The fate map of the chick forelimb-bud and its bearing on hypothesised developmental control mechanisms. *Anat. Embryol.* 179: 269–283.

Butler, J., Cosmos, E. and Cauwenberg, P. (1988) Positional signals: evidence for a possible role in muscle fibre-type patterning of the embryonic avian limb. *Development* 102: 763–772.

Chiang, C., Littingtung, Y., Lee, E., Young, K.E., Cordoen, J.L., Wetsphal, H. and Beachey, P.A. (1996) Cyclopia and axial patterning defects in mice lacking Sonic hedgehog gene function. *Nature* 383: 407–413.

Christ, B., Jacob, H.J. and Jacob, M. (1977) Experimental analysis of the origin of the wing musculature. *Anat. Embryol.* 150: 171–186.

Coates, M. (1995) Fish fins or tetrapod limbs-a simple twist of fate? *Curr. Biol.* 5: 844–848.

Cooke, J. and Summerbell, D. (1980) Cell cycle and experimental pattern duplication in chick wing during development. *Nature* 287: 697–701.

Crossley, P.H., Minowada, G., MacArthur, C.A. and Martin, G.R. (1996) Roles for FGF-8 in the induction, initiation and maintenance of chick limb development. *Cell* 84: 127–136.

Davidson, D.R., Crawley, A., Hill, R.E. and Tickle, C. (1991) Position-dependent expression of two related homeobox genes in developing vertebrate limbs. *Nature* 352: 429–431.

Fallon, J.F., Lopez, A., Ros, M.A., Savage, M.P., Olwin., B.B. and Simandl, B. (1994) FGF2: apical ridge growth signal for chick limb development. *Science* 264: 104–107.

Ferrari, D., Lichtler, A.C., Pan, Z.Z., Dealy, C.N., Upholt, W.B. and Kosher, R.A. (1998) Ectopic expression of Msx-2 in posterior limb bud mesoderm impairs limb morphogenesis while inducing BMP-4 expression, inhibiting cell proliferation and promoting apoptosis. *Dev. Biol.* 197: 12–24.

Ferretti, P. and Tickle, C. (1997) The Limbs. In: *Embryos, Genes and Birth Defects.* (P. Thorogood, ed.) pp. 101–132.

Francis, P.H., Richardson, M.K., Brickell, P.M. and Tickle, C. (1994) Bone morphogenetic proteins and a signalling pathway that controls patterning in the chick limb bud. *Development* 120: 209–218.

French, V., Bryant, P.J. and Bryant, S.V. (1997) Pattern regulation in epimorphic fields. *Science* 193: 969–981.

Green, C.R., Bowles, L., Crawley, A. and Tickle, C. (1994) Expression of the connexin43 gap junctional protein in tissues at the tip of the chick limb bud is related to the epithelial-mesenchymal interactions that mediate morphogenesis. *Dev. Biol.* 161: 12–21.

Grieshammer, U., Minowada, G., Pisenti, J.M., Abbott, U.K. and Martin, G.R. (1996) The chick limbless mutation causes abnormalities in limb bud dorso-ventral patterning: implications for the mechanism of apical ridge formation. *Development* 122: 3851–3861.

Helms, J.A., Kim, C.H., Eichele, G. and Thaller, C. (1996) Retinoic acid signalling is required during early chick limb development. *Development*, 122: 1385–1394.

Hofmann, C. and Eichele, G. (1994) Retinoids in development. In: *The Retinoids, Biology, Chemistry and Medicine* (M.B.Sporn, A.B. Roberts and D.S. Goodman eds.) pp. 387–441. Raven, New York.

Honig, L. (1981) Positional signal transmission in the developing chick limb. *Nature* 291: 72–73.

Hornbruch, A. and Wolpert, L. (1970) Cell division in the early growth and morphogenesis of the chick limb bud. *Nature* 226: 764–766.

Hornbruch, A. and Wolpert, L. (1981) Positional signalling along the antero-posterior axis: The effects of multiple polarising region grafts. *J. Embryol. Exp. Morph.* 63: 145–59.

Izpisua-Belmonte, J.C. and Duboule, D. (1992) Homeobox genes and pattern formation in the vertebrate limb. *Dev. Biol.* 152: 26–36.

Johnson, R.L. and Tabin, C.J. (1997) Molecular models for vertebrate limb development. *Cell* 90: 979–990.

Jung, H.-S., Francis-West, P.H., Widelitz, R.B., Jiang, T.-X., Ting-Berreth, S., Tickle, C., Wolpert, L. and Choung, C.-M. (1998) Local inhibitory action of BMPs and their relationships with activators in feather formation: implications for periodic patterning. *Dev. Biol.* 196: 11–23.

Kostakopoulou, K., Vargesson, N., Clarke, J.D.W., Brickell, P.M. and Tickle, C. (1997) Local origin of cells in FGF4-induced outgrowth of amputated chick wing bud stumps. *Int. J. Dev. Biol.* 41: 747–750.

Laufer, E., Nelson, C.E., Johnson, R.L., Morgan, B. and Tabin, C. (1994) Sonic hedgehog and Fgf-4 act through a signalling cascade and a feedback loop to integrate growth and patterning of the developing limb bud. *Cell* 79: 993–1003.

Li, S., Anderson, R., Reginelli, A. and Muneoka, K. (1996) FGF2 inflences cell movements and gene expression during limb development. *J. Exp. Zool.* 274, 234–247.

Loomis, C.A., Harris, E., Michaud, J., Wurst, W., Hanks, M. and Joyner, A. (1996) The mouse Engrailed-1 gene and ventral limb patterning. *Nature* 382: 360–363.

MacCabe, J.A., Errick, J. and Saunders, J.W. (1974) Ectodermal control of the dorsoventral axis in the leg bud of the chick embryo. *Dev. Biol.* 39:69–82.

Marti, E., Takada, R., Bumcrot, D.A., Sasaki, H. and McMahon, A.P. (1995) Distribution of Sonic hedgehog peptides in the developing chick and mouse embryo. *Development* 121: 2537–2547.

Martin, G.R. (1998) The roles of FGFs in early development of vertebrate limbs. *Genes Dev.* 12: 1571–1586.

Minkoff, R. and Kuntz, A.J. (1978) Cell population and cell density of mesenchyme in the maxillary process and adjacent regions during facial development in the chick embryo. *J. Embryol. Exp. Morph.* 46: 65–74.

Murray, B. and Wilson, D.J. (1997) Muscle patterning, differentiation and vascularisation in the chick wing bud. *J. Anat.* 190: 216–273.

Niswander, L. and Martin, G.R. (1993) FGF-4 and BMP-2 have opposite effects on limb growth. *Nature* 361: 68–71.

Niswander,L., Tickle, C., Vogel, A., Booth, I. and Martin, G.R. (1993) FGF-4 replaces the apical ectodermal ridge and directs outgrowth and patterning of the limb. *Cell* 75: 579–587.

Niswander, L., Tickle, C., Vogel, A. and Martin, G. (1994a) Function of FGF-4 in limb development. *Mol. Reprod. Dev.* 39: 83–88.

Niswander, L., Jeffrey, S., Martin, G.R. and Tickle, C. (1994b) A positive feedback loop coordinates growth and patterning in the vertebrate limb. *Nature* 371: 609–612.

Nogawa, H. (1981) Analysis of elongating morphogenesis of quail anterior submaxillary gland: absence of localised cell proliferation. *J. Embryol. Exp. Morph.* 62: 229–239.

Noramly, S., Pisenti, J., Abbott, U. and Morgan, B. (1996) Gene expression in the limbless mutant: Polarised gene expression in the absence of Shh and an AER. *Dev. Biol.* 179:

339–346.

Papageorgiou, S. (1998) Cooperating morphogens control hoxd gene expression in the developing vertebrate limb. *J. theor. Biol.* 192: 43–53.

Parr, B.A. and McMahon, A.P. (1995) Dorsalising signal Wnt7a required for normal polarity of D-V and A-P axes of mouse limb. *Nature* 374: 350–353.

Patel, K., Nittenberg, R., D'Souza, D., Irving, C., Burt, D., Wilkinson, D.G. and Tickle, C. (1996) Expression and regulation of Cek-8, a cell to cell signalling receptor in developing chick limb buds. *Development* 122: 1147–1155.

Robson, L.G., Kara, T., Crawley, A. and Tickle, C. (1994) Tissue and cellular patterning of the musculature in chick wings. *Development,* 120: 1265–1276.

Ros, M.A., Rivero, F.B., Hinchliffe, J.R. and Hurle, J.M. (1995) Immunohistological and ultrastructural study of the developing tendons of the avian foot. *Ant. Embryol.* 192: 483–496.

Ros, M.A., Lopez-Martinez, A., Simandl, B.K., Rodriguez, C., Izpisua-Belmonte, J.-C., Dahn, R. and Fallon, J.F. (1996) The limb field mesoderm determines initial limb bud anteroposterior aymmetry and budding independent of sonic hedgehog or apical ectodermal ridge gene expression. *Development* 122: 2319–2330.

Saunders, J.W. (1948) The proximodistal sequence of origin of the parts of the chick wing and the role of the ectoderm. *J. Exp. Zool.* 108: 363–403.

Saunders, J.W.(1977) The experimental analysis of chick limb bud development. In: *Limb and Somite Morphogenesis* (D.A. Ede, J.R. Hinchliffe and M. Balls eds.) pp. 1–24. Cambridge University Press, Cambridge.

Saunders, J.W. and Fallon, J.F. (1967) Cell death in embryonic morphogenesis. In: *Major Problems in Developmental Biology.* (M. Locke, ed.) pp. 289–314. Academic Press, New York.

Saunders, J.W. and Gasseling, M.T. (1959) The differentiation of prospective thigh mesoderm grafted beneath the apical ectodermal ridge of the wing bud in the chick embryo. *Dev. Biol.* 1: 281–301.

Saunders, J.W. and Gasseling, M.T. (1968) Ectodermal-mesenchymal interactions in the origin of limb symmetry. In: *Epithelial-mesenchymal Interactions* (R. Fleischmajer and R.E. Billingham, eds.) pp. 78–97. Williams and Wilkins, Baltimore.

Saunders, J.W., Gasseling, M.T. and Errick, J. (1976) Inductive activity and enduring cellular constitution of a supernumerary apical ectodermal ridge grafted to the limb bud of the chick embryo. *Dev. Biol.* 50: 16–25.

Schroeter, S. and Tosney, K.W. (1991) Spatial and temporal patterns of muscle cleavage in chick thigh and their value as criteria for homology. *Am. J. Anat.* 191: 325–350.

Searls, R.L. and Janners, M.Y. (1971) The initiation of limb bud outgrowth in the embryonic chick. *Dev. Biol.* 24: 198–213.

Shellswell, G.B. and Wolpert, L. (1977) The pattern of muscle and tendon development in the chick wing. In: *Vertebrate Limb and Somite Morphogenesis* (D.A. Ede, J.R. Hinchliffe and M. Balls eds.) pp. 71–86. Cambridge University Press, Cambridge.

Shubin, N.H. and Alberch, P. (1986) A morphogenetic approach to the origin and basic organisation of the tetrapod limb. *Evol. Biol.* 20: 319–387.

Smith, J.C. (1980) The time required for positional signalling in the chick wing bud. *J. Embryol. Exp. Morph.* 60: 321–328.

Smith, J.C., Tickle, C. and Wolpert, L. (1978) Attenuation of positional signalling in the chick limb by high doses of radiation. *Nature* 272: 612–613

Stark, R.J. and Searls, R.L. (1973) A description of chick wing bud development and a model of limb morphogenesis. *Dev. Biol.* 33: 138–153.

Stratford, T., Horton, C. and Maden, M. (1996) Retinoic acid is required for initiation of outgrowth in the chick limb bud. *Curr. Biol.* 6: 1124–1133.

Summerbell, D. (1974) A quantitative analysis of the effect of excision of the AER from the chick limb bud. *J. Embryol. Exp. Morph.* 32: 651–660.

Summerbell, D., Lewis, J. and Wolpert, L. (1973) Positional information in chick limb morphogenesis. *Nature* 244:492–496.

Tabin, C. and McMahon, A.P. (1997) Recent advances in hedgehog signalling. *Tren. Cell Biol.* 7: 442–446.

Tickle, C. (1981) The number of polarising region cells required to specify additional digits in the developing chick wing. *Nature* 289: 295–298.

Tickle, C. (1995) Vertebrate limb development. *Curr. Op. Genet. Dev.* 5: 478–484.

Tickle, C. and Eichele, G. (1994) Vertebrate limb development. *Ann. Rev. Cell Biol.* 10: 121–152.

Tickle, C., Summerbell, D. and Wolpert, L. (1975) Positional signalling and specification of digits in chick limb morphogenesis. *Nature* 254: 199–202.

Trelstad, R.L. (1977) Mesenchymal cell polarity and morphogenesis of chick cartilage. *Dev. Biol.* 59: 153–163.

Vargesson, N., Clarke, J., Vincent, K., Coles, C., Wolpert, L. and Tickle, C. (1997) Cell fate in the chick limb bud and relationship to gene expression. *Development* 124: 1909–1918.

Vogel, A., Rodriguez, C. and Izpisua-Belmonte, J.C. (1995) Dorsal cell fate specified by chick Lmx-1 during vertebrate limb development. *Nature* 378: 716–720.

Wolpert, L. (1969) Positional information and the spatial pattern of cellular differentiation. *J. theor. Biol.* 25: 1–47.

Wolpert, L., Tickle, C. and Sampford, M. (1979) The effect of cell killing by X-irradiation on pattern formation in the chick limb. *J. Embryol. Exp. Morph.* 50: 175–198.

Yamamoto, M., Gotoh, Y., Tamura, K., Tanaka, M., Kawakami, A., Ide, H. and Kuriowa, A. (1998) Coordinated expression of Hoxa11 and Hoxa13 during limb muscle patterning. *Development* 125: 1325–1335.

Yang, Y. and Niswander, L. (1995) Interaction between the signalling molecules WNT and SHH during vertebrate limb development: dorsal signals regulate antero-posterior patterning. *Cell* 80: 939–947.

Yang, Y., Drossopoulou, G., Chuang, P.-T., Duprez, D., Marti, E., Bumcrot, D., Vargesson, N., Clarke, J., Niswander, L., McMahon, A. and Tickle, C. (1997) Relationship between dose, distance and time in Sonic Hedgehog-mediated regulation of anteroposterior polarity in the chick limb. *Development* 124:4393–4404.

3

Pattern Formation on Butterfly Wings

Vernon French

ICAPB, University of Edinburgh, Ashworth Laboratory, Kings Buildings, West Mains Road, Edinburgh EH9 3JT, UK

3.1 Introduction

In the formation of a complex organism from the fertilised egg, cells progressively acquire distinct and diverse developmental fates. As a consequence of their interactions, the cells come to express different regulatory genes and eventually to differentiate in ways appropriate to their different positions, thereby contributing to the formation of spatially organised tissues. This integral aspect of development is usually termed 'pattern formation' (Wolpert, 1971), and a major research aim is to understand the mechanisms of signalling and response that are involved in the cell interactions.

The cuticular colour patterns displayed on adult insect wings present a novel and highly amenable system for studying the mechanisms of pattern formation. In one sense these patterns are very simple. They are accessible and two-dimensional, reflecting spatial variation in cell state within a single layer, the underlying epidermis, and their development does not involve cell migration or other complicating aspects of morphogenesis. The patterns can be very intricate and varied, however, and they are characterised by abrupt colour discontinuities between neighbouring cells. Other insects (particularly the beetles) can achieve striking arrays of coloured blotches, spots and lines, but the Lepidoptera (the butterflies and moths) are unique in their vivid, complex and highly diverse wing colour patterns. These patterns and their specification have long been the subject of comparative, experimental and theoretical study (Nijhout, 1991).

The wings of adult Lepidoptera develop from internal pockets of cells, the wing imaginal discs, that grow through larval life (Nijhout, 1991). Early in the last larval instar, the dorsal and ventral epidermal cell layers of the disc become separated along a regularly-branching pattern of haemolymph channels, the lacunae, down which grow the tracheae, forming the primary vein pattern of the wing. Late in the last larval

On Growth and Form: Spatio-temporal Pattern Formation in Biology.
Editor Mark A.J. Chaplain, G.D. Singh, J.C. McLachlan

instar, just before pupation, the disc epidermis secretes a simple cuticle and the disc is extruded to form the immobile wing of the pupa. Shortly after pupation, enlarged scale cells appear in regularly-spaced rows across the wing and the other epidermal cells undergo a period of cell division (Nijhout, 1980a; Galant *et al.*, 1998). Each scale cell gradually develops a large, flat, distal protrusion that will eventually secrete the cuticular scale. New tracheae form at the wing base and migrate out into the lacunae, adopting a slightly different branching pattern that defines the final venation of the adult wing. Very late in the pupal stage, pigments (principally melanins, ommochromes and pterins) are synthesised by the scale cells and secreted into the scale cuticle (Nijhout, 1980a). Each adult scale has a uniform colour, so the overall wing colour pattern is a finely tiled mosaic that records the spatial pattern of differentiation of the scale cells. This pattern is determined by cell interactions that occur earlier, within the epidermis of the larval wing disc and then the pupal wing.

3.2 The Diversity of Colour Patterns

Lepidopteran wings exhibit a bewildering variety of colour patterns (Figure 3.1). Almost every one of around 15,000 species of butterfly (and many more species of moth) are distinguishable by wing pattern alone, and there are usually considerable differences in pattern between the fore- and hindwings and between the dorsal and ventral wing surfaces. Furthermore, many species are genetically polymorphic (often with discrete geographical races) and, in other species, a response to environmental conditions during development produces distinct seasonal forms with different adult wing patterns (Brakefield and French, 1999).

Since the 1920s comparative studies have drawn homologies between the different wing patterns of lepidopteran species and have attempted to analyse them in relation to a basic ‘nymphalid groundplan’ of pattern elements (Schwanwitsch, 1924; Suffert, 1927). This approach has been developed and extended in recent years by Nijhout (1978, 1991). The groundplan has a set of concentric *eyespot* elements in the distal part of the wing, one centred in each wing-cell (the space between successive wing veins), plus symmetrical systems of *bands*, running anteror to posterior across the wing in central and proximal positions. *Parafocal elements* lie between the eyespots and the *marginal bands*, which run around the distal edge of the wing. Some species show almost the full array of groundplan elements, repeated in each wing-cell (Figure 3.1A), whereas other species have simple wing patterns comprising very few of the elements (Figure 3.1B). The elements may be distorted in shape and some (e.g. the eyespots or parafocals) may differ dramatically in form or in colour between the different wing-cells (Figure 3.1D). Frequently the band elements appear dislocated, with sections lying at different proximal-distal levels within the different wing-cells (Figure 3.1C). In addition to the groundplan elements, many species have large fields that are uniform in colour or are covered by repeating ripples or flecks of colour, or have stripes of colour along or between the wing veins (Figure 3.1F). Nijhout and others have used the groundplan to analyse wing patterns within the Nymphalidae and also in the other families of butterflies and in some moth families (Nijhout, 1991, 1994). By careful comparison between patterns of related species, and analysis of individual variation within a species, even very unlikely patterns can be related to the groundplan array of elements, as shown in Figure 3.1E (Nijhout and Wray, 1986, 1988).

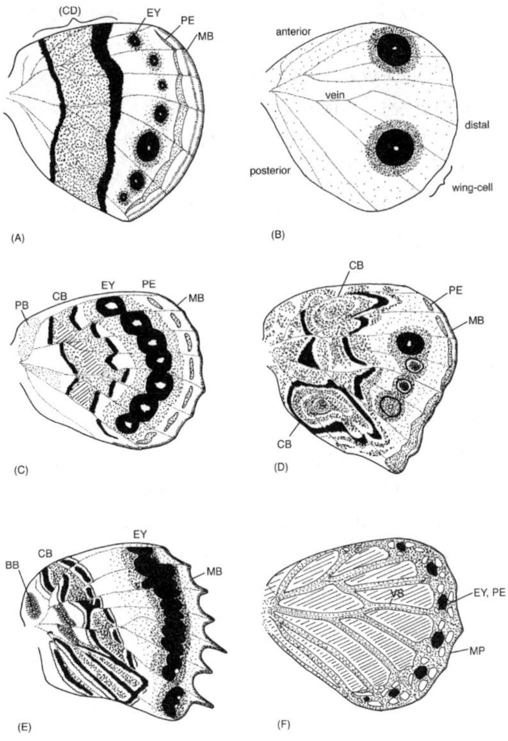

Figure 3.1 The diversity of butterfly wing patterns. The ventral hind-wing patterns of six nymphalid species, interpreted in terms of the 'nymphalid groundplan' of proximal bands (PB), central bands (CB), eyespots (EY), parafocal elements (PE) and marginal bands (MB). B-E are redrawn and adapted from Nijhout (1991). (A) *Bicyclus anynana* displays most of the groundplan array of elements (lacking only the proximal bands), with the eyespots and parafocal elements serially repeated in each wing-cell. (B) *Taenaris macrops* has only two large eyespots, each extending well beyond the wing-cell in which it is centred. (C) *Asterope occidentalium* has a full groundplan array and shows (i) fusion of adjacent eyespots to give a continuous outer band and (ii) dislocation between adjacent segments of the central bands. (D) *Smyrna blomfildia*. This pattern shows (i) major differences in pigmentation between the four eyespots and (ii) major distortions of the central band system, giving eccentric patterns on anterior and posterior parts of the wing. (E) *Charaxes brutus*. The patterns of this and related species may be interpreted in accordance with the groundplan (see Nijhout and Wray, 1986), with distortion and fusion of the eyespots, and also distortions and dislocations of the central and proximal bands. (F) *Danaus sp.* This species shows prominent venous stripes (VS) centred along the wing veins, and a complex distal pattern, interpreted as a fusion of eyespots and parafocal elements (see Nijhout, 1991).

This extensive analysis suggests that the wing patterns of the many different higher lepidopteran species are homologous (with the elements variously absent, distorted or fused): it suggests that they have evolved by modification of the mechanisms of pattern formation that operated on the ancestral lepidopteran wing (Nijhout, 1994). Furthermore, some species have eyespots that are partially fused into bands (Figures 3.1C and 3.1E), or bands that are broken up into eyespot-like swirls (Figure 3.1D), or have an apparent fusion between eyespots and the bands or the parafocal elements. These fusions indicate that the different elements of the groundplan may be specified by one basic mechanism of cell interaction that is used at different locations on the developing wing (Nijhout, 1991).

3.3 The Development of Pattern Elements: Surgical Manipulation

Surgical experiments have been the major traditional method of demonstrating the role of cell interaction during development and of exploring its properties. Experiments on wing pattern formation have concentrated on the eyespot elements and on the dorsal surface of the forewing, as this is accessible for operations at pupal stage. Nijhout (1980b) first showed that the eyespot is specified by signals from a small central focus (Figure 3.2). Ablation of the focus by cautery or damage in the early pupa of *Precis coenia*, or of *Bicyclus anynana* (Figure 3.2A(iii)), reduces or eliminates the eyespot (Nijhout, 1980b; Brakefield and French, 1995). Critically, when a small square of epidermis is transplanted from the focal position to another distal site on the early pupal wing, a concentric eyespot pattern is induced to form ectopically (Figure 3.2A(ii) and 3.2B(ii)), in the epidermis surrounding the graft (Nijhout, 1980b; French and Brakefield, 1995). The eyespot focus is the unique signalling region, as the adjacent epidermis provokes no response when transplanted.

There is a complication, however, to the interpretation of an eyespot as the epidermal response to a specific signal from the central focus, as a similar pattern may result from local epidermal damage. In *Bicyclus*, mild damage (at 12-24 hours after pupation) at a non-focal site can induce formation of an ectopic eyespot (Figure 3.2A(iv)). If epidermal damage is severe, an ectopic eyespot may result from an earlier operation (at 1 hour or 6 hours after pupation) and may also occur on the ventral wing surface (Brakefield and French, 1995). An ectopic eyespot can also occur in *Precis*, following a deep operation that damages the underlying hindwing (Nijhout, 1985). Clearly, in some situations, epidermal damage may mimic the action of an eyespot focus.

The adult eyespot pattern of scale pigmentation is specified in the early pupa, by signals from the focus and the response of the surrounding epidermal cells. Both of these properties may vary between individuals and between regions of the wing. In grafting experiments using foci taken from the anterior (small) or posterior (large) eyespot in *Bicyclus*, the size of the induced eyespot was determined by the identity of the focus (French and Brakefield, 1995). Similarly, after eyespot size was altered by generations of artificial selection in *Bicyclus*, the results of transplantation of foci between pupae from the lines with large and small eyespots, demonstrated that the size of the induced eyespot depends principally on properties of the focus (Monteiro *et al.*, 1994). After artificial selection for the relative width of the outer colour ring of the eyespot, however, transplants showed that the colour composition of the eyespot

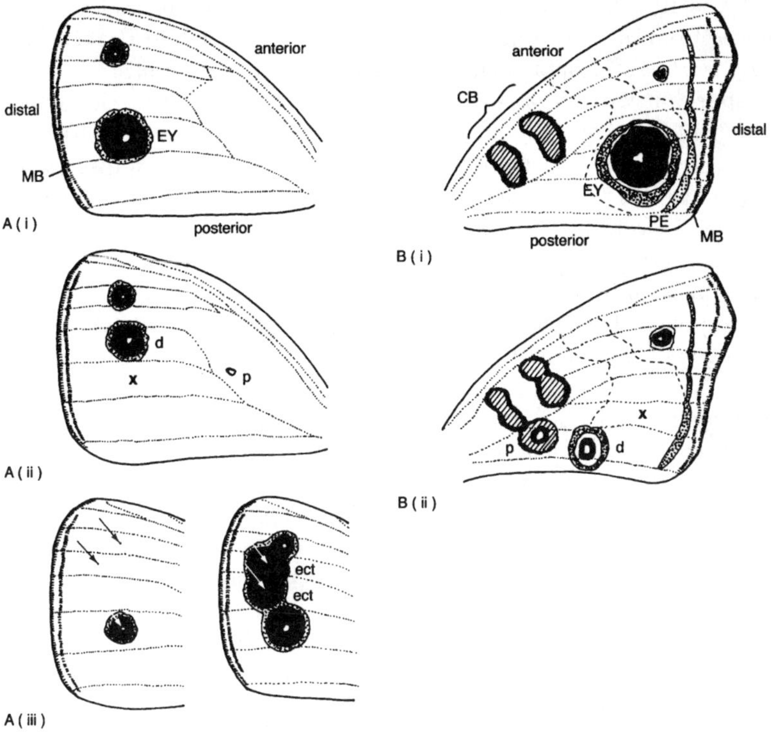

Figure 3.2 Surgical experiments on eyespot development. Camera lucida drawings of the dorsal surface of fore-wings of (A) *Bicyclus anynana* and (B) *Precis coenia*. Ai, Bi) Control colour patterns (abbreviations as in Figure 3.1). Aii) Composite drawing of the results of grafting the posterior eyespot focus on the early pupal wing of *Bicyclus* (adapted from French and Brakefield, 1995). Grafted to a distal position (d), the focus induces the formation of an eyespot; placed proximally (p), the graft differentiates normally, forming the white centre and a few dark brown scales, but the surrounding epidermis is not affected. At the donor site (X) there is a scar but no eyespot. Bii) Composite drawing of the results of focal grafting in *Precis* (adapted from Nijhout, 1980b and Brakefield *et al.*, 1996). The graft forms white and dark brown scales and, in the more distal position (d), it induces the normal buff and brown outer rings of the eyespot. Placed proximally (p), however, the graft induces the surrounding epidermis to form rings of orange and black scales. Aiii) Results of small ablations (marked by arrows) on the *Bicyclus* wing at 6 hours after pupation. Aiv) Result in *Bicyclus* of non-focal ablations (arrows) at 12 hours after pupation: dark brown and gold ectopic eyespot patterns (ect) are induced, fusing with the normal eyespots.

depends on the epidermal response to a focal signal (Monteiro *et al.*, 1997). Also the response to a focal signal may vary over the wing epidermis (Figures 3.2A(ii) and 3.2B(ii)). Hence a *Bicyclus* focus induces an eyespot when grafted distally, but in a proximal site it will heal and later form its normal white scales, yet it will elicit no response from the surrounding epidermal cells (French and Brakefield, 1995). More strikingly, a *Precis* eyespot focus grafted proximally will induce formation of a concentric eyespot, but with the orange and black scale pigmentation characteristic of the nearby bands of the central symmetry system (Brakefield *et al.*, 1996).

Together these results demonstrate that the early pupal wing has discrete eyespot foci, that these may differ in their signalling properties, and also that different regions of the wing epidermis can differ in their response. Even in adjacent wing-cells, some species have eyespots that are distinct in shape or in colour (Figure 3.1D), but it is not known whether this results from differences in focal signals or a regional difference in the epidermal response. The spatial pattern of the signalling and response properties in the pupal wing epidermis must be specified earlier, by cell interactions in the larval imaginal disc.

Comparative analysis of butterfly wing patterns suggests that the other pattern elements may share developmental mechanisms with the eyespots: they may also be specified by signalling foci (Nijhout, 1978, 1991). It is thus intriguing that the *Precis* eyespot focus can elicit a proximal pattern (Figure 3.2B(ii)), although this does not prove that the nearby bands form in response to a similar focal signal. The transplantation and ablation of proximal epidermis has not revealed any band foci (Nijhout, personal communication). Similarly, *Bicyclus* has ventral (but not dorsal) bands and these are not modified by deep ablations in putative focal positions (French, unpublished). Of course, these negative results may indicate merely that focal signalling for band specification occurs earlier and hence might be revealed by appropriate operations performed earlier, on the larval wing disc. Surgical experiments at larval stage have been limited to the removal of large distal parts of the *Precis* wing disc (Nijhout and Grunert, 1988), and have not resolved this issue. The resulting wings may be severely truncated, with missing or distorted eyespots, but have a regenerated wing margin, complete with the nearby parafocal and marginal bands. This result is consistent with, but does not prove, specification of those distal pattern elements by signalling from the edge of the wing disc (Nijhout and Grunert, 1988).

Some of the earliest surgical experiments on lepidopteran wing patterns concerned the development of bands, but in pyralid moths such as *Ephestia kuhniella* (Kuhn and von Englehardt, 1933). The results have been interpreted in terms of signalling from a line of foci, located midway between the bands (Nijhout, 1978, 1985, 1991), as microcautery of the wing epidermis can move both bands in towards this midline (Figures 3.3A and 3.3B). Surprisingly, however, this result occurs in response to damage anywhere on the pupal wing, as if the operation stops or reduces focal signalling throughout the epidermis. Earlier, just after pupation, microcautery has only a local effect on pattern, deflecting the nearest band 'inside' the site of damage (Figure 3.3C), suggesting that patterning signals may be blocked or degraded locally by the damaged epidermis (Nijhout, 1985). If signalling foci were present midway between the final band positions, early damage in that region would be expected to remove a focus and cause a symmetrical constriction in the band pattern. Instead, this operation frequently induces an ectopic ring pattern (Figure 3.3D), without affecting

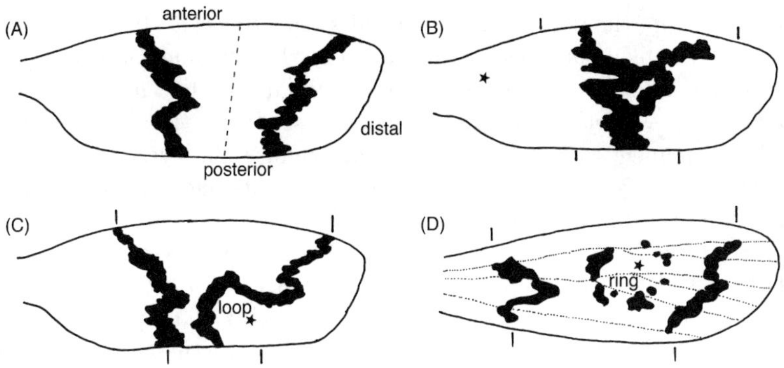

Figure 3.3 Figure 3. The band pattern on dorsal fore-wings of the moth, *Ephestia kuhniella*. (A) The control wing with prominant proximal and distal bands. The dashed line indicates the suggested medial location for the signalling foci (see text). (B) Movement of both bands towards the midline following an extreme proximal cautery (asterisk) at mid pupal stage. Lines indicate the control position of the bands. (C) Local effect of cautery at early pupal stage: the nearby band loops inside the region of damage. (D) Local effect of an early medial cautery: an ectopic ring pattern forms around the damaged area while proximal and distal bands remain in normal positions. Note: Wings A-C were dissected from pharate adults (redrawn from Kuhn and von Englehardt (1933) whereas D shows a fully-expanded adult wing (redrawn from Toussaint and French, 1988).

the position of the normal bands (Toussaint and French, 1988).

The symmetrical system of wing bands (and the other types of pattern element) may indeed resemble eyespots in being specified by signals from discrete foci, but it remains to be demonstrated unequivocally, by the effects of removal and transplantation of these regions of epidermis.

3.4 Eyespots: Foci, Gradients and Gene Expression

There is clear experimental evidence that the eyespot pattern of adult scale pigmentation is specified in the early pupal stage, as a consequence of signals from the central focal cells (see above). Nijhout (1980b, 1990) proposed that the mechanism of signalling is a simple, long-range concentration gradient which is generated by focal cells acting as the unique source of a 'morphogen', perhaps a small molecule that can pass freely through gap junctions between the epidermal cells. A point source, diffusion and decay will generate a radial concentration gradient over the epidermis. Cells close to the focus will experience a high concentration of the morphogen but, as distance increases, the concentration falls, response thresholds are crossed and the cells become specified for different pathways of pigment synthesis in the late pupa, thereby generating the eyespot pattern of concentric colour rings (Nijhout, 1980b, 1990, 1991). Alternatively, the formation of a large eyespot could occur by a cascade of short-range signals initiated by activity of the focus before propagating out across the epidermis. Each signal specifies the next peripheral state and induces those cells to produce the next signal.

Several features of the experimental results in *Bicyclus* favour the long-range gradient mechanism for eyespot specification (French and Brakefield, 1995). Attenuation of the signal will cause a gradient mechanism (but not a short-range cascade) to produce a partial pattern that consists of only the more peripheral elements. Indeed, damage to the focus often does reduce the eyespot pattern to a small patch of the outer, gold scales. Similarly, epidermal damage can mimic the focal signal and induce a pattern, ranging from a patch of gold scales (Brakefield and French, 1995) to a large ectopic eyespot (Figure 3.2A(iv)). Also, as shown clearly by the *Precis* results (Nijhout, 1980b), early ablation of the focus can abolish the eyespot, whereas damage somewhat later can still reduce its size, indicating that normal patterning depends on continued function of the focus. In a cascade mechanism the focus will become dispensible once signalling is initiated (Nijhout, 1991).

Focal grafting experiments show that the size of an induced eyespot pattern is determined by properties of the focus (see above). In a gradient mechanism, foci could differ in the level or timing of morphogen production, giving differences in the gradient profile and hence in eyespot size. In a cascade mechanism, however, the overall size of the pattern will depend on the initiating focus only if the level of each signal governs the intensity of the response and hence the level and range of the next signal in the relay (Reilly and Melton, 1996). At present, a gradient model fits most experimental results but ultimately the mechanism by which the eyespot is specified needs to be resolved by molecular investigation (Brakefield and French, 1995).

In recent years much has been revealed about the mechanisms of pattern formation through the use of genetic manipulations that alter the proteins involved in cell interaction. Mutant phenotypes can identify the genes vital to a particular developmental process and then analysis of gene sequences and expression patterns can suggest functions for their encoded proteins (e.g. as intercellular signals, receptors or components of intracellular signal transduction). Furthermore, many genetic techniques have been developed, particularly in *Drosophila*, for altering the level or the pattern of expression of a gene. Two very general conclusions from this work have significance for butterfly wing patterning: (1) intercellular signalling pathways are mostly very ancient and highly conserved, even between very different forms of animal; (2) there are relatively few such pathways and they are used repeatedly in different contexts throughout development. Thus, although complex wing colour patterns are restricted to the Lepidoptera, they may well employ developmental mechanisms already known in other contexts, in the development of other animals.

There is now good evidence for morphogen gradients in the development of *Drosophila*, not just in the syncytial early embryo but in the epidermis of the imaginal wing disc (Lawrence and Struhl, 1996). For example, *wingless* (*wg*) and *decapentaplegic* (*dpp*) are expressed in thin stripes around the edge and down the middle, respectively, of the future wing blade (Figure 3.4A). The encoded proteins (Wg and Dpp) are secreted signals that ultimately control expression of target genes in the adjacent tissue. Wg signalling, for example, activates a broad distal band of expression of *Distal-less* (*Dll*). Elegant experiments, involving the ectopic induction of the signal (at different levels) and of intracellular components of the response, have shown that both Wg (Zecca *et al.*, 1996; Neumann and Cohen, 1997) and Dpp (Nellen *et al.*, 1996; Lecuit *et al.*, 1996) are long-range signals acting directly, in concentration gradients, along the proximal-distal and anterior-posterior axes of the

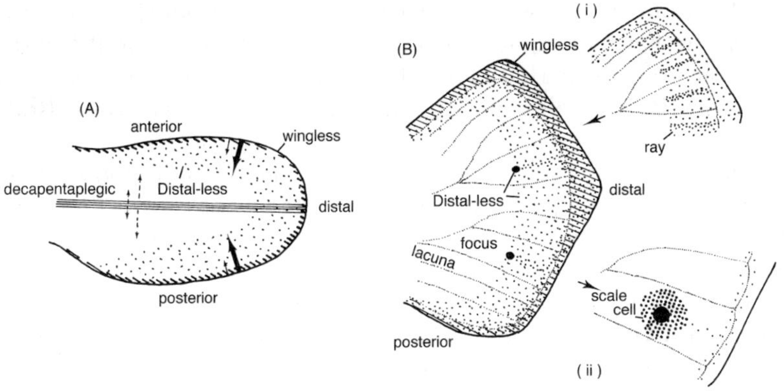

Figure 3.4 Diagrams of gene expression in *Drosophila* and butterfly wing development. (A) The *Drosophila* late larval wing imaginal disc (drawn as if it were everted, for direct comparison with the butterfly disc). Distal expression (hatched) of *wingless* results in a protein concentration gradient that controls expression of the regulatory gene *Distal-less* (large arrows) and of other genes (indicated by small arrows). Similarly, expression of *decapentaplegic* (shown by horizontal lines) gives a gradient whose concentrations define the expression zones of other genes (dashed arrows). (B) The late 5th instar wing disc of *Precis coenia*, showing expression of wingless (hatched) in the distal rim of the disc, and of Distal-less in a broad distal zone (fine stippling) and at high level in two focal spots (see Brakefield *et al.*, 1996). i) shows an earlier stage, the mid 5th instar, with Distal-less expression extending in midline rays along each wing-cell. ii) shows a later stage, the early pupal wing, with Distal-less expression in the enlarging scale cells within an area surrounding the focal spot.

wing. Another secreted signal, encoded by *hedgehog* (*hh*), is expressed posteriorly and its graded concentration controls gene expression across a central region of the wing blade (Strigini and Cohen, 1997). These studies have demonstrated the function of morphogen gradients in controlling zones of expression of target genes, and thereby in positioning morphological structures such as the veins (Sturtevant and Bier, 1995; Biehs *et al.*, 1998). Sadly, however, the *Drosophila* wing has no colour pattern!

Butterfly homologues have now been identified for some of the genes important in *Drosophila* wing development (Carroll *et al.*, 1994; Keys *et al.*, 1999). Unsurprisingly, there are many similarities in their expression patterns, but several of the butterfly genes have additional components of expression that suggest roles in specifying elements of the wing colour pattern (Figure 3.4B). Most striking is *Distal-less*, which is expressed in the butterfly wing disc in the broad distal band, but at higher levels in rays positioned between the lacunae. The rays fade in the late larval disc, except where expression is enhanced in small spots that correspond in position to the foci of the future eyespots, on both *Precis* and *Bicyclus* wings (Carroll *et al.*, 1994; Brakefield *et al.*, 1996). In *Drosophila*, *Dll* is required for normal growth and patterning in much of the leg and for normal differentiation at the wing margin (Campbell and Tomlinson, 1998). The Dll protein, as a putative transcription factor, is certainly not the focal signal in butterflies, but it may somehow control signalling from the focus. Surgical experiments indicate that focal signalling occurs in the early pupa and it is intriguing that *Dll* expression expands at that stage (Figure 3.4B(ii)), spreading out from the focal spot to the developing scale cells in a surrounding area of epidermis (Brakefield *et al.*, 1996).

The cells of the eyespot focus do not express *wg* or *dpp* (Carroll *et al.*, 1994) or *hh* (Keys *et al.*, 1999), so the corresponding secreted proteins cannot be the focal signal(s). However, these genes are expressed along the disc lacunae, or flanking the eyespot focus, suggesting that they may be involved in the earlier process of establishing the focus (Carroll *et al.*, 1994; Keys *et al.*, 1999). Furthermore, in the *Precis* forewing disc, proximal bars of *wg* expression suggest that this signal may have a role in specifying the band pattern (Carroll *et al.*, 1994; Weatherbee *et al.*, 1999). Clues to the identity of the eyespot focal signal(s) may come eventually from the expression of butterfly homologues of genes encoding other known *Drosophila* signals (such as the ligands for the Notch or EGF signalling pathways) or directly, from searches for butterfly transcripts that are specifically expressed in the early pupal focal cells. In any case, normal expression patterns are not enough: to prove a role in focal signalling and then to analyse the mechanism, mutation or manipulations of gene expression will be required.

3.5 The Development of Wing Patterns: Mechanisms and Models

A distinctive feature of butterfly wing patterns is their great diversity: even closely related species can differ markedly in the presence or absence of pattern elements and in their form and colour (Figure 3.1). Within a wing-cell the band system appears in most species as straight bars but it can form arcs, swirls or even circles; the eyespot may be circular or deformed into an ellipse or chevron; and the parafocal elements are particularly diverse in shape, ranging from bands to arcs, chevrons, double chevrons and intervenous stripes (Nijhout, 1990). All these forms of wing pattern appear on related species and are likely to be generated by similar developmental mechanisms that have diverged in various ways from a common ancestral state (Nijhout, 1990, 1991, 1994). The potential of a particular developmental mechanism for generating a range of related patterns can best be assessed by a modelling approach (Nijhout and Paulsen, 1997).

Several studies (Nijhout, 1978, 1985; Murray, 1981; Bard and French, 1984) have shown that specific diffusion or reaction-diffusion models can indeed generate patterns of morphogen concentration closely matching the band or eyespot colour patterns of particular species, when the parameters and the locations for morphogen sources (and sinks) are carefully chosen. Nijhout (1990, 1991) has gone much further, however, and has explored the diversity of morphogen concentration landscapes (and hence the potential shapes of pattern elements) that can be generated by a simple model of diffusion, given a few standard and plausible positions for the sources and sinks. Most general shapes seen in the butterfly pattern elements can be generated by diffusion operating within a standard rectangular wing-cell, with constant-level sources and sinks, as points or as lines, located only along the midline, the lateral veins or the distal edge of the wing-cell (Figure 3.5A). When the model is expanded to include two distinct gradients whose values are added together, then a very wide range of patterns

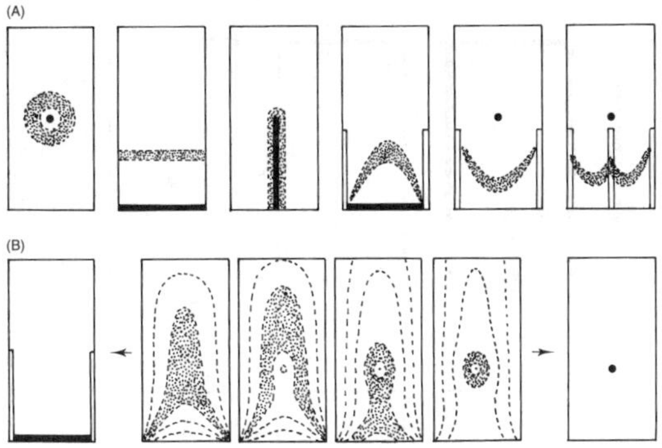

Figure 3.5 Specification of wing pattern elements by a diffusion model. (adapted from Nijhout, 1990, 1991). The wing-cell is represented by a rectangular panel with the distal edge at the bottom. Localised constant-level morphogen sources and sinks are shown by filled spots or bars, and by open bars, respectively. (A) A small set of focal positions can generate the major shapes of distal pattern element (dotted). The eyespot, band, intervenous stripe, chevron and double chevron correspond to a concentration range on the diffusion landscape resulting from the simple arrangements of foci indicated in the six panels (Note: pattern corresponding to only the distal part of the concentration landscape is indicated in the two right panels. (B) An additive two-gradient model can generate a wide range of element shapes, using the same small set of focal positions. The four central panels illustrate the efects of a change in the balance between one gradient, formed by a distal line source lateral line sinks (shown in the left panel) and the other, resulting from a midline point source (right panel). Additive morphogen concentrations are shown by dashed lines and one possible pattern element (corresponding to a particular concentration range) is shown by dots.

can be formed by subtle variation of the parameters, while retaining the same small set of potential focal (source and sink) locations (Figure 3.5B).

The diffusion model can generate patterns corresponding to at least 70% of the eyespot and parafocal patterns encountered within the Nymphalidae (Nijhout, 1990), provided that a few areas (in medial, lateral or distal positions) have been specified to act as foci. Early in the last larval instar, regularly spaced lacunae appear across the imaginal disc, and Nijhout (1990, 1991) has shown that the required set of focal positions could be established by an effect of these lacunae on reaction-diffusion operating within the enclosed wing-cell (Figure 3.6). In a specific reaction-diffusion system, Meinhardt's (1982) lateral inhibition model, with the lacunae providing an initial weak source of activator to disturb the steady state, high activator concentrations can build up along the midline of the wing-cell, and then resolve into a stable, discrete central peak (Figure 3.6A). This process shows a marked similarity to the progress of *Distal-less* expression during the formation of the eyespot focus (Figure 3.4B). Furthermore, by varying the parameters of the model, the final activator peak can be formed, as a line or as one or more points, in either lateral, medial or distal positions within the wing-cell (Figure 3.6B). If the activator peaks specify regions of the wing-cell to act later as foci, then reaction-diffusion can build on the simple

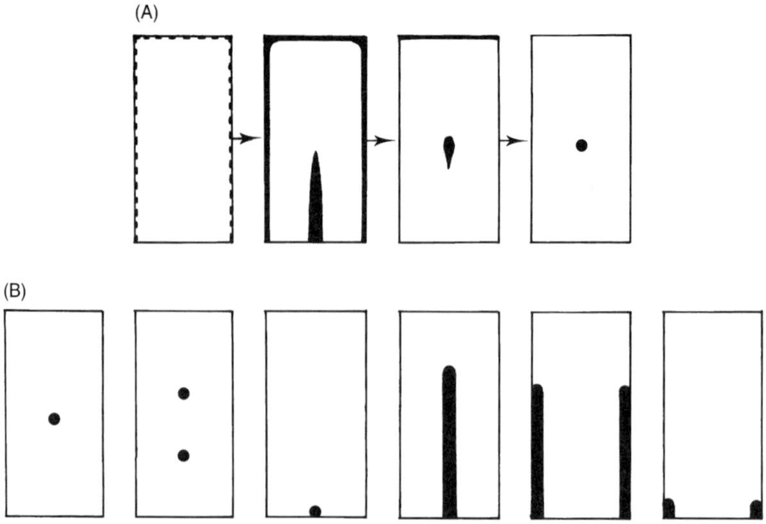

Figure 3.6 Specification of a range of focal positions by reaction-diffusion (adapted from Nijhout, 1990, 1991). The wing-cell is represented as in Figure 5. (A) The progressive establishment of a stable activator peak (filled spot), following the introduction of a small amount of activator (dashed lines) along the proximal and lateral edges of the wing-cell. With this set of parameters, activator concentration builds up along the margins, then along the midline of the wing-cell and then stabilises as a single midline peak. (B) By varying the initial activator supply and the parameters of reaction-diffusion, stable activator peaks can be produced in different midline, lateral and distal locations in the wing-cell, thus giving the 'toolbox' of potential foci for the subsequent specification of pattern elements by diffusion (see Figure 5).

spacing of the lacunae, to provide the range of initial conditions for the subsequent diffusion phase of pattern formation (Nijhout, 1990, 1991).

Nijhout's two-stage (reaction-diffusion/focus-diffusion) model is very persuasive in its ability to generate much of the observed diversity in butterfly wing pattern elements, and also in the basis of its transitions between patterns (Nijhout, 1990, 1991; Nijhout and Paulsen, 1997). Parameter changes in the first stage (resulting, for example, from mutations that alter properties of enzymes regulating the activator or inhibitor) may change the consequent pattern of foci (Figure 3.6B) and hence alter the type of wing pattern ultimately formed. In the second stage, mutation may affect the morphogen (e.g. by changing its level of synthesis or its diffusion coefficient) or change the sensitivity or nature of the response, and hence lead to changes in the details of the colour pattern (Figure 3.5B). Thus the model illustrates how the rich diversity of wing patterns may have evolved, once the basic developmental mechanism was established in ancestral higher Lepidoptera.

In principle the model has problems with size invariance. The patterns generated by reaction-diffusion systems are sensitive to scale whereas, in general, butterfly wing colour patterns are not (Nijhout, 1991). It is not clear, however, that this is a serious problem, as much of the size difference between individual butterflies may arise during late growth, after the foci have been specified in the imaginal disc. The model has achieved an excellent match with the observed eyespot and parafocal patterns, but it may have problems in accounting for the proximal band elements, particularly where straight sections of band are dislocated at the wing veins, as in Figure 3.1C.

At present, lack of empirical support must be one of the main reservations about accepting a unified model for the formation of the different types of pattern element. As explained above, there is little experimental evidence for the action of discrete signalling foci in the formation of any pattern element other than the simple eyespot. In this context, it is interesting that there are no spots of *Distal-less* expression relating to the bands, the parafocals or other forms of pattern element in any of the species examined (Brakefield *et al.*, 1996). Also, modification of the mechanism of *Bicyclus* eyespot development by artificial selection, or by gene mutation, has no effect on the other pattern elements (Brakefield and French, 1999). These observations do not suggest that common mechanisms are involved in specifying the different types of pattern element. The success of the unified model in generating pattern diversity should, however, provide a stimulus for searching (and guidance about where to look) for foci in the development of the other types of pattern.

Even in the case of the eyespot, it is far from certain that the mechanism does correspond to a focal source and long-range diffusion gradient, particularly since the spatial patterns generated by diffusion may be indistinguishable from those formed by a particular form of direct cell-cell signalling (Hammer, 1998). Nijhout (1980b, 1990) proposed a gradient formed by the diffusion of a small molecule through gap junctions within the epidermal cell layer. So far, however, there is no evidence for such a patterning mechanism in the best-known insect, *Drosophila*, although disruption of gap junctions can be associated with abnormalities of cell proliferation and patterning in the imaginal discs of some mutants (Jursnich *et al.*, 1990). Gradients are now known to be important in patterning *Drosophila* wings (see above) but they involve large, secreted proteins and they operate over a range of 10-20 cell diameters, in contrast to the approximately 100-cell range of the focal signal on the butterfly pupal epidermis (French and Brakefield, 1995). Also, these *Drosophila* gradients seem not to depend on free, extracellular diffusion of the morphogen. Wingless, for example, is secreted by the producing cells, but it then appears to be taken into adjacent cells by endocytosis and passed in vesicles across cells of the epidermis (Gumbiner, 1998). Also, the range of the Wingless gradient appears to depend on a feedback interaction with its receptor, Dfz2 (Cadigan *et al.*, 1998). Even if the butterfly foci do signal by a long-range gradient, the details of the mechanism remain obscure.

3.6 Conclusion

The Lepidoptera display spectacularly varied wing colour patterns which may function in camouflage, mimicry, warning and deflection of predator attack; in species recognition and even in thermoregulation. The system offers great opportunities for understanding how developmental mechanisms have been modulated in the evolution of adaptive phenotypic traits.

Through surgical manipulations and now molecular investigation, we are beginning to understand the eyespot and we can anticipate that this approach will reveal the developmental mechanisms that specify this simple pattern element. The extensive analysis of wing pattern variation has proposed specific homologies between very different forms of wing pattern and between the different pattern elements. Quantitative modelling is yielding plausible ideas about common mechanisms that may underlie the development of diverse patterns. We now need broadly based experimentation, to test to what extent the eyespots and other elements of wing pattern do share developmental mechanisms. When combined with a phylogenetic analysis of the pattern variation found within particular species groups, this approach may begin to indicate how developmental mechanisms may have constrained or biased the process of evolution (Brakefield and French, 1999).

Acknowledgements
I thank Paul Brakefield, Sean Carroll and Fred Nijhout for sharing their ideas, results and enthusiasms.

REFERENCES

Bard, J.B.L. and French, V. (1984) Butterfly wing patterns: how good a determining mechanism is the simple diffusion of a single morphogen? *J. Embryol. Exp. Morphol.* 84: 255–274.

Biehs, B., Sturtevant, M.A. and Bier, E. (1998) Boundaries in the *Drosophila* wing imaginal disc organise vein-specific genetic programs. *Development* 125: 4245–4257.

Brakefield, P.M. and French, V. (1995) Eyespot development on butterfly wings: the epidermal response to damage. *Dev. Biol.* 168: 98–111.

Brakefield, P.M. and French, V. (1999) Butterfly wings: the evolution of development of colour patterns. *BioEssays* 21: 391–401.

Brakefield, P.M., Gates, J., Keys, D., Kesbeke, F., Wijngaarden, P.J., Monteiro, A., French, V. and Carroll, S.B. (1996) Development, plasticity and evolution of butterfly eyespot patterns. *Nature* 384: 236–242.

Cadigan, K.M., Fish., M.P., Rulifson, E.J. and Nusse, R. (1998) Wingless repression of *Drosophila frizzled 2* expression shapes the Wingless morphogen gradient in the wing. *Cell* 93: 767–777.

Campbell, G. and Tomlinson, A. (1998) The roles of the homeobox genes *aristaless* and *Distal-less* in patterning the legs and wings of *Drosophila. Development* 125: 4483–4493.

Carroll, S.B., Gates, J., Keys, D.N., Paddock, S.W., Panganiban, G.E.F., Selegue, J.E. and Williams, J.A. (1994) Pattern formation and eyespot determination in butterfly wings. *Science* 265: 109–114.

French, V. and Brakefield, P.M. (1995) Eyespot development on butterfly wings: the focal signal. *Dev. Biol.* 168: 112–123.

Galant, R., Skeath, J.B., Paddock, S., Lewis, D.L. and Carroll, S.B. (1998) Expression pattern of a butterfly *achaete-scute* homologue reveals the homology of butterfly wing scales and insect sensory bristles. *Curr. Biol.* 8: 807–813.

Gumbiner, B.M. (1998) Propagation and localization of Wnt signaling. *Curr. Opin. Gen. Dev.* 8: 430–435.

Hammer, O. (1998) Diffusion and direct signaling models are numerically equivalent. *J. Theor. Biol.* 192: 129–130.

Jursnich, V.A., Fraser, S.E., Held, L.I., Ryerse, J. and Bryant, P.J. (1990) Defective gap-junctional communication associated with imaginal disc overgrowth and degeneration caused by mutations of the *dco* gene in *Drosophila. Dev. Biol.* 140: 413–429.

Keys, D.N., Lewis, D.L., Selegue, J.E., Pearson, B.J., Goodrich, L.V., Johnson, R.L., Gates, J., Scott, M.P. and Carroll, S.B. (1999) Recruitment of a *hedgehog* regulatory circuit in butterfly eyespot evolution. *Science* 283: 532–534.

Kuhn, A. and von Englehardt, M. (1933) Uber die Determination des Symmetriesystems auf dem Vorderflugel von *Ephestia kuhniella. Wilhelm Roux' Arch. EntwMech. Org.* 130: 660–703.

Lawrence, P.A. and Struhl, G. (1996) Morphogens, compartments and pattern: lessons from *Drosophila*? *Cell* 85: 951–961.

Lecuit, T., Brook, W.J., Ng, M., Calleja, M., Sun, H. and Cohen, S.M. (1996) Two distinct mechanisms for long-range patterning by Decapentaplegic in the *Drosophila* wing. *Nature* 381: 387–393.

Meinhardt, H. (1982) *Models of Biological Pattern Formation.* Academic Press, New York.

Monteiro, A.F., Brakefield, P.M. and French, V. (1994) The evolutionary genetics and developmental basis of wing pattern variation in the butterfly *Bicyclus anynana. Evolution* 48: 1147–1157.

Monteiro, A.F., Brakefield, P.M. and French, V. (1997) Butterfly eyespots: the genetics and development of the color rings. *Evolution* 51: 1207–1216.

Murray, J.D. (1981) On pattern formation mechanisms for lepidopteran wing patterns and mammalian coat markings. *Phil. Trans. Roy. Soc. Lond. B* 295: 473–496.

Nellen, D., Burke, R., Struhl, G. and Basler, K. (1996) Direct and long-range action of a DPP morphogen gradient. *Cell* 85: 357–368.

Neumann, C.J. and Cohen, S.M. (1997) Long-range action of Wingless organizes the dorsal-ventral axis of the *Drosophila* wing. *Development* 124: 871–880.

Nijhout, H.F. (1978) Wing pattern formation in Lepidoptera: a model. *J. Exp. Zool.* 206: 119–136.

Nijhout, H.F. (1980a) Ontogeny of the colour pattern on the wings of coenia (Lepidoptera: Nymphalidae) *Dev. Biol.* 80: 27–274.

Nijhout, H.F. (1980b) Pattern formation on lepidopteran wings: determination of an eyespot. *Dev. Biol.* 80: 27–274.

Nijhout, H.F (1985) Cautery-induced colour patterns in *Precis coenia* (Lepidoptera: Nymphalidae) *J. Embryol. Exp. Morphol.* 86: 191–203.

Nijhout, H.F. (1990) A comprehensive model for color pattern formation in butterflies. *Proc. Roy. Soc. Lond. B* 239: 81–113.

Nijhout, H.F. (1991) *The Development and Evolution of Butterfly Wing Patterns.* Smithsonian Institution Press, Washington.

Nijhout, H.F. (1994) Symmetry systems and compartments in Lepidopteran wings: the evolution of a patterning mechanism. In: *The Evolution of Developmental Mechanisms* (Eds. M. Akam, P. Holland, P. Ingham and G. Wray) *Development Supplement*, 225–233.

Nijhout, H.F and Grunert, L.W. (1988) Colour pattern regulation after surgery on the wing discs of *Precis coenia* (Lepidoptera: Nymphalidae) *Development* 102: 377–385.

Nijhout, H.F. and Paulsen, S.M. (1997) Developmental models and polygenic characters. *Am. Nat.* 149: 394–405.

Nijhout, H.F. and Wray, G.A. (1986) Homologies in the colour patterns of the genus *Charaxes* (Lepidoptera: Nymphalidae) *Biol. J. Linn. Soc.* 28: 387–410.

Nijhout, H.F. and Wray, G.A. (1988) Homologies in the colour patterns of the genus *Heliconius* (Lepidoptera: Nymphalidae) *Biol. J. Linn. Soc.* 33: 345–365.

Reilly, K.M. and Melton, D.A. (1996) Short range signalling by candidate morphogens of the TGFB family and evidence for a relay mechanism of induction. *Cell* 86: 743–754.

Schwanwitsch, B.N. (1924) On the groundplan of wing-pattern in nymphalids and certain other families of rhopalocerous Lepidoptera. *Proc. Zool. Soc. Lond. B* 34: 509–528.

Strigini, M. and Cohen, S.M. (1997) A Hedgehog activity gradient contributes to AP axial patterning of the *Drosophila* wing. *Development* 124: 4697–4705.

Sturtevant, M.A. and Bier, E. (1995) Analysis of the genetic hierarchy guiding wing vein formation in *Drosophila. Development* 121: 785–801.

Suffert, F. (1927) Zur vergleichende Analyse der Schmetterlingszeichnung. *Biol. Zentr.* 47:

385–413.

Toussaint, N. and French, V. (1988) The formation of pattern on the wing of the moth, *Ephestia kuhniella*. *Development* 103: 707–718.

Weatherbee, S.D., Nijhout, H.F., Grunert, L.W., Halder, G., Galant, R., Selegue, J. and Carroll, S.B. (1999) Ultrabithorax function in butterfly wings and the evolution of insect wing patterns. *Curr. Biol.* 9: 109–115.

Wolpert L. (1971) Positional information and pattern formation. *Curr. Top. in Dev. Biol.* 6: 183–224.

Zecca, M., Basler, K. and Struhl, G. (1996) Direct and long-range action of a Wingless morphogen gradient. *Cell* 87: 833–844.

4

Pattern Formation in Cancer

J.A. Sherratt[1], A.J. Perumpanani[2], M.R. Owen[3]

[1] *Department of Mathematics, Heriot-Watt University, Edinburgh EH14 4AS, UK*
[2] *Department of Surgery, Massachusetts General Hospital, and Shriners Burns Institute, Harvard Medical School, Bldg 1400 W, One Kendall Square, Cambridge MA 02139, USA*
[3] *Department of Mathematics, University of Utah, Salt Lake City, UT 84112, USA*

4.1 Introduction

The various stages of solid tumour formation and progression yield a wide range of pattern formation problems. Broadly speaking, these problems can be divided into two categories: those concerning the overall shape of a tumour, and those concerning internal tumour composition. The former category is important because of its potential use in prognosis: overall tumour morphology has long been known to corrolate with metastatic potential. Recently this connection has been made more precise, via the application of techniques from fractal geometry (Landini and Rippin, 1993, 1996), which shows that the fractal dimension of the tumour boundary is a powerful prognostic indicator. Mathematical modelling of the cellular and biochemical origins of variations in tumour shape have focused primarily on the initial avascular stage of tumour development. This was one of the earliest phases of tumour development to be treated mathematically (Greenspan, 1972), initially under the assumption of spherical symmetry. In the last few years, this work has been extended to study the development of asymmetries. For example, Byrne and Chaplain (1996) and Byrne (1997) studied a model consisting of conservation equations for nutrient and pressure within a tumour, the pressure coming from rapid tumour cell proliferation. Here the tumour edge enters as a free boundary, and the application of spherical harmonic analysis shows that asymmetries develop when the level of cell–cell adhesion is low.

On Growth and Form: Spatio-temporal Pattern Formation in Biology.
Editor Mark A.J. Chaplain, G.D. Singh, J.C. McLachlan

This is consistent with the fact that a reduction in homotypic adhesion is known to be a key ingredient in the onset of metastasis.

Pattern formation in the structure of solid tumours has received much more attention from mathematical modelling than tumour shape. The earliest work in this area focused on the division into a viable rim of proliferating cells around a necrotic core (Greenspan, 1972). Recently, this issue has been studied in detail by Ward and King (1997); their calculations give formulae for the thickness of the viable rim as a function of key model parameters. Many other aspects of internal patterning have also been studied mathematically. For example, the ingrowth of blood vessels into a solid tumour (angiogenesis) involves a complex series of patterning steps, with capillary sprouts advancing into the tumour and then merging to form loops. This has been modelled by Chaplain and coworkers in a series of papers (Chaplain and Sleeman, 1990; Chaplain and Stuart, 1993; Orme and Chaplain, 1996): Orme and Chaplain (1996) have shown how the development of capillary loops depends on an interplay between chemotaxis and haptotaxis of endothelial cells.

In this chapter we focus on two aspects of the internal patterning of tumour cells: the formation of a capsule around solid tumours, and the ability of macrophages to induce tumour heterogeneity.

4.2 Tumour Encapsulation

The presence of a capsule around a tumour is the most significant gross morphological feature determining the clinical outcome. Tumours that are encapsulated (i.e. having a dense band of surrounding connective tissue) have a favourable prognosis and only produce symptoms related to pressure effects on surrounding tissue (Bailey *et al.*, 1995). Such tumours are known as benign. Malignant tumours on the other hand do not have a well circumscribed capsule; the cancerous cells invade neighbouring tissue and are carried far from their primary site by the metastatic cascade (Aznavoorian *et al.*, 1990). Malignant tumours are potentially lethal and may either arise de novo or in existing benign tumours which may be encapsulated. In the latter case the cancerous cells have to disrupt the capsular barrier before spreading further.

Any normal tissue in the body may be viewed simplistically as a collection of different types of cells anchored into position by the presence of intervening extracellular matrix. Among other elements, this extracellular matrix is composed of strands of connective tissue fibres such as collagen, elastin, fibronectin, etc. (William and Gray, 1995). A tumour arises when one of the cells (or sometimes a few) proliferates more rapidly than its neighbours. This aggressive proliferation may be the result of a mutation which encourages division, or one that makes the cell deaf to the inhibitory growth signals from its neighbours (Robbins *et al.*, 1984). In any case the result is a group of cells proliferatively outpacing the neighbouring cells and producing a localised growth - a benign tumour. The pathological hallmark of such a benign tumour is the presence of a dense band of connective tissue around it - the capsule. The capsule is composed chiefly of matrix fibres; we use the term "connective tissue" throughout, following convention. Here we begin by considering the mechanism of capsule formation, using a mathematical formulation of existing ideas on encapsulation. We will use a combination of analytical and numerical techniques to describe the mathematical analogue of encapsulation as a bifurcation from a constant shape wave to an aggregating wave of connective tissue. The work in this section has been previously reported in Perumpanani *et al.* (1997).

4.2.1 Formulation of the model

We derive a model for the growth of a benign tumour based on a continuum approach, in which $m(x,t)$ and $c(x,t)$ represent the concentrations of the tumour cells and connective tissue respectively. Here x and t are the space and time coordinates. The model studies the averaged behaviour of the tumour cells in the direction of expansion only, and ignores variations in the plane perpendicular to the direction of growth.

In the absence of any surrounding connective tissue we describe the local proliferation of the benign tumour cells by $f(m)$ and the flux of the cells by J_m, say. However, the presence of connective tissue will influence the movement of the benign tumour cells. We describe the effects of connective tissue on the motility of tumour cells by $\theta(c)$, and model the overall flux of the tumour cells by the product $\theta(c)J_m$. We base our model on the "expansive growth hypothesis" for capsule formation, which suggests that the capsule may form by the passive convection of extracellular matrix (ECM) around the tumour as this grows (Berenblum, 1970). This is supported by some pathological evidence, although the alternative "foreign body hypothesis" has some support: this proposes that the capsule forms via an inflammatory response to the tumour (Barr *et al.*, 1988). Based on the expansive growth hypothesis stated earlier, we model the connective tissue flux as being proportional to the flux of the cells, that is, $\theta(c)J_m$. This is the appropriate flux term since the density of connective tissue c is measured per unit volume rather than per cell. In practice there should be a separate saturation term, in addition to $\theta(c)$, representing the limitation on matrix reorganisation potential per cell. However, since we already have a saturation effect via $\theta(c)$, we omit this term for simplicity. Under these assumptions we write the model for benign tumour growth as

$$\frac{\partial m}{\partial t} = f(m) + \frac{\partial}{\partial x}[\theta(c)J_m], \tag{4.1}$$

$$\frac{\partial c}{\partial t} = k\frac{\partial}{\partial x}[c\,\theta(c)J_m]. \tag{4.2}$$

Our approach is based on the prediction of Sherratt and Nowak (1992) that the early growth of a tumour has the form of a travelling wave moving outwards from an initial site of disease into the surrounding normal tissue. The increase in tumour size is primarily driven by cell division; cell movement is just local random motion at the tumour edge. We use simple functional forms of f and J_m which give travelling wave solutions reminiscent of those demonstrated by Sherratt and Nowak (1992). The advantage of such a generic model approach is that the ensuing simplicity allows an analytical appraisal of the equations while retaining the necessary qualitative features. Throughout, we describe the proliferation of tumour cells using the logistic growth model which Vaidya and Alexandro (1982) established as an appropriate model for tumour cell growth. Under an appropriate rescaling of tumour cell density m and time t, this gives $f(m) = m(1-m)$.

We assume that the boundary does not have any significant bearing on the evolution of the tumour and hence study the case of an infinite domain, which in our numerical simulations is represented by a large finite domain with zero derivative conditions on the boundary. We assume that at time $t = 0$ the tumour cell population is zero except in a small region centred at $x = 0$. We take the initial connective tissue density $c(x, 0)$ to be zero near this region, and a non-zero constant (set arbitrarily to 0.2) away from it. The results of simulations are not sensitive to the details of these initial conditions. To reduce simulation time, we solve only on the region $x \geq 0$, using a symmetry boundary condition at $x = 0$.

4.2.2 Connective tissue waves

We begin by considering equations (4.1) and (4.2) in the simple case of $\theta(c) \equiv 1$; that is, changes in connective tissue density have no effect on cell motility. This case will be relevant during the early stages of tumour growth, in which connective tissue accretion on the surface of the tumour is too low to significantly restrain its expansion. Moreover, study of this simple case gives important mathematical insights into the behaviour for more general forms of $\theta(c)$.

We have solved equations (4.1) and (4.2) numerically with $\theta(c) \equiv 1$ for two different forms for the cell flux. The first case is $J_m = \partial m/\partial x$, in which case equation (4.1) becomes the very well-studied Fisher equation (Murray, 1991). Standard theory shows that for this equation, initial conditions of the form we are using for m evolve to a travelling wave, moving with constant shape and with speed 2. Secondly, we have considered $J_m = m\,\partial m/\partial x$, giving the non-linear Fisher equation, for which our initial conditions evolve to a sharp-front travelling wave, moving with speed $1/\sqrt{2}$ (Sanchez-Garduno and Maini, 1995; Sherratt and Marchant, 1996). Our interest is in the way in which connective tissue is convected by these waves of cell density.

Numerical simulations show that, for both flux terms, the solution for c depends critically on the parameter k, which reflects the strength of connective tissue convection. When k is small, a wave of connective tissue moves in parallel with the wave of tumour cells, with the effect that a constant band of connective tissue is pushed ahead by the growing tumour (Figure 4.1). However, when k increases above a critical value, the form of the solution alters significantly, and the peak in the c-wave increases with time, corresponding to a gradually increasing intensity of the predicted capsule (Figure 4.2).

Before launching into a detailed mathematical analysis of these results, we present a simple intuitive explanation for the change in behaviour. In equation (4.2) with $\theta(c) \equiv 1$, kJ_m is the velocity at which connective tissue is being convected, while a is the speed of the tumour cell wave. When the velocity of the connective tissue wave is slower than the velocity of the cell wave, we expect only a fleeting ripple to be produced in the solution for c. However, when $kJ_m > a$ at some points, the ECM at these points is actually convected faster than the speed at which the cell wave moves, causing a build-up of connective tissue. An analogous physical situation is the difference between the rippling of the surface of a cornfield by wind (small k) and a gust of wind blowing loose strands of hay into a pile (large k). In the analysis that follows, we will show that the bifurcation value of k, at which the c-wave changes from a constant shape wave to an aggregating wave, is indeed $k = a/\max\{J_m\}$.

We will study this case of $\theta(c) \equiv 1$ analytically, with two key objectives – to

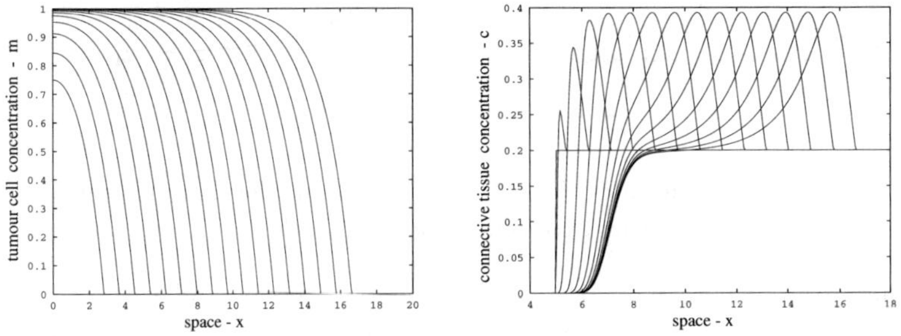

Figure 4.1 Numerical solutions of equations (4.2) with $k = 3.8$, $\theta(c) \equiv 1$, and $J_m = m\,\partial m/\partial x$. In this case the bifurcation value of k is 4. We plot m and c as functions of space at equal intervals of time. Both the m and the c waves eventually evolve into constant shape travelling waves.

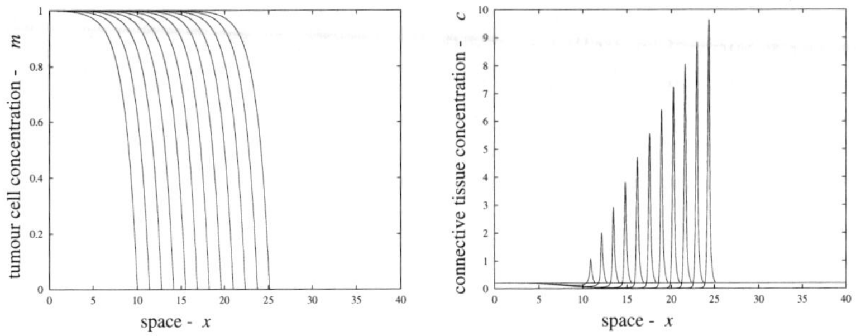

Figure 4.2 Numerical solutions of equations (4.2) with $k = 4.05$, $\theta(c) \equiv 1$, and $J_m = m\,\partial m/\partial x$. In this case the bifurcation value of k is 4. We plot m and c as functions of space at equal intervals of time measuring 1.9. The solution has the form of a constant shape travelling wave for m and an aggregating wave for c.

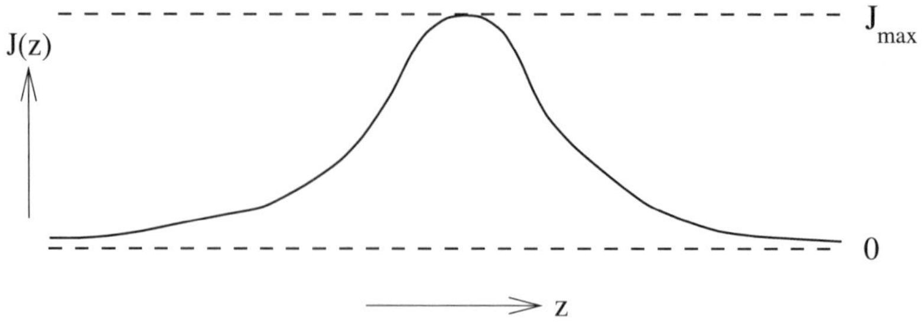

Figure 4.3 The qualitative form of the cell flux $J_m = J(z)$ when cell movement is not hindered by extracellular matrix levels.

determine the value of k at which the c-wave changes form, and the rate at which the peak of c increases when k is above this value. We denote by $M(z)$ the sharp-front travelling wave solution of (4.1) that evolves from our initial conditions; here $z = x - at$ and a is the speed of this cell wave. In the case of the non-linear Fisher equation, a closed-form expression for $M(z)$ exists, but not for the Fisher equation. In either case J_m depends only on M and its derivatives, and can thus be written as a function of z, say $J(z)$.

Rewriting (4.2) in terms of coordinates $\tau = t$ and $z = x - at$ gives

$$\frac{\partial c}{\partial \tau} + [kJ(z) - a]\frac{\partial c}{\partial z} = -kJ'(z)c\,. \tag{4.3}$$

This first-order partial differential equation can be solved exactly using the method of characteristics, and we will use this method to determine the qualitative form of the solution for $c(z, \tau)$ for general $J(.)$. Of course, in practice the solution for $m(x, t)$ only approaches $M(z)$ asymptotically as $t \to \infty$, and thus our solution for c will also only be exact in this limit.

The qualitative form of $J(z)$ is illustrated in Figure 4.3, and is the same for any realistic flux expression. Crucially, we will assume that $J(z) \geq 0$ for all z, with $J(z) \to 0$ as $z \to \pm\infty$, and $J'(z)$ non-zero except at the unique local maximum, at which $J(z) = J_{max}$, say. In the case of non-linear cellular diffusion, $J(z)$ is identically zero for sufficiently large z, since the m-wave is of sharp-front type, but this will not be significant in our calculations.

The characteristics of (4.3) are given by solving

$$d\tau = \frac{dz}{kJ(z) - a} = \frac{-dc}{kJ'(z)c}\,,$$

which gives the two characteristic functions

$$\mathcal{C}_1(c, z) = c/c_{tw}(z) \quad \text{and} \quad \mathcal{C}_2(\tau, z) = \tau - G(z)$$

$$\text{where} \quad c_{tw}(z) = 1/[a - kJ(z)] \quad \text{and} \quad G(z) = -\int c_{tw}(z)\,dz\,.$$

Thus the solution for c is given by eliminating the parameter θ, say, from

$$\frac{c}{c_{tw}(z)} = \frac{c_0(\theta)}{c_{tw}(\theta)} \quad \text{and} \quad G(z) - \tau = G(\theta)\,. \tag{4.4}$$

Here $c_0(x) \equiv c(x, t = 0)$. The key determinant of the form of this solution is the sign of $kJ_{max} - a$. If this expression is negative then $\mathcal{C}_1(c, z)$ is finite for all z, so that a possible solution is $c(z, \tau) = c_{tw}(z)$, independent of τ. This is a travelling wave solution for c and it is straightforward to show, using the method of characteristics, that all bounded initial conditions evolve to this solution, multiplied by $c_0(\theta)$, as $\tau \to \infty$.

The case $kJ_{max} > a$ is more complicated, however (Figure 4.4). Then $c_{tw}(z)$ has the qualitative form illustrated in Figure 4.4b, with infinities at the points, z_1 and z_2 say ($z_1 < z_2$), at which $kJ(z) = a$. Thus the solution for c cannot be of travelling wave form, and must be determined from (4.4). Integrating c_{tw} shows that $G(z)$ has the qualitative form illustrated in Figure 4.4c, also with infinities at z_1 and z_2. In order to use the solution (4.4), we require the parameter θ, which is given by $\theta(z, \tau) = G^{-1}[G(z) - \tau]$; here G^{-1} denotes the local inverse. The qualitative form of θ as a function of z at successively increasing τ is illustrated in Figure 4.4d. Crucially $\theta(z_1, \tau) \equiv z_1$ and $\theta(z_2, \tau) \equiv z_2$.

For simplicity, we begin by considering the case $c_0(x) \equiv 1$, so that (4.4) implies $c(z, \tau) = c_{tw}(z)/c_{tw}(\theta(z, \tau))$. The qualitative form of $c_{tw}(\theta)$ as a function of z at increasing τ is illustrated in Figure 4.4e, and compared to $c_{tw}(z)$. However, from these sketches alone it is not possible to determine the form of the ratio $c_{tw}(z)/c_{tw}(\theta)$, since the behaviour near $z = z_1$ and $z = z_2$ is not clear, and requires detailed analysis.

For z close to z_1, $kJ(z) - a \approx \lambda(z - z_1)$, where $\lambda = kJ'(z_1)$; our assumption that $J'(z) \neq 0$, except at the unique maximum, implies that $\lambda > 0$. Thus to leading order near $z = z_1$,

$$
\begin{aligned}
G(z) &= \lambda^{-1} \log |z - z_1| \\
\Rightarrow G^{-1}(\xi) &= z_1 + \exp(\lambda\xi) \cdot \mathrm{sign}(z - z_1) \\
\Rightarrow \theta &\equiv G^{-1}[G(z) - \tau] \\
&= z_1 + \exp\left\{\lambda\left[\lambda^{-1} \log |z - z_1| - \tau\right]\right\} \cdot \mathrm{sign}(z - z_1) \\
&= z_1 + (z - z_1)e^{-\lambda\tau} \\
\Rightarrow c(z, \tau) &\equiv c_{tw}(z)/c_{tw}(\theta) \\
&= \left[\frac{1}{-\lambda(z - z_1)}\right]\left[\frac{1}{-\lambda(\theta - z_1)}\right]^{-1} \\
&= e^{-\lambda\tau}.
\end{aligned}
$$

Thus the solution for c near $z = z_1$ decreases to zero exponentially in time. Moreover, this calculation is valid whenever both θ and z are close to z_1, so that the region in which c is close to zero remains finite as τ increases.

A similar calculation shows that $c(z_2, \tau) = e^{+\mu\tau}$, where $\mu = -kJ'(z_2) > 0$, so that c has a maximum at $z = z_2$, whose height increases exponentially with time. However, the width of the peak decreases at a corresponding exponential rate, so that the net amount of connective tissue within this peak ($\propto$ its area) is roughly constant. This is because the calculation of the solution form near $z = z_2$ is only valid provided $1 \gg |\theta - z_2| = |z - z_2|e^{\mu\tau}$; that is, provided $|z - z_2| \ll e^{-\mu\tau}$.

Putting these various calculations together implies that the qualitative form of $c(z, \tau)$ when $kJ_{max} > a$ and $c_0(x) \equiv 1$ is as illustrated in Figure 4.4e. There is a peak in c at $z = z_1$, whose height and width increase and decrease exponentially in

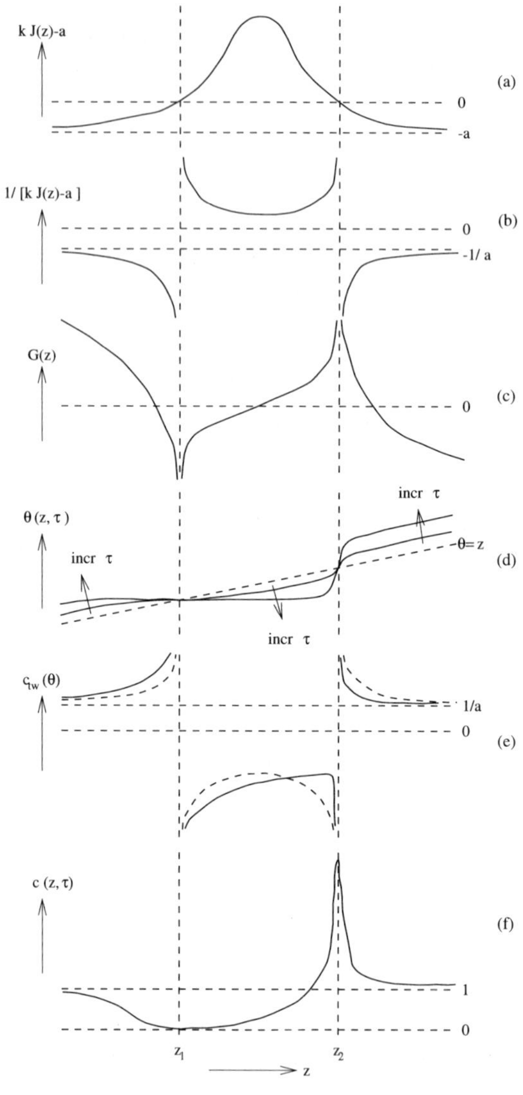

Figure 4.4 The qualitative form of the various functions involved in the solution of (2) using the method of characteristics. The explanation for these forms is given in the main text. (a) illustrates the form of $kJ(z) - a$. (b) shows $1/[kJ(z) - a]$, and (c) shows its integral, $G(z)$. (d) illustrates the form of $\theta(z, \tau)$. (e) shows $c_{tw}(\theta)$ (full curve) compared with $c_{tw}(z)$ (dashed curve). Finally, (f) shows the ratio $c_{tw}(z)/c_{tw}(\theta)$ at one value of τ

time, respectively, and a trough in c centred at $z = z_1$, whose width increases with time.

For illustration, we consider the particular case of non-linear cellular diffusion, $J_m = m\,\partial m/\partial x$. In this case, standard analysis (Sanchez-Garduno and Maini, 1995) shows that the travelling wave solution for cell density that evolves from our initial conditions has the form

$$\begin{aligned} M(z) &= \begin{cases} 1 - \exp\left[\frac{z-z_c}{\sqrt{2}}\right], & z < z_c \\ 0, & z > z_c \end{cases} \\ \Rightarrow J(z) &= \begin{cases} \frac{1}{\sqrt{2}}\left(\exp\left[\frac{z-z_c}{\sqrt{2}}\right] - \exp\left[(z-z_c)\right]\right), & z < z_c \\ 0, & z > z_c\,. \end{cases} \end{aligned}$$

Solving the equation $kJ(z) = 1/\sqrt{2}$ shows that

$$z_1,\, z_2 = \sqrt{2}\log\left(\frac{1 \pm \sqrt{1-4/k}}{2}\right).$$

When $k < 4$ these roots are complex, and the travelling wave solution for c exists and is the long-term solution: this solution is

$$c(z) = \begin{cases} \dfrac{\sqrt{2}}{\sqrt{2} - k\sqrt{2}\exp\left(\dfrac{z-z_c}{\sqrt{2}}\right) + k\sqrt{2}\exp(\sqrt{2}(z-z_c))}, & z < z_c \\ 1, & z > z_c\,. \end{cases}$$

(for $c_0(x) \equiv 1$). When $k > 4$, however, the solution for c has the aggregating waveform described above; straightforward calculation shows that

$$\begin{aligned} \lambda &= (k/4)\sqrt{1-4/k}\left[1 - \sqrt{1-4/k}\right] \\ \mu &= (k/4)\sqrt{1-4/k}\left[1 + \sqrt{1-4/k}\right] \end{aligned}$$

These results are confirmed by numerical simulations of (4.1); the example of $k = 4.05$ is illustrated in Figure 4.2.

When $c_0(x)$ is not constant, the solution for c is given by multiplying $c_{tw}(z)/c_{tw}(\theta)$ by $c_0(\theta)$. Moreover, this solution is then only exactly valid when $m(x, 0)$ is exactly in travelling wave shape, although since m evolves to the travelling wave shape quite quickly (Figures 4.1a and 4.2a), we expect the analysis to determine the key qualitative features of the solution for c. For the step function form of $c_0(.)$, this multiplication by $c_0(\theta)$ has the effect of removing the trough in the solution for $c(z, \tau)$, although the growing peak remains.

This analysis applies only when $\theta(.) \equiv 1$, implying that the extracellular matrix density does not affect cell movement. Allowing $\theta(.)$ to vary is extremely difficult because the equations no longer decouple, so the method used above cannot be applied. However, numerical simulations suggest that provided the variation in $\theta(.)$ is not too great, the qualitative behaviour of the solutions remains similar to when $\theta(.)$ is constant.

4.2.3 Extension to model transcapsular spread

The modelling in this section has demonstrated a novel bifurcation of a travelling wave into a more general waveform, not of constant shape. It also presents a succinct model for the accretion of connective tissue on the surface of a growing tumour.

Equations (4.1) and (4.2) are of course extremely simplistic and neglect a great many features of real tumour growth. A tumour cell interacts with the connective tissue in several ways, depending on the range of mutations it has undergone (Aznavoorian *et al.*, 1990). Benign tumour cells merely advect and compress surrounding connective tissue, thereby causing encapsulation (Carter *et al.*, 1985). Further mutations can make the cells invasive. These mutations can cause the cells to produce large amounts of proteases leading to digestion of the connective tissue, or can make the cells sensitive to connective tissue gradients (chemotaxis and haptotaxis) (Aznavoorian *et al.*, 1990) or alter their adhesiveness (Parsons, 1993). These phenotypic changes in a cell allow it to invade neighbouring tissue and to be carried away far from its primary site (Stetler-Stevenson *et al.*, 1993).

When malignant transformation occurs in an encapsulated benign tumour, the capsule is at first disrupted by the action of proteases and the cells can then escape into the surrounding tissue (Carter *et al.*, 1985). Clinically this transcapsular spread represents a crucially important point in the evolution of the tumour. Before transcapsular spread the prognosis for a malignant tumour is favourable, since the tumour can be completely removed by surgery. However, once the capsule has been disrupted, more aggressive forms of surgery (e.g. wide dissection) will be required, often in combination with other forms of treatment such as chemotherapy or radiotherapy. More importantly, the final outcome of treatment in patients with transcapsular spread compares poorly with tumours that are confined by the capsule (Bailey *et al.*, 1995).

We conclude this section by describing the results of simulations of an improved version of equations (4.1) and (4.2) which incorporates some of the features discussed above. The model includes a study of the behaviour of normal cells. Tumour cells differ from normal cells both in their local behaviour (e.g. the loss of contact inhibition and the ability to produce proteases) and in their spatial behaviour (e.g. enhanced chemokinesis, chemotaxis and haptotaxis). Here we describe the consequence of a mutation resulting in the loss of contact inhibition and another resulting in the production of a protease by the cancer cells.

When a malignant tumour produces proteases (e.g. matrix metalloproteases), this typically occurs at the growing edge of the tumour which is in contact with the surrounding connective tissue (Xie *et al.*, 1994). The body produces other proteins called antiproteases that have the ability to neutralise the effects of the proteases. Also, being small biochemicals, proteases can diffuse into the surrounding tissue. This results in the digestion of surrounding connective tissue, providing a mechanism for the possible disruption of the capsule. To study the occurrence of such capsular disruption we expand equation (4.2) to incorporate the dynamics of the proteases. The new model has the form

$$\frac{\partial n}{\partial t} = k_1 n(k_2 - n - m) + k_3 \frac{\partial}{\partial x}\left[n\frac{\partial n}{\partial x}\right], \tag{4.5}$$

$$\frac{\partial m}{\partial t} = k_4 m(k_5 - n - m) + k_6 \frac{\partial}{\partial x}\left[m \frac{\partial m}{\partial x}\right], \tag{4.6}$$

$$\frac{\partial c}{\partial t} = -k_7 pc + k \frac{\partial}{\partial x}\left[cm \frac{\partial m}{\partial x}\right], \tag{4.7}$$

$$\frac{\partial p}{\partial t} = k_8 uc - k_9 p - k_{10} up - k_{11} pc + k_{12} \frac{\partial^2 p}{\partial x^2}. \tag{4.8}$$

where $n(x,t)$ is the concentration of the normal cells and $p(x,t)$ is the concentration of the protease. Here the k_i's are all positive constants. k_1 and k_4 represent the linear growth rates of the normal cells and the tumour cells, and k_2 and k_5 their maximum carrying capacities. Thus the difference $k_5 - k_2$ is a measure of the loss of contact inhibition. k_3 and k_6 represent the effective diffusion coefficients of the normal cells and tumour cells. The interpretation of the various terms in equations (4.7) and (4.8) are as follows:

$-k_7 pc$	represents the degradation of connective tissue by the protease
$k_8 uc$	represents the production of the protease at the interface of the tumour and the surrounding tissue
$-k_9 p$	models the natural degradation of the protease
$-k_{10} up$ and $-k_{11} pc$	represent the neutralisation of the protease by the action of antiproteases.

The constant k_{12} is the diffusion coefficient of the protease.

Figure 4.5 illustrates a typical numerical solution of the improved model (4.8). In this simulation, we begin solution with $k_8 = 0$; that is, no protease is produced. This results in the formation of a capsule in a manner very reminiscent of the solutions discussed in Section 4.2.2. But there is also the dynamics of the normal cell population to consider; the receding wave of these cells corresponds to their out-competition by the tumour cells. Partway through the simulation, we alter k_8 to have a positive value, which crudely simulates the occurrence of a mutation resulting in the production of protease by the tumour cells. Once this mutation has occurred and protease production starts, the capsule is disrupted, as shown by the absence of increasing waves of c. In this second part of the solution, the cancer cells can spread unhindered by the capsule.

The simulation illustrated in Figure 4.5 has two key implications. Firstly, it demonstrates capsule formation via the mechanism described earlier in the section, but in an enlarged and somewhat more realistic model, adding to the credibility of this mechanism. Secondly, it shows that the mechanism is consistent with postulated methods of capsule disruption. However, it must be stressed that for cancer cells to spread aggressively, they need additional mechanisms of motility such as chemotaxis and haptotaxis, and alterations in adhesivity. Work is currently in progress in our group to understand the contribution of these various interactions to cancer invasion.

4.3 Macrophage Induction of Tumour Heterogeneity

Macrophages form an important part of the immune response to cancer. They are the mature form in tissue of a type of white blood cell known as a monocyte, and are

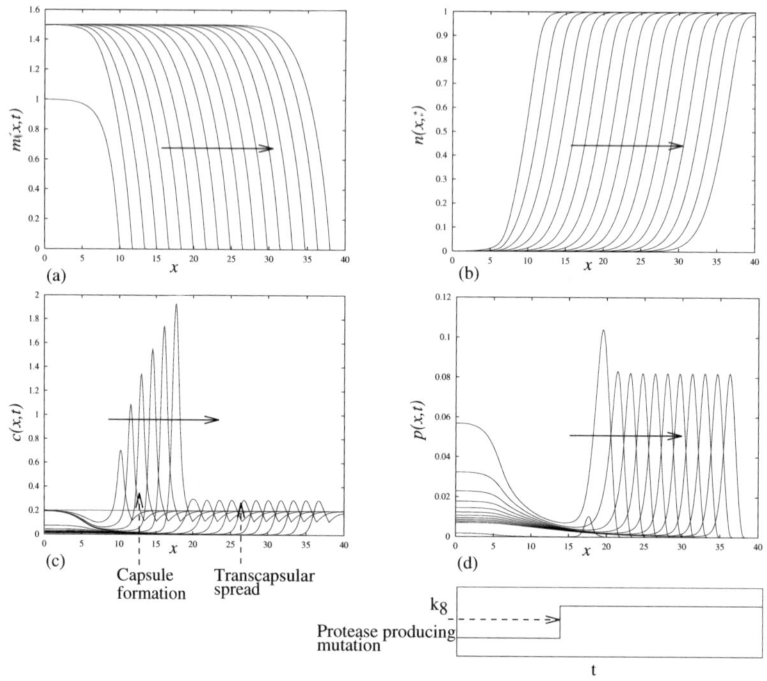

Figure 4.5 Numerical solution of equation (4.8) showing the replacement of normal tissue (b) by tumour cells (a) and an incipient capsule (c) being degraded by the action of a protease (d) produced by a mutation in the tumour cell. The parameter values used in this simulation are $k_1 = 5$, $k_2 = 1$, $k_3 = 1$, $k = 5$, $k_4 = 1$, $k_5 = 1.5$, $k_6 = 1$, $k_7 = 5$, $k = 5$, $k_9 = 3.1$, $k_{10} = 0.1$, $k_{11} = 0.1$, $k_{12} = 0.1$. The arrows show the direction of increasing time.

recruited to tumour sites by attracting gradients of chemicals produced by tumour cells. This recruitment can be very effective: for example, in many cases of breast carcinoma, macrophages constitute over 50% of the total cell mass (O'Sullivan and Lewis, 1994). The role of macrophages within tumours is complex. They are known to selectively kill tumour cells (Mantovani *et al.*, 1992) but they can also promote tumour growth by encouraging vascularisation (Lewis *et al.*, 1995). These various processes will be explained more fully later in the chapter; for more detailed biological reviews, see Hamilton and Adams (1987), Esgro *et al.* (1990) and Mantovani (1990).

Here we consider the effect of macrophage-mediated killing of tumour cells during the early, avascular stage of tumour growth. This process is initiated when a cell in the tissue undergoes a mutation, giving it a proliferative advantage over its normal counterparts (Weinberg, 1989); mathematical models for this process are given in Wheldon (1975) and Sherratt and Nowak (1992). The avascular phase of growth terminates because of the limited rate at which nutrients can diffuse into a solid tumour (Groebe and Mueller-Klieser, 1991) and thereafter the tumour becomes quiescent; this diffusion-limited cessation of growth was the first area of tumour biology to be modelled mathematically (Burton, 1966; Greenspan, 1972; Adam, 1986) and recent modelling has clarified the effects of tumour shape on the process (Byrne and Chaplain, 1995; Ward and King, 1997). Tumour growth is then resumed via the onset of angiogenesis (Folkman, 1995) in which tumours acquire their own blood supply, leading to unconstrained (and potentially lethal) growth; for models of the angiogenic process, see Chaplain and Sleeman (1990); Chaplain and Stuart (1993); Anderson and Chaplain (1998). Subsequent metastasis, leading to secondary tumours, requires the development of invasive tumour cell phenotypes; for modelling of tumour invasion, see Perumpanani *et al.* (1996).

Previous mathematical models of tumour immunology fall into two main categories, using either generic representations of the immune response (Adam, 1993; Albert *et al.*, 1980; Hiernaux and Lefever, 1988; Sherratt and Nowak, 1992) or detailed models focusing on particular aspects of immune cell–tumour cell interactions (Kuznetsov *et al.*, 1993; Kuznetsov, 1996; Adam and Bellomo, 1996). Our approach falls in between these categories of model: we focus specifically on macrophages, and use equations that are based at least qualitatively on known details of macrophage–tumour interactions; however, we keep our modelling sufficiently simple that we can consider the effects of macrophages within the context of a tumour that is growing in space. The work presented in this section has been previously reported in Owen and Sherratt (1997, 1998a, b).

4.3.1 Model development

Our model restricts attention to the interactions between mutant cells, normal tissue, macrophages, and their chemical regulators. We make the simplifying assumption that there is a single chemical regulator responsible for activation of macrophages, control of their proliferation, and stimulation of their influx from the bloodstream. In reality, many different chemicals are involved, including tumour necrosis factor (Mace *et al.*, 1988), macrophage colony stimulating factor (Sampson-Johannes and Carlino, 1988), and members of the monocyte chemotactic protein family (Bottazzi *et al.*, 1983; van Damme *et al.*, 1992); however, all these regulators derive primarily from mutant cells, and can be treated together to a first approximation.

Activation and inhibition of mutant and normal cell growth are not included in our model; instead we represent the proliferative advantage of mutant cells in terms of an increased basic growth rate. Thus the variables we consider are the densities of macrophages, $l(\mathbf{x}, t)$; mutant cells, $m(\mathbf{x}, t)$; normal cells, $n(\mathbf{x}, t)$; macrophage–mutant cell complexes, $c(\mathbf{x}, t)$; and the concentration of the chemical regulator, $f(\mathbf{x}, t)$. In the following sections we will outline the assumptions and interactions relevant to each species, and give mathematical descriptions in terms of differential equations.

Macrophages destroy tumour cells by binding to form a complex and then lysing the mutant cell; direct cell–cell contact is an essential part of this process (Jonjic *et al.*, 1992). Quiescent macrophages must be biochemically activated before complex formation and lysis are possible (Hamilton and Adams, 1987; Mantovani, 1990). We assume that the rate of complex formation is linear with respect to chemical concentration, and macrophage and tumour cell densities. In addition we assume that the complex returns viable macrophages after lysis of the mutant cell (Esgro *et al.*, 1990). Note that the source of regulators is mutant cells. Schematically this can be represented in the following form:

$$f + l + m \xrightarrow{k_1} c + f \qquad c \xrightarrow{k_2} l + \text{debris}.$$

Here k_1 and k_2 are positive constants. It is important to stress that there is no definitive experimental data on the details of this tumour cell destruction; we have considered various alternative formulations for this process, such as separating the activation and complex formation steps, but such changes do not significantly alter the qualitative model behaviour.

The remaining assumptions that we make with regard to macrophages are (i) they have spatially random motion combined with movement up gradients in the concentration of chemical regulator; (ii) they proliferate only in the presence of the chemical regulator (Bottazzi *et al.*, 1990), and such proliferation increases linearly with the concentration; (iii) their proliferation is limited by the crowding effect of all cell types; (iv) there is an influx from capillaries, which, due to chemotaxis, increases linearly with regulator concentration; and (v) they die with some constant rate per cell. In the absence of chemical regulator, the normal background level of tissue macrophages is maintained by a constant influx, at a rate denoted by I.

We use a term for proliferation of macrophages of the form

$$(\text{chemical concentration}) \times (\text{macrophage density}) \times (\text{crowding term}).$$

The most common representation of crowding effects on cell division is the logistic term, $r(x) = (k - x)$, which has been used in modelling a wide range of biological systems (Fisher, 1937; Sherratt and Murray, 1991). In this case it is inappropriate, since the logistic crowding term can be negative, which does not make sense when it is multiplied by the chemical concentration – macrophage death should not be promoted. Therefore we use the term $r(x) = (N + N_e)/(N + x)$, where N_e is the equilibrium cell density in normal tissue and N is a measure of crowding response. There is no particular significance in this choice of functional form; in the absence of detailed biological data, only the qualitative form of the function is known.

Combining these terms gives the following conservation equation for macrophages:

$$\frac{\partial l}{\partial t} = \overbrace{D_l \nabla^2 l}^{\text{cell migration}} + \overbrace{\frac{\alpha f l (N + N_e)}{N + l + m + n}}^{\text{proliferation}} + \overbrace{I(1 + \sigma f)}^{\text{influx}} - \overbrace{k_1 f l m + k_2 c}^{\text{lysis}} - \overbrace{\delta_l l}^{\text{death}} . \qquad (4.9)$$

Most tumours arise initially from a mutation that affects the control of cell division, giving the mutant cell a proliferative advantage over its peers (Weinberg, 1989). Such mutations have been the subject of previous detailed models (Wheldon, 1975; Sherratt and Nowak, 1992), but here our focus is on the role of macrophages during the early stages of tumour growth, and thus we use a very simple model. Specifically, we assume that the dynamics of mutant and normal cells are alike except for the removal of mutant cells by macrophages, and a scaling of the mutant cell growth rate by $\xi > 1$ to model their proliferative advantage. We use the same crowding term as discussed above, and a growth rate in normal tissue of δ, balanced by an equal rate of cell death, so that combining these elements with diffusion to simulate random cell migration gives the following conservation equations for mutant and normal cells:

$$\begin{aligned} \frac{\partial m}{\partial t} &= \overbrace{D_m \nabla^2 m}^{\text{cell migration}} + \overbrace{\frac{\xi \delta m (N + N_e)}{N + l + m + n}}^{\text{proliferation}} - \overbrace{\delta m}^{\text{death}} - \overbrace{k_1 f l m}^{\text{lysis}} \\ \frac{\partial n}{\partial t} &= D_n \nabla^2 n + \frac{\delta n (N + N_e)}{N + l + m + n} - \delta n. \end{aligned} \qquad (4.10)$$

Chemotaxis, macrophage proliferation, and mutant cell lysis are the important interactions considered in this model, and all are stimulated by chemicals produced by mutant cells, which we are representing as a single generic regulator $f(\mathbf{x}, t)$, whose sole source is mutant cells. We assume a constant secretion rate β per unit of mutant cell density, and a linear natural decay with rate δ_f. The complex density is also assumed to decay linearly, with rate δ_c. Including random motion and the binding and lysis terms discussed above completes the derivation of our model:

$$\begin{aligned} \frac{\partial f}{\partial t} &= \overbrace{D_f \nabla^2 f}^{\text{diffusion}} + \overbrace{\beta m}^{\text{prod}^n} - \overbrace{\delta_f f}^{\text{decay}} \\ \frac{\partial c}{\partial t} &= \overbrace{D_c \nabla^2 c}^{\text{migration}} + \overbrace{k_1 f l m - k_2 c}^{\text{lysis}} - \overbrace{\delta_c c}^{\text{death}} . \end{aligned} \qquad (4.11)$$

Equations (4.9) to (4.11) are non-dimensionalised using the following rescalings, where L is half a typical cell length, and f_0 is a typical concentration of chemical regulator:

$$t^* = \delta t, \; x^* = x/L, \quad l^* = l/N_e, \; m^* = m/N_e, \; n^* = n/N_e, \; c^* = c/N_e, \quad f^* = f/f_0,$$

$$D_l^* = \frac{D_l}{\delta L^2}, \quad D_m^* = \frac{D_m}{\delta L^2}, \quad D_n^* = \frac{D_n}{\delta L^2}, \quad D_f^* = \frac{D_f}{\delta L^2}, \quad D_c^* = \frac{D_c}{\delta L^2},$$

$$\alpha^* = \alpha f_0/\delta, \quad N^* = N/N_e, \quad I^* = I/(N_e \delta), \quad \sigma^* = f_0 \sigma, \quad k_1^* = k_1 f_0 N_e/\delta,$$

$$k_2^* = k_2/\delta, \quad \delta_l^* = \delta_l/\delta, \quad \delta_f^* = \delta_f/\delta, \quad \delta_c^* = \delta_c/\delta, \quad \beta^* = \beta N_e/(f_0\delta).$$

Recall that N_e is the equilibrium cell density in normal tissue. Applying these rescalings, and dropping the asterisks for notational simplicity, gives

$$\frac{\partial l}{\partial t} = D_l\nabla^2 l + \frac{\alpha f l(N+1)}{N+l+m+n} + I(1+\sigma f) - k_1 f l m + k_2 c - \delta_l l \qquad (4.12)$$

$$\frac{\partial m}{\partial t} = D_m\nabla^2 m + \frac{\xi m(N+1)}{N+l+m+n} - m - k_1 f l m \qquad (4.13)$$

$$\frac{\partial n}{\partial t} = D_n\nabla^2 n + \frac{n(N+1)}{N+l+m+n} - n \qquad (4.14)$$

$$\frac{\partial f}{\partial t} = D_f\nabla^2 f + \beta m - \delta_f f \qquad (4.15)$$

$$\frac{\partial c}{\partial t} = D_c\nabla^2 c + k_1 f l m - k_2 c - \delta_c c. \qquad (4.16)$$

The experimental data available for parameter estimation is limited, but enables order of magnitude estimates to be obtained for most of the parameters, and details of this are given in Owen and Sherratt (1998a) for the kinetic parameters, and in Owen and Sherratt (1998b) for motility parameters.

4.3.2 Spatially independent solutions

Although we are primarily concerned with spatially varying solutions, as an essential precursor to this work, we discuss solutions of the spatially homogeneous system given by setting $D_l = D_m = D_n = D_f = D_c = 0$. In the resulting system of ordinary differential equations, there are four types of steady state to consider. The first two are somewhat trivial: a macrophage-only state $l = I/\delta_l$, $m = n = f = c = 0$, and the normal tissue state, consisting of normal cells and macrophages, $l = I/\delta_l$, $m = 0$, $n = 1 - I/\delta_l$, $f = c = 0$. Straightforward linear stability analysis shows that both these steady states are unstable to the introduction of mutant cells.

The third type of steady state is that of a tumour with no normal cells remaining, denoted the mutant-only steady state, and the fourth steady state is of mixed type, with all cell species coexisting. Real tumours certainly contain a variety of cell types (Mantovani, 1990), and other models (Wheldon, 1975; Sherratt and Nowak, 1992) also predict this type of solution. If the growth advantage of mutant cells is too high, coexistence is not possible, and steady states of this type cannot exist. Figure 4.6 illustrates the way in which these last two steady states change with σ, the macrophage influx parameter. We have identified a number of bifurcation points, at which the number and stability of steady states change. This is discussed in some detail in Owen and Sherratt (1998a); for the purposes of this chapter the key result is that, with respect to homogeneous perturbations, at least one tumour steady state is stable, and that for realistic parameter ranges there is only a very small region of parameter space in which both coexistence and mutant-only steady states are stable.

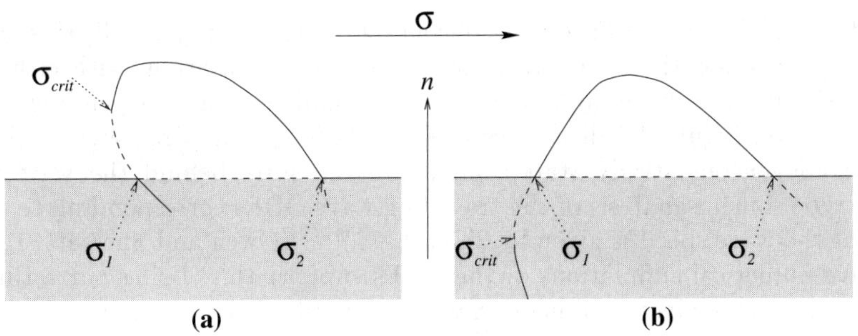

Figure 4.6 Schematic representation of the bifurcation structures for the homogeneous tumour steady states of the dimensionless model (4.16). Solid lines indicate stability, dashed lines instability. The bifurcation parameter σ is the measure of the increase in macrophage influx from the bloodstream in response to chemoattractants secreted by tumour cells.

Numerical simulations of the ordinary differential equation model given by setting $D_l = D_m = D_n = D_f = D_c = 0$ suggest that when only one steady state is locally stable, it is also globally stable for positive solutions; in the small regions of parameter space where two tumour steady states are both locally stable, solutions evolve to one or other of these, depending on initial conditions. In particular, this implies that introduction of a small mutant cell density to the normal tissue state causes the solution to evolve to a tumour steady state. This occurs despite the presence of cytotoxic macrophages, because the macrophage immune response is second order in the density of mutant cells. By this we mean that as well as having an explicit dependence on the mutant cell density m, the immune response term $-k_1 flm$ also depends on the concentration of chemical regulator f, whose production rate is itself dependent on the mutant cell density.

4.3.3 Wave front solutions and spatial patterns

Real tumours in vivo are nearly always spatially heterogeneous, often with no apparent order whatsoever. This should be no surprise given the varied environment in which tumours grow, and the wide variety of interactions with different cell types and tissue structures. Such spatial variations can arise from a variety of biological mechanisms, which include convection and mechanical stress due to movement of the underlying tissue structures, and heterogeneities in the underlying tissue itself. These features do not relate specifically to the macrophage–tumour interaction, and are not considered in our model, although they may be very significant. However, we will show that spatial variations can arise simply as a consequence of cell movement and chemical diffusion, within the context of the macrophage–tumour interaction. In this section we discuss the implications of purely random motility of cells and diffusion of the chemical regulator; we also demonstrate the existence of travelling wave solutions, and the possibility of spatial patterning within the tumour.

We are concerned with the evolution of model solutions following a spatially localised introduction of mutant cells to an infinite domain on which the system is at the normal tissue steady state. In a wide range of numerical simulations, we have found that provided the chemical diffusion coefficient D_f is sufficiently small, this results in a travelling wave solution originating from the site of mutation, with either the coexistence or mutant-only tumour steady state behind the wave, depending on the kinetic parameters. Figure 4.7 illustrates this simulation, with parameters such that the stable homogeneous steady state, and hence the state behind the wave, is of coexistence type. Linear analysis of the travelling wave ODEs corresponding to (4.16) suggests that the wave speed is given by $2[D_m(\xi - 1)]^{1/2}$ (Owen and Sherratt, 1998b), and extensive numerical simulations of the PDEs support this. Using our estimated parameter values, this gives a dimensional wave speed of the order of 10^{-10}ms^{-1}, indicating that it would take about 100 days for a tumour to grow to a size of 1mm.

These results show that the macrophage population has no effect on the speed of tumour growth; however, the macrophages do have a significant effect on tumour composition, because of their effect on the homogeneous steady states. Moreover, when the chemical diffusion coefficient D_f increases above a critical value, numerical simulations of the model predict a much more significant effect on tumour composition, with a spatial pattern developing behind the advancing wavefront, as illustrated in Figure 4.8. Note that the pattern is stationary behind the wave, growing in extent but not changing in form or location. Such regular patterns are not observed in the stochastic inhomogeneous environment of a real tumour, but their observation in model solutions suggests that the macrophage–tumour interaction may be a first step towards the more irregular spatial inhomogeneities seen in the real situation.

This pattern formation is an example of a Turing mechanism (Murray, 1991). However, there is no prepattern involved: in this case the mutant cells are the local activator, and the chemical regulator is the long-range inhibitor. Thus, an intuitive explanation for this spatial instability, illustrated in Figure 4.9 is that given a local perturbation increasing the density of mutant cells, chemical regulator production will also increase locally. Then if the chemical diffuses fast enough, it will act non-locally to activate macrophages to the tumoricidal state and to stimulate an additional influx of macrophages. This will suppress non-local mutant cell growth, while enhancing local mutant cell growth, due to the relative suppression of local macrophage activation, so the original perturbation will grow in time. Thus it is the inclusion of chemical diffusion, which occurs at a much faster rate than random cell migration, that is responsible for this diffusion-driven instability. The diffusion coefficient D_f for the chemical regulator is expected to be substantially larger than the random motility coefficients for cells, but we are not aware of data enabling precise values to be calculated. Such data would in fact be relatively easy to obtain via in vitro experiments, and our results suggest this would be a valuable line of experimental investigation.

We analyse the patterning phenomenon by linearising the model (4.16) about the stable homogeneous steady state. A general perturbation can then be expressed as a linear combination of spatial modes, and so substituting solutions proportional to $e^{i\kappa x + \lambda t}$ into the linearised equations and looking for non-trivial solutions gives the dispersion relation for the real part of λ (i.e. the growth rate) as a function of wave number κ. The homogeneous steady state is then stable to all spatial perturbations if $\text{Re}(\lambda) < 0$ for all $\kappa \in \text{R}$, whereas if $\text{Re}(\lambda) > 0$ for some κ then the mode corresponding to that wave number will grow exponentially in time, destabilising the homogeneous steady state.

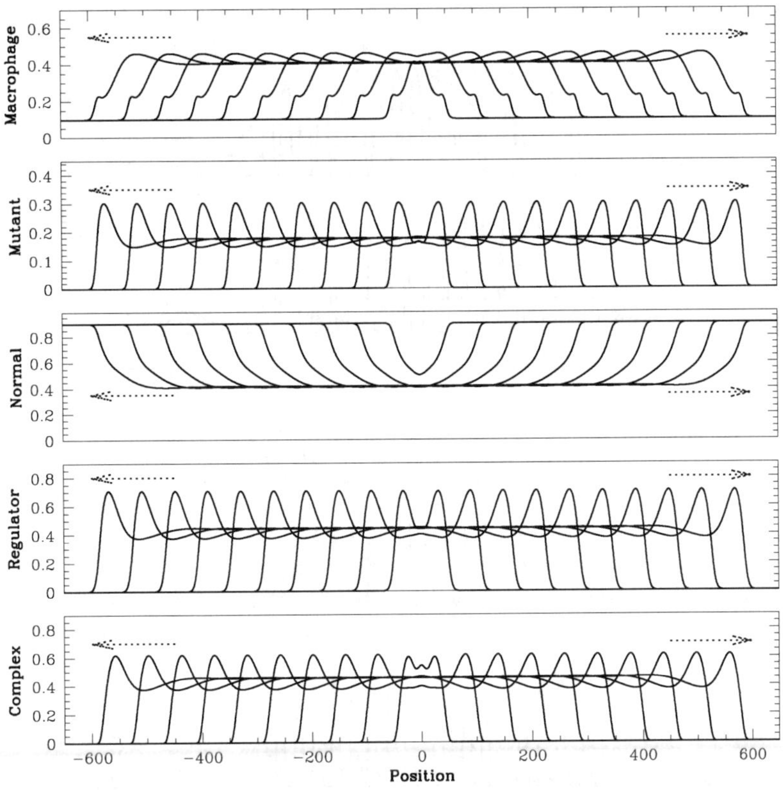

Figure 4.7 Numerical simulation in one dimension of the dimensionless model (4.16), showing the evolution from a localised mutation at the centre. The initial conditions were the normal tissue steady state $l = 0.1, m = 0, n = 0.9, f = 0, c = 0$ on the whole domain, apart from the mutation $l = 0, m = 1, n = 0, f = 0, c = 0$ for $-0.5 < x < 0.5$. A travelling wave is established as the tumour grows, with the composition behind the wave front that of the homogeneous stable steady state, which is dependent on the kinetic parameters. The parameters are: $D_l = D_m = D_n = 5,\ D_f = 30,\ D_c = 2.5, \alpha = 0.01,\ \beta = 5.0,\ \delta_c = 0.5,\ \delta_f = 2.0,\ \delta_l = 0.1,\ N = 1,\ I = 0.01,\ k_1 = 10.0,\ k_2 = 0.2,\ \sigma = 58,\ \xi = 2.8$.

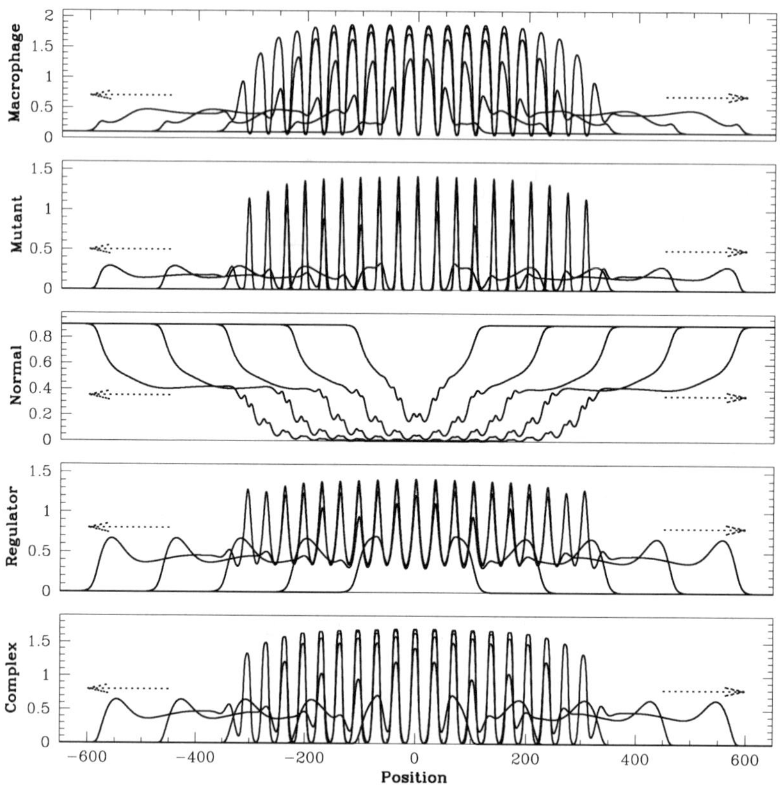

Figure 4.8 Numerical simulation in one dimension of the model (4.16), showing the evolution from a localised mutation at the centre. The initial conditions were the normal tissue steady state $l = 0.1, m = 0, n = 0.9, f = 0, c = 0$ on the whole domain, apart from the mutation $l = 0, m = 1, n = 0, f = 0, c = 0$ for $-0.5 < x < 0.5$. The parameters are: $D_l = D_m = D_n = 5$, $D_f = 100$, $D_c = 2.5$, $\alpha = 0.01$, $\beta = 5.0$, $\delta_c = 0.5$, $\delta_f = 2.0$, $\delta_l = 0.1$, $N = 1$, $I = 0.01$, $k_1 = 10.0$, $k_2 = 0.2$, $\sigma = 58$, $\xi = 2.8$.

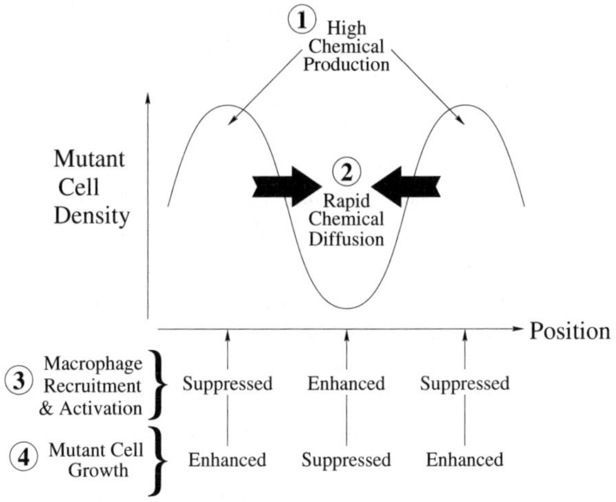

Figure 4.9 An intuitive explanation for the development of spatial inhomogeneities within a tumour, due to interactions with macrophages. A locally elevated mutant cell density is reinforced by the consequently elevated chemical regulator production (step 1) – if the chemical diffuses sufficiently fast (step 2), then local macrophage recruitment and activation will be suppressed (step 3), and hence mutant cell growth enhanced (step 4). Correspondingly, non-local recruitment and activation of macrophages will be enhanced, and mutant cell growth suppressed.

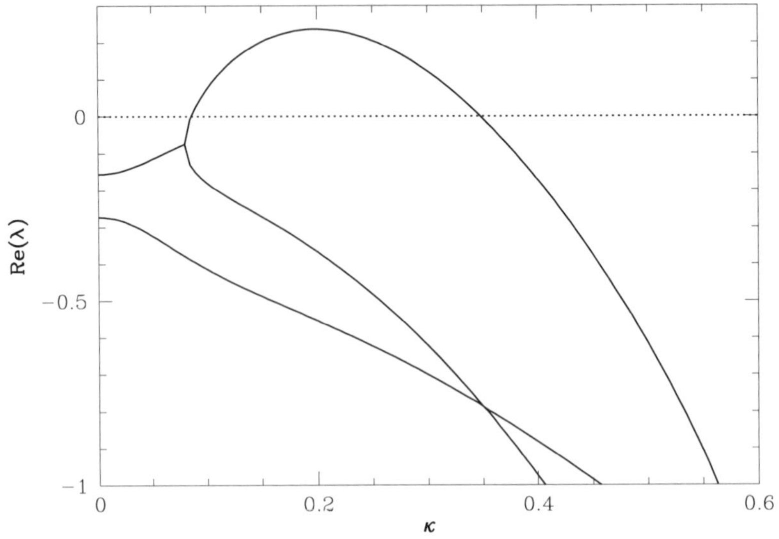

Figure 4.10 Numerically calculated dispersion relation for the model (4.16), showing Re(λ) as the wavenumber κ varies. The different curves are for different eigenvalues. For κ between about 0.09 and 0.345 the real part of one of the eigenvalues is positive, indicating that the homogeneous steady state is unstable to perturbations with those wavenumbers.

Figure 4.10 illustrates the numerically calculated dispersion relation for a parameter set for which a range of wavenumbers are unstable; Figure 4.8 shows the corresponding numerical simulation of the partial differential equations. Although the wavefront is moving, the pattern which grows behind the front is stationary in space, with a wavelength of 33 cell lengths. The wavenumber of the fastest-growing mode in the dispersion relation is approximately $\kappa = 0.2$, giving a wavelength of 31 cell lengths, in good agreement with the above observation from numerical simulations.

4.3.4 Numerical simulations in two dimensions

So far we have considered solutions in one space dimension only, but real tumours reside in a three-dimensional tissue. For any mathematical model, the range of potential solution behaviours is much greater in higher dimensions, and while numerical solution of the equations in three dimensions would be too computationally intensive for our current resources, two-dimensional simulations are feasible, and show a number of features of interest. Perhaps a central theme in the biological background to this work, and indeed in much of tumour biology, is the heterogeneity between and within tumours. It is heterogeneity within tumours that is the focus of this chapter, and we have demonstrated a possible mechanism for the development of spatial heterogeneity. The more realistic two-dimensional setting and the possibility of still greater complexity, such as an instability of the wavefront itself, is the focus of interest in an investigation of two-dimensional solutions.

The dispersion relation as calculated in Section 4.3.3, and illustrated in Figure 4.10, is still valid, so that whenever patterns appear in one dimension they will also appear in two, but the precise nature of those patterns requires investigation. In one dimension the unstable wave number κ is a scalar, and a corresponding instability can only give rise to a single pattern, but in two dimensions the corresponding wave number (κ_x, κ_y) is two-dimensional, with $\kappa = \{\kappa_x^2 + \kappa_y^2\}^{1/2}$. If one of κ_x and κ_y is zero then the pattern will be just simple stripes, but if they are both non-zero then the pattern will be two-dimensional, tiling the plane. This introduces the possibility of more than one pattern being stable for a given parameter set, depending on the domain and initial conditions.

The model equations were solved in two dimensions using an alternating direction implicit method. This means that each time step consists of an explicit half-time step in the x direction and an implicit half-time step in the y direction, and then similar half-time steps exchanging explicit for implicit and vice versa. Figure 4.11 shows a shaded plot of the macrophage density that develops following a localised introduction of mutant cells into normal tissue. A circular travelling wave is established, behind which the tumour steady state is unstable to spatial perturbations, which leads to the formation of a stationary target pattern. The concentric rings in the circular geometry of the simulation correspond to stripes in a planar geometry; the break-up of the pattern at the centre seems likely to be due to curvature effects. Numerical simulations for the same parameters, but on a finite rectangular domain with randomly generated initial conditions, indicate that for the parameters used in the figure, both spotted and striped patterns stably coexist, with the initial conditions determining which state is reached. This is a well-known phenomenon away from a primary Turing bifurcation (Murray, 1991); note that sufficiently close to a Turing bifurcation, only one of the striped and spotted patterns can be stable (Ermentrout, 1991). However, our simulations indicate that, for the parameters in Figure 4.11, stripes always develop in the simulation of the growing tumour. In real tumours such patterns would be significantly modified by the effects of environmental noise, so the results illustrated in Figure 4.11 are not expected to represent real situations that might arise in practice. The essential implication of our results is rather that macrophage–tumour cell interactions can act as a first step in the initiation of both spatial and spatio-temporal heterogeneity.

4.3.5 Summary of model predictions

The ability of the immune system to spontaneously eliminate tumours was widely debated in the 1950s and 60s. In recent years, attention has switched to the more complex regulatory effects of the immune system on tumour progression. Our work suggests that these two philosophies are not separate, but rather have an important overlap. Our model predicts that macrophages are unable to spontaneously eliminate whole tumours, in keeping with the conventional view that, without intervention, the immune system is an ineffective weapon against the majority of cancers. Nevertheless, our model predicts that the ability of macrophages to selectively kill tumour cells has an extremely significant effect on tumour development, since it is able to induce spatial inhomogeneities. This new prediction is consistent with observations of Leek *et al.* (1996), which demonstrate the existence of macrophage hot spots within breast carcinomas, in a manner highly reminiscent of our results. Moreover, the work of Leek

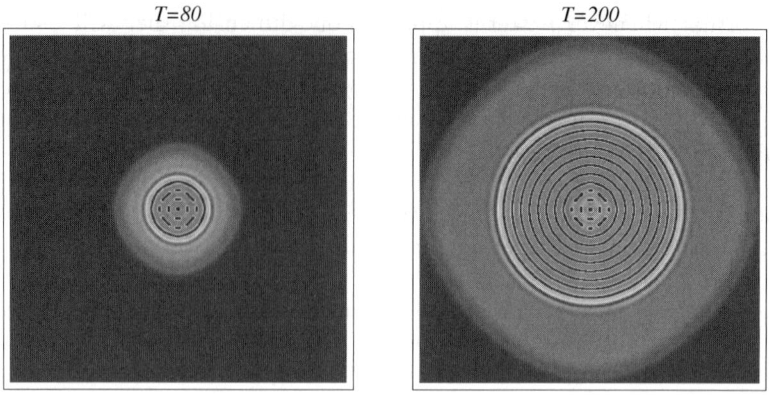

Figure 4.11 Numerical simulation in two dimensions, showing the macrophage density after 80 and 200 dimensionless time units of evolution from a single unit seeding region with a mutant cell density of 1, located at the centre of the domain. Macrophage density is represented on a grey scale, with darker colouring indicating lower cell density. The domain size is 2000×2000, and the parameters are $D_l = D_m = D_n = 5$, $D_f = 500$, $D_c = 2.5$, $\alpha = 0.01$, $\beta = 5.0$, $\delta_c = 0.5$, $\delta_f = 2.0$, $\delta_l = 0.1$, $N = 1$, $I = 0.01$, $k_1 = 10.0$, $k_2 = 0.2$, $\sigma = 40$, $\xi = 2.0$.

et al. (1996) suggests that these hot spots may play a key role in inducing tumour angiogenesis. This implies that high macrophage densities actually favour tumour progression, despite their antitumour activities, an observation made previously by Lewis *et al.*, (1995). Our work suggests that this antitumour activity may nevertheless be crucial in establishing the spatial structure that is later able to induce angiogenesis.

Experimental investigation of the predictions of our model is potentially possible using cell spheroid assays, which could in principle be seeded with macrophage cells. This line of inquiry provides the exciting possibility of providing a detailed, quantitative link between the results of a model of the type we are considering, and experimental data. This would provide a natural route towards detailed applications of our model to in vivo tumour dynamics.

REFERENCES

Adam, J.A. (1986) A simplified mathematical model of tumor growth. *Math. Biosci.*, 81: 229–244.

Adam, J.A. (1993) The dynamics of growth-factor-modified immune response to cancer growth: One dimensional models. *Mathl. Comp. Modelling*, 17: 83–106.

Adam, J.A. and Bellomo, N. (1997) *A Survey of Models for Tumor-Immune System Dynamics*. Birkhäuser.

Albert, A., Freedman, M. and Perelson, A.S. (1980) Tumors and the immune system: The effects of a tumor growth modulator. *Math. Biosci.*, 50: 25–28.

Anderson, A.R.A. and Chaplain, M.A.J. (1998) Continuous and discrete models of tumor-induced angiogenesis *Bull. Math. Biol.*, 60: 857–899.

Aznavoorian, S., Stracke, M.L., Krutzsch, H., Schiffman, E. and Liotta, L.A. (1990) Signal transduction for chemotaxis and haptotaxis by matrix molecules in tumour cells. *J. Cell Biology*, 110: 1427–1438.

Bailey, H., Love, R.J., Mann, C.V. *et al.* (1995) *Bailey and Love's Short Practice of Surgery.* Chapman and Hall.

Barr, L.C., Carter, R.L. and David, A.J.S. (1988) Encapsulation of tumors as a modified wound healing response. *The Lancet,* II: 135–137.

Berenblum, I. (1970) The nature of tumour growth. In: *General Pathology,* (Florey HEW, ed.), Lloyd-Luke.

Bottazzi, B., Polentarutti, N., Acero, R., Balsari, A., Boraschi, D., Ghezzi, P., Salmona, M. and Mantovani, A. (1983) Regulation of the macrophage content of neoplasms by chemoattractants. *Science,* 220: 210–212.

Bottazzi, B., Erba, E., Nobili, N., Fazioli, F., Rambaldi, A. and Mantovani, A. (1990) A paracrine circuit in the regulation of the proliferation of macrophages infiltrating murine sarcomas. *J. Immunol.,* 144: 2409–2412.

Burton, A.C. (1966) Rate of growth of solid tumours as a problem of diffusion. *Growth,* 30: 159–176.

Byrne, H.M. (1997) The importance of intercellular adhesion in the development of carcinomas. *IMA J. Math. Appl. Med. Biol.,* 14: 305–323.

Byrne, H.M. and Chaplain, M.A.J. (1995) Growth of nonnecrotic tumors in the presence and absence of inhibitors. *Math. Biosci.,* 130: 151–181.

Byrne, H.M. and Chaplain, M.A.J. (1996) Modelling the role of cell-cell adhesion in the growth and development of carcinoma. *Mathl. Comp. Modelling,* 24: 1–17.

Carter, R.L., Barr, L.C. and O'Brien, C.J. *et al.* (1985) Transcapsular spread of metastatic squamous cell carcinoma from cervical lymph nodes. *Am. J. Surgery,* 150: 495–499.

Chaplain, M.A.J. and Sleeman, B.D. (1990) A mathematical model for the production and secretion of tumour angiogenesis factor in tumours. *IMA J. Math. Appl. Med. Biol.,* 7: 93–108.

Chaplain, M.A.J. and Stuart, A.M. (1993) A model mechanism for the chemotactic response of endothelial cells to tumor angiogenesis factor. *IMA J. Math. Appl. Med. Biol.,* 10: 149–168.

Ermentrout, B. (1991) Stripes or spots - non-linear effects in bifurcation of reaction-diffusion equations on the square. *Proc. R. Soc. Lond. A,* 434: 413–417.

Esgro, J.J., Whitworth, P. and Fidler, I.J. (1990) Macrophages as effectors of tumor immunity. *Immunol. Allergy Clin. N. Am.,* 10: 705–729.

Fisher, R.A. (1937) The wave of advance of advantageous genes. *Ann. Eugenics,* 73: 353–369.

Folkman, J. (1995) Angiogenesis in cancer, vascular, rheumatoid and other disease. *Nature Med.,* 1: 27–31.

Greenspan, H.P. (1972) Models for the growth of a solid tumour by diffusion. *Stud. Appl. Math.,* 51: 317–340.

Groebe, K. and Mueller-Klieser, W. (1991) Distribution of oxygen, nutrient and metabolic waste concentrations in muticellular spheroids and their dependence on spheroid parameters. *Eur. J. Biophys.,* 19: 169–181.

Hamilton, T.A. and Adams, D.O. (1987) Mechanisms of macrophage-mediated tumor injury. In *Tumor Immunology – Mechanisms, Diagnosis, Therapy* (den Otter, W. and Ruitenberg, E.J. eds), pp. 89–107. Elsevier Science.

Hiernaux, J.R. and Lefever, R. (1988) Population dynamics of tumors attacked by immunocompetent killer cells. In *Theoretical Immunology,* (Perelson, A.S. ed.), volume.2, pp. 19–36. Addison Wesley.

Jonjic, N., Alberti, S., Bernasconi, S., Peri, G., Jilek, P., Anichini, A., Parmiani, G. and Mantovani, A. (1992) Heterogeneous susceptibility of human-melanoma clones to monocyte cytotoxicity – role of ICAM-1 defined by antibody blocking and gene transfer. *Eur. J. Immunol.,* 22: 2255–2260.

Kuznetsov, V.A. (1996) Harpoon model for cell-cell adhesion and recognition of target cells by the natural killer cells. *J. Theor. Biol.,* 180: 321–342.

Kuznetsov, V.A., Zhivoglyadov, V.P. and Stepanova, L.A. (1993) Kinetic approach and estimation of the parameters of cellular interaction between the immunity system and a tumor. *Arch. Immunol. Ther. Exp.*, 41: 21–31.

Landini, G. and Rippin, J.W. (1993) Fractal dimensions of the epithelial–connective tissue interfaces in premalignant and malignant epithelial lesions of the floor of the mouth. *Anal. Quant. Cytol. Histol.* 15: 144-149.

Landini, G. and Rippin, J.W. (1996) How important is tumour shape? Quantification of the epithelial–connective tissue interface in oral lesions using local connected fractal dimension analysis. *J. Pathol.* 179: 210-217.

Leek, R.D., Lewis, C.E., Whitehouse, R., Greenall, M., Clarke, J. and Harris, A.L. (1996) Association of macrophage infiltration with angiogenesis and prognosis in invasive breast carcinoma. *Cancer Res.*, 56: 4625–4629.

Lewis, C.E., Leek, R., Harris, A.L. and McGee, J.O.D. (1995) Cytokine regulation of angiogenesis in breast cancer – the role of tumor-associated macrophages. *J. Leukoc. Biol.*, 57: 747–751.

Mace, K.F., Ehrke, M.J., Hori, K., Maccubbin, D.L. and Mihich, E. (1988) Role of tumor necrosis factor in macrophage activation and tumoricidal activity. *Cancer Res.*, 48: 5427–5432.

Mantovani, A. (1990) Tumour-associated macrophages. *Curr. Opin. Immunol.*, 2: 689–692.

Mantovani, A., Bottazzi, B., Colotta, F., Sozzani, S. and Ruco, L. (1992) The origin and function of tumor-associated macrophages. *Immunol. Today*, 13: 265–270.

Murray, J.D. (1991) *Mathematical Biology.* Springer Verlag, Berlin.

Orme, M.E. and Chaplain, M.A.J. (1996) A mathematical model of the first steps of tumour-related angiogenesis: Capillary sprout formation and secondary branching. *IMA J. Math. Appl. Med. Biol.*, 13: 73–98.

O'Sullivan, C. and Lewis, C.E. (1994) Tumor-associated leukocytes - friends or foes in breast-carcinoma. *J. Path.*, 172: 229–235.

Owen, M.R. and Sherratt, J.A. (1997) Pattern formation and spatiotemporal irregularity in a model for macrophage-tumour interactions. *J. Theor. Biol.* 189: 63–80.

Owen, M.R. and Sherratt, J.A. (1999) Mathematical modelling of macrophage dynamics in tumours. *Math. Mod. Meth. Appl. Sci.* 9: 513–539.

Owen, M.R. and Sherratt, J.A. (1998b) Modelling the macrophage invasion of tumours: effects on growth and composition. *IMA J. Math. Appl. Med. Biol.* 15: 165–185.

Parsons, D.F. (1993) Tumor cell interaction with stromal elastin and type-I collagen: The consequences of specific adhesion and proteolysis. *Tumor Biol.*, 14(3): 137–143.

Perumpanani, A.J., Sherratt, J.A., Norbury, J. and Byrne, H.M. (1996) Biological inferences from a mathematical model for malignant invasion. *Invasion and Metastasis*, 16: 209–221.

Perumpanani, A.J., Sherratt, J.A. and Norbury, J. (1997) Mathematical modelling of capsule formation and multinodularity in benign tumour growth. *Nonlinearity*, 10: 1599-1614.

Robbins, S.L., Cotran, R.S. and Kumar, V. (1984) *Pathologic Basis of Disease.* Saunders, Philadelphia; London.

Sampson-Johannes, A. and Carlino, J.A. (1988) Enhancement of human monocyte tumoricidal activity by recombinant M-CSF. *J. Immunol.*, 141: 3680–3686.

Sanchez-Garduno, F. and Maini, P.K. (1995) Travelling wave phenomena in some degenerate reaction-diffusion equations. *J. Diff. Eq.* 117: 281–319.

Sherratt, J.A. and Marchant, B.P. (1996) Non-sharp travelling waves in the Fisher equation with degenerate non-linear diffusion. *Appl. Math. Lett.*, 9: 33-38.

Sherratt, J.A. and Murray, J.D. (1991) Mathematical analysis of a basic model for epidermal wound-healing. *J. Math. Biol.*, 29: 389–404.

Sherratt, J.A. and Nowak, M.A. (1992) Oncogenes, anti-oncogenes and the immune response to cancer: A mathematical model. *Proc. R. Soc. Lond. B*, 248: 261–271.

Stetler-Stevenson, W.G., Aznavoorian, S. and Liotta, L.A. (1993) Tumor cell interactions with the extracellular matrix during invasion and metastasis. *Ann. Rev. Cell. Biol.*, 9: 541–73.

Vaidya, V.G. and Alexandro, F.J. Jr. (1982) Evaluation of some mathematical models for tumour growth. *Int. J Biomed. Comput*, 13: 19.
van Damme, J., Proost, P., Lenaerts, J.P. and Opdenakker, G. (1992) Structural and functional identification of two human, tumor-derived Monocyte Chemotactic Proteins (MCP-2 and MCP-3) belonging to the chemokine family. *J. Exp. Med.*, 176: 59–65.
Ward, J.P. and King, J.R. (1997) Mathematical modelling of avascular tumour growth. *IMA J. Math. Appl. Med. Biol.*, 14: 39–69.
Weinberg, R.A. (1989) Oncogenes, anti-oncogenes, and the molecular bases of multistep carcinogenesis. *Cancer Res.*, 49: 3713–3721.
Wheldon, T.E. (1975) Mitotic autoregulation of normal and abnormal cells: Alternative mechanisms for the derangement of growth control. *J. Theor. Biol.*, 53: 421–433.
William, P.L. and Gray, H. (1995) *Gray's Anatomy.* Churchill Livingstone.
Xie, B., Bucana, C.D., and Fidler, I.J. (1994) Density-dependent induction of 92-kd type type-IV collagenase activity in cultures of A431 human epidermoid carcinoma cells. *Am. J. Pathol.*, 144(5): 1958–67.

5

Turing Structures of the Second Kind

Axel Hunding

Department of Chemistry C116, H.C. Ørsted Institute, University of Copenhagen, Universitetsparken 5, DK 2100 Copenhagen Ø, Denmark

5.1 Introduction

Spontaneous pattern formation in biology is an intriguing problem. Until a few decades ago, such processes were perplexing as they seemed to run counter to intuition based on the second law of thermodynamics. The last decades, however, have seen a vast effort in the field of non-linear dynamics, which has provided ample evidence, theoretical as well as experimental, of physicochemical mechanisms capable of producing reliable patterns spontaneously from an initially homogeneous state.

The same decades have experienced huge improvements in genetic research, and many genes involved in early embryogenesis are now known. Highly conserved gene clusters have been found, yet any idea of a common basis for pattern formation as such, rooted in an ancestral 'proto-pattern gene' or gene cluster is not supported (Martinez *et al.*, 1998).

Rather the basis for biological pattern formation seems to have emerged well before the Cambrian explosion some 600 million years ago. Genes involved in axis specification in early embryos belong to gene classes which arose as much as a billion years before the emergence of multicellular organisms. These genes must thus have originated with different essential roles. In Wolpert's words: 'Cells are more complicated than embryos in the sense that the intracellular network of signals is more complex than the intercellular signals in the developing embryo. Most of the features required for development are already present in single cells and in single cell organisms like the Protozoa. In a curious sense development and multicellular animals were inevitable once the cell was invented' (Wolpert, 1990).

Genes involved in embryonic pattern formation may code for proteins, which are able to bind to the promoter region of other genes, or bind to proteins already bound to promoters, thus creating a three-dimensional jigsaw puzzle, which determines whether the gene under control from the promoter is activated or not. An essential property

On Growth and Form: Spatio-temporal Pattern Formation in Biology.
Editor Mark A.J. Chaplain, G.D. Singh, J.C. McLachlan

of such genes is then to provide an interlinked, highly non-linear system.

A highly non-linear response of such control systems may be an important feature already selected during the evolution of single cells. Several processes need a fine-tuned control and thus off/on response in narrow ranges. Genes are already highly interdependent in single cells. High non-linearity is a necessity for multistability in connected reaction networks, and multistability is a prerequisite for multifunctionality. Indeed many species classified as unicellular organisms have recently been shown to have a number of cell–cell self-organizing properties, and collective behavior displaying regulative phenomena in colony morphogenesis (Shapiro and Dworkin, 1997).

The highly non-linear response found in unicellular systems may then be advantageous in other contexts as well. It seems to be a requirement for decoding positional information along gradients. Several gradient decoding systems have been implied in prephylotypic embryogenesis.

Highly non-linear response is advantageous for pattern formation processes other than gradient decoding, however, and well-known dynamical properties of such systems in general may be the common basis for the observed spatiotemporal self-organization in early embryos.

Indeed the theoretical and experimental basis for highly non-linear systems is already well established. Oscillations and waves, multistability, and pattern formation are all well known properties of such systems. I will mainly discuss the emergence of stationary patterns in biochemical reaction-diffusion systems (Turing structures). I will argue that pattern formation based on this mechanism is likely to have evolved simultaneosly with the capability of biosystems to read positional information along gradients. It follows that observed pattern formation processes in biology may not rely exclusively on gradient control; other features from non-linear dynamics, like Turing stuctures or waves may play an equally important role without any further essential requirements of the system.

5.2 Pattern Formation in Reaction-Diffusion Systems

An example of a reaction-diffusion system capable of spontaneous pattern formation is Gierer and Meinhardt's reactions

$$\frac{dx}{dt} = A - Bx + \frac{x^\gamma}{y} \tag{5.1}$$

$$\frac{dy}{dt} = x^\gamma - y \tag{5.2}$$

where A and B are positive constants. Since x increases dx/dt, and y decreases it (due to the x^γ/y term), the system is known as an activator-inhibitor system. Also, x increases dy/dt through the x^γ term and thus acts as an activator as well in this rate. In order to obtain pattern formation, the above rates are completed with extra terms to describe diffusion in space.

It is well known that pattern formation by Turing's mechanism does not need rate terms with explicit activation or inhibition. The interplay between the two reactants according to the rate laws may be less obvious, but some of these systems may be classified as activator-inhibitor systems nonetheless, as discussed below.

In more generality, a Turing system of the first kind is defined as

$$\partial c/\partial t = F(c) + D\Delta c \tag{5.3}$$

where vector c now contains the concentrations (say, the two components x and y above), F is a vector function defining the rates, and Δc is the Laplacian operator describing Fickian diffusion. For model systems with diffusion in one spatial direction only, this term for component x, say, is simply $\partial^2 x/\partial s^2$.

Bifurcations, such as onset of oscillations or build-up of Turing structures, are discussed in terms of linear stability theory. To leading order, the behavior is determined by the Jacobian J. For a two-component system, we may define J as

$$J = \partial F(c)/\partial c = \begin{pmatrix} a & b \\ c & d \end{pmatrix} \tag{5.4}$$

evaluated at the homogeneous stationary state. The system without diffusion is usually required to be stable and thus

$$a + d = Tr < 0 \tag{5.5}$$

$$ab - cd = \det > 0 \tag{5.6}$$

When diffusion is added, the system may develop patterns with a characteristic wavelength $\lambda = 2\pi/\kappa$, with

$$\kappa^2 = \frac{aD_y + dD_x}{2D_x D_y} \tag{5.7}$$

provided

$$0 < \det < \frac{(aD_y + dD_x)^2}{4D_x D_y} \tag{5.8}$$

These inequalities are derived in many sources. They may be used to show that the diffusion coefficients usually have to be different by an order of magnitude or more. For details and references see Hunding and Engelhardt (1995).

For the particular system above, we find the stationary state (x^*, y^*) by setting both rates to zero; this immediately yields $y^* = (x^*)^\gamma$, which inserted in the x-rate yields $A - Bx^* + 1 = 0$ and thus

$$x^* = \frac{A+1}{B}; \qquad y^* = (\frac{A+1}{B})^\gamma \tag{5.9}$$

By differentiation we find the elements of the Jacobian:

$$a \quad = \quad \partial x'/\partial x = -B + \frac{\gamma}{x^*}\frac{(x^*)^\gamma}{y^*} = -B + \frac{\gamma B}{A+1} \tag{5.10}$$

$$b \quad = \quad \partial x'/\partial y = -\frac{(x^*)^\gamma}{(y*)^2} = -(\frac{B}{A+1})^\gamma \tag{5.11}$$

$$c \quad = \quad \partial y'/\partial x = \gamma(x*)^{\gamma-1} = \gamma(\frac{A+1}{B})^{\gamma-1} \tag{5.12}$$

$$d \quad = \quad \partial y'/\partial y = -1 \tag{5.13}$$

Thus it is easy to calculate det $= ad - bc = B$, which is always positive, as required by equation (5.6). The condition for the trace Tr from equation (5.5) then becomes

$$0 > Tr = a + d = -B + \frac{\gamma B}{A+1} - 1 \tag{5.14}$$

which may be rewritten as

$$B < B_c = \frac{A+1}{\gamma - (A+1)} \tag{5.15}$$

Usually a simplified version of this reaction scheme is discussed, obtained with $A = 0$ and the Hill coefficient $\gamma = 2$, in which case the Jacobian elements are simplified to $a = B$, $b = -B^2$, $c = 2/B$ and $d = -1$.

The condition for the trace, (5.14) now becomes $0 > B - 1$, and for the characteristic wave number κ we find from (5.7) that

$$\kappa^2 = \frac{BD_y + (-1)D_x}{2D_x D_y} = \frac{1}{2}\left(\frac{B}{D_x} - \frac{1}{D_y}\right) \tag{5.16}$$

If the activator is removed by a term in the rate of the form $-\mu x$, one may introduce the activator range as $r_x = \sqrt{D_x/\mu}$ and an analogous inhibitor range $r_y = \sqrt{D_y/\nu}$. For the simplified system under discussion, we have $\mu = B$ and $\nu = 1$. Equation (5.16) for κ is often discussed with these terms, since it may be written

$$\kappa^2 = \frac{1}{2}\left(\frac{1}{r_x^2} - \frac{1}{r_y^2}\right) \tag{5.17}$$

If the range of y is much larger than the range of x, the last term may be neglected and we get

$$\kappa \sim \frac{1}{\sqrt{2}}\frac{1}{r_x} \tag{5.18}$$

and thus the wavelength of the pattern

$$\lambda = \frac{2\pi}{\kappa} = 2\sqrt{2}\pi r_x \tag{5.19}$$

In the more general case with nonvanishing A and Hill constant γ different from the particular value 2, the reader may substitute the elements from (5.10) to (5.13) and convince themself that the wavelength is not quite as simple as above. One may still define a general range of x with the Jacobian element a substituted for μ and the element $|d|$ for ν but this no longer reflects simple first-order rate constants in the original rate equations. Indeed, in general reaction schemes, there may not be any such first-order rate constant terms present at all.

Another feature of Turing systems is a certain structure in the Jacobian elements a to d. It is found that a and d must have different signs, say positive a and negative d. The two off-diagonal elements, b and c, must have different signs as well, and thus it is a necessary condition that the Turing Jacobian has one of the following two structures:

$$J = \begin{pmatrix} + & - \\ + & - \end{pmatrix} \quad \text{(I)} \qquad\qquad J = \begin{pmatrix} + & + \\ - & - \end{pmatrix} \quad \text{(II)} \tag{5.20}$$

Two further Jacobians may be obtained from the trivial operation of interchanging x and y:

$$J = \begin{pmatrix} - & + \\ - & + \end{pmatrix} \quad (\mathrm{I}') \qquad J = \begin{pmatrix} - & - \\ + & + \end{pmatrix} \quad (\mathrm{II}') \tag{5.21}$$

The first class (I) is known as an activator-inhibitor system. The second class (II) has no obvious classification of activators or inhibitors. Both x and y enhance the first rate, and both decrease the second rate. One may define the self-activator to be the substance which has a positive element (+ in the above matrices) in the diagonal, i.e. the component which enhances its own formation. The other component is generally self-inhibitory (minus sign in the diagonal). According to inequality (5.5), the numerical value of this self-inhibition must be larger than the value for the self-activator (i.e. with $a > 0$, d must be sufficiently negative to yield the required $a + d < 0$). The presence of sufficient self-inhibition is therefore just as important as the presence of a self-activator.

This is a subtle point when referred directly to the rates of the original system, rather than the small deviation terms in the Jacobian. The reader may not notice any obvious self-inhibitory term in the rate (5.2), but this role is played by the simple first-order reaction term $-y$, or in slightly more generality $-\nu y$. The fact that this term is most important to satisfy the requirement for the Turing Jacobian matrix may then be compared to equation (5.1). Here a similar expression of form $-Bx$ is present but the net positive self-activation term in the Jacobian arises from the larger magnitude of the last term x^γ/y in the stationary state. It is the elements in the Jacobian matrix, defining what happens with small deviations from the homogeneous stationary state, which are important for the onset of pattern formation, not particular rate terms in the original rate equations.

This statement is true for any discussion of the onset of pattern formation from homogeneity. However, once a pattern has emerged, the above linear analysis has to include strong non-linearities, and specific kinetics terms may then be important for the discussion.

5.3 Slowing Down the Effective Diffusion of the Self-activator

It took quite some time from the theoretical discovery of Turing stuctures in 1952 to their experimental realization by De Kepper's group in 1990; see for example Dulos *et al* (1996).

One reason for this was the theoretical requirement for substantially different diffusion constants predicted with the standard theory as outlined above. Real diffusion constants in water do not exhibit such large variation, even with small molecules compared to large ones, say a globular protein.

One may observe, however, that transport of matter in living systems often occurs in environments which are poorly described as pure water. If the self-activator binds reversibly to a (quasi) stationary third component (perhaps a gel like substance), its effective diffusion is slowed down.

The theoretical analysis of such a 2+1 system (Pearson, 1992) shows that Turing patterns may indeed emerge there, even with equal diffusion constants. An expression for the characteristic wave number κ, similar to (5.17) for simple systems, reemerges, but here the large range of the self-inbihitor is not due to a comparatively large diffusion constant. Rather it obtains because the generalization of ν is now required to be small. In this situation inequality (5.5) cannot be fulfilled, which is to say that to obtain Turing patterns in the 2+1 system, the original two-state system must exhibit strong time-periodic solutions. When the self-activator is then coupled to the third component, the time oscillatory solutions are replaced with a pattern, i.e. a stationary solution with periodic variations in space (Figures 5.1 to 5.3).

Turing stuctures were realized experimentally in such an oscillator–gel system, and it is now felt that these 2+1 systems are much more likely candidates for real pattern formation than those originally discussed. Numerous time oscillatory systems are known, experimentally as well as theoretically, and they are abundant in biosystems. The additional requirement of slowing down one component by binding to a quasi-stationary third component poses little problems in such systems.

5.4 Turing Patterns of the Second Kind

In the previous systems all rates and diffusion constants were independent of position, and the formation of Turing patterns of the first kind, as discussed above, is thus a spontaneous symmetry-breaking event.

The substances involved may act as morphogens, as concentration patterns formed may influence rates in other systems, setting them on or off depending on their position in space. Indeed, rates in biosystems are often catalyzed by enzymes, which may be inhibited to varying degree by other substances. Thus concentration gradients or more involved patterns may influence the rates of new pattern-forming systems, but this time with ingoing rates depending on position.

In general, such a system is a Turing system of the second kind:

$$\partial c/\partial t = F(k(r), c) + \nabla \cdot D(r)\nabla c \tag{5.22}$$

with vector r the position.

If the rates are spatially varying, perhaps along a gradient, it is trivial that solutions to (5.22) may show the same spatial geometry, i.e. they also vary in a gradient manner. However, just as symmetry-breaking takes place when a homogeneous system develops into a Turing pattern of the first kind, the patterns arising in these new systems may show symmetry-breaking as well and have geometry unrelated to the wavelength of the imposed gradient.

An example of this has been demonstrated for pattern formation in a spherical region (a cell or early blastula). If a gradient from the animal pole to the vegetal pole is built into the Turing rates, bipolar Turing patterns may form perpendicular to the imposed axis. Thus the wavelength of this pattern is not that of the imposed gradient, and its spatial variation is along another axis.

Another extension to first-order systems may be present in systems where diffusion is not restricted to self-diffusion and where cross interactions among the substances are important. Indeed, pure diffusion may be replaced by simple or more non-linear cell–cell communication, but if such communication involves active transport, $D(r)$ in (5.22) may have nondiagonal elements, which may even be negative. Such terms may strongly widen the parameter region for which pattern formation is possible.

In even more generality, once a pattern has formed within a certain region, it may start cell growth or cell migration in well-defined regions. This would change the overall geometry and in turn might trigger transitions to new patterns with yet another geometry, etc. The result would be an autonomous physicochemical system developing through a succession of patterns on a well-defined prescribed route in space and time.

5.5 Size Regulation of Pattern Forming Processes

Early investigations on the original Turing mechanism suggested that the parameter range in which a particular pattern emerged was fairly narrow for a number of particular models. Only about 30% variation of rate constants was allowed. But rates in biosystems are not determined with rate constants like well-studied inorganic reactions; they are given by enzyme activities, which may be controlled by activator or inhibitor substances. Indeed, many biochemical processes even in unicellular systems are kept in narrow ranges under strict control of highly non-linear reaction networks. Such control may be based on allosteric enzymes or kinetic cooperative systems. Indeed series, cascades or more complicated networks of such building blocks are known experimentally to create biosystems with very steep off/on control. Maintaining parameters within very narrow limits is thus already achieved in a number of quite complicated processes, even in single cells. So Turing pattern formation needs control that is no narrower than obtained in other existing systems.

Size regulation for pattern-forming mechanisms is usually discussed in terms of a characteristic length L of the embryo. Linear stability analysis shows that L enters the equations through the term $\kappa^2 D/L^2$, i.e. through the squared characteristic length. Maintaining parameters independent of L thus requires some sort of control, which varies with L^2. It has been suggested that a simple control system of this type may be based on a substance which is created (or destroyed, or simply lost) proportional to the surface area of the region in question. Such a crude system may give sufficient initial size regulation, to be refined later in evolution to more sophisticated control.

One may conclude with a general remark on size control of any pattern-forming mechanism, Turing or otherwise. Repetitive structures often show a remarkably well-controlled number of units (segments, somites, etc.), but any mechanism capable of discerning among 20 or 22 such units (say) must be able to respond to a variation of about 10%, which also imposes strict off/on control. And such narrow limits are not easy to achieve by other proposed simple size-regulating models, like dilution of a substance by cell division.

5.6 Pair-rule Patterns in *Drosophila*

The genes regulating the initial pattern-forming events in the *Drosophila* embryo were characterized during the 1980s. A remarkably small number of genes determine the major axis and initial subdivision of the embryo (Jäckle and Sauer, 1993; Akam *et al.* 1994; Tautz and Schmid, 1998). The discovery of a seven-stripe repetitive pattern of the gene *fushi tarazu ftz* under control of a single all-or-none element on the promoter evoked quite some interest, from theoreticians and experimentalists alike, for the possible presence of a Turing pattern forming mechanism. A few years later, this idea was largely abandoned by experimentalists. Other pair-rule genes also formed seven stripes, and some of these genes seemed to exert control on *ftz*, which was thus not the primary organizer. These primary pair-rule genes also affect each other's expression. Two of these primary pair-rule genes (*hairy* and *even-skipped*) were shown to have specific control elements for at least a few stripes, and on this basis it was inferred that the generation of seven stripes for each of the genes *h, eve, runt* and *ftz* were under specific single-stripe control, generated by combinations (cues) of previously expressed maternal and gap genes.

To most experimentalists, the problem of formation of periodicity (pair-rule stripes) from nonperiodic positional information is thus largely settled, with remaining questions of minor importance, although occasionally some researchers admit or state that the formation of 28 stripes with precise phase relationships is still poorly understood. Theoreticians argue that the cue model should be prone to unacceptably large deviations in the final pair-rule pattern from only small variations in the maternal and gap protein locations. It is suggested that the stripe-specific cue control is only one aspect of the total system, and that a Turing pattern component may be present nonetheless to insure precise phase relationships (Hunding, 1993); see Figure 5.4.

Recently the promoter region of the primary pair-rule gene *runt* has been found to contain both stripe-specific elements and a disperse region controlling all seven stripes. It has been speculated that such disperse regulatory elements may have escaped detection for *eve* as well. (Klingler *et al.*, 1996). Another recent addition to the mutual pair-rule control system has been found in the form of a protein that interacts with the Hairy protein, and has been shown to be essential for proper segmentation. (Poortinga *et al.*, 1998). Such examples indicate that the originally established working model for pair-rule stripe formation is still open for substantial extensions.

The focus has changed from stripe formation in the evolutionarily recent *Drosophila* to research on other, older species like the flour beetle *Tribolium* and grasshoppers. It has been found that the control system established for early *Drosophila* morphogenesis is not general. In *Drosophila*, a so-called long-germband insect, the pair-rule stripes are generated roughly simultaneously in a syncytial unicelluar but multinuclear embryo. More primitive short-germband insects initiate development in a syncytium as well, but segments are generated one after another in a cellularized embryo, and laid down sequentially from a terminal growth zone. This, at least in part, implies cell–cell interactions. Interestingly, some insects (polyembryonic wasps) are evolutionary descendants from species with a syncytial blastoderm (like *Drosophila*), but seem to have lost the pair-rule (double segment) expression state. Some segments are initiated quite rapidly, but not from a posterior growth zone, whereas the remaining segments are laid down sequentially. (Grbić *et al.*, 1998).

Sequential segment formation may be modeled by a molecular clock mechanism that is some autonomous biochemical oscillation, combined with posterior growth. Interestingly, experimental evidence for such a mechanism has recently been presented for chick segmentation and somitogenesis, and a *hairy* homolog is involved. (Palmeirim *et al.*, 1997).

Nevertheless, some of the pair-rule genes may not have been evolutionary original pattern-forming genes in the early embryo. They seem to have their initial role in the nervous system. In vertebrates, numerous homologs to *Drosophila* segmentation genes have been found, but they are not involved in somitogenesis. On the other hand, homologs of some other neurogenic genes from the fruit fly are presumably involved in somitogenesis.

A tentative conclusion may then be that several of the genes involved in early (so-called prephylotypic) embryo formation are versatile gene control units, which may be borrowed and reused in different control processes, not necessarily with the same pattern-forming function. Or, in a theoretician's vocabulary (citing the introduction), '... an essential property of such genes is then to provide an interlinked, highly non-linear system. Well-known dynamical properties of such systems in general may be the common basis for the observed spatio-temporal self-organization in early embryos.'

5.7 Clocks and Patterns from the Same Mechanism

The experimental result that a molecular clock mechanism is present in chick segmentation and somitogenesis demonstrates there is a non-linear control system creating chemical oscillations. The nature of this clock is unknown. This section demonstrates that a chemical oscillator, extended with a mechanism for slowing down the spatial movement of the self-activator, has a rich repertoire of possible dynamic modes. Variation of a single parameter in such a system is enough to change the system from displaying time oscillations to create spatial patterns instead. This sort of mechanism is easy to achieve in real systems, once autonomous oscillations are already present, and it is suggested that such a control system has been exploited during evolution.

The precise form of the oscillator rates is not essential for the argument here, but to be specific, a modification of the well-known Sel'kov oscillator is used:

$$\frac{du}{dt} = \nu - k_1 \frac{uv^\gamma}{L + uv^\gamma} \qquad (5.23)$$

$$\frac{dy}{dt} = k_1 \frac{uv^\gamma}{L + uv^\gamma} - k_2 v + k_3 \qquad (5.24)$$

It is customary to simplify and normalize these equations to the form

$$\frac{\partial x}{\partial \tau} = 1 - xy^\gamma + D_x \frac{\partial^2 x}{\partial s^2} \qquad (5.25)$$

$$\frac{\partial y}{\partial \tau} = \alpha x y^\gamma - \alpha y + k + D_y \frac{\partial^2 y}{\partial s^2} \qquad (5.26)$$

where we have introduced diffusion as well in one spatial coordinate s. Linear stability theory on this system shows that it has a Jacobian of the form (II′) in equation (5.21) so this system is not a standard activator-inhibitor system (I). The component with self-enhancement (+ in the diagonal of the Jacobian) is y, and we introduce a mechanism for slowing down this component's net spatial spread by means of a third immobile component P. Thus add a reaction of the type $Y + P \rightleftharpoons Z$. The total system then becomes

$$\frac{\partial x}{\partial \tau} = f(x, y) + D_x \frac{\partial^2 x}{\partial s^2} \qquad (5.27)$$

$$\frac{\partial y}{\partial \tau} = g(x, y) - k_f p y + k_r z + D_y \frac{\partial^2 y}{\partial s^2} \qquad (5.28)$$

$$\frac{\partial z}{\partial \tau} = k_f p y - k_r z \qquad (5.29)$$

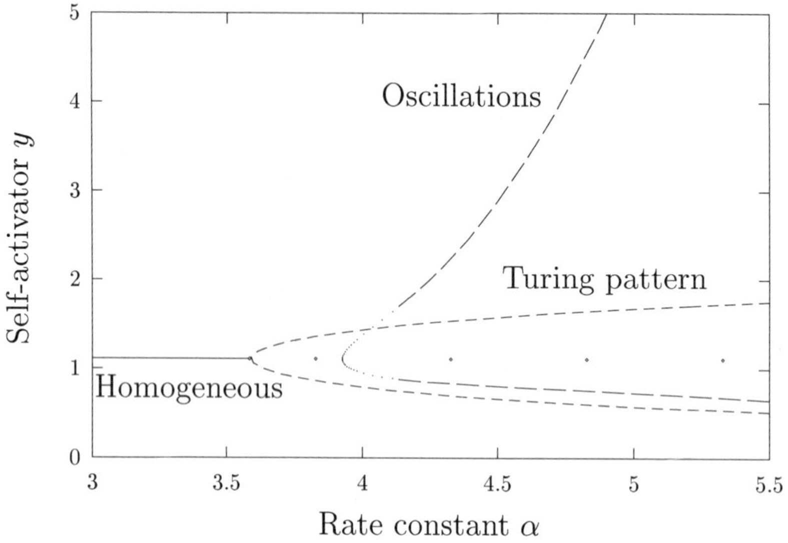

Figure 5.1 Bifurcation diagram for oscillator-gel mechanism, Eqs(5.28-5.29). Ordinate is the self-activator y, abscissa the generalized rate-constant α. For α less than about 3.6, the homogeneous state is stable. For larger α, a spatial pattern (Turing structure) emerge with increasing amplitude. For α larger than about 4.1, a homogeneous time oscillator also may appear. This solution is unstable below $\alpha = 4.1$ and disappears for $\alpha = 3.9$. Total immobile component $w = 25.0$ here. Rate constants k_f and k_r are 4.0 and 1.0 respectively. Thus both Turing structures or high amplitude, relaxation type time oscillations are possible with a change of a parameter α.

Put $w = p + z$ for the total amount of the immobile component, free and bound. Bifurcations, such as transitions from an initial homogeneous state to oscillations, have been studied systematically for various values of w, and increasing values of the rate constant α. Such a bifurcation diagram may be recorded for different values of w, different ratios of the in- and outgoing rate constants k_f and k_b, or different absolute magnitudes of these rate constants; see Figures 5.1 and 5.2.

Variation of the relative magnitude of the slowing-down effect from w changes the properties of the solutions obtained. For low or vanishing values of w, time oscillations emerge (Figure 5.3). With increasing w, Turing patterns take over, but a parameter region is still present, where time oscillations are dominant. Eventually, for sufficiently large w, homogeneous time oscillations disappear, and Turing structures are dominant.

One may suggest that such a control system is easy to exploit by nature. The oscillatory features may be involved in sequential segment formation as seen in short-germband insects, and incorporation of a slowing-down mechanism may be exploited to create Turing patterns for fast all-at-once stripe formation as in long-germband insects like *Drosophila*. Many theoreticians have argued that the presence of a syncytium points to a diffusion-related control mechanism. The presence of an oscillator-delay mechanism, as discussed above, should incorporate some means for slowing down diffusion.

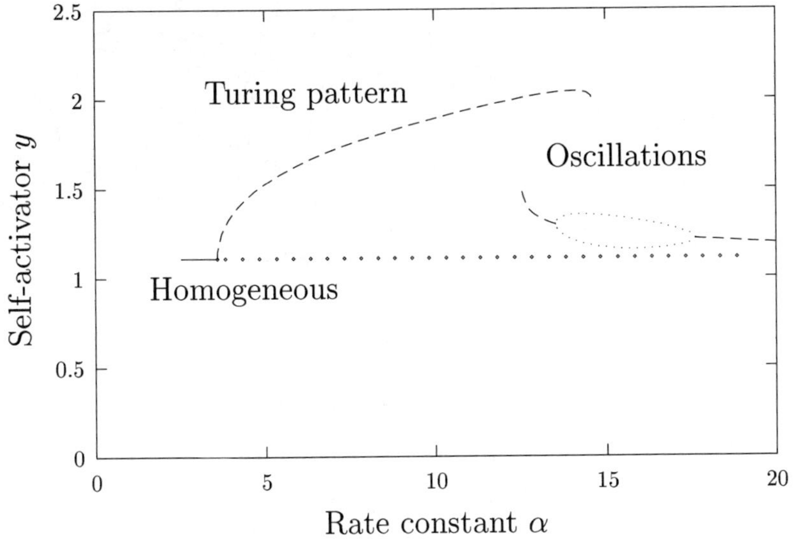

Figure 5.2 Bifurcation diagram for Eqs(5.28-5.29), as in previous figure, but here with increased total immobile component $w = 150$. $k_f = 100$, $k_r = 25$. Now, large amplitude time oscillations are no longer present, but spatial patterns emerge as before. For α larger than about 14, these patterns become unstable, and new low amplitude inhomogeneous time oscillations appear, which eventually disappear as well, to give rise to a low amplitude Turing pattern.

Such a delay component in the pair-rule system may have been experimentally found in the form of asymmetric localization (Davis and Ish-Horowicz, 1991; Francis-Lang *et al.*, 1996). Contrary to the transcripts of the gap genes, the pair-rule gene transcripts localize exclusively to a thin layer, the apical periplasm. This process limits diffusion.

The presence of both a syncytium and diffusion-limiting processes specific for the pair-rule system may be seen as supporting the above model for Turing pattern formation.

5.8 Conclusion

Although concepts from non-linear dynamics are alien to most experimental biologists, nonetheless, the mathematics of complex systems may be essential for understanding biological pattern formation and self-organization, thereby extending the heritage of D'Arcy Thompson in a manner not foreseen.

5.9 Numerical Treatment of 3D Pattern Formation

Efforts to develop efficient software for the calculation of spontaneous pattern formation in biological systems have led to fast codes for vector supercomputers. The numerical methods I have used are largely the same as described earlier (Hunding, 1993). The method of lines was used and thus the system of non-linear partial differential equations was converted to a large system of ordinary differential equations

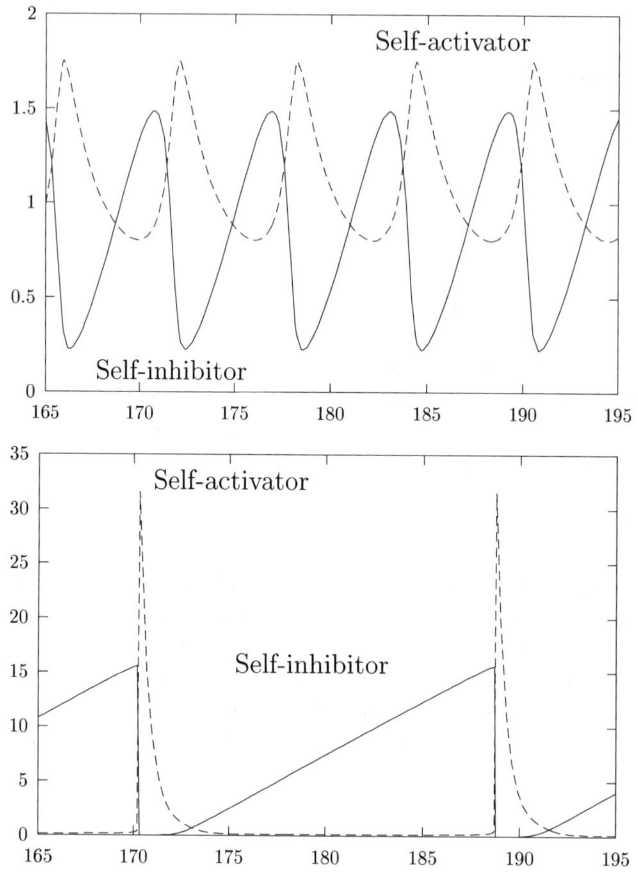

Figure 5.3 Time-oscillations in oscillator-gel mechanism, Eqs(5.28-5.29). For small or vanishing slowing down (small w), oscillations in time (limit cycles) are dominant. Abscissa is time, ordinate is concentrations of the two reactants shown, the self-inhibitor and the self-activator. Here, $w = 0.25$, $k_f = 0.4$ and $k_r = 0.1$. For moderate α, normal oscillations are present (top) ($\alpha = 0.9$), for increasing α the form changes to spike-like relaxation type oscillations (bottom) ($\alpha = 2.0$).

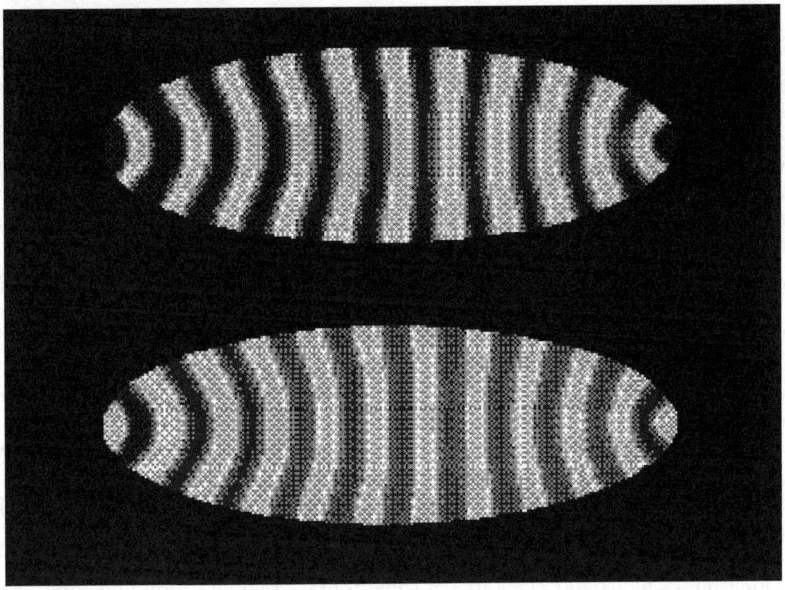

Figure 5.4 Numerical solution of Eqs(5.28-5.29). Formation of Turing stripes. Note that the self-activator y (bottom) and self-inhibitor x (top) make stripes which are out of phase. It is suggested that such a pattern is a much more reliable control unit to position the 4 times 7 pair-rule stripes with correct phase relations than a pure cue model.

by discretization of the Laplacian in three curvilinear coordinates. The resulting system is stiff and solved accordingly (modified Gear code). A factor 500 in CPU time is gained in scalar mode over nonstiff codes.

The Jacobian used in the corrector step is a sparse-banded matrix which is rearranged (chessboard numbering of meshpoints) to yield large blocks within which the solution vector elements may be iterated in parallel (RBSOR method). Implementation on vector computers results in a huge speed-up: The RBSOR code runs efficiently and close to the top speed of vector machines. Routine production runs yield 750 Mflops (single processor) and 1320 Mflops have been recorded on a two-processor CRAY C90.

Acknowledgements
This work was supported by the Danish Research Council, grant 9600657. I have used the CONT continuation package developed by Igor Schreiber, Department of Chemical Engineering, Prague Institute of Chemical Technology.

REFERENCES

Akam, M., Holland, P., Ingham, P. and Wray, G. eds., 'The Evolution of Developmental Mechanisms', *Development 1994 Supplement.*

Davis, I. and Ish-Horowicz, D. (1991) Apical localization of pair-rule transcripts requires 3′ sequences and limits protein diffusion in the *Drosophila* blastoderm embryo. *Cell* 67:

927–940.

Dulos, E., Boissonade, J., Perraud, J.J., Rudovics, B. and De Kepper, P. (1996) Chemical Morphogenesis - Turing Patterns in an Experimental Chemical System. *Acta Biotheoretica* 44: 249–261.

Francis-Lang, H., Davis, I. and Ish-Horowicz, D. (1996) Asymmetric localization of *Drosophila* pair-rule transcripts from displaced nuclei: evidence for directional nuclear export. *EMBO J.* 15: 640–649.

Grbić, M., Nagy, L.M. and Sand, M.R. (1998) Development of polyembryonic insects: a major departure from typical insect embryogenesis. *Dev. Genes Evol.* 208: 69–81.

Hunding, A. (1993) Modeling spatial pattern formation in *Drosophila. Comments Theor. Biol.* 3: 141–167.

Hunding, A. and Engelhardt, R. (1995) Early biological morphogenesis and nonlinear dynamics. *J. Theor. Biol.* 173: 401–413.

Jäckle, H. and Sauer, F. (1993) Transcriptional cascades in *Drosophila. Curr. Opinion Cell Biol.* 5: 505–512.

Klingler, M., Soong, J., Butler, B. and Gergen, J.P. (1996) Disperse versus compact elements for the regulation of *runt* stripes in *Drosophila, Dev. Biol.* 177: 73–84.

Martinez, D.M., Bridge, D., Masuda-Nakagawa, L.M. and Cartwright, P. (1998) Cnidarian homeoboxes and the zootype. *Nature* 393: 748–749.

Palmeirim, I., Henrique, D., Ish-Horowicz, D. and Pourquié, O. (1997) Avian *hairy* Gene expression identifies a molecular clock linked to vertebrate segmentation and somitogenesis. *Cell* 91: 639–648.

Pearson, J.E. (1992) Pattern formation in a (2+1)-species activator-inhibitor-immobilizer system. *Physica A* 188: 178–189.

Poortinga, G., Watanabe, M. and Parkhurst, S.M. (1998) *Drosophila* CtBP: a Hairy-interacting protein required for embryonic segmentation and Hairy-mediated transcriptional repression. *EMBO J.* 17: 2067–2078.

Shapiro, J.A. and Dworkin, M., eds. (1997) *Bacteria as Multicellular Organisms.* Oxford University Press, Oxford.

Tautz, D. and Schmid, K.J. (1998) From genes to individuals: developmental genes and the generation of the phenotype. *Phil. Trans. R. Soc. Lond. B* 353: 231–240.

Wolpert, L. (1990) The evolution of development. *Biol. J. Linn. Soc.* 39: 109–124.

6

Pattern Formation Mechanisms in Skin and Hair: Some Experimental Tests

Barry N. Nagorcka and David A. Adelson

CSIRO, Division of Animal Production, Locked Bag 1, Delivery Centre, Blacktown, NSW 2148, Australia

6.1 Introduction

At first sight, mammalian hair follicles may seem to be located in the skin in an irregular manner apart from a tendency to be approximately equidistant from each other. The hair fibres they produce are often assumed to be round or elliptical. A closer examination reveals a much higher degree of order and pattern in the location of follicles, and in the shape or form of their fibres. Taking sheep as an example (Figure 6.1), we see (i) that some follicles are arranged in trio groups with a strong tendency for the trio to be near linear, (ii) that the trio groups themselves frequently show a strong tendency to be aligned forming rows, (iii) that each trio group appears to be positioned on the anterior margin of a larger group of follicles, (iv) that follicles on the posterior margin of the larger follicle group are often isolated or single follicles, (v) that follicles within the group tend to be branched forming "bundles" of follicles. In addition, the follicles forming trio groups frequently have attached sweat glands. The follicles in mice also tend to group and show some of these characteristics. However, none of the mouse follicles has sweat glands.

Follicles produce fibres during a hair growth cycle, which is very short in mice and rabbits compared to the cycle in sheep. In mice and rabbits the cross-sectional shape of some fibres is not round or elliptical but bilobed, trilobed or even tetralobed (Figure

On Growth and Form: Spatio-temporal Pattern Formation in Biology.
Editor Mark A.J. Chaplain, G.D. Singh, J.C. McLachlan

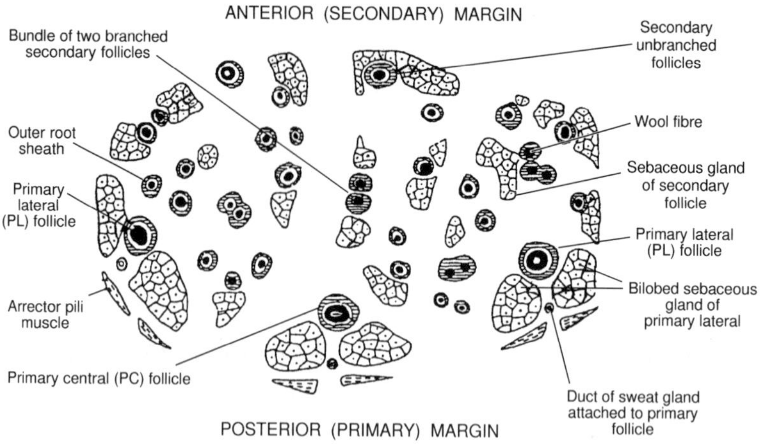

Figure 6.1 Transverse section through an adult merino wool follicle group at the level of the sebaceous gland (Hardy and Lyne, 1956). A near-linear trio of primary follicles (two PL and one PC follicles) can be seen on the posterior margin of the group. Unbranched secondary follicles can be seen on the opposite side (anterior margin) of the group. Several bundles of branched secondary follicles along with some unbranched secondary follicles can be seen in between the two margins. Reproduced by permission of CSIRO publishing.

6.2) (Dry, 1926; Straile and Mann, 1973). That part of the fibre which exhibits these multilobed cross-sectional shapes corresponds to the most rapid growth phase of the hair growth cycle. These fibres are usually among the largest of the fibres in the pelage and are referred to as "guard hairs". They are often produced by those follicles which form the central member of a trio group. Other fibres have kidney-shaped cross-sections and a bend at the proximal end of the shield region (Figure 6.2) of the fibre (they are called awls and auchenes). The great majority of fibres, however, are short and form the underfur. They are called zigzag fibres because they have a zigzag shape along their length.

Multilobed cross-sectional shapes are also found in wool fibres in coarse-wool sheep. The fibres in fine-wool sheep, on the other hand, have circular or elliptical cross-sections, but even in these it is possible to find bilateral, bilobed and trilobed groupings of different cortical cell types within the fibres.

Even though there are many differences between species in both follicle arrangement and fibre structure, there are also patterns which appear to be common across species. The differences can be viewed as variations on a theme. This has certainly been pointed out by Lyne (1966) in relation to follicle arrangement. Unfortunately, no common theme has been described for fibre form and structure apart from the hair growth cycle and fibre cell types. The differences in fibre form and structure have been described and used as a basis for species identification (Brunner and Coman, 1974).

This chapter summarizes the results of some theoretical and associated experimental research which point to one pattern formation mechanism, namely a Turing type reaction-diffusion (RD) system, as having the capacity to generate the spatial patterns required to establish the common themes in both follicle arrangement in the skin and

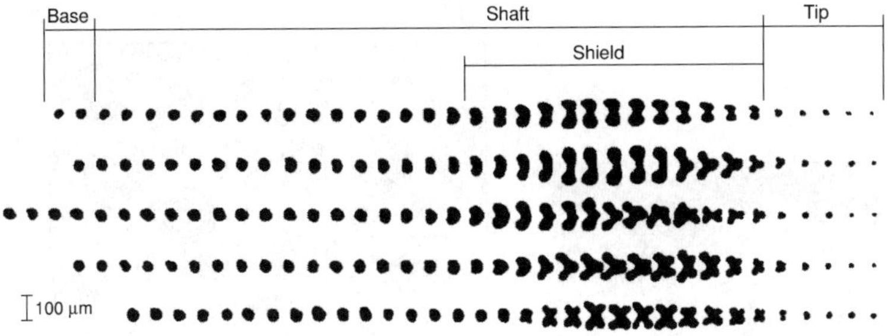

Figure 6.2 Each row consists of cross-sections at 1mm intervals along individual telogen hairs from Dutch rabbits and is typical of the results of Straile and Mann (1973). Each fibre is divided into a tip, shaft and base. In the examples shown here, the shaft contains a broad section which immediately follows the tip and is known as the shield.

in the form and structure of the fibre. Other mechanisms have also been suggested. It has also been proposed that the dermal papilla in the follicle bulb functions as a source of a substance which then establishes a diffusion gradient around the dermal papilla. The gradient in this case is able to control the formation of the fibre. The predictions based on both RD systems and gradients in follicle initiation and fibre formation are compared with each other and with experimental observations.

6.2 Follicle Initiation and Development

6.2.1 Equidistant follicles

Wool follicles first appear in fetal sheep skin on the mid-side at about 60 days post conception (Carter, 1943). These earliest follicles can be seen in transverse skin sections (i.e. in planes cut parallel to the skin surface) to be approximately equidistant from each other (Figure 6.3a). There appears to be no other regularity in their relative location. The approximately equidistant but irregular array of the earliest follicles seems to be common to all mammalian species. It is as if there is an inhibitory field associated with each new follicle, a field which prevents other follicles from initiating within a certain distance of the new follicle. Wigglesworth (1940) first suggested the possible existence of inhibitory fields in relation to the initiation of insect bristles. The relative location of insect bristles is the same as for the earliest hair follicles. An inhibitory field would arise if at the time a follicle (or bristle) initiates, it produces an inhibitory substance which diffuses (radially) away (Wigglesworth, 1940: figure 13). Where the concentration of the inhibitor is above a certain threshold value, it prevents the initiation of new follicles (bristles). New follicles will only initiate where there is sufficient distance between neighbouring follicles. Based on inhibitory fields it is possible to predict that new follicles will initiate only in between existing follicles, and to predict this will occur at a rate which maintains the density of follicles at a constant level as the skin grows in area.

A substrate depletion field was also suggested by Wigglesworth (1940) as an alternative mechanism to an inhibitory field. A substrate depletion field would arise if a

(B)

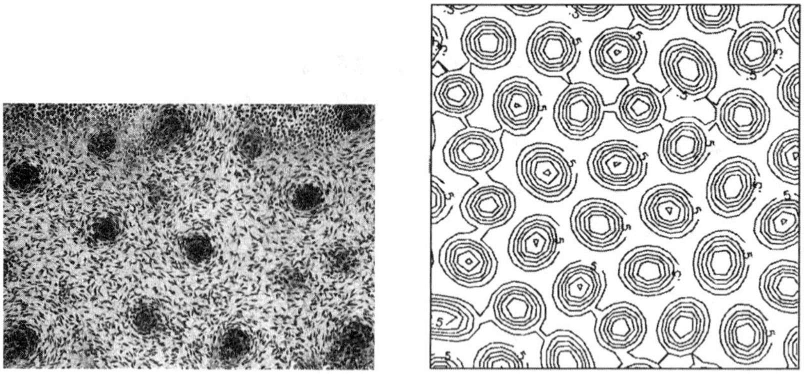

Figure 6.3 (a) A transverse section just below the epidermis in fetal merino sheep skin at ~70 days gestation, showing plugs of epithelial cells, i.e. primordial wool follicles, growing down into the dermis. (b) A contour plot showing the distribution of U produced by the RD system (6.1) in a two-dimensional sheet. Localized maxima in the concentration of U (spots) can be seen to be approximately equidistant from each other. (The contours are at intervals of 0.5 in arbitrary units chosen so that $U_0 = V_0 = 1.0$).

substrate, which is required for follicle initiation to occur, is used up by existing follicles and diffuses towards existing follicles where it is required. The spatial distribution of the substrate would tend to be the inverse of the expected distribution for an inhibitor. Wigglesworth felt that it would not be possible to distinguish between an inhibitory and a substrate depletion field.

6.2.2 Increasing density of follicles

Following the appearance of the earliest follicles, the skin continues to grow in area and new follicles do indeed initiate between existing follicles in fetal sheep skin (Carter 1943: figure 1, Plate 5; Nagorcka 1995a: figure 2b). This would be entirely consistent with the existence of inhibitory (depletion) fields (Claxton, 1964) were it not for the fact that the total density of follicles is observed to increase substantially despite a large increase in skin area (Carter and Hardy, 1947). If inhibitory (depletion) fields are controlling follicle initiation then it would be necessary to propose that the diffusion rate of the inhibitor (substrate) decreases with developmental stage or time, in order to explain the increase in follicle density observed during follicle initiation and development. In the sheep fetus, the total follicle density increases by a factor of 4 over a period of about 60 days. Follicle initiation occurs almost continuously during this period and it has been classified into a series of stages based on observations of follicle grouping which builds up in very small localized regions of the skin towards its mature form (Hardy and Lyne, 1956; Figure 6.1, this chapter). We refer to these stages as a time series of follicle generations.

6.2.3 *Spatial patterns in follicle location*

The earliest follicles to initiate in the skin are regarded here as the first generation of follicles. These follicles are normally denoted as PCX, or primary central early follicles (Hardy and Lyne, 1956). Looking a short time later at a particular small region of the fetal skin (e.g. the midside), a second generation of follicles can be seen initiating in between existing follicles. Follicles in the second generation (normally denoted as PCY, or primary central late follicles) are still located approximately equidistant from existing (neighbouring) follicles. The first and second generations are the follicles mentioned in Section 6.2.2 and they lead to an increase in the total follicle density.

A short time later the third generation (denoted as PLx, or primary lateral early follicles) appears further increasing the total density, but this time the new follicles show a strong preference to be located on either side of the first generation of follicles, forming a near linear trio group of follicles. Slightly later still, the fourth generation of follicles (denoted as PLy, or primary lateral late follicles) appear, and they too show a strong preference to be located on either side of the second generation of follicles. The first and second generations form the central follicles of the trio groups. At the end of the third and fourth follicle generations, a high proportion of follicles form trio groups of follicles, and the trio groups show a strong tendency to be aligned with each other (Nagorcka, 1995a: figure 2).

At an early stage of follicle development in all four generations of follicles mentioned thus far, sweat glands begin by budding from the posterior side of the follicles (Hardy and Lyne, 1956). The sweat glands are frequently used to identify the early or primary follicles in the adult population. The trios and sweat glands are arranged in such a way that the line passing through the centre of a sweat gland and the associated follicle tends to be at right angles to the near-linear trio group of follicles.

After the primary follicles have appeared in fetal sheep skin, several more generations of follicles initiate in the epidermis. These later generations of follicles are called secondary follicles. Unlike primary follicles, they do not develop sweat glands. The first generation of secondary follicles (i.e. the fifth generation of follicles) initiates on the anterior side of the trio groups. The next two or three generations of secondary follicles arise in between the trio group of follicles and the first generation of secondary follicles. Up to this stage all follicles initiate from the epidermis, and we call them epidermal follicles. Follicle initiation from the epidermis then ceases.

Further generations of follicles do appear, but they form by budding from the side of the last two or three generations of epidermal (secondary) follicles. The budding follicles are called derived or branched follicles; and each cluster of branched follicles, together with the epidermal follicle from which they originally branch, forms a bundle of follicles. This establishes the characteristic grouping of follicles seen in adults (Figure 6.1).

To date, no mechanisms involving inhibitory or substrate depletion fields, or other gradients, have been suggested as an explanation for (i) the formation of the primary trio group of follicles, (ii) their tendency to align into rows with a particular orientation in the integument, (iii) the building up of a group of follicles based on the primary trio group, (iv) the capacity of some epidermal follicles to branch, or (v) how some follicles develop sweat glands and others do not. Only Claxton and Scholl (1973) attempted an explanation for (i) using an interaction between a wavefront and an inhibitory field.

6.2.4 Spatial patterns produced by an RD system

It has been suggested (Nagorcka and Mooney, 1985) that an RD system of the type originally proposed by Turing (1952) may be present in the epidermis where it functions as a prepattern mechanism generating patterns which control follicle initiation and development. An RD system has the capacity to generate, quite spontaneously, stable stationary wavelike spatial patterns in the distribution of its chemical components. The range of patterns which can be generated in the epidermis (a two-dimensional sheet of cells) has been calculated by Nagorcka and Mooney (1992) using a two-component RD system of the form

$$
\begin{aligned}
\frac{\partial U(\mathbf{r},t)}{\partial t} &= \overbrace{D_U \nabla^2 U(\mathbf{r},t)}^{diffusion} + \overbrace{AU(\mathbf{r},t) - BV(\mathbf{r},t) + P_U}^{netproduction} - \overbrace{T_U\left[U(\mathbf{r},t), V(\mathbf{r},t)\right]}^{nettransport} \\
\frac{\partial V(\mathbf{r},t)}{\partial t} &= D_V \nabla^2 V(\mathbf{r},t) + CU(\mathbf{r},t) - DV(\mathbf{r},t) + P_V - T_V\left[U(\mathbf{r},t), V(\mathbf{r},t)\right]
\end{aligned}
\tag{6.1}
$$

where $U(\mathbf{r},t)$ and $V(\mathbf{r},t)$ are the concentrations of the two chemical components of the RD system at the point $\mathbf{r}$ and time t in the epidermis. D_U and D_V are the diffusion rates of U and V. P_U and P_V are constant source terms for U and V, and A, B, C and D are constant reaction coefficients. The terms defining the net production of U and V in equation (6.1) are linear in U and V. T_U and T_V specify the net transport of U and V across the basement membrane from the epidermis into the underlying dermis, and are defined as follows:

$$
\begin{pmatrix} T_U \\ T_V \end{pmatrix} = \begin{pmatrix} q_U t_U \\ q_V t_V \end{pmatrix}
$$

where

$$
\begin{pmatrix} t_U \\ t_V \end{pmatrix} = \begin{pmatrix} \tau_{U1}\left[U - U_0\right] + \tau_{U3}\left[U - U_0\right]^3 + \cdots \\ \tau_{V1}\left[V - V_0\right] + \tau_{V3}\left[V - V_0\right]^3 + \cdots \end{pmatrix}
$$

and q_U and q_V are defined to represent any blocking or saturation effects on the transport of U and V, respectively. For the results presented here, q_U is assumed to have the form $q_U = \epsilon_{U0}/\left(\epsilon_{U1} U V^2 + 1\right) + \epsilon_{U2}$, where ϵ_{U0}, ϵ_{U1} and ϵ_{U2} are constants, and q_V is set to zero (Nagorcka and Mooney, 1992). The constants U_0 and V_0 are chosen such that $AU_0 - BV_0 + P_U = 0$ and $CU_0 - DV_0 + P_V = 0$; in which case (U_0, V_0) corresponds to one solution of the RD system equations (6.1), namely the spatially uniform steady state, i.e. $U = U_0$ and $V = V_0$ throughout the epidermis.

Although the same RD system given by equations (6.1) is proposed to exist in the dermis (as well as the epidermis), values for the parameters defining the reaction, diffusion and transport terms are chosen to lie outside the Turing space (Murray,

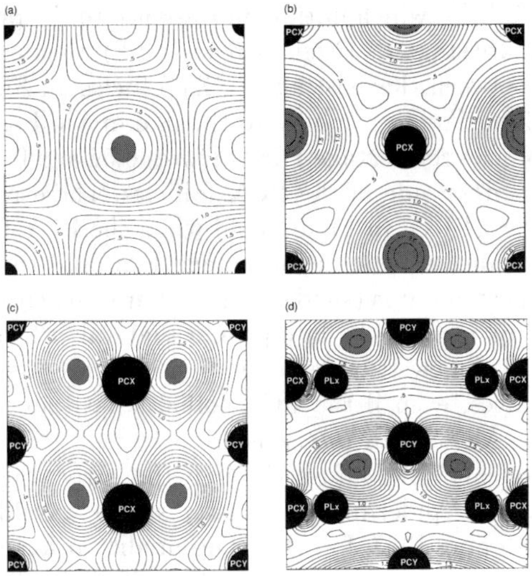

Figure 6.4 Contour plots of the distribution of U obtained by solving the RD system equations (6.1) in a two-dimensional sheet representing the basal epidermal cell layer. The black discs indicate holes in the domain where existing follicles join the epidermis. Fixed boundary conditions are applied around the edges of the discs corresponding to the distributions (U and V) arising in the upper part of the follicles depending on their stage of development (Figures 6.5c and e). Reproduced by permission of CSIRO publishing.

1989). This means that U and V will be maintained at the constant values U_0 and V_0 in the dermis.

Values for the parameters defining the reaction, diffusion and transport terms in the epidermis are chosen to lie within the Turing space, which means that the uniform state will be unstable and wavelike patterns will tend to arise spontaneously. It has been shown (Nagorcka and Mooney, 1992: figure 4) that the RD system equations (6.1) will generate spatial patterns ranging from stripes, to fingerprint patterns, to regular hexagonal arrays of spots (localized maxima), to irregular arrays of approximately equidistant spots shown in Figure 6.3b. The degree of directional asymmetry in the net transport terms determines which of these patterns will be generated. In regions of the skin where the directional asymmetry in transport across the basement membrane is substantial, an irregular array of spots with large amplitude will be generated by the RD system. It is proposed (Nagorcka, 1995a) that such large-amplitude spots will increase the mitotic activity of the epidermal cells in localized areas covered by the spots causing a plug of cells to grow down into the underlying dermis to form the PCX early stage F1 (Figure 6.5a) primordial follicles. In this way the initial spotted pattern (Figure 6.3b; also represented by a regular pattern in Figure 6.4a) is able to function as a prepattern for the initiation of PCX follicles (Figure 6.3a).

Figure 6.4a shows the first prepattern in U, represented here as a small-scale periodic array rather than the prepattern in Figure 6.3b. The areas where U exceeds the threshold of 1.9 concentration units (units chosen so that $U_0 = V_0 = 1.0$) are shaded.

In Figure 6.4b, when the follicles which initiate in response to the prepattern in Figure 6.4a, PCX follicles, reach late stage F1 (Figures 6.5a and b), then a new set of maxima (shaded regions) arise in between the existing (PCX) follicles. In Figure 6.4c when the new follicles (PCY) which initiate in response to the maxima in Figure 6.4b reach late stage F1, and the PCX follicles reach stage F2a (Figures 6.5e and f), then new maxima (shaded regions) arise on either side of the PCX follicles. These maxima initiate the first primary lateral follicles (PLx). In Figure 6.4d when the PCY follicles reach stage F2a, the PCX follicles advance a little beyond stage F2a, and the new PLx follicles reach late stage F1, then new maxima (shaded regions) arise on either side of the PCY follicles. These new maxima initiate the late primary lateral (PLy) follicles on either side of the PCY follicles. The new trios tend to align with the first trios, which have PCX follicles at their centres, even if the PCY follicles have no alignment with the first trios, as shown here. (All contour intervals are 0.1.)

The RD system components now diffuse through the early stage PCX follicles as well as the basal layer of the epidermis. Stationary stable wavelike prepatterns are now produced in the developing PCX follicles. The predicted form of the prepattern depends on the stage (i.e. the size and shape) of the follicle. The first prepattern expected to arise is a bilateral pattern (Figure 6.5a), which causes the mitotic activity to be higher on one side of the follicle than on the other so that the follicle grows at an angle to the skin surface (late stage F1, Figure 6.5b). It has also been proposed that the localized minimum of the bilateral pattern triggers a sweat gland bud to develop on one side of the follicle. The sweat gland bud causes the shape of the upper part of the follicle to change (stage F2a), and this in turn leads to the appearance of a bilobed pattern in the upper part of the follicle (Figure 6.5c).

At the same time as causing the follicle to grow at an angle to the skin surface and initiating a sweat gland bud, the bilateral pattern also causes the pattern in the distribution of the RD system components in the epidermis to change. Therefore, at the time the PCX follicles reach stage F1, it is predicted that new localized maxima will appear in the basal epidermal cell layer in between the existing PCX follicles (Figure 6.4b). This new pattern also functions as a prepattern, causing a second generation of follicles (PCY) to initiate. Not only are the PCY follicles predicted to be approximately equidistant from the PCX follicles, but total follicle density is also predicted to increase (Figure 6.6a).

Another pattern change in the epidermis is predicted to occur when the bilobed pattern appears in the upper part of the PCX follicles, i.e. as they reach stage F2a (Figure 6.5c). At this time the PCY follicles have developed to stage F1 (Figures 6.5a and b). The bilobed pattern causes two new localized maxima to arise on either side of the PCX follicles, forming trios (Figure 6.4c); the two lateral follicles are also predicted to be located on either side of the line joining the sweat gland bud to the centre of the follicle. Once again the predicted total density of follicles is increased by the appearance of the PLx follicles (Figure 6.6a).

A further pattern change is predicted to occur in the epidermis when the PCY follicles reach stage F2a. This time localized maxima appear on either side of the PCY follicles (Figure 6.4d). The trios formed by this fourth generation of follicles (PLy) tend to be aligned with the early trios, giving the appearance of rows of follicles. The total follicle density is predicted to be increased by these PLy follicles (Figure 6.6a).

A further three pattern changes are expected to occur, giving rise to three more

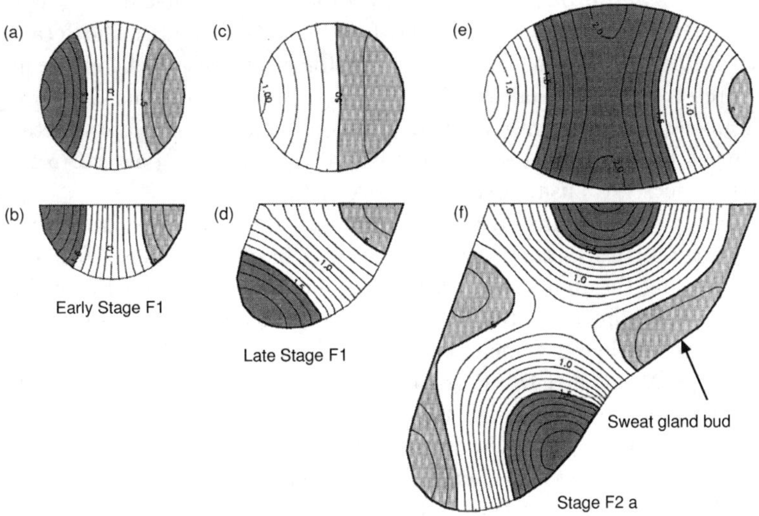

Figure 6.5 The prepatterns of U predicted to arise in early stage F1 (a,b), late stage F1 (c,d) and stage F2a (e,f) primordial follicles. These distributions were obtained by solving the RD system equations (6.1) in domains representing the different stages of primordial follicles. (Contour intervals are 0.1.) Regions where the concentration of U is above 1.5 units are darkly shaded. Regions where U is below 0.5 units are lightly shaded. Reproduced by permission of CSIRO publishing.

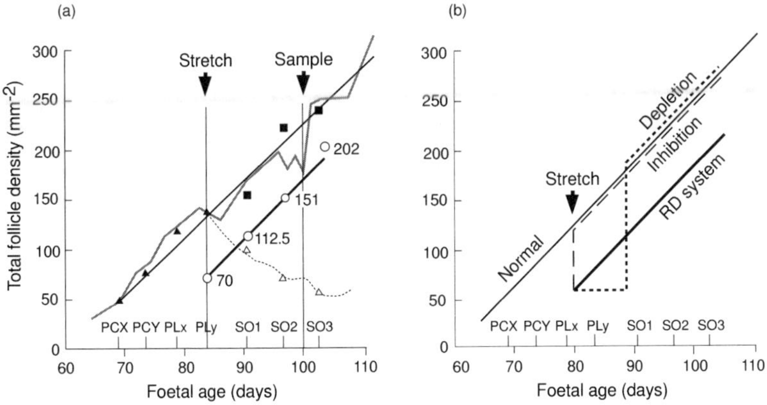

Figure 6.6 (a) The predicted follice densities following each of four generations of primary follicles (PCX, PCY, PLx, PLy (▲)) and three generations of secondary follicles (SO1, SO2, SO3 (■)) are compared to observed densities (solid grey curve) and the effect of skin stretching (○). (b) The expected changes in follicle density caused by skin stretching depending on mechanism.

generations of follicles (Nagorcka, 1995b). Each new generation tends to be equidistant from existing follicles and produces an increase in the total follicle density consistent with observation (Figure 6.6a), despite the fact that the area of the fetal skin is growing quite rapidly. These new generations of follicles are observed to have no sweat glands and are referred to here as the SO1, SO2 and SO3 generations. The loss of the sweat gland following the fourth generation can be explained if the diameter of the primordial follicles has decreased to a value where the bilateral pattern in the early stage F1 follicles effectively disappears (Nagorcka, 1995b).

The RD theory of follicle initiation therefore predicts that new generations are caused by existing follicles as they develop. On this basis six generations of follicles are expected following the first (PCX) follicle generation. Each generation increases the total follicle density in the fetal skin despite the fact that the fetal skin area is also increasing. The predicted follicle densities following each of four generations of primary follicles (PCX, PCY, PLx, PLy) are plotted (▲). The predicted total follicle densities as a consequence of the next three generations of secondary follicles (SO1, SO2, SO3) are also plotted (■). The results are consistent with the observed total follicle density (solid grey curve). When the sweat gland ceases to initiate on follicles (i.e. primary follicles cease to initiate) the primary follicle density is predicted to decline with increasing fetal skin area (△) consistent with observation (dotted curve).

One feature of follicle initiation not explained by the RD system is the asymmetry observed in follicle groups (all secondaries initiating closest to the trios are observed to be located on their anterior side, Figure 6.1). It has been suggested that this asymmetry may be the result of the arrector pili musculature attached to the primary follicles of the trio group being attached to the epidermis on the anterior side of group and blocking follicle initiation from this region of the epidermis (Nagorcka, 1995b 373). One possibility is that the muscle attached to the basement membrane may prevent the dermis from responding to any overlying maxima in U. The consequences of this have not been included in the current calculations.

6.3 Experiments on Follicle Initiation

6.3.1 Skin stretching

In the previous section it can be seen that the RD system has the capacity to explain many aspects of follicle initiation and development. On the other hand, gradients established by diffusion and interaction between expression patterns of DNA binding proteins are a powerful mechanism for providing positional information during the earliest stages of embryonic development (Pick, 1988). Therefore, despite the fact that the RD system appears to have a greater capacity to explain many morphological features than the gradients, it is desirable to design experiments to directly test between these mechanisms. It is possible to do this because the RD system and gradients offer completely different explanations for the increasing density of follicles during the period of follicle initiation.

Gradients in the form of inhibitory or substrate depletion fields maintain follicle density at a particular level determined by the diffusion rate of the substance forming the gradient. In contrast, an RD system generates new generations of follicles at a rate determined by the rate of follicle development. In terms of the RD theory, pattern changes in the epidermis are forced by pattern changes in the developing follicles which occur in response to changes in the size and shape of developing follicles. This major difference between gradients and the RD prepattern mechanism makes it possible to directly test between these ideas. Consider for the moment the consequences of suddenly stretching the skin to double its area at some time during follicle initiation (e.g. following the PLy generation in Figure 6.6b). The follicle density would be immediately be halved, falling to 70mm^{-2}, and assuming for the moment that the epidermis recovers quickly so that its thickness returns to normal and the skin volume increases, then

(i) if an inhibitory field is present, the concentration of the inhibitor would suddenly decrease and follicles would immediately initiate to fill the increased area (long dashed curve, Figure 6.6b) to the expected (normal) follicle density (heavy solid curve, Figure 6.6b), or,

(ii) if a substrate depletion field is present, the concentration of the substrate would decrease and there would be a delay while more substrate is synthesized, after which follicles would initiate to fill the increased area (short dashed curve, Figure 6.6b) to the expected (normal) follicle density, or,

(iii) if an RD system is present, follicle development would continue to cause new generations of follicles to initiate at the normal rate. There would be no tendency to fill the available area but simply to add new generations, of the same density characteristic of the RD system, at the normal rate. The follicle density would continue to increase at the normal rate and so follicle density would remain permanently depressed below the normal level (heavy solid curve, Figure 6.6b). This prediction has been confirmed by recalculating the sequence of prepatterns generated by an RD system following stretching immediately after the PLy generation. In this case the three generations of secondary follicles (SO1, SO2, SO3) are still predicted to initiate (Figure 6.6a, (○)), increasing the follicle density at about the normal rate, but remaining below the normal level of follicle density.

The predicted outcomes for (i), (ii) and (iii) are shown in Figure 6.6b. It does not indicate the fact that there will be some immediate decrease in skin thickness and cell density per unit area as a consequence of stretching. Some delay for recovery of the skin following stretching must be expected. After recovery, however, the follicle density is expected to return to normal if initiation is controlled by a gradient ((i) and (ii)), but to remain permanently depressed below normal in the case of an RD system (iii) even though the follicle density continues to increase after stretching.

These hypotheses have now been tested experimentally. Using fetal surgery hemispheres were inserted under the epidermis in sheep foetuses at about 84 days, i.e. at a time when primary follicle initiation is (approximately) completed. Discs were also placed under the skin in the equivalent site on the opposite midside to act as control sites. The foetuses were then placed back in their mother's womb and allowed to develop to about day 103. They were then removed again and skin samples taken from above the apex of the hemisphere, and from above the centre of the discs. Measurements have confirmed that despite several new generations of follicles appearing after stretching, the follicle density increases at the normal rate, so it remains permanently depressed, i.e. below the normal (control) density level, in the stretched skin above the dome. These results are consistent with the predictions based on the RD system (Figure 6.6b).

6.3.2 *Follicle density and follicle diameter*

The mean distance between neighbouring localized maxima in the first prepattern produced by the RD system in a two-dimensional sheet is called the wavelength λ of the RD system. It is a characteristic of the reaction and diffusion rates defining the RD system. Each new prepattern in the epidermis produces a new generation of follicles at a density of λ^{-2}. For a given rate of follicle development and after a specified number of follicle generations, the accumulated follicle density is also proportional to λ^{-2} (Nagorcka, 1995a, b).

The area covered by each localized maximum, i.e. the area over which U exceeds a certain threshold, determines the number of epithelial cells and the number of dermal cells which will contribute to an early stage follicle. On this basis it is expected that the diameter of primordial follicles will be proportional to λ.

It follows that the RD theory predicts the existence of an inverse relationship between follicle density and follicle diameter. Measurements have now been carried out to test this prediction. Fetal sheep skin was sampled at day 81 of gestation from a number of sheep breeds covering a wide range of follicle densities. The sample time was chosen to include approximately the four generations of primary follicles. No allowance was made for any possible differences in the rate of follicle development between breeds. Even so, the results in Figure 6.7 appear to be essentially consistent with the prediction.

In a further experiment providing a test of the relationship between follicle density and diameter, murine epidermal growth factor (mEGF) was injected subcutaneously into newlyborn bandicoots (an Australian marsupial) contained in their mother's pouch (Adelson *et al.*, 1997). The injections were carried out during the period of primary central follicle initiation, but before the initiation of the primary lateral follicles. The effect of mEGF was to increase the diameter of the primordial follicles and to decrease the total follicle density. The magnitude of the effect is consistent with the inverse relationship between these quantities predicted by the RD theory. Other effects of the mEGF treatment are outlined in the next section.

The size of the dermal prepapilla is determined at the same time as the diameter of the primordial follicle. The size of the dermal prepapilla is expected to be one factor determining the size of the dermal papilla in the mature follicle. The dermal papilla is a major factor controlling the diameter of the fibre produced. It follows that the inverse relationship between follicle density and follicle diameter may provide an explanation for the strong inverse genetic correlation known to exist between wool follicle density and fibre diameter (Brown and Turner, 1968).

6.3.3 *The sweat gland and the trio group*

The RD system generates prepatterns which place maxima on either side of the early primary follicles of sheep (Figure 6.4c and d), leading to the initiation of lateral follicles and the formation of the trio group. According to the RD theory, these prepatterns are a consequence of the shape change in the upper part of the early primary follicles

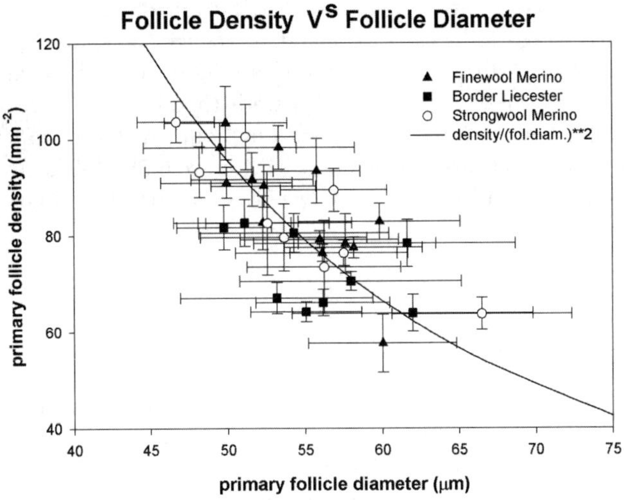

Figure 6.7 Measurements were made of the total (primary) follicle density and the follicle diameter in samples of fetal skin taken at 81 days of gestation. Fetuses from a range of sheep breeds were sampled to extend the range of follicle densities and diameters sampled. Results for individual fetuses are plotted. There appears to be an inverse relationship between follicle density and follicle diameter consistent with expectations based on the RD theory (solid line).

caused by the growth of sweat gland buds (Figure 6.5c). Assuming this is true, it follows that if it were possible to prevent the initiation of the sweat gland, or to substantially reduce the shape change, then the trio grouping would also disappear. This direct link between the sweat gland and the trio group in sheep, predicted only by the RD theory, provides a basis for testing the theory.

The predicted link between sweat gland and trio group led us to examine skin samples taken from mutant sheep without sweat glands. These samples showed that the loss of the sweat gland was indeed accompanied by the loss of the trio group. Unfortunately, the same mutation caused a very large change in the total follicle density, and this may complicate the interpretation.

The predicted association between sweat gland buds and trios was also tested in the experiment involving the subcutaneous injection of mEGF, the same experiment noted in the previous section. In this experiment it was observed that sweat glands, along with sebaceous glands and dermal papillae, were "inhibited or retarded" (Adelson *et al.*, 1997). It was also noted that the trio grouping was lost in the skin of bandicoot young. In terms of the RD system, the loss or reduced physical size of the sweat gland bud could explain the disappearance of the trio grouping.

6.3.4 Patterns of mitotic activity

In terms of the RD theory, one of the steps leading to the initiation of follicles is an increase in mitotic activity in the epidermal tissue in response to a localized maximum in U and the establishment of a dermal condensation (Nagorcka and Mooney, 1985).

If this is a general response, then the spatial distribution of mitotic activity in the epithelial cells of early stage follicles is expected to reflect the spatial patterns in U which are predicted to arise in primordial follicles (Figure 6.5). Spatial patterns of mitotic activity in early stage primary follicles were therefore measured to test the predicted patterns in Figure 6.5. The observations confirmed the appearance of a bilateral pattern in mitotic activity consistent with Figure 6.5a (Adelson *et al.*, 1992). Non-uniform distributions of mitotic activity were also observed in stage F2a follicles (Figure 6.5c).

6.4 Fibre Formation

6.4.1 Concentric patterns of cell type

Primordial follicles grow down into the dermis and soon reach a stage where the epithelial cells at the base of the follicle begin to grow around the cells of the prepapilla to form the dermal papilla and the follicle bulb. At this stage the rapidly dividing epithelial cells near the papilla begin to migrate away from the basal membrane, now adjacent to the dermal papilla as well as the dermis, and up the follicle. The epithelial cells in the bulb differentiate into a range of cell types, associated with the inner root sheath (IRS), outer root sheath (ORS) and the fibre, as they migrate out of the bulb. The cells in the bulb migrate up the follicle bulb at different rates depending on their location in the bulb (Figure 6.8), such that the older cells are always higher in the bulb. Cells higher in the bulb near the apex of the dermal papilla may retain an association with the basement membrane for longer than those lower in the bulb.

A differentiation scheme has been proposed (Nagorcka and Mooney, 1982) which depends on the age of the epithelial cells (i.e. the time since the cells become committed to a path of differentiation while still near to or attached to the basement membrane), and the concentration of dermal factor Z they experience at particular ages. Z is produced in the dermal papilla and diffuses radially away, setting up a gradient through the epithelial cells of the bulb (Figure 6.8). Such a differentiation scheme is able to account for all the cell types which arise in the mature follicle bulb. These include all cell types making up the IRS and ORS, which surround the fibre in the follicle, and the cell types such as the ortho- and paracortical cells which make up the fibre itself (Figure 6.8). Although this differentiation scheme can account for all the cell types observed in the bulb, it can only produce them in spatial patterns which are concentric about the dermal papilla.

6.4.2 Non-concentric patterns

A differentiation scheme based on a gradient with its source in the dermal papilla would be quite satisfactory were it not for the fact that many non-concentric patterns are observed in fibres. For example, the guard hairs of many mammalian species exhibit multilobed cross-sectional shapes along much of their length (Figure 6.2). Even the zigzag fibres in mice which are essentially round or elliptical in cross-section show both bilateral and concentric patterns in the distribution of the cortical cell types within the fibres (Priestley and Rudall, 1965). In wool fibres too, which in the fine-wool breeds are normally round or slightly elliptical, a range of non-concentric patterns in the

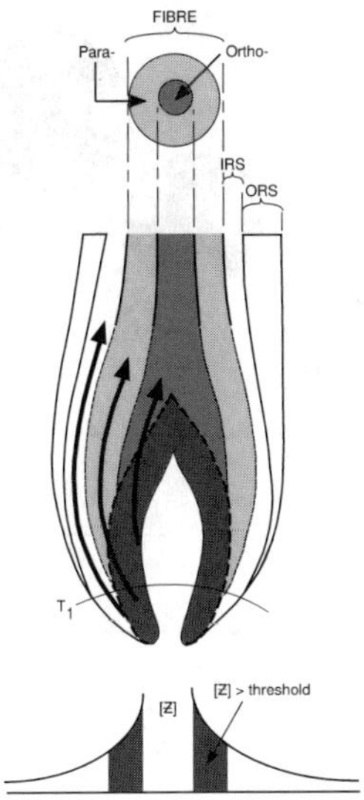

Figure 6.8 Epithelial cells divide rapidly in the follicle bulb, shown here in longitudinal section, and migrate out of the bulb up the follicle at rates dependent on the paths they follow as indicated by the relative lengths of the lines with arrowheads. These cells differentiate into various cell types (see text) controlled, in part, by a factor, Z, produced in the dermal papilla (the invaginated region at the centre of the bulb). Z diffuses radially away.

distribution of para- and orthocortical cells is frequently observed (Ahmad and Lang, 1957).

A radial gradient of a diffusible dermal factor is unable to account for these non-concentric patterns. Instead it has been proposed that the same RD system suggested as a prepattern mechanism for follicle initiation and development, is also present in the mature follicle bulb, where it continues to function as a prepattern mechanism (Nagorcka and Mooney, 1982; Nagorcka, 1984; Nagorcka *et al.*, 1993). It has been proposed that the epithelial cells differentiate as inner root sheath cells (IRS) or presumptive fibre cells at age T_1 (days) depending on the concentration of the dermal factor Z and the RD system component U which the cells experience above the T_1 level (Figure 6.9 and Nagorcka *et al.*, 1993). Epithelial cells aged T_1 which experience a very low concentration of Z differentiate as outer root sheath (ORS) cells. Presumptive fibre cells older than T_1, and hence located higher in the bulb, differentiate as ortho- and paracortical cells depending on the concentration of Z and U they experience. Z is produced in the dermal papilla and diffuses radially away, as indicated in the bottom plot of the concentration of Z as a function of radial distance from the central axis at the level in the bulb corresponding to age T_1. The region where Z exceeds the threshold required for differentiation as presumptive fibre cells is darkly shaded. If U is constant everywhere then different cell types always form as concentric rings about the central axis. In general, the particular prepattern in U which arises in the bulb will now, along with Z, regulate cell differentiation during fibre formation. In essence the quantity ZU controls fibre cross-sectional shape; and $aZ + bU^{-1}$, where a and b

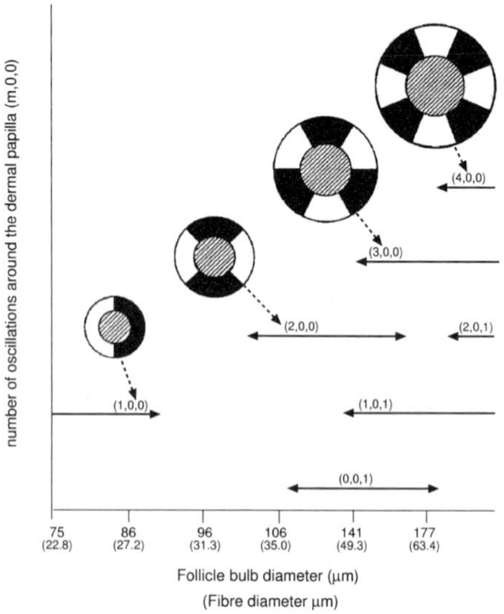

Figure 6.9 The horizontal lines delimited by arrowheads show the range of follicle bulb diameters (plotted on the horizontal axis) for which the linear wave functions, indicated here as an ordered triplet (m, n, j) and fully defined by Nagorcka and Mooney (1982), are expected to contribute to a non-uniform distribution of U (and V) in the bulb.

are constants, controls the spatial distribution of cortical cell type within the fibre.

To gauge the range of prepatterns which may arise in the bulb, a linearized form of the RD system (6.1) was solved (Nagorcka and Mooney, 1982), and the range of bulb sizes was determined for which the uniform steady-state solution of the RD system is expected to be unstable. The results are summarized in Figure 6.9 in the form of a periodic table of linear wave functions which indicate what spatial patterns may contribute to the instabilities for a given bulb size. The linear wave functions are denoted by the ordered triplets (m, n, j): m (plotted on the vertical axis) in effect specifies the number of full-wave variations exhibited by the wave function around the dermal papilla; j specifies the number of half-wave variations along the length of the bulb, i.e. in the proximodistal direction. Threshold plots showing the distribution of maxima (dark areas) and minima (light areas) in transverse sections through the bulb containing the dermal papilla at the centre (hatched area) have been plotted for some wave functions of the form $(m, 0, 0)$ (Figure 6.9). Spatial variations in the form of these linear wave functions are likely to contribute to a stable non-uniform spatial distribution which arises. In fact, this has been confirmed through numerical solutions of the non-linear RD system equations (6.1) in three-dimensional shapes representing the follicle bulb (Nagorcka *et al.*, 1993). The stable non-uniform distributions in U and V obtained numerically demonstrate that the RD system is able to generate non-concentric wave-like prepatterns in the follicle bulb.

6.5 Experiments on Fibre Formation

In general, a mixture of the patterns summarized in Figure 6.9 will dictate the final form of the prepattern which arises in the bulb. The relative weighting of the patterns in the mixture will depend on the size and shape of the bulb, the non-linear form of the RD system equations, and to some extent the developmental history in the follicle bulb. Introducing the RD-generated prepattern into the differentiation scheme has the effect of imposing a wavelike spatial pattern on the (predicted) distribution of various cell types within the follicle bulb. The dependence of the non-concentric wavelike spatial pattern on the size and shape of the bulb is a characteristic of an RD system functioning as a prepattern mechanism in the bulb. No other mechanism has been proposed which predicts a periodic relationship between spatial patterns in the fibre and bulb size and shape. Experiments have now been proposed to test for this periodic relationship between the distribution of cell type and bulb size and shape.

6.5.1 The relationship between pattern and follicle bulb size

The periodic table in Figure 6.9 was calculated for follicle bulbs of increasing size. The scaling parameter determining size is plotted on the horizontal axis. Figure 6.9 predicts that the prepattern of U (and V) is expected to range from bilateral in "small" bulbs, to uniform in "medium" bulbs, to bilobed in "large" bulbs (i.e. a bulb diameter range of 75 to 150 μm). Numerical solution of the non-linear RD system equations (6.1) in a finite element representation of three-dimensional bulbs confirms that stable prepatterns of these forms do arise (Nagorcka *et al.*, 1993). The situation is more complicated in "very large" bulbs where complex mixtures of the patterns in Figure 6.9 are potentially involved. The spatial patterns of ZU and $aZ + bU^{-1}$ have been calculated in bulbs representing each of the bulb size classes. According to the proposed cell differentiation scheme, ZU determines mitotic activity and the distribution of ZU low in the bulb determines the cross-sectional shape of the fibre. The pattern of the quantity $aZ + bU^{-1}$ higher in the bulb determines the pattern of ortho- and paracortical cells within the fibre. Therefore, the numerical results predict an association between the pattern of mitotic activity in the bulb, the fibre cross-sectional shape and the pattern of ortho- and paracortical cells within the fibre.

To test these predictions, samples of skin 1cm in diameter and the associated fibre samples were taken from the midside of sheep known to contain a mixture of follicle bulb sizes throughout the skin. Serial sections were taken parallel to the skin surface so that it was possible to follow follicles down to the bulb. In this way it was possible to look at the fibre cross-section, the pattern of ortho- and paracortical cells, and the pattern of mitotic cells in the bulb in which the fibre is formed. The results, which are discussed in more detail by Nagorcka *et al.*, (1993), appear to be in agreement with the theoretical predictions outlined above.

A fundamental characteristic of Turing type RD systems, however, is the dependence of the periodic pattern on tissue size. In the range of bulb sizes considered here the prediction is that the prepattern in the bulb will change from bilateral to bilobed as the bulb size increases from "small" to "large". To test this size dependence a lock of fibres were sampled by cutting close to the skin surface and separating them into length classes. Each length class of fibres was embedded in wax, sectioned and stained

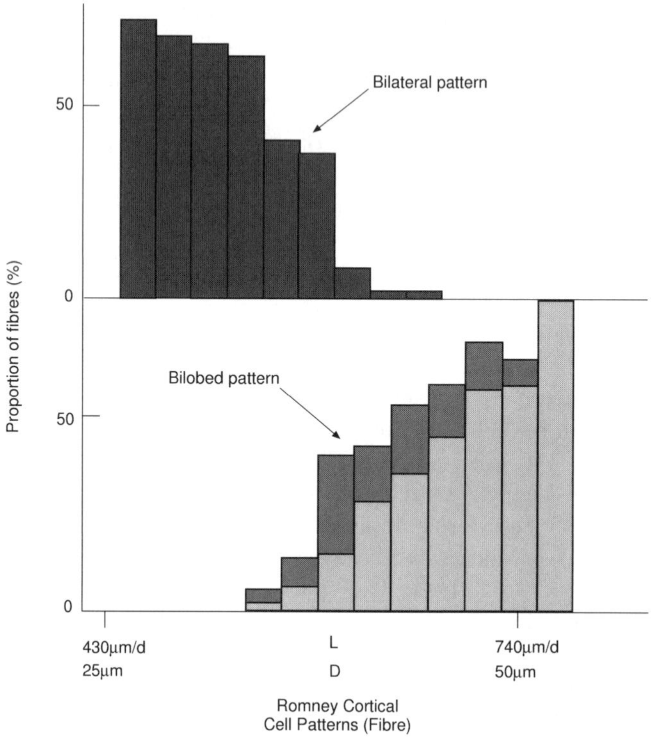

Figure 6.10 A lock of fibres was taken from the midside of an adult Romney sheep and separated into length classes. The estimated length growth rate (L) of each length class is plotted on the horizontal axis. The corresponding fibre diameter (D) is also shown. The proportion of fibres in each length class containing (a) a bilateral pattern of ortho/paracortical cells, and (b) a bilobed pattern (lightly shaded), or a pattern mixture containing a strong bilobed component (intermediate shading) of ortho/paracortical cells, is plotted on the vertical axis. The change from a bilateral pattern in fine fibres (small bulbs) to thicker fibres (large bulbs) is consistent with the predictions in Figure 6.9.

to observe the ortho- and paracortical cell distribution. The proportion of fibres in each length class containing a bilateral or bilobed pattern of ortho- and paracortical cells is plotted against fibre length in Figure 6.10. There is known to be a linear relationship between fibre length, fibre diameter and follicle bulb diameter (Henderson, 1965). The results in Figure 6.10 are therefore consistent with the predicted change in prepattern with increasing size of the follicle bulb.

6.5.2 Change of pattern within a fibre

The results in Figure 6.10 are based on observations of the patterns present in different fibres and their follicle bulbs. The follicle bulbs have a range of sizes and are assumed to have a similar shape. To further test for a direct link between follicle bulb size/shape and prepattern in the bulb, it would be desirable to change the size/shape of a particular bulb, leaving all other factors unchanged, and observe a change in prepattern. Fortunately, the fibre itself contains a permanent and continuous record of the prepatterns existing in the follicle bulb, in the pattern of ortho- and paracortical cells and/or in the shape of the cross-section.

Cortical Cell Distribution

One technique for changing bulb size would be to apply a small depilatory force to a small number of wool fibres by attaching them to a spring under tension, as demonstrated by Lyne and Jolly (1969). This would need to be applied for about two weeks to allow time for the bulb to change its size, due to a lag in the response of wool/hair growth to changes such as nutritional changes (Nagorcka, 1977). In this way, if the RD theory is correct, large-diameter fibres containing a bilobed pattern of ortho/paracortical cells should be seen to reduce their diameter and to change the ortho/paracortical pattern from bilobed to bilateral. Upon release of the depilatory force, the pattern and diameter change should be observed to reverse. Unfortunately, the fibres examined and reported by Lyne and Jolly were too fine for this test; they contained only a bilateral pattern.

Fibre Cross-sectional Shape

Numerical solutions of the RD system equations in follicle bulbs of a fixed shape (Nagorcka *et al.*, 1993: figure 32.1b) have demonstrated that the amplitude of the bilateral or bilobed wavelike prepattern, and other prepatterns reflecting the linear wave functions of the form $(m, 0, 0)$, decreases as you move lower in the bulb towards its base (Nagorcka, unpublished). This reduces the influence of the prepattern in U through ZU (Section 6.4.2) on fibre cross-section. In theory this is the reason why most wool fibres appear almost circular. If the shape of the bulb is changed by increasing the diameter of the dermal entrance into the dermal papilla, it can be shown that the amplitude of the wavelike prepattern increases near the base of the bulb. This situation actually occurs at the beginning of a hair cycle as the epithelial cells of an early stage (anagen) follicle begin to grow around the dermal papilla. The change in bulb shape in this situation may pass through two or more prepatterns having the form $(m, 0, 0)$ (Nagorcka, unpublished). In which case the cross-sectional shape of the fibre would be expected to show a number of multilobed cross-sections. Although this has not been confirmed experimentally, it is suggested as the explanation for some guard hairs having a sequence of different multilobed cross-sections in the tip and shield region of the fibre, as can be seen in Figure 6.2.

6.6 Conclusion

Hair and wool follicles are located in the skin in specific patterns which exhibit a great deal more order than simply a tendency to be approximately equidistant from each other. To date, no one has been able to explain how a gradient in the form of an inhibitory field or a substrate depletion field, for example, is able to provide an explanation for this observed order. The periodic prepatterns produced by an RD system, on the other hand, are able to provide an explanation for the development of the observed order in follicle location. In terms of an RD system, the prepatterns which arise in existing follicles not only force prepattern changes in the epidermis leading to new generations of follicles, but they can also determine the location of new follicles relative to the existing, developing follicles causing the follicles to become grouped in particular ways. In contrast, inhibitory or depletion fields will allow new follicles to initiate only as space becomes available, so that they fill the available space to a density dictated by the diffusion rates of the substances involved. In order to test between these two mechanisms, the skin of sheep foetuses was stretched during follicle initiation to determine whether subsequent follicles would fill the available space, or initiate at a rate dictated by the rate of development of existing follicles as predicted on the basis of an RD system. The results were consistent with the predictions of an RD system.

A direct link between the development of a sweat gland bud on the central primary follicle and the appearance of two lateral follicles to form a linear trio group, non-uniform distributions of mitotic activity in early stage follicles, and an inverse relationship between follicle density and follicle diameter, have all been predicted using the tissue size and shape dependence of the RD system prepatterns. All of these predictions have also been supported by experimental observations.

In the case of the mature follicle bulb, the prepatterns produced by an RD system present in the bulb are once again dependent on the size and shape of the follicle bulb. The dependence is indicated by the bulb size ranges for which various wave funtions are unstable (Figure 6.9), which was confirmed by numerical solution of the RD system equations (6.1). Although a gradient of a factor diffusing away from the dermal papilla can account for all cell types forming multicell layers in the bulb, it cannot explain non-concentric patterns of the multicell layers. The RD system not only provides an explanation for the non-concentric patterns, but it also predicts a dependence of pattern on follicle bulb size (and shape). Experimental observations obtained thus far appear to be consistent with the bulb size dependence of the prepatterns (Figure 6.10). A more extensive testing of the periodic table in Figure 6.9 has yet to be carried out.

At this stage it is concluded that the tissue size and shape dependence characteristic of RD systems of the Turing type is necessary to explain the initiation and development of follicles and the formation of fibres.

Acknowledgements

We would like to acknowledge the financial support given by the Australian Wool Research and Promotion Organisation.

REFERENCES

Adelson, D.L., Kelley, B.A. and Nagorcka, B.N. (1992) Increase in dermal papilla cells by proliferation during development of the primary wool follicle. *J. Agric, Res.* 43: 843–56.

Adelson, D.L., Hollis, D.E., Merchant, J.C. and Kelley, B.A. (1997) *In vivo* effects of epidemal growth factor on epidermal pattern formation and hair follicle initiation in the marsupial bandicoot *Isoodon macrourus. Reprod. Fertil. Dev.* 9: 493–500.

Ahmad, N. and Lang, W.R. (1957) Ortho-para cortical differentiation in "anomalous" Merino wool. *Aust. J. Biol. Sci.* 10: 118.

Brown, G. and Turner, H.N. (1968) Response to selection in Australian merino sheep. II. Estimates of phenotypic and genetic parameters for some production traits in merino ewes and an analysis of the possible effects of selection on them. *Aust. J. Agric. Res.* 19: 303–22.

Brunner, H. and Coman, B.J. (1974) *The Identification of Mammalian Hair.* Melbourne: Inkata Press.

Carter, H.B. (1943) Studies in the biology of the skin and fleece of sheep. 1–3. *Coun. Sci. Industr. Res. Aust. Bull.* No. 164.

Carter, H.B. and Hardy, M.H. (1947) Studies in the biology of the skin and fleece of sheep. 4. The hair follicle group and its topological variations in the skin of the Merino foetus. *Coun. Sci. Industr. Res. Aust. Bull.* No. 164.

Claxton, J.H. (1964) The determination of patterns with special reference to that of the primary central skin follicles in sheep. *J. Theor. Biol.* 7: 302–17.

Claxton, J.H. and Scholl, C.A. (1973) A model for pattern formation in the primary skin population of sheep. *J. Theor. Biol.* 40: 353–67.

Dry, F.W. (1926) The coat of the mouse (*Mus Musculus*). *J. Genet.* 26: 287–340.

Hardy, M.H. and Lyne, A.G. (1956) The pre-natal development of wool follicles in sheep. *Aust. J. Biol. Sci.* 9: 423–21.

Henderson, A.E. (1965) Relationship of wool follicle and wool fibre dimensions. In *Biology of the skin and hair growth.* Eds. A.G. Lyne and B.F. Short. pp. 447–60. Sydney: Angus and Robertson.

Lyne, A.G. (1966) The development of hair follicles. *Aust. J. of Sci.* 28: 370–77.

Lyne, A.G. and Jolly, M. (1969) In *Advances in biology of the skin.* Eds. W. Montagna and R.L. Dobson. Vol 9. p. 247. Oxford: Pergamon Press.

Murray, J.D. (1989) *Mathematical Biology.* Berlin: Springer Verlag.

Nagorcka, B.N. (1977) The description and analysis of wool growth. *Aust. J. Agric. Res.* 28: 737–46.

Nagorcka, B.N. (1984) Evidence for a reaction-diffusion system as a mechanism controlling mammalian hair growth. *Biosystems* 16: 323–332.

Nagorcka, B.N. (1995a) The reaction-diffusion theory of woll (hair) follicle initiation and development. I. Primary follicles. *Aust. J. Agric. Res.* 46: 333–55.

Nagorcka, B.N. (1995b) The reaction-diffusion theory of woll (hair) follicle initiation and development. II. Secondary original follicles. *Aust. J. Agric. Res.* 46: 357–78.

Nagorcka, B.N., Adelson, D.L., Mooney, J.R. and Kelley, B.A. (1993) Predicted and observed spatial prepatterns in (hair) wool follicle bulbs. In *Exerimental and Theoretical Advances in Biological Pattern Formation.* Eds. H.G. Othmer, P.K. Maini and J.D. Murray. pp. 255-266. New York: Plenum Press.

Nagorcka, B.N. and Mooney, J.R. (1982) The role of a reaction-diffusion system in the formation of hair fibres. *J. Theor. Biol.* 98: 575–607.

Nagorcka, B.N. and Mooney, J.R. (1985) The role of a reaction-diffusion system in the initiation of primary hair follicles. *J. Theor. Biol.* 114: 243–72.

Nagorcka, B.N. and Mooney, J.R. (1992) From stripes to spots: Prepatterns which can be produced in the skin by a reaction-diffusion system. *IMA J. Math. Appl. Med. and Biol.* 9: 249–67.

Pick, L. (1998) Segmentation: Painting stripes from flies to vertebrates. *Dev. Genet.* 23: 1–10.

Priestley, G.C. and Rudall, M. (1965) Modifications in the Huxley layer associate with changes in fibre diameter and output. In *Biology of the skin and hair growth.* Eds. A.G. Lyne and B.F. Short. pp. 35–57. Sydney: Angus and Robertson.

Straile, W.E. and Mann, S.J. (1973) Characteristics of hairs and related mechanoreceptors in the mouse and rabbit. *J. Exp. Zoo.* 185: 189–208.

Turing, A.M. (1952) The chemical basis of morphogenesis. *Phil. Trans. R. Soc.* B237: 37–72.

Wigglesworth, V.B. (1940) Local and general factors in the development of 'pattern' in *Rhodnius prolixus* (hemitera). *J. Exp. Biol.* 17: 180–99.

7

Some Mathematical Models for Biological Pattern Formation

Philip K. Maini

Centre for Mathematical Biology, Mathematical Institute, 24–29 St Giles', Oxford OX1 3LB, UK

7.1 Introduction

The study of growth and form is central to the understanding of many areas in the physical, chemical and biological sciences. D'Arcy Thompson was one of the key figures in the early studies of this area. He recognized that although classification in biology was extremely important, it was also crucial to develop theories to explain observed traits. In the latter half of this century, advances in this area have been made by understanding a few systems in depth, which serve as paradigms for the broader picture in which Thompson was interested. Of course, while pattern is not form, both are closely related.

Perhaps the most famous example of spatio-temporal pattern formation, and one of the earliest studied, is the Belousov–Zhabotinsky reaction, in which bromate ions oxidize malonic acidic in a reaction catalysed by iron which, depending on its state, can assume two different colours – reddish orange or blue (Johnson and Scott, 1996). By observing the colour changes, one can see that this reaction exhibits a wide range of spatio-temporal patterning, such as propagating fronts, spiral waves, target patterns and toroidal scrolls. Such oscillatory and wave-like patterns also arise in physiology, for example in the propagation of electrical activity in the heart, and in biology, for example, waves of cAMP that initiate cell aggregation in the slime mould. Spatial patterns have recently been shown to arise in the CIMA reaction (Maini *et al.*, 1997).

The formation of spatial pattern in developmental biology is a central issue and

On Growth and Form: Spatio-temporal Pattern Formation in Biology.
Editor Mark A.J. Chaplain, G.D. Singh, J.C. McLachlan

is still not resolved. Although genes play a crucial role in embryology, a study of genetics alone will not be sufficient to help us understand how the physicochemical properties of embryonic material interact to produce the complex spatio-temporal signalling cues which ultimately determine cell fate. An ideal tool for studying such complex non-linear interactions is mathematical modelling, and there is now a vast literature on such models (Murray, 1993). In Section 7.2 some models for biological pattern formation will be classified and their applications discussed. In Section 7.3 the coupling of such models will be illustrated by two examples: the formation of feather germs in dorsal chick skin, and aggregation in cellular slime mould. In Section 7.4 the role of domain growth will be analysed in the context of the spatio-temporal patterning of tooth primordia in alligators and in pigmentation patterning in fish. Finally, Section 7.5 contains a discussion of future goals and areas of research.

7.2 Simple Models

A number of models have been proposed to account for spatio-temporal patterning phenomena and regeneration in several areas of developmental biology. In this section, I broadly classify these models. I term them *simple models* because in later sections I will consider the coupling of such models. But from a mathematical viewpoint, they can be far from simple!

7.2.1 Chemical prepattern models

Chemical prepattern models assume that a spatial pattern in some chemical (termed a morphogen) is set up and cells respond to this pattern by differentiating accordingly. Wolpert (1969) proposed a simple model in which a source–sink mechanism, coupled with diffusion and degradation, led to a spatial gradient in a single morphogen. He proposed that this provided *positional information* for cells, which differentiated according to a series of threshold values.

A more complicated spatial pattern can be set up by the interaction of two chemicals. Turing (1952) showed that a system of two chemicals reacting and diffusing could evolve to a spatial pattern in chemical concentration. Such a pattern consisted, in its simplest one-dimensional form, of a series of peaks and troughs. These models may be couched in terms of partial differential equations, or discretized and analysed as cellular automata (Bard, 1981).

These models differ in the sense of cell interpretation of the chemical prepattern. In Wolpert's model, the chemical prepattern is set up by a simple process which can only produce a simple gradient. To use this gradient to generate a complicated pattern, it is hypothesized that a complex series of thresholds exist and cells have the machinery to interpret multiple thresholds. In Turing's model, complex spatial patterns arise due to a complex chemical interaction, but the interpretation of the prepattern is via a single threshold (Nagorcka, 1989).

Turing's model takes this form:

$$\frac{\partial u}{\partial t} = D_u \nabla^2 u + f(u, v) \tag{7.1}$$

$$\frac{\partial v}{\partial t} = D_v \nabla^2 u + g(u, v) \tag{7.2}$$

where $u(\underline{x}, t), v(\underline{x}, t)$ denote chemical concentrations at position $\underline{x}$ and time t, D_u, D_v are diffusion coefficients and f, g are reaction kinetics. These equations are solved subject to certain initial conditions, typically random perturbations about a uniform steady state, and boundary conditions, which are typically periodic, zero flux, or fixed. The forms of the kinetic terms may be derived in a number of ways.

Phenomenological Models

The kinetic terms are chosen such that one of the chemicals is an *activator*, the other an *inhibitor*. An example is the Gierer–Meinhardt model (1972):

$$\begin{aligned} f(u, v) &= \alpha - \beta u + \gamma u^2/v, & (7.3)\\ g(u, v) &= \delta u^2 - \mu v, & (7.4) \end{aligned}$$

where $\alpha, \beta, \gamma, \delta$ and μ are constants.

Hypothetical Models

Derived from a hypothetically proposed series of chemical reactions. For example, Schnakenberg (1979) proposed a series of trimolecular autocatalytic reactions involving two chemicals:

$$X \underset{k_2}{\overset{k_1}{\rightleftharpoons}} A, \quad B \xrightarrow{k_3} Y, \quad 2X + Y \xrightarrow{k_4} 3X.$$

Using the law of mass action and denoting the concentrations of X, Y, A and B by u, v, a and b, respectively, we have

$$\begin{aligned} f(u, v) &= k_2 a - k_1 u + k_4 u^2 v, & (7.5)\\ g(u, v) &= k_3 b - k_4 u^2 v & (7.6) \end{aligned}$$

where $k_1, \ldots, k_4$ are (positive) rate constants. Assuming that there is an abundance of A and B, a and b can be considered to be approximately constant.

Empirical Models

The kinetics are fitted to experimental data. For example, the Thomas (1975) immobilized-enzyme substrate-inhibition mechanism involves the reaction of uric acid (concentration u) with oxygen (concentration v). Both reactants diffuse from a reservoir maintained at constant concentrations u_0 and v_0, respectively, onto a membrane containing the immobilized enzyme uricase. They react in the presence of the enzyme with empirical rate $V_m uv/(K_m + u + u^2/K_s)$, so that

$$f(u,v) = \alpha(u_0 - u) - \frac{V_m uv}{K_m + u + u^2/K_s}, \tag{7.7}$$

$$g(u,v) = \beta(v_0 - v) - \frac{V_m uv}{K_m + u + u^2/K_s} \tag{7.8}$$

where α, β, V_m, K_m and K_s are positive constants.

Actual Chemical Reactions

Although it was in 1952 that Turing predicted the spatial patterning potential of chemical reactions, this phenomenon has only recently been realized in actual chemical reactions. Therefore, it is now possible in certain cases to write down detailed reaction schemes.

The first Turing patterns were observed in the chlorite–iodide–malonic acid starch reaction (CIMA reaction) (Castets *et al.*, 1990; De Kepper *et al.*, 1991). The model proposed by Lengyel and Epstein (1991) stresses three processes: the reaction between malonic acid (MA) and iodine to create iodide, and the reactions between chlorite and iodide and between chloride and iodide. These reactions take the form

$$\mathrm{MA} + \mathrm{I_2} \rightarrow \mathrm{IMA} + \mathrm{I^-} + \mathrm{H^+}$$

$$\mathrm{ClO_2} + \mathrm{I^-} \rightarrow \mathrm{ClO_2^-} + \frac{1}{2}\mathrm{I_2}$$

$$\mathrm{ClO_2^-} + 4\mathrm{I^-} + 4\mathrm{H^+} \rightarrow \mathrm{Cl^-} + 2\mathrm{I_2} + 2\mathrm{H_2O}.$$

The rates of these reactions can be determined experimentally. By making the experimentally realistic assumption that the concentrations of malonic acid, chlorine dioxide and iodine are constant, Lengyel and Epstein derived the following model:

$$\frac{\partial u}{\partial t} = k_1 - u - \frac{4uv}{1+u^2} + \nabla^2 u$$

$$\frac{\partial v}{\partial t} = k_2 \left[k_3 \left(u - \frac{uv}{1+u^2} \right) + c\nabla^2 v \right]$$

where u, v are the concentrations of iodide and chlorite, respectively, and k_1, k_2, k_3 and c are positive constants.

General results on the patterning properties of reaction-diffusion equations can be found in the books by Britton (1986), Edelstein-Keshet (1988), Fife (1979), Grindrod (1996), Murray (1993) and Segel (1984).

7.2.2 Cell movement models

Cell movement models assume that a spatial pattern arises in cell density, and cells then differentiate in a density-dependent manner. There are a number of ways in which cell aggregation can occur. For example, cells can move in response to gradients in certain chemicals (chemotaxis) and it has been shown by a number of authors that such cell–chemical interactions can lead to spatial pattern formation (Keller and Segel, 1971; Maini *et al.*, 1991). These models involve reaction and diffusion, but spatial patterning arises due to the advective term introduced by chemotaxis. The typical model takes this form:

$$\frac{\partial n}{\partial t} = D_n \nabla^2 n - \nabla.(\chi(u) n \nabla u) + f(n, u) \tag{7.9}$$

$$\frac{\partial u}{\partial t} = D_u \nabla^2 u + g(n, u), \tag{7.10}$$

where $n(\underline{x}, t), u(\underline{x}, t)$ denote cell density and chemoattractant concentration, respectively, at position $\underline{x}$ and time t, D_n, D_u are diffusion coefficients, f, g are terms incorporating production and degradation and $\chi(u)$ is the chemotactic sensitivity.

Cells can also respond to mechanical cues, such as movement in response to advective forces exerted by cells on a common substratum, or due to movement up gradients in adhesive activity (haptotaxis). The first model incorporating these ideas was proposed by Oster *et al.* (1983) and since then such models have been extensively studied (Murray, 1993). The mathematical equations of these models are quite different. They consist of conservation equations for cells and extracellular matrix, which take the general form of the equations above, but the main difference is the force-balance equation, which describes the behaviour of a viscoelastic material.

It has also been proposed that cells move to minimize energy (Cocho *et al.*, 1987a, b; Steinberg, 1970; Sulsky *et al.*, 1984). Such models can be set up mathematically and solved to show cell sorting and patterning behaviour consistent with a number of experimentally observed phenomena.

7.2.3 Applications

Turing assumed that the chemicals in his reaction-diffusion system were growth hormones, so the spatial pattern in chemical concentrations would result in spatially non-uniform growth and hence pattern. He applied his theory to account for whorled leaves and to growth-induced shape changes in the early embryo which he proposed could account for gastrulation. Since his seminal paper, reaction-diffusion models have been proposed to account for a vast number of patterning processes in Nature, too great to be completely reviewed here, so here are a few examples to give a flavour of the applications.

Geirer and Meinhardt (1972) used their model to account for pattern formation in *Hydra* and their model was also consistent with several of its regenerative properties. Reaction-diffusion models have been used to account for compartmentalization in insect development and to provide an explanation for the occurrence of various mutants (Meinhardt, 1982). However, for *Drosophila* it now appears that patterning is due to a cascade of protein interactions rather than to a reaction-diffusion mechanism.

Reaction-diffusion models have been applied to shell patterns (Meinhardt, 1995) and to butterfly wing pigmentation patterns (Nijhout, 1990). Reaction-diffusion and cell movement models have been applied to animal coat markings (Bard, 1981; Cocho *et al.*, 1987a, b; Murray, 1981; Murray and Myerscough, 1991; Murray *et al.*, 1990) and to skeletal patterning in the limb (Maini and Solursh, 1991).

Although these models are based on very different biological assumptions, many of them share common properties. For example, the patterning in reaction-diffusion and in many cell movement models arises from the interaction of *short-range activation* and *long-range inhibition*. On the one hand, this has the disadvantage of making it very difficult to use models to distinguish between mechanisms; on the other hand, it does mean that one can make general conclusions and predictions that are *mechanism-independent*. This leads to the idea of *developmental constraints*: only certain patterns are possible, regardless of the mechanism (Oster and Murray, 1989).

7.3 Coupled Models

Many pattern formation phenomena arise as the result of the coupling of mechanisms. In some cases the mechanisms on their own do not produce pattern, it is the interaction that leads to pattern; in other cases each subsystem can exhibit patterns, and the coupling leads to complex patterning. In this section I consider an example from each class of model.

7.3.1 Feather germ formation

The formation of skin organs, e.g. hair, teeth, feathers and sweat glands, has been widely studied. The skin is composed of two basic layers, the epidermis and the dermis, which are separated by a thin layer of tissue called the basal lamina. In many cases these layers interact during organ formation. An example of this is the formation of feather germs, which determine the future sites of feathers. A feather germ consists of an epidermal thickening, termed a placode, overlying a condensation of cells in the dermis, termed a papilla. On the chick back, these form along the dorsal midline and then propagate out on either side, eventually forming a hexagonal lattice of feather germs.

A number of models have been proposed to account for this tissue interaction. For example, Nagorcka *et al.* (1987) considered a model of two coupled reaction-diffusion systems, while Shaw and Murray (1990) used a coupled mechanochemical system. Both models were shown to exhibit spatial patterns. These tissue interaction models have the common property of being able to produce patterns independently in the dermis and the epidermis. However, experimental studies show that the dermis cannot produce coherent patterns without the presence of the epidermis and vice versa. In other words, tissue interaction is crucial for the patterning process.

In an attempt to address this issue, Cruywagen and Murray (1992) proposed a mechanochemical model in which epidermal cells produce a chemical that diffuses into the dermis, where it acts as a chemoattractant for dermal cells, while the dermis secretes a chemical that diffuses into the epidermis, where it serves to enhance the tractional forces exerted by epidermal cells. The full model consists of seven partial differential equations, but using some approximations and quasi-steady-state assumptions on the chemical equations, it may be reduced to the following system of two equations (I refer the reader to the original paper for full details of the experimental motivation and mathematical derivation of the model equations):

$$\nabla^2\theta - \beta\nabla^4\theta + \nabla^2\left\{\frac{\tau n^2}{1+cn^2}\right\} = \rho\theta \tag{7.11}$$

$$\frac{\partial n}{\partial t} = D\nabla^2 n - \nabla\cdot(n\nabla\alpha(1-\theta)) . \tag{7.12}$$

The first equation is derived by considering the epidermis as a viscoelastic sheet and then assuming that the viscous terms are set to zero. This leads to an equation for the dilation θ, where β, τ, c and ρ are constants. A further approximation yields the relation $N = 1 - \theta$, for the epidermal cell density N. Dermal cell density is denoted by n and the third term in the first equation incorporates the enhancement of cell traction in the epidermis due to the chemical produced by the dermal cells. The second equation models dermal cell diffusion and chemotaxis, where the effect of the epidermally secreted factor on dermal chemotaxis is incorporated into the second term on the right-hand side. This model has been extensively studied (Cruywagen *et al.*, 1992; 1994a, b) and shown to exhibit many of the properties of the full model. Crucially for this model, neither submodel on its own can exhibit spatial patterning. It is the coupling of the submodels that leads to pattern formation. It has been shown that this model can exhibit propagating patterns consistent with those observed experimentally (Figure 7.1).

7.3.2 Cell aggregation in slime mould

The life cycle of the cellular slime mould *Dictyostelium discoideum* serves as an excellent paradigm for morphogenesis and has for many years attracted the interest of developmental biologists and theoreticians. Under favourable conditions, the amoebae feed on bacteria in the soil and divide. Starvation triggers a developmental programme which is initiated by cell–cell signalling via the extracellular messenger cyclic 3′5′-adenosine monophosphate (cAMP). The chemotactic response to this signal leads, through the phenomenon of cell streaming, to the formation of a multicellular organism composed typically of 10^4 to 10^5 cells. This organism passes through an intermediate motile (slug) phase during which cells differentiate into prespore and prestalk types, before developing a fruiting body, aiding the dispersal of spores from which new amoebae develop under favourable conditions. The comparative simplicity of morphogenesis in *Dictyostelium* has made it an attractive model system for the study of self-organization, and many of the molecular and cellular mechanisms which are involved in cell aggregation, collective movement and differentiation have now been identified.

A large body of experimental and theoretical work exists on the molecular mechanisms of the cAMP signalling dynamics, and models incorporate the detailed biochemical dynamics of the binding of extracellular cAMP to its surface receptors (Devreotes, 1989; Goldbeter, 1996; Tang and Othmer, 1994).

The incorporation of diffusion of extracellular cAMP into these models yields a description of the signalling dynamics in a stationary cell layer, which is a valid approximation for the situation at the beginning of cell aggregation (Tyson *et al.*, 1989; Monk and Othmer, 1990; Tang and Othmer, 1995). This leads to a coupled system of reaction-diffusion equations in which the variables are extracellular cAMP

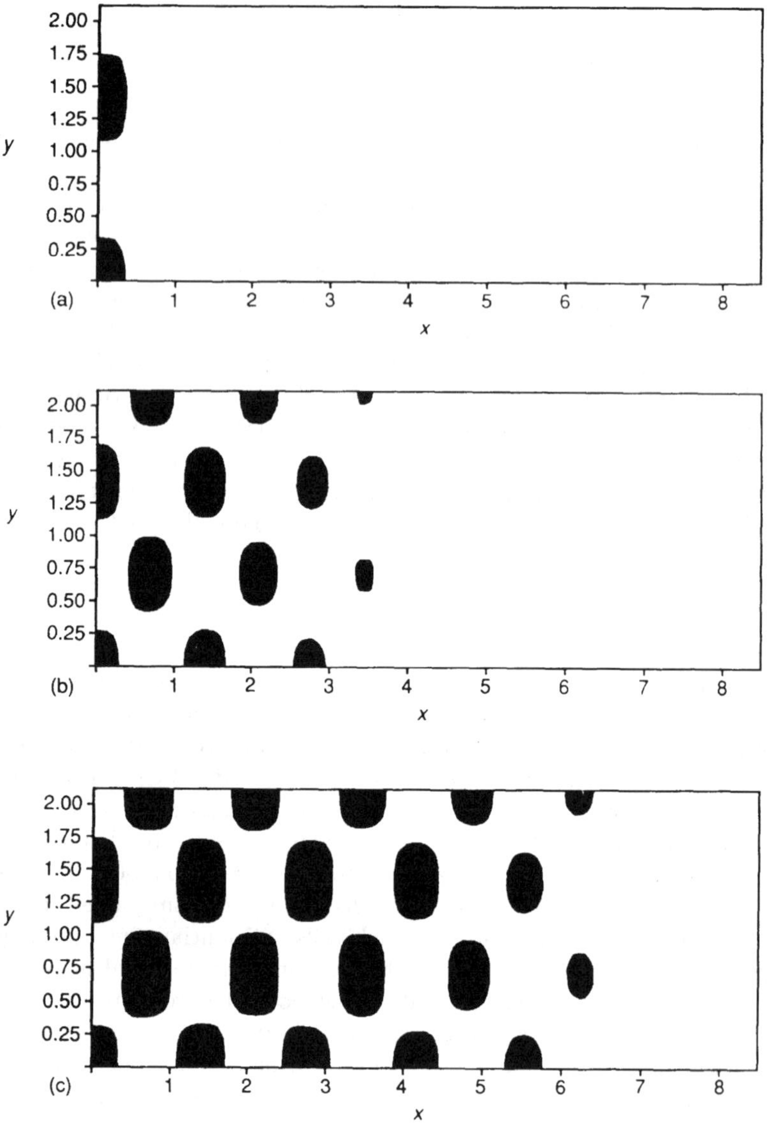

Figure 7.1 Sequential pattern formation in the caricature model (7.11)-(7.12) for tissue interaction. (a) Initially a single row consisting of spots of high cell density is specified at one end of a rectangular domain. (b, c) As the system evolves, the pattern propagates with more rows added sequentially. In this way the mechanism converts a simple one-dimensional pattern into a complicated two-dimensional form. Regions of high cell density are shaded. (Reproduced with permission from Cruywagen *et al.*, 1992)

concentration and density of cAMP receptors. This model is of excitable type and exhibits spiral waves and target patterns of cAMP concentration.

To model cell streaming, one must couple cell movement and chemotaxis to this excitable system. So far, this has been modelled in two different ways. The first approach consists of modelling discrete cells equipped with cAMP-dependent movement rules coupled to a finite-difference approximation for the continuous cAMP dynamics (Dallon and Othmer, 1997; Kessler and Levine, 1993; Van Oss *et al.*, 1996). A second approach is to approximate the cell distribution by a continuous density, resulting in a system of coupled partial differential equations for the cell density and the cAMP dynamics (Vasiev *et al.*, 1994; Höfer *et al.*, 1995a, b). The former approach has the advantage that one can follow distinct cells and compare their detailed motion with that of amoebae in experiments. The latter modelling approach allows one to use the extensive analytical machinery of partial differential equation theory to fully understand how each process and parameter affects model behaviour. However, both models rely on the same mechanism – a chemotactic cell response to an excitable system involving extracellular cAMP and cAMP receptors.

Consider the continuum model of Höfer *et al.* (1995). The nondimensionalized model takes the following form:

$$\frac{\partial n}{\partial t} = \nabla \cdot (\mu \nabla n - \chi(v) n \nabla u) \tag{7.13}$$

$$\frac{\partial u}{\partial t} = \lambda[\phi(n) f_1(u,v) - (\phi(n)+\delta) f_2(u)] + \nabla^2 u \tag{7.14}$$

$$\frac{\partial v}{\partial t} = -g_1(u)v + g_2(u)(1-v), \tag{7.15}$$

where n, u and v denote cell density, extracellular cAMP concentration and fraction of active cAMP receptors, respectively. We motivate each equation in turn. The first equation is the conservation equation for cell density. It is assumed that cell movement has a random component, with cell diffusion coefficient μ, and a chemotactic component, with sensitivity denoted by $\chi(v)$. To account for adaptation, the latter is taken to be of the form $\chi(v) = \chi_0 v^m/(A^m + v^m)$, $m > 1$, where χ_0 and A are positive constants. Hence an appreciable chemotactic response requires a minimal fraction of active receptors, yet the response saturates for a large fraction of active receptors. Note that many models of chemotaxis assume χ to be a constant. Under that assumption, cells would respond to a pulse of chemoattractant by moving towards the wave in the wavefront, then moving with the wave in the waveback, resulting in a net movement away from the source of attractant, rather than towards it. This is the so-called chemotactic wave paradox (Soll *et al.*, 1993). The form of $\chi(v)$ chosen above resolves this paradox (Höfer *et al.*, 1994).

The second and third equations are a simplified model of the cAMP–cell receptor dynamics modified by cell density effects (Martiel and Goldbeter, 1987). The first term on the right-hand side of the second equation accounts for cAMP production by assuming that the rate of cAMP synthesis per cell is $f_1(u,v) = (bv + v^2)(a + u^2)/(1+u^2)$. This models autocatalytic cAMP production with saturation, mediated by cAMP binding to active receptors. The rate of cAMP degradation per cell is taken to be $f_2(u) = du$. The cell density dependence is reflected in the factor

$\phi(n) = n/(1 - \rho n/(K + n))$, and δ accounts for cAMP degradation in the absence of cells (Höfer *et al.*, 1995a). The parameters a, b, d, ρ, K, δ and λ are all positive constants.

The first term on the right-hand side of the third equation accounts for receptor densitization and, assuming the law of mass action, $g_1(u) = k_1 u$. The second term on the right-hand side models resensitization of the desensitized $(1 - v)$ fraction of receptors, at the constant rate $g_2(u) = k_2$. The parameters k_1 and k_2 are positive constants.

Good estimates of most of the parameters are available from the experimental literature (Höfer *et al.*, 1995b), and substituting these values into the model, we find that the model captures the key features of the aggregation process. A typical aggregation sequence is shown in Figure 7.2.

It is found that the initial uniform state is unstable from the outset, suggesting that the coupled dynamics of cAMP wave propagation and cell movement exhibit a patterning instability perpendicular to the direction of wave propagation. An analytically tractable caricature of the above model has been proposed by Höfer *et al.* (1995b) from which it is possible to derive explicit conditions on the parameters for the uniform steady state to go unstable. The dispersion relation predicts the wavelength of the fastest-growing unstable mode, and agrees closely with the results of numerical simulations and experimental observations.

A more detailed numerical study of the model reveals that this minimal model, based on biochemical and mechanical mechanisms established at the cellular level, can account for a number of other important properties. For example, our numerical simulations show that low initial cell densities lead to the formation of a central hole. The formation of a central hole is observed experimentally, if the system is treated with caffeine (Steinbock and Müller, 1995). Caffeine treatment lowers the excitability of the system, and in the model this corresponds to lowering λ. This is equivalent, in some sense, to lowering the cell density n. Another key experimental prediction of the model is discussed in Section 7.5.

7.4 The Role of Domain Growth

In many developmental systems, domain growth does not play a role in pattern *formation*, it simply results in already established patterns growing in size. However, in some cases, domain growth is crucial for the pattern *selection* process. Here are two such examples.

7.4.1 Alligator tooth primordia

Experimental evidence suggests that jaw growth plays a crucial role in the spatial positioning of tooth primordia in *Alligator mississippiensis*. These primodia are either resorbed or differentiate to form functioning teeth. It also appears that existing primordia inhibit neighbouring primordium formation. Although there is substantial literature on dentition development in vertebrates, the mechanism of primordium morphogenesis is still not understood. In this case the role of mathematical modelling is to determine whether the processes thought to be of most significance (here domain growth and inhibition) are sufficient to produce the observed patterns.

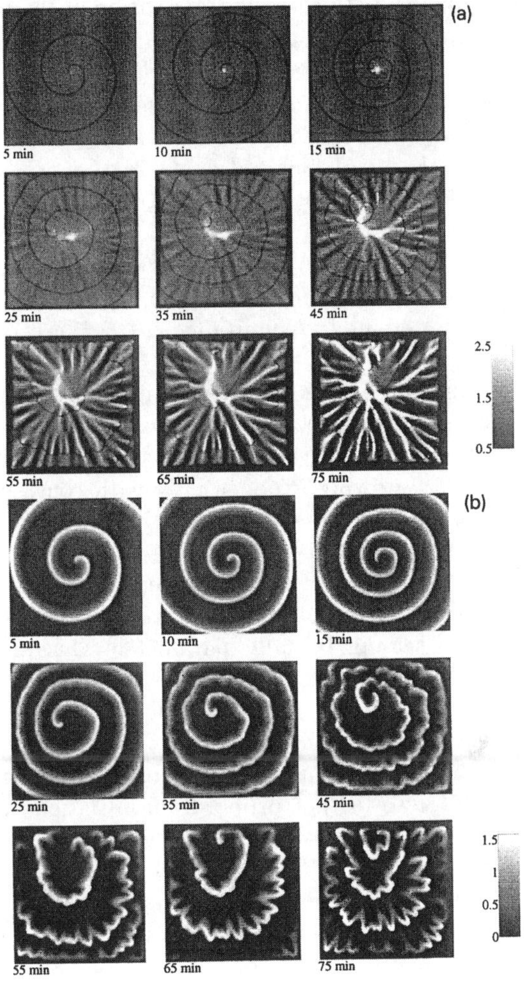

Figure 7.2 Spatio-Temporal evolution of (a) cell density, and (b) cAMP concentration in a numerical simulation for equations (7.13) to (7.15). (Reproduced with permission from Höfer *et al.*, 1995b)

The formation of primordia in the lower jaw is symmetric about the midline, so one can consider only half the lower jaw. The first seven primordia are initiated in the spatio–temporal sequence $7-3-6-2-5-1-4$ where the 4 denotes the anterior-most primordium (the 4th primordium to form) and 7 denotes the posterior-most primordium (the 7th to form). Kulesa *et al.* (1996) investigated the spatial patterning properties of the following model (I refer the reader to the original paper for full details on the experimental background, motivation and derivation of the model):

$$\frac{\partial u}{\partial t} = \gamma(hc - u + u^2 v) - ru + \exp(-2rt)\frac{\partial^2 u}{\partial x^2} \tag{7.16}$$

$$\frac{\partial v}{\partial t} = \gamma(b - u^2 v) - rv + d\ \exp(-2rt)\frac{\partial^2 v}{\partial x^2} \tag{7.17}$$

$$\frac{\partial c}{\partial t} = -\delta c - ru + p\ \exp(-2rt)\frac{\partial^2 c}{\partial x^2}. \tag{7.18}$$

This is the non-dimensionalized version of the model on the spatial domain $[0, 1]$, where u and v satisfy zero flux boundary conditions, $c(0, t) = c_0(t)$, with c satisfying zero flux at $x = 1$.

In this model, u and v are components satisfying a modified Schnakenberg (1979) kinetics with the source of u depending on the chemical c, which itself diffuses and is degraded linearly. The other linear degradation terms represent dilution due to the assumption that the domain is growing exponentially with constant growth rate r. Domain growth also modifies the diffusion coefficients by the factor $\exp(-2rt)$ when the domain is non-dimensionalized to a fixed domain.

In this model, the source term can be taken as the bifurcation variable. As this decreases below a critical value, diffusion-driven instability occurs and a spatially varying pattern in u and v is set up. Initially, parameter values are chosen so that the source profile is such that everywhere in space it is above the critical bifurcation value. As the domain grows, the c profile becomes sufficiently low for diffusion-driven instability to occur in the u, v system and u evolves to a one-peaked pattern. It is then assumed that this peak marks a tooth primordium and initiates the production of c, which acts as an inhibitor to further primordium initiation. This ensures that the second primordium forms sufficiently far away from primordium 1. Kulesa *et al.* showed that this mechanism could lead to the correct spatio-temporal initiation of the first seven tooth primordia. Hence they conclude that the interaction of inhibition and domain growth is a plausible mechanism for generating this complex sequence. Such a mathematical model can be used to make predictions about the outcome of experimental manipulations.

7.4.2 Pigmentation patterning in fishes

Unlike mammalian skin patterns, which simply enlarge with body growth, patterns on certain fishes change *qualitatively* with domain growth. A striking example of this is the stripe pattern on the marine angelfish *Pomacanthus*. As the fish grows, its stripes grow wider apart until they are about twice as far apart as in the juvenile case. At this stage, new stripes appear between the existing stripes in such a way as to restore the original interstripe spacing. The new stripes are thinner than existing stripes but they gradually broaden.

Kondo and Asai (1995) considered a simple reaction-diffusion model on a one-dimensional growing domain and showed that it exhibited similar mode-doubling behaviour to that observed in an one-dimensional anterior–posterior cross-section of *Pomacanthus* but there was one crucial drawback – their model predicted that all stripes had the same width.

Recently, Painter *et al.* (1999) have considered a generalized Turing model for this phenomenon on a two-dimensional domain which represents the surface of the fish. In their model, the pigmentation pattern is assumed to arise as a result of cell movement (this is consistent with results on zebrafish). They consider a typical reaction-diffusion system coupled to a cell movement equation of the form

$$\frac{\partial n}{\partial t} + \nabla.n\underline{v} = D_n\nabla^2 n - \nabla.[\chi_u(u)n\nabla u + \chi_v(v)n\nabla v] + n(r_x + r_y) \tag{7.19}$$

where $\underline{v} = (r_x x\mathbf{i} + r_y y\mathbf{j})$ is the uniform velocity field generated by growth, r_x and r_y are growth rates in the x and y directions, respectively, u and v are the concentrations of the chemicals in the reaction-diffusion system, n is pigment cell density and $\chi_u(u)$ and $\chi_v(v)$ are the chemotactic sensitivities.

They show that this system can produce the mode-doubling phenomenon observed in *Pomacanthus*, with the inserted stripes being narrower than the existing stripes. The reason behind this behaviour is that as the domain grows, the pattern in the underlying reaction-diffusion system undergoes mode doubling with stripes all of the same size. However, under the biologically realistic assumption that cell movement is on a longer timescale than chemical diffusion, the cell rearrangement induced by this new pattern occurs very slowly, resulting in the new inserted stripes being thinner than the existing stripes, broadening only gradually. More generally, this model system is using chemotaxis as a novel positional information interpreter.

7.5 Discussion and Future Directions

Development of spatial pattern and form is unquestionably one of the central areas in biology, as was recognized by D'Arcy Thompson. It has led to genuine interdisciplinary research which has advanced many areas of science and mathematics. For example, the study of parabolic systems of partial differential equations was largely motivated by Turing's paper. The rich behaviour of these systems has resulted in the development of a large body of mathematical theory. Cell movement models pose different mathematical questions, such as blow-up and collapse (Othmer and Stevens, 1997), while mechanical models are a system of mixed hyperbolic/parabolic/elliptic equations for which very little theory exists.

While the study of pattern formation has been the source of a great number of mathematical problems, it is reasonable to ask what role mathematics has to play in pattern formation. It is clear that pattern formation in many areas arises as the result of a complex interaction of mechanisms. The language of mathematics is ideally suited to understanding non-linear interactions and it is only through mathematical and computational approaches that one can predict the outcome of such complex interactions. This allows one to make predictions that are experimentally testable. Of course, no mathematical model can take into account *every* process:

approximations and simplifications have to be made. In this respect, mathematics is like any other scientific tool – it has its limitations, but it can be used in conjunction with experimental tools to help elucidate the underlying mechanisms involved in pattern formation.

Presently, biology offers mathematics a range of challenges. At one extreme are the areas such as slime mould, in which many of the processes are known and data on parameters available. Here the role of modelling is to make precise experimental predictions on the effects of manipulating small details in the system. At the other extreme, in which most areas lie, are systems in which very little is known. In such cases the role of modelling is to see what processes are sufficient to produce the patterning; for example, in the case of alligator teeth it has been shown that domain growth, coupled with inhibition, is sufficient to produce the required sequence of pattern. This does not mean that this *is* the mechanism, it only says that it is a candidate mechanism. It also makes experimentally testable predictions.

For a model to be of use, it must provoke experimentation. An example of this is in the skeletal development of the chick limb. Both reaction-diffusion and cell movement models predict that the complexity of pattern is intimately linked to the size of the domain. To test this theory, Wolpert and Hornbruch (1990) constructed double anterior chick limbs which had the same size as a normal chick limb. They found that the double anterior limbs gave rise to two humeral bones, even though the limb was the same size as a normal limb, which only produces a single humerus. This contradicts the model predictions. Maini *et al.* (1992), however, showed that if there was a spatially varying diffusion coefficient, then a reaction-diffusion system could predict such a pattern. At the same time as this work was being carried out, Brümmer *et al.* (1991) showed that there was indeed a gradient in diffusivity across the limb bud.

Another use of mathematical models is to provide alternative explanations for observed phenomena. For example, in slime mould, it is observed in situ that as streaming proceeds, the wave speed and wavelength of the spiral patterns decrease (Gross *et al.*, 1976). Previously this has been explained by assuming that biochemical changes must be occurring in the cell–cAMP system. However, an alternative, and much simpler explanation comes out of the modelling. As cell streams grow, they alter the propagation conditions for the cAMP waves. This is initially equivalent to increasing λ in equation (7.14), leading to an increase in excitability of the medium. This in turn leads to an increase in the rotation frequency of the spiral core. As a result (Tyson and Keener, 1988) the wave speed and wavelength of the spiral patterns decrease. Hence the model shows that complex biochemical changes are not necessary to account for the observations.

One of the key areas of future study is domain growth. An understanding of how this can be incorporated into the modelling and the effects that it has on the patterns exhibited by models is an exciting new area to be investigated. Preliminary studies show that the mode-doubling phenomenon mentioned in Section 7.4 is very robust. If one wishes to generate a certain number of stripes on a large domain using reaction-diffusion theory then, due to the multiplicity of stable solutions on large domains, the pattern produced may be very sensitive to initial conditions. However, if we start off with a small domain and let it grow, then the correct stripe pattern can be generated robustly. Alternatively, boundary conditions can be used to select certain patterns

(Dillon *et al.*, 1994).

Another important area for study is the effect of domain geometry on the patterns predicted. Kauffman *et al.* (1978) solved a reaction-diffusion system on an ellipse and showed that the nodal lines of chemical concentration closely matched the compartmental division lines observed in *Drosophila* wing imaginal disc. However, Bunow *et al.* (1981) showed that if the equations were solved on a more realistic geometry, the nodal lines differed significantly from those on the ellipse. It is surprising that so little work has been done on the role of domain shape. However, with increased computer power, this is becoming easier to do. The effect of domain shape on patterns in fish is presently being analysed (Varea *et al.*, 1997, Madzvamuse, 1999). The advances in computer technology mean that such studies are now more feasible.

Until now most models have been couched in terms of coupled systems of partial differential equations, in which quantities such as cell density are modelled as continuum variables. This allows one to use the methods of applied mathematics to investigate solution behaviour. However, the continuum approximation is not a good one in the limit of low density, when cells really behave as discrete entities. Furthermore, advances in biotechnology now enable single cell trajectories to be followed in some systems (e.g. slime mould). Therefore, in such areas, models considering cells as discrete entities are more appropriate. This leads to coupled systems of discrete ordinary partial differential equations and a whole new area of mathematics.

In most cases thus far, modelling has been phenomenological. Recently Othmer and Stevens (1997) have investigated in detail how one can derive models at the macroscopic cell population level by considering processes for movement at a microscopic cellular level. In this way they were able to show how different movement rules at a local level could result, for example, in the different phenomenological forms of chemotactic sensitivity commonly used. Such a modelling approach leads to a greater degree of insight about the processes which are present.

Although I have focused here on differential equations, integral equations can also be used to model pattern formation. In this framework, a convolution kernel is used to model the patterning interaction. Recently Sekimura *et al.* (1999) used a novel model of this form to account for the formation of stripe patterns in scale cells in the butterfly wing. In particular, they showed how a spatial gradient in a parameter can robustly select rows of the correct orientation over other rows or spots.

In conclusion, growth and form provide a seemingly endless source of interesting and novel mathematical problems, while mathematics can be used as a tool to explore different mechanisms and processes underlying these phenomena. For the area to advance, we must have truly interdisciplinary research efforts involving biological experimentation, mathematical modelling and computational investigations. The recent advances in biotechnology, coupled with the rapid increases in computational power, now make such progress a real possibility.

REFERENCES

Bard, J.B.L. (1981) A model for generating aspects of zebra and other mammalian coat patterns. *J. Theor. Biol.* 93: 363–385.

Britton, N.F. (1986) *Reaction-Diffusion Equations and Their Applications to Biology.*

London: Academic Press.

Brümmer, F., Zempel, G., Buhle, P., Stein J.-C. and Hulser, D.F. (1991) Retinoic acid modulates gap junction permeability: A comparative study of dye spreading and ionic coupling in cultured cells. *Exp. Cell. Res.* 196: 158–163.

Bunow, B., Kernevez, J.P., Joly, G. and Thomas, D. (1981) Pattern formation by reaction-diffusion instabilities: Application to morphogenesis in *Drosophila*. *J. Theor. Biol.* 84: 629–649.

Castets, V., Dulos, E., Boissonade J. and De Kepper, P. (1990) Experimental evidence of a sustained Turing-type equilibrium chemical pattern. *Phys. Rev. Lett.* 64(3): 2953–2956.

Cocho, G., Pérez-Pascual, R. and Rius, J.L. (1987a) Discrete systems, cell-cell interactions and color pattern of animals. I. Conflicting dynamics and pattern formation. *J. Theor. Biol.* 125: 419–435.

Cocho, G., Pérez-Pascual, R., Rius J.L. and Soto, F. (1987b) Discrete systems, cell-cell interactions and color pattern of animals. II. Clonal theory and cellular automata. *J. Theor. Biol.* 125: 437–447.

Cruywagen G.C. and Murray, J.D. (1992) On a tissue interaction model for skin pattern formation. *J. Nonlinear Sci.* 2: 217–240.

Cruywagen, G.C., Maini P.K. and Murray, J.D. (1992) Sequential pattern formation in a model for skin morphogenesis. *IMA J. Math. Appl. Med. and Biol.* 9: 227–248.

Cruywagen, G.C., Maini P.K. and Murray, J.D. (1994a) Travelling waves in a tissue interaction model for skin pattern formation. *J. Math. Biol.* 33: 193–210.

Cruywagen, G.C., Maini P.K. and Murray, J.D. (1994b) Bifurcating spatial patterns arising from travelling waves in a tissue interaction model. *Appl. Math. Lett.* 7: 63–66.

Dallon, J.C. and Othmer, H.G. (1997) A discrete cell model with adaptive signalling for aggregation of *Dictyostelium discoideum*. *Phil. Trans. Roy. Soc. Lond. B* 352: 391–417.

De Kepper, P., Castets, V., Dulos E. and Boissonade, J. (1991) Turing-type chemical patterns in the chlorite-iodide-malonic acid reaction, *Physica D* 49: 161–169.

Devreotes, P.N. (1989) *Dictyostelium discoideum*: a model system for cell-cell interactions in development. *Science* 245: 1054–1059.

Dillon, R., Maini P.K. and Othmer, H.G. (1994) Pattern formation in generalised Turing systems: I. Steady-state patterns in systems with mixed boundary conditions. *J. Math. Biol.*, 32: 345–393.

Edelstein-Keshet, L. (1988) *Mathematical Models in Biology.* New York: Random House.

Fife, P.C. (1979) *Mathematical Aspects of Reacting and Diffusing Systems, Lect. Notes in Biomath.* New York: Springer-Verlag.

Gierer, A. and Meinhardt, H. (1972) A theory of biological pattern formation. *Kybernetik* 12: 30–39.

Goldbeter, A. (1996) *Biochemical Oscillations and Cellular Rhythms.* Cambridge: Cambridge University Press.

Grindrod, P. (1996) *The Theory of Applications of Reaction-Diffusion Equations: Pattern and Waves.* Oxford: Oxford University Press.

Gross, J.D., Peacey M.J. and Trevan, D.J. (1976) Signal emission and signal propagation during early aggregation in *Dictyostelium discoideum*. *J. Cell Sci.* 22: 645–656.

Höfer, T., Maini, P.K., Sherratt, J.A., Chaplain, M.A.J., Chauvet, P., Metevier, D., Montes P.C. and Murray, J.D. (1994) A resolution of the chemotactic wave paradox. *Appl. Math. Lett.* 7: 1–5.

Höfer, T., Sherratt, J.A. and Maini, P.K. (1995a) *Dictyostelium discoideum*: cellular self-organization in an excitable biological medium. *Proc. Roy. Soc. Lond. B* 259: 249–257.

Höfer, T., Sherratt, J.A. and Maini, P.K. (1995b) Cellular pattern formation during *Dictyostelium* aggregation. *Physica D* 85: 425–444.

Johnson, B.R. and Scott, S.K. (1996) New approaches to chemical patterns. *Chem. Soc. Rev.* 25: 265–273.

Kauffman, S.A., Shymko, R.M. and Trabert, K. (1978) Control of sequential compartment formation in *Drosophila. Science* 199: 259–270.

Keller, E.F. and Segel, L.A. (1971) Travelling bands of bacteria: a theoretical analysis. *J. Theor. Biol.* 30: 235–248.

Kessler, D.A. and Levine, H. (1993) Pattern formation in *Dictyostelium* via the dynamics of cooperative biological entities. *Phys. Rev. E* 48: 4801–4804.

Kondo, S. and Asai, R. (1995) A reaction-diffusion wave on the skin of the marine angelfish *Pomacanthus. Nature* 376: 765–768.

Kulesa, P.M., Cruywagen, G.C., Lubkin, S.R., Maini, P.K., Sneyd, J., Ferguson, M.W.J. and Murray, J.D. (1996) On a model mechanism for the spatial patterning of teeth primordia in the alligator. *J. Theor. Biol.* 180: 287–296.

Lengyel, I. and Epstein, I.R. (1991) Modeling of Turing structures in the chlorite-iodide-malonic acid-starch reaction system, *Science* 251: 650–652.

Madzvamuse, A. (1999) D. Phil. Thesis, Oxford (in preparation).

Maini, P.K. and Solursh, M. (1991) Cellular mechanisms of pattern formation in the developing limb. *Int. Rev. Cytology* 129: 91–133.

Maini, P.K., Myerscough, M.R., Winters, K.H. and Murray, J.D. (1991) Bifurcating spatially heterogeneous solutions in a chemotaxis model for biological pattern formation. *Bull. Math. Biol.* 53: 701–719.

Maini, P.K., Benson, D.L. and Sherratt, J.A. (1992) Pattern formation in reaction diffusion models with spatially inhomogeneous diffusion coefficients. *IMA J. Math. Appl. Med. and Biol.* 9: 197–213.

Maini, P.K., Painter, K.J., and Chau, H.N.P. (1997) Spatial pattern formation in chemical and biological systems. *Faraday Trans.* 93(20): 3601–3610.

Martiel, J.-L. and Goldbeter, A. (1987) A model based on receptor desensitization for cyclic AMP signaling in *Dictyostelium* cells. *Biophys. J.* 52: 807–828.

Meinhardt, H. (1982) *Models of Biological Pattern Formation.* New York: Academic Press.

Meinhardt, H. (1995) *The Algorithmic Beauty of Sea Shells.* New York: Springer-Verlag.

Monk, P.B. and Othmer, H.G. (1990) Wave propagation in aggregation fields of the cellular slime mould *Dictyostelium discoideum. Proc. R. Soc. Lond. B* 240: 555–589.

Murray, J.D. (1981) A pre-pattern formation mechanism for animal coat markings. *J. Theor. Biol.* 88: 161–199.

Murray, J.D. (1993) *Mathematical Biology.*New York: Springer-Verlag.

Murray, J.D. and Myerscough, M.R. (1991) Pigmentation pattern formation on snakes. *J. Theor. Biol.* 149: 339–360.

Murray, J.D., Deeming, D.C. and Ferguson, M.W.J. (1990) Size dependent pigmentation pattern formation in embryos of Alligator mississippiensis: time of initiation of pattern generation mechanism. *Proc. Roy. Soc. Lond.* B 239: 279–293.

Nagorcka, B.N. (1989) Wavelike isomorphic prepatterns in development. *J. Theor. Biol.* 137: 127–162.

Nagorcka, B.N., Manoranjan, V.S. and Murray, J.D. (1987) Complex spatial patterns from tissue interactions — an illustrative model. *J. Theor. Biol.* 128: 359–374.

Nijhout, H.F. (1990) A comprehensive model for colour pattern formation in butterflies. *Proc. R. Soc. Lond. B* 239: 81–113.

Oster, G.F. and Murray, J.D. (1989) Pattern formation models and development. *Zoology* 251: 186–202.

Oster, G.F., Murray, J.D. and Harris, A.K. (1983) Mechanical aspects of mesenchymal morphogenesis. *J. Embryol. Exp. Morph.* 78: 83–125.

Othmer, H.G. and Stevens, A. (1997) Aggregation, blowup, and collapse: The ABC's of taxis in reinforced random walks, *SIAM J. Appl. Math.* 57: 1044–1081.

Painter, K.J., Othmer, H.G. and Maini, P.K. (1999) Stripe formation in juvenile *Pomacanthus* explained by a generalised Turing mechanism with chemotaxis. *Proc. Natl. Acad. Sci. USA* 96: 5549–5554.

Schnakenberg, J. (1979) Simple chemical reaction systems with limit cycle behaviour. *J.*

Theor. Biol. 81: 389–400.

Segel, L.A. (1984) *Modelling Dynamic Phenomena in Molecular and Cellular Biology.* Cambridge: Cambridge University Press.

Sekimura, T., Zhu, M., Cook, J., Maini, P.K. and Murray, J.D. (1999) Pattern formation of scale cells in Lepidoptera by differential origin-dependent cell adhesion. *Bull. Math. Biol.* (in press).

Shaw, L.J. and Murray, J.D. (1990) Analysis of a Model for complex skin patterns. *SIAM J. Appl. Math.* 50(2): 628–648.

Soll, D.R., Wessels, D. and Sylwester, A. (1993) The motile behavior of amoebae in the aggregation wave in *Dictyostelium discoideum.* In *Experimental and Theoretical Advances in Biological Pattern Formation*, Othmer, H.G., Maini, P.K. and Murray, J.D. (eds.) London: Plenum Press. pp. 325–328.

Steinberg, M.S. (1970) Does differential adhesion govern self-assembly processes in histogenesis? Equilibrium configurations and the emergence of a hierarchy among populations of embryonic cells. *J. Exp. Zool.* 173: 395–434.

Steinbock, O. and Müller, S.C. (1995) Spatial attractors in aggregation patterns of *Dictyostelium discoideum. Z. Naturforschung* 50: 275–281.

Sulsky, D., Childress, S. and Percus, J.K. (1984) A model of cell sorting. *J. Theor. Biol.* 106: 275–301.

Tang, Y.H. and Othmer, H.G. (1994) A G-protein based model of adaptation in *Dictyostelium discoideum. Math. Biosci.* 120: 25–76.

Tang, Y.H. and Othmer, H.G. (1995) Excitation, oscillations and wave propagation in a G-protein based model of signal transduction in *Dictyostelium discoideum. Phil. Trans. R. Soc. Lond. B* 349: 179–195.

Thomas, D. (1975) Artifical enzyme membranes, transport, memory and oscillatory phenomena. In *Analysis and Control of Immobilized Enzyme Systems*, Thomas, D. and Kernevez, J.-P. (eds.). New York: Springer-Verlag. pp. 115–150.

Turing, A.M. (1952) The chemical basis of morphogenesis. *Phil. Trans. Roy. Soc. Lond. B* 327: 37–72.

Tyson, J.J. and Keener, J.P. (1988) Singular perturbation theory of traveling waves in excitable media (a review). *Physica D* 32: 327–361.

Tyson, J.J., Alexander, K.A., Manoranjan, V.S. and Murray, J.D. (1989) Spiral waves of cyclic AMP in a model of slime mold aggregation. *Physica D* 34: 193–207.

Van Oss, C., Panfilov, A.V., Hogeweg, P., Siegert, F. and Weijer, C.J. (1996) Spatial pattern formation during aggregation of the slime mould *Dictyostelium discoideum. J. Theor. Biol.* 181: 203–213.

Varea, C., Aragón, J.L. and Barrio, R.A. (1997) Confined Turing patterns in growing systems. *Phys. Rev. E* 56(1): 1250–1253.

Vasiev, B.N., Hogeweg, P. and Panfilov, A.V. (1994) Simulation of *Dictyostelium discoideum* aggregation via reaction-diffusion model. *Phys. Rev. Lett.* 73: 3173–3176.

Wolpert, L. (1969) Positional information and the spatial pattern of cellular differentiation. *J. Theor. Biol.* 25: 1–47.

Wolpert, L. and Hornbruch, A. (1990) Double anterior chick limb buds and models for cartilage rudiment specification. *Development* 109: 961–966.

8

On Pattern and Growth

Hans Meinhardt

Max-Planck-Institut für Entwicklungsbiologie, Spemannstraße 35, D-72076 Tübingen, Germany

8.1 Introduction

The development of a higher organism from a single cell is a highly complex process requiring the coupling together of many reactions. These reactions are frequently non-linear and contain many feedback loops. Such processes are difficult to understand by intuition alone. In many branches of science, mathematically based modelling has provided substantial help for the understanding of complex systems. All biological processes are assumed to be based ultimately on the interaction of molecules. However, the development of a higher organism may appear too complex to be partitioned into a set of differential equations that describe the production, distribution, interaction and decay of the molecules involved. However, it is possible to separate development at least partially into discrete elementary steps that are accessible to modelling. Models have been proposed for several different elementary steps. In the following sections, some of them will be discussed and compared with recent findings obtained by molecular genetic investigations.

The famous book by D'Arcy Thompson is entitled *On Growth and Form.* In the models to be described below, a step preceding the generation of form will be discussed: the generation of patterns. It is assumed that during development a change in the physical shape is preceded by signals that determine where these changes should occur and what type they have to be. In these pattern-forming reactions, growth plays a dual role. Growth can provide the precondition to insert new structures. An example is the insertion of new leaves at the tip of a growing shoot. For other systems, the converse is important: new structures should not be inserted despite the fact of

On Growth and Form: Spatio-temporal Pattern Formation in Biology.
Editor Mark A.J. Chaplain, G.D. Singh, J.C. McLachlan

growth. For instance, in a growing embryo a single anteroposterior axis should be maintained, although we know from classical experiments that an inclination exists to form secondary axes. As will be discussed below, the maintenance of a single organizing region requires special mechanisms. Thus, emphasis will be given to the role of growth in pattern formation.

8.2 The Generation of Organizing Regions by Local Self-enhancement and Long-Range Inhibition

Although many eggs have pronounced asymmetries, the final pattern cannot already be present in the egg in a hidden form. At the early stages of development, many embryos can be fragmented and each part forms a complete organism. The sea urchin embryo at the 16-cell stage is an example. In some vertebrates, including mice and humans, this can also occur spontaneously, giving rise to identical twins. This indicates that at least in these cases the early embryo is not a mosaic-like arrangement of differently determined cells that have a fixed fate. Instead, a communication must exist between different parts, such that the removal of some parts becomes detected and the missing parts replaced.

Some small specialized regions obviously play a decisive role for the overall organization of the developing organism. Such organizing regions direct pattern formation in the surrounding tissue. One such region is the dorsal lip of the amphibian embryo (Spemann and Mangold, 1924). The embryo first forms a blastula, a partially hollow sphere. The embryo proper becomes shaped in the process of gastrulation, an invagination of cells that starts at a particular position. This process leads to a tube-like structure within the cavity touching at one side the outer surface. In this way the embryo obtains its basic cell sheets, the ecto-, meso-, and endoderm. The cells at the dorsal border of the first invaginating tissue have organizing properties as mentioned above. A transplantation to the opposite (ventral) side causes the formation of a second embryonic axis. By using tissues for transplantation that differ in their pigmentation, it has been shown that the secondary embryo is formed mostly from host tissue, not from the transplant. Obviously the transplanted cells have an instructive influence on the surrounding tissue. The resulting pattern has no discontinuity at the border between the transplanted and the host cells. Many genes involved in the formation of the organizing regions are now known (DeRobertis, 1995).

Pattern formation requires that originally equivalent cells become different from each other. We have proposed that biological pattern formation is based on local self-enhancement and long-range inhibition (Gierer and Meinhardt, 1972). It can be shown that the mechanisms proposed by Turing (1952) are also based on this principle (Meinhardt, 1992), although Turing did not interpret his mathematically abstract concept in this way. Pattern formation from initially homogeneous conditions is not a privilege of living systems: formation of sand dunes, patterns of erosion, lightning, stars and galaxies are examples form the inorganic world. Also in these cases, patterns arise from the amplification of small fluctuations. The emerging patterns becomes globally stabilized by long-range antagonistic effects to testifying the universal character of the proposed mechanism.

A molecular realization of this concept requires a substance (or a cascade of substances) that has an autocatalytic feedback on its own synthesis. A simple molecular interaction with pattern-forming capability could consist of an 'activator' $a(x)$ whose autocatalysis is slowed down by a long-ranging 'inhibitor' $h(x)$ (Gierer and Meinhardt, 1972). The following system of equations provides an example:

$$\frac{\partial a}{\partial t} = \frac{\rho a^2}{h} - \mu_a a + D_a \frac{\partial^2 a}{\partial x^2} + \sigma_a \tag{8.1}$$

$$\frac{\partial h}{\partial t} = \rho a^2 - \mu_h h + D_h \frac{\partial^2 h}{\partial x^2} + \sigma_h \tag{8.2}$$

Equation (8.1) states that a concentration change of the activator per unit time ($\partial a/\partial t$) is proportional to an autocatalytic production term (a^2), and that the autocatalysis is slowed down by the action of the inhibitor ($1/h$). As any other biological substance, the activator molecules become degraded. It is natural to assume that the number of disappearing activator molecules is proportional to the number of activator molecules present. This is expressed by the term $-\mu_a a$. The autocatalysis must be non-linear since the production rate must overcome the disappearance by the linear decay. A non-linearity of the reaction will result if the active component is not the activator itself but a dimer of two activator molecules.

The concentration of a in a given cell may also vary due to an exchange of molecules with neighbouring cells. Diffusion is the simplest way to take this exchange into account. Although diffusion is a good approximation, the real process is often more complex, requiring a chain of several molecules: molecules that are exposed on the cell surface, molecules that act as receptors on neighbouring cells and molecules that transmit the signal from the cell surface to the nucleus. Other modes of redistribution of molecules are conceivable as well. In plants, for instance, active transport plays a major role.

The last term in equation (8.1), σ_a, describes a small basic (i.e. an activator-independent) activator production. This term insures that the concentration of the activator never sinks to zero. This is important for the initiation of the autocatalytic reaction at low activator concentrations. The term ρ, the *source density*, describes the general ability of the cells to perform the autocatalytic reaction. Slight asymmetries in the source density may have a strong influence on the *orientation* of the emergent pattern. A necessary condition for the formation of a stable pattern is that the inhibitor diffuses much faster than the activator and has a shorter half-life, i.e., $D_h \gg D_a$ and $\mu_h > \mu_a$ must be satisfied. In other ranges of parameters, oscillations and travelling waves can occur. These modes will be discussed further in connection with the patterns on the shells of molluscs.

The system generates patterns whenever the size of the field exceeds the range of the activator. A small local fluctuation or any asymmetry will grow further due to its self-enhancement. The concomitantly produced inhibitor spreads into the surroundings and suppresses the activator production there. A stable situation is reached when the activator maximum is in a dynamic equilibrium with the surrounding cloud of inhibition. As shown in Figure 8.1, graded concentration profiles – more or less periodically arranged maxima or stripe-like patterns – can be generated in this way. In fields with a size comparable to the activator range, the high activator concentration is formed at one end of the field. This is a very important aspect for the understanding of pattern formation in an early embryo. One side of the embryo becomes different from the other. The graded activator and/or inhibitor distribution can activate different genetic information in different parts of the tissue in an ordered sequence (see Figure 8.4). Thus, the system described by equations (8.1) and (8.2) is able to generate 'positional information' (Wolpert, 1969). Important for biological applications is that these patterns are self-regulating and in a wide range independent of the details of initiation. A most extreme case is the re-establishment of a normal pattern during regeneration. In the model, after removal of the activated region, the remaining inhibitor decays until a new maximum is formed via autocatalysis, restoring in this way a gradient. Due to this insensitivity, a deviation from the normal pattern at one level can frequently be compensated at the next level. Therefore, errors do not propagate — this important feature insures the robustness of the complex network of coupled reactions and thus development as a whole.

Pattern-forming systems have recently been uncovered that follow this type of regulation. In the fly *Drosophila* the precursor cells of the peripheral nervous system (neuroblasts) and the ectoderm form from two lateral bands of ectodermal cells. Crucial for neuroblast formation are the genes of the *achaete-scute* complex. Both genes have a direct autocatalytic feedback on their own synthesis (Van Doren *et al.*, 1992; Culi and Modolell, 1998). Cells that express these genes activate the gene *Delta* that codes for a secreted molecule. This message is received by Notch, a receptor molecule ubiquitously present on all cells in that region. Its stimulation leads to a repression of the *achaete-scute* complex. In this way, particular cells are singled out to participate in the formation of the peripheral nervous system while the remaining cells form the ectoderm.

The interaction given in equations (8.1, 8.2) is only an example. To illustrate an alternative possibility, an interaction of three substances in which only inhibitions are involved is presented and it is shown that this can satisfy our general criterion for pattern formation (Meinhardt, 1989). Meanwhile it has turned out that this scheme is presumably close to the interactions that are actually involved in the formation of organizing regions. First the scheme: let us assume two substances, a and b, that inhibit each other mutually. A small increase of a above an equilibrium leads to a stronger repression of the b production and thus to a further increase of a, and so on in the same way as if a would be autocatalytic. The same holds for b. Taken together, a and b form a switching system in which either a or b is high. To allow pattern formation, a long-ranging signal is required which makes sure that a wins the a, b competition only in a particular region, whereas b wins in the remaining region. In other words, regions with high a and with high b concentrations have to obtain a balanced ratio. This is the case, for instance, if the a molecules control the production of a substance c that acts in an inhibitory manner on the a production. The following system of equations is an example for such a scheme:

$$\frac{\partial a}{\partial t} = \frac{\rho}{\kappa + b^2} - \mu_a a + D_a \frac{\partial^2 a}{\partial x^2} + \sigma_a \tag{8.3}$$

$$\frac{\partial b}{\partial t} = \frac{\rho}{\kappa + a^2/c^2} - \mu_b b + D_b \frac{\partial^2 b}{\partial x^2} + \sigma_b \tag{8.4}$$

$$\frac{\partial c}{\partial t} = \gamma a - \mu_c c + D_c \frac{\partial^2 c}{\partial x^2} \tag{8.5}$$

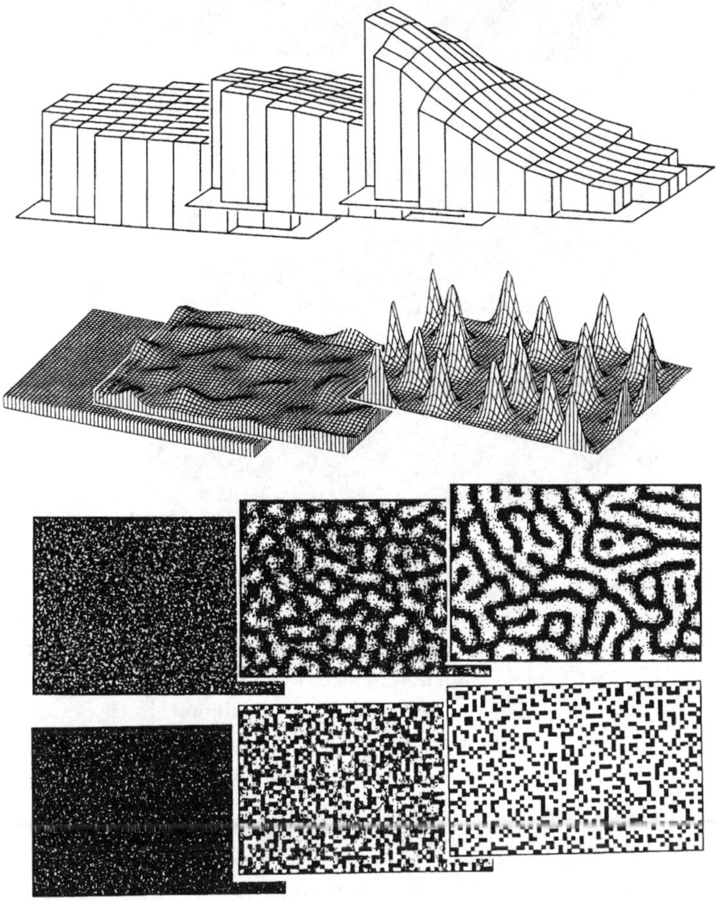

(a)

(b)

(c)

(d)

Figure 8.1 Stages in the generation of elementary patterns by local self-enhancement and long-ranging inhibition: the initial distribution, an intermediate distribution and the final activator distributions. (a) Monotonic gradients are formed if the range of the activator is comparable to the size of the field. This is appropriate to supply the tissue with positional information. (b) A more or less regular arrangement of peaks results in fields that are large compared to the range of the inhibitor. (c) Stripe-like distributions result if the autocatalysis saturates (Meinhardt, 1989). (d) The same parameter but without activator diffusion leads to a segregation into two cell types, as required for instance in the early patterning of *Dictyostelium discoideum*

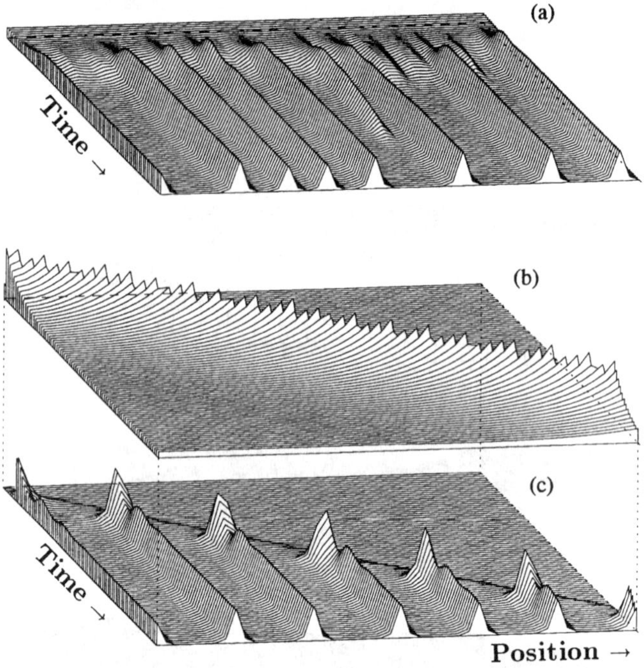

Figure 8.2 Formation of a regular pattern behind a wavefront. (a) If pattern formation takes place in a large field, the resulting pattern is irregular since some maxima become suppressed by others that are too close. (b, c) Regularly spaced maxima appear if cells become competent after a wave has passed. New maxima emerge at a defined distance from the last established maximum. In this example, a small baseline inhibitor production, σ_h in equation (8.2), maintains the cells in a low non-patterned state. The wave is assumed to increase temporarily a baseline activator production, σ_a in equation (8.1), initiating in this way the onset of pattern formation.

In the amphibian organizer, the proteins Noggin and Chordin are present in high concentrations (Harland and Gerhart, 1977). At the opposite ventral side of the marginal zone, a high BMP-4 concentration is formed (Thomsen, 1997). Proteins Noggin/Chordin and BMP-4 downregulate each other mutually, as required to generate the instability mentioned above. For the balanced expression of both groups at opposite positions, a long-ranging substance is expected that has its highest expression in the organizing region and that has an inhibitory influence on the expression of marker genes characteristic for the organizer. A candidate with such antidorsalizing properties has been found by Moos *et al.* (1995): a TGF-β homologue related to BMP-3. As expected from the model, it has the highest concentration in the organizer region but its overexpression leads to a down regulation of all organizer-specific genes. A similar function has been ascribed to a secreted metalloprotease Xolloid that cleaves Chordin and the Chordin/BMP-4 complex (Piccolo *et al.*, 1997). The authors have proposed that Xolloid restricts the size of the organizer region, in agreement with the model proposed.

8.3 Regular and Irregular Periodic Patterns

In fields larger than the range of the antagonist, several activator maxima will emerge. If pattern formation starts when the field already has a substantial size, the resulting activator maxima will have a somewhat irregular arrangement, but a maximum and a minimum distance will be maintained (Figure 8.1b). The distributions of stomata and hairs have corresponding patterns. The irregularity results because, at an early stage, the mutual inhibition of maxima is relatively weak and incipient maxima can appear close together. With increasing peak height, some of these maxima lose the mutual competition and disappear, leaving larger interstices (Figure 8.2a).

The model also describes the insertion of new peaks during growth. With increasing distance, the inhibitor concentration becomes lower and lower until the onset of autocatalysis is triggered. A corresponding process is the insertion of new stomata into the largest interstices (Bünning and Sagromsky, 1948) or the insertion of new bristles during the moulting of some insects (Wigglesworth, 1940).

Regular structures are formed if the field grows at its boundaries. The existing fully developed maxima allow the formation of new maxima only at a well-defined distance. The initiation of new leaf primordia behind the growing shoot of a plant is a prominent example from botany. This requirement of growth to form regular periodic structures has a severe consequence: it requires time. However, there are cases where very regular patterns are formed although the domain has already reached a substantial size. The formation of feather primordia on birds (Davidson, 1983; Jung *et al.*, 1998) or the differentiation of ommatidia in insect eyes (Tomlinson, 1988) are examples. This indicates that nature has found a way to circumvent the problem. In these cases the regularity is achieved by a 'simulated' growth (Koch and Meinhardt, 1994). Cells become competent in a wave-like manner. Although many cells are already present, pattern formation can take place only in a narrow zone that sweeps over the field, and existing maxima direct the position of new ones as if physical growth were at work. Figure 8.2b and c show the generation of a regular pattern behind a moving wavefront.

8.4 The Wavelength Problem: Stabilization of a Monotonic Gradient by a Feedback on the Source Density

As a rule, the size of a morphogenetic field increases during the growth of the embryo. In a usual reaction-diffusion system, a graded concentration profile can be maintained only over a range of about a factor 2. With increasing field size, a tendency exists to change from a monotonic distribution into a symmetric and ultimately into a periodic distribution, either by insertion of new maxima or by splitting of existing maxima. This is inappropriate if the graded concentration should be used in the growing embryo as positional information for the determination of the primary body axes since, for instance, several heads would appear instead of one.

Some observations provide hints of how this wavelength problem has been solved by nature. *Hydra*, a small freshwater polyp, maintains its polar structure over substantial growth. Nevertheless, a fragment one-tenth of the normal body size is still able to regenerate. In such a fragment the regeneration of the head occurs always at the side pointing towards the original head. Morgan (1904) interpreted similar observations

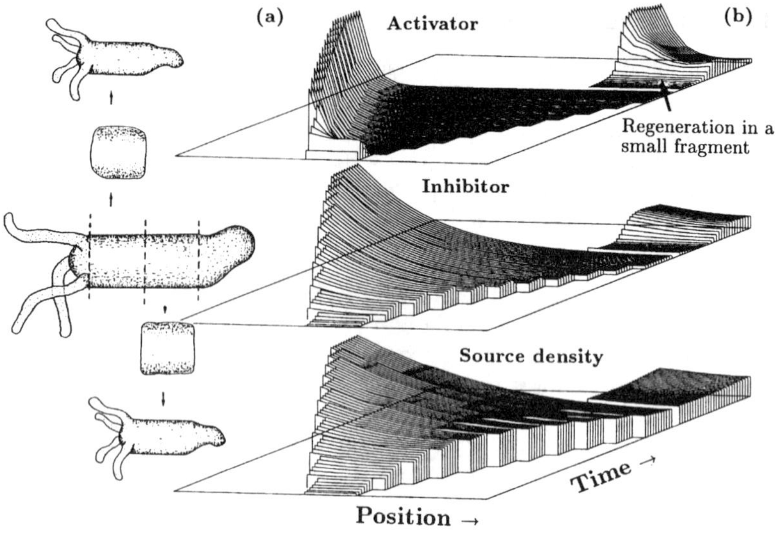

Figure 8.3 Maintenance of a polar distribution in a growing field. (a) Biological observation: a hydra maintains its polar structure over a large range of sizes. Small fragments regenerate with the original polarity. (b) Model: a polar structure can be stabilized upon growth if the activator has a feedback on the ability of the cell to synthesize the activator, i.e. on the source density ρ in (8.1). The source density obtains a graded distribution. In regions of low source density, the initiation of secondary maxima is unlikely. Nevertheless, a small fragment regenerates even when derived from a region of low source density.

with *Tubularia* by assuming there exists a systematic change in the ability for head regeneration and by assuming that a competition takes place. The tissue has an intrinsic asymmetry. Cells originally closer to the head have an advantage and win the competition. Therefore, it is the *relative* position that is decisive in determining which side will win.

In terms of our model, the range of dominance of the activated over the non-activated region (the so-called apical dominance) can be increased by an order of magnitude if there exists a feedback of the activator or inhibitor on the source density i.r. on ρ in equation 8.1 (Meinhardt and Gierer, 1974). Due to this feedback, the source density becomes graded too. In a region of low source density, the initiation of secondary maxima is unlikely. In contrast, the rising source density at an existing maximum reinforces this maximum. Thus, the polar distribution remains stable despite the fact of growth (Figure 8.3). The graded source density provides the long-lasting information about the polarity of the system. A small fragment regenerates a pattern according to the original polarity since the graded source density provides a systematic head start for cells at the apical end to outcompete the others. Since the source density has a long time constant, it remains essentially unchanged during regeneration. Although secondary maxima are successfully suppressed, regeneration is not impaired. Regeneration can be fast since no symmetry breaking is required nor communication by diffusion over the total field (Figure 8.3).

The source density must have a more shallow distribution than the activator. This may be achieved by a diffusible intermediate molecule. Alternatively, it may not be the activator but the inhibitor that controls the source density. Anyway, due to its higher diffusion rate, the inhibitor has a shallower distribution. No further diffusion of the sources would be required. In equation (8.1) the activator a is nearly independent of the absolute ρ concentration. Therefore a feedback of a on ρ does not create stability problems. This is, however, not true if the inhibitor accomplishes this feedback. Equation (8.6) provides an example in which the inhibitor causes the increase of the source density. The system is nevertheless stable since this increase slows down at higher levels of the source density.

$$\frac{\partial \rho}{\partial t} = \mu_\rho(h/\rho - \rho) \tag{8.6}$$

In an interaction according to equations (8.1), (8.2) and (8.6) the inhibitor has a dual role (Figure 8.3). Firstly, it localizes the activator maximum. Secondly, it increases the local source density. In this way it is also responsible for the generation of polarity within the tissue. Although, in the steady state, inhibitor and source density would have similar distributions, they have very different properties otherwise. The inhibitor must have a high diffusion rate and a short time constant to enable a stable pattern. In contrast, the source density must be non-diffusible and must have a long time constant. Since the wavelength of the system is no longer determined by the activator diffusion, the activator diffusion can be very small. Therefore, the pattern can regenerate in small fragments. Since after regeneration the ρ-gradient will be restored in the course of time (Figure 8.3), this gradient will not be diluted out in repeated rounds of regeneration.

Since the source density concept is very important to explain many biological observations, a somewhat anthropomorphic analogy should provide a better intuition of the mechanism envisaged. A king, a president (or any other local hero) usually has a strong tendency to suppress others from becoming a king too - a long-range inhibition. On the other hand, he promotes individuals in his surroundings to obtain different levels in a hierarchy, to become ministers, etc. In this way the centre of power generates a hierarchy. The central figure has only to suppress a relatively small region of the total field to remain unique. If, due to this non-uniformity, the centre of power becomes void, it is usually clear who will win the subsequent competition.

8.5 Space-dependent Activation of Genes by a Morphogen

The generation of signals by the exchange of molecules via diffusion works only in small fields. In larger fields the time required to exchange information by randomly moving molecules would be much too long. Indeed, the sizes of embryonic fields in which developmental decisions take place are always small, smaller than 1 mm and less than 100 cells across (Wolpert, 1969). Therefore, the signals generated at small scales have to be translated into more permanent states that can be maintained upon further growth, independent of the evoking signal. The obvious means is a stable concentration-dependent (and thus space-dependent) activation of genes. The choice of a particular pathway under the influence of a morphogenetic signal requires the activation of particular genes and the suppression of alternative genes. This situation has formal similarities with pattern formation. As discussed above, pattern formation in space requires activation at a particular position and inhibition in the remaining part. The selection of a particular pathway requires the activation of a particular gene and the suppression of the alternative genes. Cell determination can be regarded as pattern formation among alternative genes. Based on this similarity, I have proposed that gene activation requires a direct or indirect feedback of gene products on the activation of their own gene and their mutual competition such that only one of the alternative genes can remain active in a particular cell (Figure 8.4a and b; Meinhardt, 1978). Meanwhile many such genes have been found; examples are *deformed* (Regulski *et al.*, 1991), *hunchback* (Simpson-Brose *et al.*, 1994) and *twist* (Leptin, 1991).

Such a dynamic regulation of gene activation has many properties that are essential for a developing organism. Firstly, such a system has a threshold. The activation of a gene becomes an all-or-nothing event. If a threshold is surpassed, due to the autoregulation, the response of the cell becomes independent of the exact level of the morphogen concentration. Therefore, small deviations from a desired value are not propagated into the hierarchically next level of gene activation. Secondly, the cells obtain a long-term memory in respect of the signals they have seen. The signal is required only for the initiation of gene activity. Feedback makes the activity self-maintaining.

Based on ligation experiments with early insect embryos (Sander, 1976), I have proposed that the cells measure a particular morphogen concentration not all at once but that, starting from a default gene activity, other genes become activated in a stepwise manner. Each further step requires a higher morphogen concentration. This process comes to rest if the actual gene activation corresponds to the local morphogen concentration (Figure 8.4). To require higher and higher morphogen concentrations for such a promotion, the system has to be less and less sensitive for the morphogen. In order that these transitions nevertheless occur, the 'higher' genes must be more efficient in the autoregulation to compensate the reduced sensitivity (Meinhardt, 1978, 1998).

Due to the autocatalysis of the gene activation and the unidirectional promotion, the process is essentially irreversible. A reduction of the signalling molecules is without effect since the morphogen is not required for the maintenance of the gene activity. This mode of stepwise promotion has a very essential function during development. As mentioned, the range of the signalling molecules is usually small. The initial activation of genes takes place in small fields that grow afterwards. During growth the distance between a cell and the morphogen source necessarily increases, so the morphogen concentration decreases. Due to the irreversible character of the stepwise promotion, the cells maintain their once obtained state of determination even if they escape from the influence of a morphogen source. This stability is possible without the determination being absolutely fixed. For instance, if during regeneration a new morphogen source is generated, the cells can still be 'distally transformed'. In the regeneration of a cockroach leg, this process plays a decisive role. The stepwise and irreversible promotion enables on the one hand the required stability to cope with the fading influence of the morphogen during growth. On the other hand, it provides some (unidirectional) flexibility if regulation is required. Although the actual morphogen concentration changes permanently, this avoids having to confine the measurements to a narrow time window.

Meanwhile several systems have been described that show this behaviour. Gurdon *et*

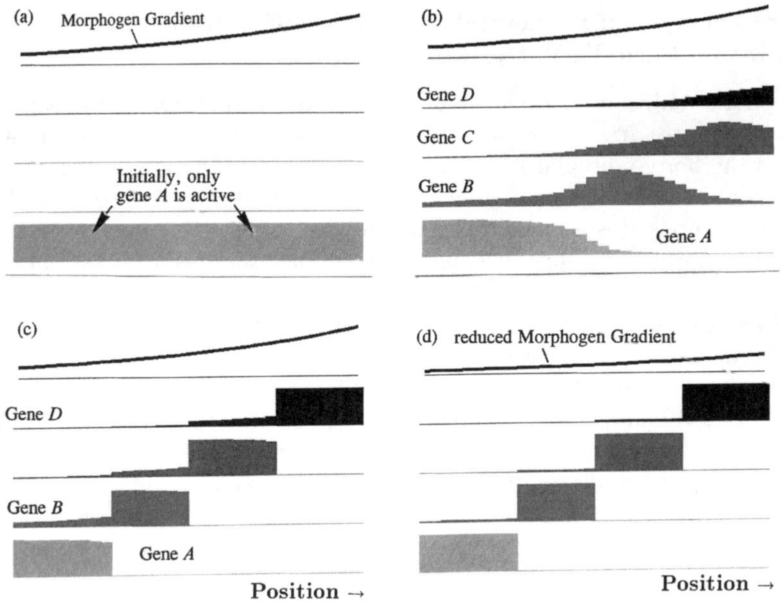

Figure 8.4 Space-dependent gene activation by a morphogen gradient. A set of genes $A, B, \ldots$ is assumed whose gene products feed back on the activation of the corresponding genes. In addition, all genes compete with each other for activity. This has the consequence that only one of these genes can be active within one cell. Depending on the local level of the morphogen, a promotion from gene A to gene B, and so on, takes place until the local morphogen concentration is insufficient for a further step. Sharply confined regions of gene activities emerge despite the fact that the initiating signal is graded. After a decrease of the morphogen concentration, the gene activation maintains the level it has achieved.

al. (1995) found that in *Xenopus* cells from the blastula stage, a low concentration of Activin (a member of a family of growth factors) will activate the *Xbra* gene, whereas a high concentration is required to activate the gene *Xgsc*. These genes remain activated after removal of Activin. However, applying Activin first in low and later in high concentrations leads to a reprogramming of cells from *Xbra* to *Xgsc*. Gurdon *et al.* (1995) called this a ratchet-like switching, in full agreement with the mechanism I have proposed.

The model has predicted further that if, upon a transplantation, cells become exposed to a higher morphogen concentration, they adapt to the new situation. In contrast, after transplantation into a region of low morphogen concentration, the cells should maintain their original state of determination. Such a regulatory behaviour has been found by Grapin-Botton *et al.* (1998) for the *Kreisler* gene in the segmentation of the chick hindbrain and by Udolf *et al.* (1995) in the determination of *Drosophila* neuroblasts.

8.6 Somite Formation: Sequential Conversion of a Periodic Pattern in Time into a Periodic Pattern in Space

The formation of somites is a crucial step in the primary anteroposterior patterning of vertebrates. Somites are the precursors of many essential structures, including the vertebrae and the body musculature. During such a metamerization two patterns are laid down. There is a periodic pattern consisting of a serial repetition of homologous structures. Superimposed is a sequential pattern which makes the repetitive subunits different from each other. For instance, only the thoracic somites form vertebrae that bear ribs.

Classic observations restrict possible models for somite formation. Although the somites are formed in an anterior-to-posterior sequence, the formation of a somite does not require a trigger of its anterior neighbour (Deuchar and Burges, 1967). Therefore, this mechanism is fundamentally different from the formation of a strictly periodic pattern behind a wavefront (Figure 8.2). In amphibians a heat shock of 30min will nevertheless lead to the malformation of several somites (Pearson and Elsdale, 1979); this is much shorter than the 120min required to form a single somite. It suggests that the malformation of one somite has an effect on several subsequent somites, and perhaps a limited cell-to-cell communication is involved.

To be compatible with this and other observations, a model has been proposed that has recently found strong experimental support (Figure 8.5; Meinhardt, 1982, 1986a). It is assumed that each somite consists of an anterior half and a posterior half. Consider two non-overlapping cell states called A and P; how they can be generated has been described above (Figure 8.4). If two such cell states not only exclude each other but activate each other on a longer range, these cell states stabilize each other in a symbiotic manner. A stable arrangement would consist of alternating A and P states, since neighbouring cell state need each other close by to be maintained while the local exclusiveness makes sure these states do not merge or overlap. Molecular genetic observations on segment polarity genes in *Drosophila* have confirmed this general scheme (Meinhardt, 1994).

While A and P cells stabilize each other in the region of a common boundary, it is a property of such an interaction that a group of cells consisting of one type only (A or P) can oscillate back and forth between the two states. For instance, if all cells are in state A, then state P obtains substantial help while state A is not supported. Therefore, after a certain time, the cells switch from A to P. Later, for the same reason, the cells switch back to A, and so on.

This spatially homogeneous oscillating system can convert into a pattern which is stable in time if, at any location, an A–P border has been formed (Figure 8.2d). Imagine that all cells are in the state P except the most anterior cells that are in state A. Cells close to this border stabilize each other while the P cells distant to this border switch from P to A, forming in this way a second A–P border, and so on. But what gives rise to the first border? To account for several observations, it is assumed that some sort of positional information provides the driving force for this oscillation. Imagine a field of cells exposed to a graded morphogen concentration which has its high point at the posterior end, and further, that all cells are in the state P. Cells above a certain threshold concentration switch from P to A, a process that leads to the first P–A border. Again, cells distant to the new P/A border will switch back to

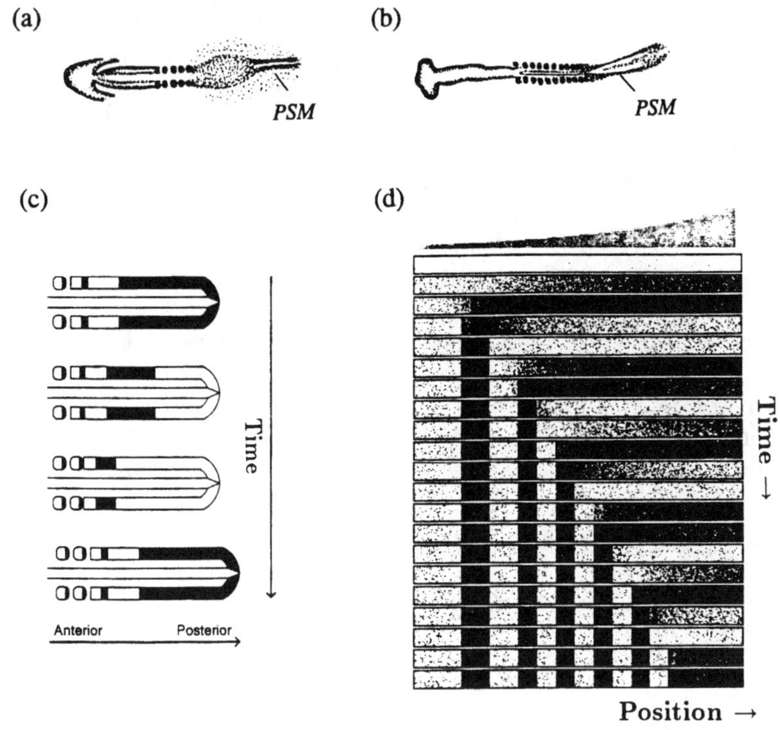

Figure 8.5 Somite formation. (a, b) In chicken, somites separate from the non-segmented presomitic mesoderm (PSM) in an anterior-to-posterior sequence. At 25 h of incubation about 5 somites are visible. Ten hours later, about 12 somites are present. (c) Observation of Palmeirim *et al.* (1997): the activity of the *c-hairy* gene oscillates in the presomitic mesoderm. (d) Model: under the influence of positional information (top), cells oscillate between two states, A (white) and P (black). Only cells above a threshold can participate. This leads to a first A/P border. At such a border, the two states stabilize each other mutually. The remaining cells switch to the alternative state, forming a further border, and so on. Each further cycle leads to one additional A/P pair (Meinhardt, 1982, 1986a). In the course of time, the region with a stable pattern in space enlarges at the expense of the region where oscillations take place.

A, forming the next A/P border, and so on. After each full cycle, one pair of A–P stripes is added. In the course of time, the region of the stable spatially alternating A–P pattern enlarges at the expense of the cells which still oscillate between A and P. The borderline between the stable and oscillating cells moves over the field in a wave-like manner. Since the oscillation is driven by the gradient, the transition occurs a bit earlier at the posterior end and sweeps towards the anterior in a wave-like manner. It comes to rest at a certain distance from the last-formed border. Thus, although the somites separate from each other in an anterior-to-posterior sequence, the non-trivial prediction of the model was that this is based on oscillations with a wave-like phase shift causing an apparent wave in the posterior-to-anterior direction. The dynamics of the *c-hairy1* gene activation, as observed by Palmeirim *et al.* (1997) in chickens, corresponds precisely to the predicted dynamics of gene activation for the state P (Figure 8.2c).

The somites are not only a periodic structure, they are also different from each other. Keeping in mind that the cells forming a more posterior somite have made precisely one additional A/P transition in their history, this suggests a convenient and reliable mechanism: each oscillation drives a transtion to a more posterior specification. The system may be compared with the pendulum escapement mechanism of a grandfather clock. The primary event is the periodic movement of the pendulum. The stepwise advancement of the pointer provides a count of the periodic events (Meinhardt, 1982, 1986a). Whether such a mechanism is realized is still an open question.

8.7 Patterns on Shells of Tropical Molluscs

The shapes of mollusc shells and phyllotaxis - the arrangement of leaves on a growing shoot - were central subjects in the research of D'Arcy Thompson. His interest was focused more on the final geometrical arrangement than on the molecular basis of its emergence. In the last sections of this chapter I will show that the colour patterns on seashells and the arrangement of leaves on a growing shoot share essential elements: both are formed in narrow zones and the final pattern is a record of what happens in these zones in the course of time. Both patterns depend, therefore, on growth. Phyllotaxis and some of the mollusc patterns require closely related mechanisms: an autocatalytic reaction that is counteracted by two antagonists. The first is responsible for a pattern in time, the second for a pattern in space. That such very different patterning processes can be simulated by closely related models suggest a universality of the underlying mechanism - a result that perhaps would have found D'Arcy Thompson's interest.

The pigment patterns on mollusc shells are presumably without functional significance. And since nature was able to play, this has enabled an incredible pattern diversification in the different species. The basic patterns are lines perpendicular, parallel or oblique to the direction of growth and they can be accounted for by the same mechanism as described above: autocatalysis coupled with an antagonistic reaction (Meinhardt and Klingler, 1987). Activated cells are assumed to produce pigment. Keeping the space-time character in mind, it is clear that lines oblique to the direction of growth result from travelling waves. As is well known for waves of an epidemic, travelling waves arise if a slowly spreading self-enhancing process is coupled with a non-spreading antagonistic reaction. Neighbouring regions become infected, and the

Figure 8.6 Crossings of pigmentation lines: shell of *Tapes literatus* in front of a computer simulation. Crossings suggest that two waves do not annihilate each other but penetrate each other at a collision – an unusual behaviour for waves in excitable media. Model: a three-component system is assumed. An activator–antagonist system is tuned to produce a steady-state activation. An additional diffusible inhibitor causes the delayed extinction of activation in each cell due to the subsequently triggered neighbouring cells. At the point of collision, no new neighbours are available which could provide the signal for this extinction. The cells remain in the steady state until two backwards waves are initiated; for details see Meinhardt and Klinger (1987); Meinhardt, (1998).

slower antagonistic reaction brings the system back to the original state.

8.7.1 Formation of crossings by an additional diffusible inhibitor

On many species the patterns document an unusual behaviour of the generating system. For instance, on some species the oblique lines frequently branch. Branches result from a sudden trigger of a backwards-running wave. On other shells, crossings are visible, resulting from a penetration of waves (Figure 8.6). Again, this is unusual since waves in excitable media usually annihilate each other upon a collision. Since the waves don't penetrate into a region that is in the refractory phase, due to the counter wave, a simulation of this behaviour is possible under the assumption that two antagonists are involved. In the correct parameter range, the first (essentially non-diffusible) antagonist would keep a cell, once activated, in a steady state. Due to the activator diffusion, the activation can spread. The additional diffusible inhibitor extinguishes the activation of the preceding cell. Therefore, travelling waves result which appear normal. However, if two waves collide, the situation is different. No newly activated cell is available to extinguish the activation at the point of collision.

These cells remain in the steady state until the refractory period of the neighbouring region is over. These cells will become reinfected. Via the diffusible inhibitor, the newly activated cells extinguish the steady-state activation of the cells at the point of collision.

This mode of travelling wave formation is a remarkable reversal of the mechanism of stabilization of a single activator maximum by positive feedback as discussed above (Figure 8.3). For the penetration of travelling waves, it is necessary local *self-destabilization* by negative feedback that leads to a permanent displacement into a neighbouring region and ultimately to travelling waves.

8.8 Arrangement of Leaves and Staggered Dots on Shells: Two Corresponding Patterns

Some shells are decorated with pigment dots that are arranged along oblique lines (Figure 8.7). This pattern can be simulated again by assuming two antagonists: one causes the separation of the dots along the space-coordinate and the other along the time-coordinate.

The successive initiation of leaves behind the tip of a growing shoot, called *phyllotaxis*, seems to have nothing in common with any of the patterns on the seashells, but this impression is misleading. As in shells, the new structures on a growing shoot appear over the course of time. In plants, instead of a growing edge, there is a narrow zone next to the tip of the shoot. The tip itself, the so-called meristem, consists of undifferentiated, rapidly dividing cells. Cells just leaving this zone are able to form new leaves. According to classical models, the initiation of a new leaf is inhibited by existing leaves (Schoute, 1913). Therefore, a new leaf can be initiated only when a certain number of non-leaf cells have appeared after the last leaf. In this way, a certain distance is maintained between the sites of leaf initiation. According to the model, lateral inhibition on its own would be insufficient. The leaf, being in the centre of inhibition, would inhibit itself. Therefore, the lateral inhibition has to be complemented by the local self-amplification that is strong enough to cope with this cloud of inhibition.

In many plants, leaves are initiated along spirals. Seeds also frequently have a corresponding arrangement (Figure 8.7). Such patterns result if the last leaf and the penultimate leaf have a repelling influence on new leaves. This can be simulated with the basic model. However, since the inhibitor has to diffuse rapidly, the spatial information of the penultimate leaf on the positioning of a new leaf is minute. Therefore, this mechanism would not be robust against small perturbations, in contrast to the observations. There is a tendency in the simulations to revert to an alternating (distichous) or pairwise, 90^0-rotated (decussate) arrangement. Helical patterns can emerge very reliably if an inhibition in space *and* and an inhibition in time is assumed (Figure 8.7; Bernasconi, 1994; Meinhardt *et al.*, 1998). According to this model, it is not the inhibiting influence of the penultimate leaf that leads to the helical arrangement, but the long-lasting memory of cells or their progeny in the leaf-forming zone that their ancestors had once formed a leaf at this position. Since this memory is based on a non-diffusing substance, it remains localized.

Of course, the initiation of leaves is more complex. A local signal on its own would be insufficient to describe leaf outgrowth since this would lead to a rotation-symmetric

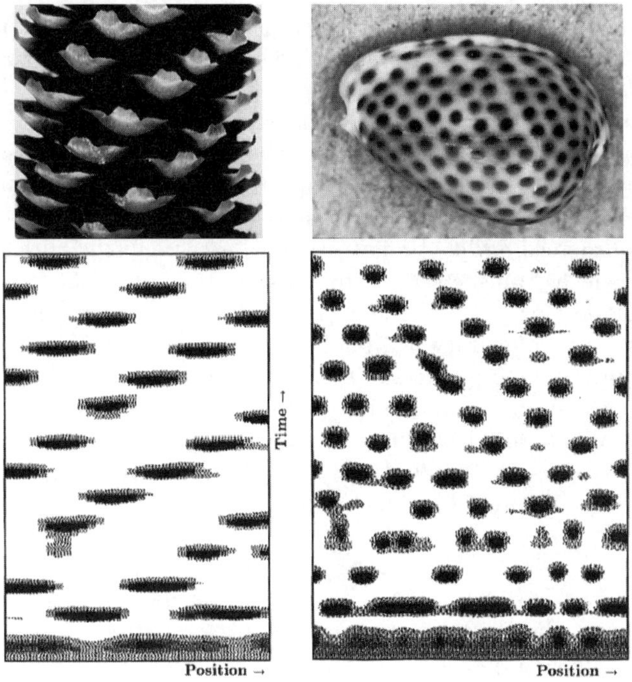

Figure 8.7 Helical arrangement of seeds on a fir cone and oblique rows of pigmented dots on the shell of *Persicula persicula*. Both patterns require a pulse-like activation in a short time interval. Each must occur at a somewhat displaced position in comparison with the preceding activation. It assumes the action of two inhibitors, one with a high diffusion rate and one with a long time constant. The plot shows the activator concentration in a linear array of cells as a function of time, on the right, with cyclic boundary conditions (Meinhardt, 1998). The calculation starts with homogeneous initial concentrations and small random fluctuations.

structure like a needle but not to a plane as given in a leaf. In extending a model for segmentation and limb initiation in insects (Meinhardt, 1983b, 1986b), it has been proposed that leaf formation requires a preceding subdivision (Koch and Meinhardt, 1994; Meinhardt, 1996). In this way the leaves would have two different surfaces from the outset, termed adaxial and abaxial (upper/lower or U/L). A similar proposal has been made by Waites and Hudson (1995). They described the gene *phantastica* in *Antirrhinum majus* that is required for dorsal (U) cell identity. Two solutions are conceivable: one is that during outgrowth, a periodic pattern is laid down, for instance a serial repetition of the cell specification $\ldots A, B, C, A, B, C \ldots$. A precondition for leaf initiation might be a C/A border, then both sites of the incipient leaf would have necessarily different determinations, C versus A. This mechanism would require a determination pattern with a ring-shaped geometry around the growing shoot. It would appropriate for plants such as bamboo with a nodal structure that surrounds the shoot like a belt. However, for a helical arrangement such as shown in Figure 8.7 the mechanism would require an unrealistically narrow sequence of such rings. An alternative would be that the signals shown in Figure 8.7 determine the lower side of the leaf and induce the signals for the upper side. Again, this would lead to a confrontation of two different determinations with a common border and thus to proper signals for leaf initiation.

A recent observation by Schichnes *et al.* (1997) provides support for such a model. In maize plants with a certain genetic background, supernumerary leaf blades sprout out from normal leaves at right angles, so-called flaps. The supernumerary blades appear in pairs at the abaxial side and have a normal adaxial/abaxial epidermal and internal organization. In the models outlined above, this will occur if cells with abaxial (L) specification are prone to transdetermine to adaxial (U) due to the genetic defect. This would lead to two new confrontations (LLL/UUU/LLL) at the lower side of a leaf and thus to two supernumerary leaf blades. If this view turns out to be correct, the process of leaf initiation on plants and the initiation of appendages in insects and vertebrates would become very similar in that borders play a decisive role (Meinhardt, 1983a, b; Koch and Meinhardt, 1994). In this way it is possible to achieve a leaf or a hand that remains flat in spite of the initial cone-shaped outgrowth, and the two faces of the plane can become different from each other.

8.9 Conclusion

Relatively simple molecular interactions can account for pattern formation during the development of higher organisms. The main steps that are postulated include the generation of positional information by a system of short-range autocatalysis and long-range inhibition and, under its control, the regional activation of different autoregulatory genes at particular positions. Segmentation requires cell states that exclude each other locally but activate each other over a long range. This generates a self-regulating neighbourhood. As in somite formation, this process may involve a stepwise conversion of an oscillating pattern into a stable periodic pattern.

Models are available for other important processes: the formation of filamentous branching structures such as the tracheae or blood vessels; and the initiation of new substructures such as legs and wings at particular positions within the organism. Although the discussion of these models goes beyond the scope of this article, it is fair to say that important steps in development have been correctly predicted by mathematically based models using cascades of relatively simple molecular interactions.

Many theoretical problems remain open. To mention only a few, the models describe only how existing cells can become different from each other. Cell-specific and position-specific control of cell division have to be included. Cells have been regarded as homogeneous building blocks. But they have a complex internal structure. How is this complexity achievable without local stabilization by autoregulatory genes? Perhaps biological pattern formation was invented during evolution for the orientation of cell division - an archetype of pattern formation. But how is this achieved? Enough problems to be getting on with!

REFERENCES

Bernasconi, G.P. (1994) Reaction-diffusion model for phyllotaxis. *Physica D* 70: 90–99.
Bünning, E. and Sagromsky, H. (1948) Die Bildung des Spaltöffnungsmusters in der

Blattepidermis. *Z. Naturforsch.* 3b: 203–216.
Culi, J. and Modolell, J. (1998) Proneural gene self-stimulation in neural precursors - an essential mechanism for sense organ development that is regulated by notch signaling. *Genes Dev.* 12: 2036–2047.
Davidson, D. (1983) The mechanism of feather pattern development in the chick. II. Control of the sequence of pattern formation. *J. Embryol. Exp. Morphol.* 75: 261–273.
De Robertis, E.M. (1995) Developmental biology - dismantling the organizer. *Nature* 374: 407–408.
Deuchar, E.M. and Burges, A.M.C. (1967) Somite segmentation in amphibian embryos: is there a transmitted control mechanism? *J. Embryol. Exp. Morphol.* 17: 349–358.
Gierer, A. and Meinhardt, H. (1972) A theory of biological pattern formation. *Kybernetik* 12: 30–39.
Grapin-Botton, A., Bonnin, M.A., Sieweke, M. and Le Douarin, N.M. (1998) Defined concentrations of a posteriorizing signal are critical for *MadB/Kreisler* segmental expression in the hindbrain. *Development* 125: 1173–1181.
Gurdon, J.B., Mitchell, A. and Mahony, D. (1995) Direct and continuous assessment by cells of their position in a morphogen gradient. *Nature* 376: 520–521.
Harland, R. and Gerhart, J. (1997) Formation and function of Spemanns organizer. *Ann. Rev. Cell Dev. Biol.* 13: 611–667.
Jung, H.S., Franciswest, P.H., Widelitz, R.B., Jiang, T.X., Tingberreth, S., Tickle, C., Wolpert, L. and Chuong, C.M. (1998) Local inhibitory-action of BMPs and their relationships with activators in feather formation - implications for periodic patterning. *Dev. Biol.* 196: 11–23.
Koch, A.J. and Meinhardt, H. (1994) Biological pattern-formation - from basic mechanisms to complex structures. *Rev. Modern Phys.* 66: 1481–1507.
Leptin, M. (1991) *twist* and *snail* as positive and negative regulators during *Drosophila* mesoderm development. *Genes Dev.* 5: 1568–1576.
Meinhardt, H. (1978) Space-dependent cell determination under the control of a morphogen gradient. *J. Theor. Biol.* 74: 307–321.
Meinhardt, H. (1982) *Models of Biological Pattern Formation.* Academic Press, London.
Meinhardt, H. (1983a) A boundary model for pattern formation in vertebrate limbs. *J. Embryol. Exp. Morphol.* 76: 115–137.
Meinhardt, H. (1983b) Cell determination boundaries as organizing regions for secondary embryonic fields. *Dev. Biol.* 96: 375–385.
Meinhardt, H. (1986a) Models of segmentation. In: *Somites in developing embryos* (R. Bellairs, D. A. Edie, J. W. Lash, Eds), Nato ASI Series A, Vol. 118, pp. 179–189. Plenum Press, New York.
Meinhardt, H. (1986b) Hierarchical inductions of cell states: a model for segmentation in *Drosophila. J. Cell Sci. Suppl.* 4: 357–381.
Meinhardt, H. (1989) Models for positional signalling with application to the dorsoventral patterning of insects and segregation into different cell types. *Development* (Supplement, 1989), pp. 169–180.
Meinhardt, H. (1992) Pattern-formation in biology - a comparison of models and experiments. *Reports Progr. Physics* 55: 797–849
Meinhardt, H. (1994) Biological pattern-formation - new observations provide support for theoretical predictions. *BioEssays* 16: 627–632.
Meinhardt, H. (1996) Models of biological pattern-formation - common mechanism in plant and animal development *Int. J. Dev. Biol.* 40: 123 – 134.
Meinhardt, H. (1998) *The Algorithmic Beauty of Sea Shells.* 2nd edn. Springer, New York.
Meinhardt, H. and Gierer, A. (1974) Applications of a theory of biological pattern formation based on lateral inhibition. *J. Cell Sci.* 15: 321–346.
Meinhardt, H. and Klingler, M. (1987) A model for pattern formation on the shells of molluscs. *J. theor. Biol.* 126: 63–69.

Meinhardt, H., Koch, A.J. and Bernasconi, G. (1998) Models of pattern formation applied to plant development. In: *Symmetry in Plants*, (D. Barabe and R. V. Jean, Eds). World Scientific, Singapore, pp. 723–758.

Moos, M., Wang, S.W. and Krinks, M. (1995) Anti-dorsalizing morphogenetic protein is a novel tgf-beta homolog expressed in the Spemann organizer. *Development* 121: 4293–4301.

Morgan, T.H. (1904) An attempt to analyse the phenomena of polarity in tubularia. *J. Exp. Zool.* 1: 587–591.

Palmeirim, I., Henrique, D., Ishhorowicz, D. and Pourquie, O. (1997) Avian *hairy* gene-expression identifies a molecular clock linked to vertebrate segmentation and somitogenesis. *Cell* 91: 639–648.

Pearson, M. and Elsdale, T. (1979) Somitogenesis in amphibian embryos. *J. Embryol. Exp. Morphol.* 51: 27–50.

Piccolo, S., Agius, E., Lu, B., Goodman, S., Dale, L. and De Robertis, E.M. (1997) Cleavage of chordin by *xolloid* metalloprotease suggests a role for proteolytic processing in the regulation of Spemann organizer activity. *Cell* 91: 407–416.

Regulski, M., Dessain, S., McGinnis, N. and McGinnis, W. (1991) High-affinity binding-sites for the deformed protein are required for the function of an autoregulatory enhancer of the deformed gene. *Genes Dev.* 5: 278–286.

Sander, K. (1976) Formation of the basic body pattern in insect embryogenesis. *Adv. Insect Physiol.* 12: 125–238.

Schichnes, D., Schneeberger, R. and Freeling, M. (1997) Induction of leaves directly from leaves in the maize mutant Lax midrib1-0. *Dev. Biol.* 186: 36–45.

Schoute, J.C. (1913) Beiträge zur Blattstellung. *Rec. trav. bot. Neerl.* 10: 153–325.

Simpson-Brose, M., Treisman, J. and Desplan, C. (1994) Synergy between the *hunchback* and *bicoid* morphogens is required for anterior patterning in *Drosophila*. *Cell* 78: 855–865.

Spemann, H. and Mangold, H. (1924) Über Induktion von Embryonalanlagen durch Implantation artfremder Organisatoren. *Wilhelm Roux' Arch. Entw. mech. Org.* 100: 599–638.

Thomsen, G.H. (1997) Antagonism within and around the organizer: BMP inhibitors in vertebrate body patterning. *Trends in Genetics* 13: 209–211.

Tomlinson, A. (1988) Cellular interactions in the developing *Drosophila* eye. *Development* 104: 183–193.

Turing, A. (1952) The chemical basis of morphogenesis. *Phil. Trans. Roy. Soc. Lond. B.* 237: 37–72.

Udolf, G., Lüer, K., Bossing, T. and Technau, G.M. (1995) Commitment of the CNS progenitors along the dosroventral axis of *Drosophila* neuroectoderm. *Science* 269: 1278–1281.

Van Doren, M., Powel, P.A., Pasternak, D., Singson, A. and Posakony, J.W. (1992) Spatial regulation of proneural gene activity: auto- and cross-achtivation of *achaete* is antagonized by *extramacrochaetae*. *Genes Dev.* 6: 2592–2605.

Waites, R. and Hudson, A. (1995) *phantastica*: a gene required for dorsoventrality in leaves in *Antirrhinum majus*. *Development* 121: 2143–2154.

Wigglesworth, V.B. (1940) Local and general factors in the development of 'pattern' in *Rhodnius prolixus*. *J. Exp. Biol.* 17: 180–200.

Wolpert, L. (1969) Positional information and the spatial pattern of cellular differentiation. *J. Theor. Biol.* 25: 1–47.

9

Diversity in Pattern and Form of Biological Systems and Evolution: A Theoretical Approach

Toshio Sekimura

College of Engineering, Chubu University, Kasugai, Aichi 487-8501, Japan

9.1 Introduction

Diversity in biological patterns and forms is a really fascinating issue. People ask the reason why diversity was generated during the history of evolution, but nobody really understands and knows about it. Through recent progress in biological theory and biological experiments, some specific aspects of the diversity have been well described. It is generally known that there are some key factors to help understand it. In this chapter, I review some of them briefly; I also summarise what progress has been made on understanding the diversity so far and what still remains unsolved.

We can see a rich variety of patterns and forms in both plants and animals. One of the most fascinating and unsolved problems in biology is to understand how and why the diversity in pattern and form was generated during the long history of evolution. Because of its complexity, progress in modelling this problem had been unremarkable until recently. However, due to striking progress both in theory and computing facilities, a theoretical approach now seems to be developing very rapidly. Here I present an overview of recent progress on the theoretical approach to diversity in pattern and form of biological systems and its relation to evolution.

On Growth and Form: Spatio-temporal Pattern Formation in Biology.
Editor Mark A.J. Chaplain, G.D. Singh, J.C. McLachlan

9.2 Possible Factors in Determining Pattern and Form

It is generally known that there are some key factors which underpin pattern and form formation in biological systems (Seilacher, 1970). I now consider briefly some of the key factors or aspects.

9.2.1 Gene activity

Homoeotic genes have recently been found in many animals and plants from studies of gene expression in molecular biology, developmental genetics e.g. the hedgehog gene in *Drosophila* and floral homoeotic genes in *Arabidopsis* (Meyerowitz *et al.*, 1991). These genes encode secreted signals which finally result in protein gradients that are responsible for pattern and form formation in specific regions. Based on these findings, what was previously a hypothetical "morphogen" has become a realistic substance.

According to Waddington's definition (Waddington, 1975), " morphogenetic change in organisms result from the change in developmental programmes of genes in phylogeny". Of course, changes in the programme here do not only mean changes in gene expression, but also include changes in self-organisation of cells, cell movement, etc. A theoretical approach to the diversity in pattern and form should be made based on the new findings on homoeotic genes mentioned above.

With respect to gene activity, we must not discard the so-called developmental constraints which refer to a historical aspect of gene expression. No organism could appear abruptly without former ancestors. This means that all organisms have restrictions on morphological changes both in ontogeny and phylogeny. This factor seems to suggest that the problem is one of uncertain events during the history of evolution.

9.2.2 Physico-chemical aspect

This captures biological patterns and forms based on the laws of physics and chemistry. Organisms on the earth are certainly under physico-chemical restrictions: efficiency in biochemical reactions, materials to construct bodies, gravity, temperature, solar radiation, etc.

The idea of D'Arcy Thompson, author of *On Growth and Form*, is very remarkable and noteworthy (Thompson, 1917). He emphasised that present patterns and forms can be explained reasonably by well-known physico-chemical laws without taking into account historical factors. Along this line, Goodwin (Webster and Goodwin, 1996) has developed his idea that morphogenesis is characterised by symmetry-breaking cascades that generate forms which fall into hierarchically nested equivalence classes. He further notes, "This leads to a morphogenetic definition of homology that is based solely on generative criteria and is independent of evolutionary history" (Webster and Goodwin, 1996).

Recently, rapid progress in modelling biological pattern formation has been based on physico-chemical laws, both qualitatively and quantitatively. This is mainly due to Turing's discovery of diffusion-driven instability in reaction-diffusion equations (Turing, 1952). Murray (1981) applied the theory successfully to colour pattern formation on animal skin surfaces such as the leopard, tiger, etc. In his models he showed that non-uniform stable patterns of chemicals such as pigments could be generated from random perturbations around the initial spatially uniform steady-state pattern in two-dimensional domains, as a result of physico-chemical processes (Murray, 1981).

Odell *et al.* (1981) presented a mechanical model for the folding and movement of cell sheets based on hypothetical mechanical properties of the cellular cytoskelton, which is applicable to certain elementary types of gastrulation and neurulation (Odell *et al.*, 1981). And in 1997 Weijer's group presented a hydrodynamical model for aggregation and mound formation in *Dictyostelium discoideum* (Vasiev *et al.*, 1997; chapter 11 in this book). In the model, cell movement is modelled as the flow of a compressible fluid controlled by cAMP-induced chemotactic forces, frictional and adhesion forces and internal pressure. Direct comparisons of simulations with experimental images of successive aggregation stages show striking agreement.

9.2.3 Ecological aspect

This is an attempt to capture morphological changes in organisms through processes of natural selection, in which only organisms adapted to their surrounding environments could survive. This means that any unsuited organisms could not survive for a long time, even if they could exist from purely physico-chemical and gene activity aspects.

Until recently the ecological approach to biological pattern and form has not been well developed. However, based on modern theories developed in evolutionary ecology, social biology, etc., new progress in this approach is strongly expected in the near future. Modern ecological theories such as *inclusive fitness* (Hamilton, 1964) or *evolutionary stable strategy* (Maynard-Smith, 1974), which are based on game theory and optimization theory, have so far been applied successfully to quantitative problems such as altruistic behaviour and mating behaviour in animals. These theories are candidates for application to spatial problems such as morphological evolution in plants and animals, although mainly they deal with the dynamics of gene frequency for maximal reproduction.

9.2.4 Biological function

This is an attempt to capture morphological changes in pattern and form from the biological function of organisms. It is based on the idea that any tissue, organ or structure of an individual should play its role as part of a well-organised biological system. Morphological changes in ontogenesis and any phylogenetic change also need to be well coordinated. Biological function in this case also includes an increase in adaptability to the environment.

In the book *The Triumph of the Embryo* (Wolpert, 1991), Wolpert has described the reason why gastrulation in multicellular organisms evolved successfully. As shown in Section 9.2.2, it is certainly true that we could explain gastrulation from the aspect of Newtonian mechanics. On the other hand, Wolpert has assumed that by having a hole available in which to keep food, the efficiency of organisms to digest the nutrition is remarkably improved, and as a result gastrulation is programmed in gene expression. This example seems to show how collaboration between key factors (here physical and functional factors) does work to produce morphological changes in evolution.

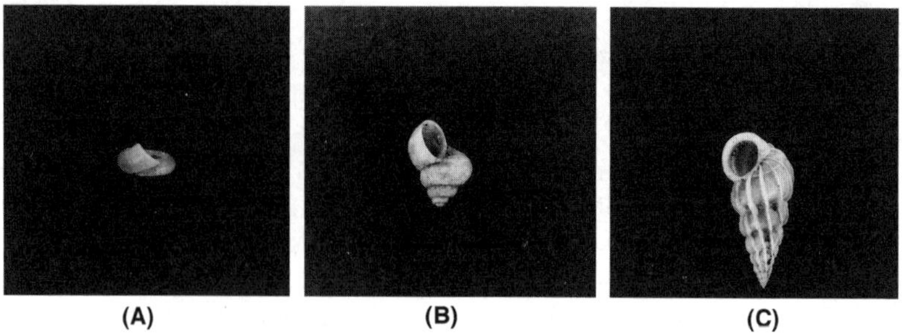

Figure 9.1 Examples of coiled shells (A) *Spirostoma nakadai*, (B) *Tropidophora legata*, (C) *Epitonium (Gyroscala) perplexum*. (courtesy of Dr R. Morita of the Natural History Museum and Institute, Chiba)

9.3 Pioneering Work

Due to the interdisciplinary character of the problem, there are very few articles to deal with diversity in biological pattern and evolution. However, we can find some pioneering works from a theoretical or computational viewpoint, which I review briefly in the following.

9.3.1 Shell shape

Using a logarithmic growth model, Raup (1966) demonstrated computationally that a spirally coiled shell shape can be produced by four parameters, W, D, T and S, where W is the expansion rate of the generating curve per revolution, D is the distance between the generating curve and the coiling axis, T is the translation rate of the generating curve along the axis per revolution, and S is the shape of the generating curve. These parameters are expected to relate to gene activity regarding growth rate of shell. He showed successfully that change in parameters in three-dimensional space (W, D and T) with fixed S value could regenerate four taxonomic groups having a coiled shell. This suggests that a large group of animals with a coiled shell have almost the same way to grow with different growth rates (Raup, 1966). Figure 9.1 shows some photographs of coiled shells.

9.3.2 Plant evolution

Niklas (1986) how it is possible to simulate plant evolution by using a parametric model. He succeeded in constructing plants on a computer by using three parameters that represent three particular morphological characteristics: the branching angle (the angle between adjacent branches), the probability of branching, and the permitted range of rotation angle between the plane defined by a pair of new axes and the plane formed during the previous generation of branching. He has hypothesised that plants with branching patterns that gather the most light can be predicted to be the most successful. Consequently, changes in the plant's shape that increase the ability to gather light should confer competitive advantages. A three-dimensional or cubical space of these three parameters represents a "universe" of possible branching patterns.

Each point within the cube represents a "species" of plant with a unique combination of characteristics. In this way, he showed that a desktop computer can recreate the major trends in plant evolution (Niklas, 1986)!

Cummings and Strickland (1998) presented a model of phyllotaxis based on physical laws. The initial pattern of the leaf primordia is known to arise at the meristem. Cells of an epithelial sheet at the meristem are envisaged as joined at their lateral surfaces by adhesive connections, biochemicals. They estimated theoretically the energy of the cell surface molecules, and by using a minimisation principle with respect to the energy, they derived the resultant equations of the model which are the familiar Helmholtz and Laplace equations of mathematical physics. And by using these equations, they have reproduced theoretically diverse phyllotacic patterns, including three main types – spiral, distichous and decussate – in one simple algorithm (Cummings and Strickland, 1998).

9.3.3 Animal skin and insect wing

Murray (1981) examined the effect of animal body surface scale on the spatial patterns formed by a reaction-diffusion system. He showed that a change in a parameter corresponding to body scale generates a rich variety of patterns which remarkably correspond to real patterns on animal skin surfaces in nature (Murray, 1981). In this case the parameter corresponding to body scale in the reaction-diffusion equation is supposed to relate to genes that control animal growth.

Nijhout (1998) suggested that in the theoretical study of colour patterns in butterfly wings is incorporated into an analytical model (based on differential equations) by assuming that the reaction constants represent gene activity. Genes control the rate of a reaction or, in a threshold system, whether a reaction will proceed or not. This is reasonable since many genes code for proteins which have enzymatic activity affecting biochemical reaction rates within the cell (Nijhout, 1998). The day-flying butterfly *Papilio dardanus* is widely distributed across sub-Saharan Africa, and it is well known that a spectacular polymorphism has evolved through differences between seasonal forms, between males and females, and between different geographic isolates. The females have evolved wing colour patterns that mimic several different species of unpalatable danaids. They are profoundly different and hardly recognisable as members of a single species (Figure 9.2).

9.4 Summary and Conclusion

As noted in Section 9.2, certain key elements are required to understand diversity in pattern and form of biological systems and its relation to evolution. It is obvious that these factors are not independent of the others; rather, they are closely related as a whole. However, researches in theory and experiment seem to have proceeded almost independently, in spite of the comprehensive or interdisciplinary characteristic of the problem. As a result, there have been very few works or articles on this fascinating problem.

Living organisms are generally overcoming some key conditions (genetic, physico-chemical, ecological, etc.) to survive in a given environment. Regarding diversity in biological patterns and forms, it would be necessary to consider the problem along

Figure 9.2 Female wing patterns of *Papilio dardanus*: (A) *cenea*, (B) *planemoides*, (C) male-like female. (Courtesy of Drs A.P. Vogler and A. Cieslak of the Natural History Museum, London)

this line, i.e. what conditions the organisms satisfy for living. It no longer seems sufficient to consider the problem only from a specific view-point. Recently, there has been rapid progress in modelling biological pattern formation, both qualitatively and quantitatively. It seems to me that each discipline (molecular biology, mathematical biology, ecology, paleontology, etc.) is now mature enough to interact with the others.

Acknowledgments
I thank Professors B. Goodwin and P. K. Maini for their critical reading of the manuscript and their helpful comments. At the D'Arcy Thompson meeting in memory of the 50th anniversary of his death, held in Dundee during September 1998, I had the opportunity to organize a workshop on theoretical approach to diversity in biological pattern and evolution. It was an attempt to promote information exchange among various scientists working in different fields. I also thank Professor H. Honda and Dr S. Kondo for agreeing to give their talks as part of the workshop. This work was supported in part by a grant from the Human Frontier Science Program (RG 0323/1999-M) and an EPSRC Visiting Fellowship Grant GR/M81878.

REFERENCES

Cummings, F.W. and Strickland, J.C. (1998) A model of phyllotaxis. *J. Theor. Biol.* 192: 531–544.

Hamilton, W.D. (1964) The genetical evolution of social behaviour I and II. *J. Theor. Biol.* 7: 1–16 and 17–32.

Maynard Smith, J. (1974) The theory of games and the evolution of animal conflicts. *J. Theor. Biol.* 47: 209–221.

Meyerowitz, E.M., Bowman, J.L., Brockman, L.L., Drews, G.N., Jack, T., Sieburth, L.E.,

and Weigel, D. (1991) *Development Supplement* 1: 157–168.
Murray, J.D. (1981) On pattern formation mechanisms for lepidopteran wing patterns and mamalian coat markings. *Phil. Trans. Roy. Soc. Lond.* 295: 473–496.
Nijhout, F. (1998) private communication. See also *The Development and Evolution of Butterfly Wing Patterns* (1991) Smithsonian Institution Press, Washington.
Niklas, K.J. (1986) Computer-simulated plant evolution, *Scientific American* 254: 68–75.
Odell, G., Oster, G.F., Burnside, B., and Alberch, P. (1981) The mechanical basis for morphogenesis. *Dev. Biol.*,85: 446–462.
Raup, D.M. (1966) Geometric analysis of shell coiling: general problems. *J. Paleontol.* 40: 1178–1190.
Seilacher, A. (1970) Arbeitskonzept zur Konstruktion Morphologie. *Lethaia* 3: 393–396.
Thompson, D.W. (1917) *On Growth and Form.* Second edition, 1970, Cambridge University Press, Cambridge.
Turing, A.M. (1952) The mechanical basis of morphogenesis. *Phil. Trans. Roy. Soc. Lond.* 237: 37–72.
Vasiev, B., Siegert, F., and Weijer, C.J. (1997) A hydrodynamic model for *Dictyostelium discoideum* mound formation. *J. Theor. Biol.* 184: 441–.
Waddington, C.H. (1975) *The Evolution of an Evolutionist.* Cornell Univ. Press, Ithaca, NY.
Webster, G. and Goodwin, B. (1996) *Form and Transformation: Generative and Relational Principles in Biology.* Cambridge University Press, Cambridge.
Wolpert, L. (1991) *The Triumph of the Embryo.* Oxford University Press, Oxford.

10

Developmental Morphologies Not Directly Specified by the Genome of the Individual

John C. McLachlan

School of Biomedical Sciences, University of St Andrews, Bute Building, University of St Andrews, St Andrews KY16 9TS, Scotland, UK

10.1 Introduction

Heredity in general and genes in particular were of little interest to D'Arcy Thompson. His view was that there were underlying processes in morphology which were universal and independent of the medium in which they appeared. Yet in recent times, the discovery of segment polarity, segment identity, and paired genes in *Drosophila*, and their conserved equivalents in all animals, has demonstrated clearly a strong relationship between genes, morphologies and evolution. In an overview of Thompson's contribution to modern biology, it is therefore necessary to address this problem directly. Was D'Arcy simply wrong on this matter? Or are there aspects of morphogenesis which have a non-genetic component? Genes are the major units of heredity, and stable morphologies are heritable. However, it does not necessarily follow from this that morphologies are specified in their entirety by the genome. Knowledge of the complete sequence of the genome of an individual may still leave it quite unclear how that individual attained its adult morphology. Understanding of many other factors and mechanisms may be required. Conventionally, it often seems to be taken for granted that the instructions for development are specific (i.e. complex, direct and deterministic), and that this specification is carried out by the genome of the individual during development. Individual variation, arising from environmental influences is widely appreciated as being important in evolution (Stearns, 1989; J.D. Thompson, 1991). However, such plasticity of response is generally ascribed to the interaction of the environment directly with the genotype of the organism, giving rise to "norms of reaction" (Schmalhausen, 1949). I wish to argue that, on the contrary, the instructions which result in individual morphologies may be non-specific (i.e. simple, indirect and stochastic), and may not arise from the genome of the individual. Variation may

On Growth and Form: Spatio-temporal Pattern Formation in Biology.
Editor Mark A.J. Chaplain, G.D. Singh, J.C. McLachlan

arise as an immediate consequence of the operation of these indirect and stochastic mechanisms. The environment, including cell–cell interactions, is a component of the morphogenetic mechanism, rather than a separate factor acting to blur the genotype. One advantage of this approach is that it makes variability an inevitable consequence of developmental mechanisms, rather than a secondary phenomenon imposed later in the course of morphogenesis. Mathematical modellers may welcome release from the requirement that their models must have robust and consistent outcomes, or may even incorporate stochastic elements into future models. Study of human anatomy reveals there are considerable variations from individual to individual, especially in tissues like blood vessels.

I attempt to describe briefly all the different kinds of mechanisms which lead to morphologies not specified in their entirety by the individual's genome, identifying eight such classes in all. Although some of the examples are familiar, I have marshalled them together so they may lend mutual support to the overall argument.

Four of these classes do not derive from the genome, though they may act upon them. These are (1) physical and geometrical constraints, Section 10.2, (2) non-genetic methods of inheritance, Section 10.3, (3) environmental specification, Section 10.4 and (4) self-generating patterns arising from initially homogeneous starting states, Section 10.5. One mechanism (5) derives from the parental genome, rather than the genome of the individual, Section 10.6. Three derive from the genome, but do not directly specify developmental morphologies; in other words, the mechanism is defined by the genome but not the outcome. These are (6) tissue self-assembly, Section 10.7, (7) tissue interactions, Section 10.8 and (8) specification of morphologies by use, Section 10.9.

While it is a historical commonplace to note that with a few exceptions the genome is identical in all cells in an individual, and that the genome itself cannot therefore initiate differences between cells, the implications of this seem largely neglected. Differential gene expression must be initiated, at least in the first instance, by factors other than the genes. However, I do not argue that these factors act independently of the genome: they act on the genome or with the genome. They are merely not controlled by the genome.

The mechanisms I discuss operate in phylogeny as well as ontogeny. The morphologies produced by the interaction of the environment with the developmental programme are the raw material on which selection operates, and therefore these mechanisms play a vital part in the evolutionary process. As long as the mechanisms establishing morphogenesis remain constant, as they always do in the case of geometrical constraints, and frequently do in the case of environmental constraints, then the product of the interaction will be stably heritable, as required for the operation of natural selection. Where they change, new morphological variants may be generated rather quickly.

In 1894 William Bateson published *Materials for the Study of Variation*, in which he argued that such study would lead to elucidation of important biological phenomena. It was indeed study of the anomalous phenomenon of homoeosis (a term coined by Bateson) that led to the discovery of homoeotic genes, which has done so much to demonstrate the common genetic mechanisms underlying the development of all living things. Bateson's studies of variation might therefore be interpreted as the ultimate triumph of genetic determinism in development. It would be pleasing if, in addition, studies of variation proved to be the pathway to elucidating non-genetic mechanisms which also exert profound influences on development and evolution.

10.2 Physical and Mathematical Constraints

As D'Arcy Thompson (1942) indicated, organisms are subject to various physical and mathematical laws, such as close packing and minimum energy considerations. These may significantly influence adult form. The outcomes of interaction between genes and such constraints may be predictable, but they are not determined by the genes. For example, in some instances, an observer from another planet would be unable to predict morphologies from genes without intimate knowledge of the magnitude of physical forces in operation on Earth. In the extreme case, evolution may only be able to select between discrete permitted morphologies: it may be unable to alter their form in any way.

Thompson paid particular attention to close packing, as seen in eggs after early cleavage events (and also in the scutes of tortoise shells; Stevens, 1974). There are many other examples of such phenomena, however. Euler classified polyhedral solids on the basis on the numbers of vertices and faces they possessed. D'Arcy Thompson pointed out that the external skeletons of radiolarians abide by this classification, and it has subsequently become clear that brochosomes - particles excreted by plant-eating insects - also adopt the same morphologies, even though they are far smaller. At a smaller scale still, viruses and molecular structures adopt the same deployments, acting under the same physical, non-genetic rules (Galloway, 1988).

Another of D'Arcy Thompson's interests lay in the coiling of shells (Cook, 1914). He was fascinated by the equiangular spiral, that most elegant of mathematical forms, which is generated in many biological systems as a consequence of constant proportional growth. This constructs a form which is identical in morphology at any scale. Thompson believed that any shell could be described by a few mathematical parameters, although his original listing has been superseded by that of Raup (1962, 1966). Raup indicated that shell coiling can be described by four parameters. These are (i) the shape of the generating curve (essentially the opening of the shell), (ii) the rate of expansion of the generating curve, (iii) the degree of overlap between successive rotations of the curve, and (iv) the distance between the axis and the generating curve. These are constants for an individual species.

Complex mathematical relationships also appear in plants (Stevens, 1974). Physical constraints, such as considerations of close packing, may explain their form. But there may be more complex interactions. In sunflowers (*Compositae*) the disc florets grow in two sets of spirals, one clockwise and one anticlockwise. The number of spirals in one set relates to the number in the other as members of the Fibonacci series. Douasy and Couder (1992) have demonstrated that magnetisable oil drops falling in a uniform magnetic field (and hence becoming magnetised themselves) can be deployed in just such an arrangement at particular field and timing parameters. This does not prove of course that sunflower florets are patterned by magnetism, but it does demonstrate that a purely physical system can establish exactly the same complex pattern as a biological one. The key component in determining the distribution of the drops is energy minimisation. Of course, this does not establish the exact nature of the interaction in plants, but it is pregnant with possibilities.

This class of explanation may be contrasted with adaptive explanations; for instance, that leaf deployment patterns are determined by considerations of maximisation of insolation. Energy considerations are of course also adaptive, but at the level of selection of developmental programmes, not of adult forms.

Branching patterns (highly ordered) appear in many physical and biological settings, as in blood vessels, lung, trees and bushes, lightning forks, and river tributary beds. These have many underlying features in common (Stevens, 1974), which reveal that they are consequences of non-genetic factors.

Another physical system generating patterns which seem familiar to biologists is that of stripes in stone fields (Werner and Hallet, 1993). Here slopes on Arctic and alpine hills subject to periodic freezing generate patterns in which bands of large stones alternate with bands of small stones in stripes running up and down the field. Sorted stone circles are also a possible outcome of physical mechanisms (R.M. Anderson, 1993). A conventional biologist looking at such patterns might be confident that they were (a) specified by DNA and (b) of survival value to the adult landscape.

Fractals offer another way of approaching the problem of form which would surely have fascinated D'Arcy Thompson (Meinhardt, 1979; West and Goldberger, 1987). Their power lies in their ability to generate complex morphologies from very simple algorithms. It may be no coincidence that one of the major practical applications of fractals is in describing complex morphologies (e.g. in storing photographs on computer disks). It would be satisfying if this turned out to be related to the mechanisms by which the form was generated in the first place.

10.3 Morphologies Arising from Non-genetic Mechanisms of Inheritance

Extrachromosomal inheritance is well documented (Jinks, 1964), and many forms of this are not mediated by the genome. These include inheritance of cellular organelles, cytoskeletal elements and membrane structures. One example is the inheritance of induced phenotypic changes in single-celled organisms such as *Tetrahymena* and *Paramecium* (Beisson and Sonneborn, 1965; Nelsen *et al.*, 1989). These have organised rows of cilia on their surface, which are asymmetric. Handling manipulations can result in the disturbances of the pattern of these rows: for example, a membrane patch may be effectively rotated through 90° in particular individuals. This rotated patch is then stably inherited through subsequent generations, even though the DNA of the organism is unaffected. Indeed, the nature of genetic exchange in these organisms is such that the original DNA is soon diluted; however, the phenotype remains unaffected.

This phenomenon is believed to arise from direct membrane replication on the template of existing membranes, so that alterations can be propagated directly. A similar effect may underlie those diseases apparently transmitted by self-replicating proteins or prions, without the intervention of DNA or RNA. This indicates that, at least in principle, membrane-template reproduction is not confined to unicellular organisms. These diseases include the sheep disease scrapie and its analogues in cattle and humans, bovine spongiform encephalopathy and Creuzfeldt-Jacob disease, respectively (Pruisner, 1991).

10.4 Environmental Specification

Factors in the environment may provide cues to development. These may determine polarity, or determine the choice between different possible differentiation states. In these cases the information necessary for development is not present on the genome, and development cannot proceed without information imparted by the environment. Such environmental influences may derive from physical or cellular interactions.

The spherical egg of the marine alga *Fucus* settles on the sea floor, and extends a process in one particular direction (Jaffe, 1968). An oriented cell division then occurs to create two different cells, one of which gives rise to the rhizoid (holdfast) and the other to the thallus (photosynthetic organ) progenitors. Extension of the initial process is therefore the key axis-determining step. The direction of extension is determined by the direction of light, current flow, and the proximity of other eggs, in ways that plausibly have adaptive value. However, in experimental situations, chemical and electrical gradients can also determine the axis, so it appears extremely responsive even in an artificial environment. It has been postulated that membrane particles migrate in the plane of the membrane to establish this polarity.

Polarity also arises in *Xenopus laevis* eggs from environmental factors. The "grey crescent", signalling the future site of gastrulation, appears after fertilisation. Since gastrulation marks the formation of the main body axis, this is the key step in going from a radially symmetrical structure to the bilateral symmetry of the adult animal. However, the grey crescent forms at 180° to the point of sperm entry (Gerhart *et al.* 1981). So the main body axis is determined by a cue from the environment, which is indeed random and indeterminate since it can occur anywhere along the equator. In addition, fertilised *Xenopus* eggs also adopt a particular orientation with respect to gravity. If the egg is rotated so that the sperm entry point is up, then reversal or twinning of the axis can occur (Gerhart *et al.*, 1981). Hence the position of the main body axis can be reset by the physical environment in the form of gravity. In "low gravity" rotating devices, where the net gravity vector is close to zero, the eggs develop normally (Neff *et al.*, 1985). This indicates, perhaps, that several separate mechanisms operate, giving multiple redundancy during such an important step.

A similar phenomenon occurs in early mammalian embryos, where the first differentiation decision involves the appearance of the centrally placed inner cell mass (ICM) cells, and the peripheral trophectoderm cells, from the undistinguished morula. This first differentiation decision is made under environmental influence. Cells with many cell contacts become inner cell mass, and cells with few cell contacts become trophectoderm (Hillman *et al.*, 1972). Cells removed from all cell contacts make no differentiation decision, but re-enter division until an appropriate number is reached to provide the required cues for differentiation.

A special version of environmental influence is where the developmental programme has pre-established a choice of forms, and the environment selects between them. It is customary to distinguish between the continuous variation of reaction norms, and discontinuous variation of polyphenisms, but there is no particular reason to assume that different mechanisms need be involved.

Examples of such alternatives are familiar. Studies of environmental determination of sex are well known (Woodward and Murray, 1993). Many other examples exist. The geometrid moth *Nemoria arizonaria* has a spring "catkin-like" form and a summer "twig-like", initiated by dietary tannin levels (Green, 1989). Many examples of this type of developmental response to environmental factors are found in insects. Wing polyphenisms in grasshoppers, locusts, aphids and others are particularly well marked (Dingle, 1985; Green, 1989; Ritchie *et al.*, 1987; Wagner and Liebherr, 1992). Physical factors influencing morphology include temperature, photoperiod, population density, moisture and food quality. In Crustacea, *Daphnia retrocurva* exists in a rounded spring form and a spined summer form. The summer form, although supporting fewer eggs through reduction in the size of the brood chamber, is probably more resistant to the abundant summer predators (Dodson, 1988).

Plants also show environmental plasticity (Clausen *et al.*, 1948). For instance, arrowleaf plants growing on land have rigid leaves covered with a waxy cuticle that reduces water losses, whereas those under water have flexible leaves lacking this cuticle, enabling the plant to absorb mineral nutrients directly.

A less extreme example at the whole animal level is found in those salamanders that metamorphose in warm ponds but are neotenous in cold water, such as mountain lakes. This may be because the hypothalamus/pituitary/thyroid axis is retarded by cold (Jenkin, 1970) or because thyroxin is less effective in the cold (Dent, 1968). Neoteny may be advantageous in cold environments, so this phenomenon may have adaptive value. (On the other hand it may not.)

Environmental factors may demonstrably lead to discontinuous variation. For instance, vertebral number in mice might be thought of as a constant for mathematical modelling purposes. However, not only is there a small degree of natural variation, but vertebral numbers will increase their range markedly under the influence of an environmental factor which affects growth (Gregg and Snow, 1983).

10.5 Self-establishing Interactive Mechanisms

The competing claims of environmental and maternal factors in development have almost acquired the mythic status of the roles of Nature and Nurture in promoting human intelligence. In particular, they are sometimes seen as constituting a Universe of Possibilities, so that disproof of one counts as proof of the other. However, in 1952 the mathematician Alan Turing suggested an alternative means by which order could be generated from initially homogeneous, non-polar, systems. He described reaction-diffusion models (Turing, 1952), in which a diffusible autocatalytic activator, also producing its own diffusible inhibitor, could move from a homogeneous apolar state to a consistent polar stable state following *any* perturbation. The key criterion for such a system is that the inhibitor diffuses faster than the activator. Turing also indicated that physical forces could serve as equivalents of diffusion and inhibition, so that non-chemical versions were also possible.

Such mechanisms could be viewed as establishing a polar prepattern which can then be interpreted by cells to give rise to varying classes of behaviour. As the field over which the mechanism operates increases in size, more complex outcomes arise, such as stripes, spots and peaks. These outcomes show generic similarities with each other, though there is a degree of indeterminacy in detail.

Although these mathematical models require an initial perturbation in departing from the initial state, they differ from environmental establishment of polarity as seen in *Xenopus* and *Fucus* eggs (see above), in that the outcome takes a consistent form which is independent of the place or size of the initial stimulus. At first, this seems to defy the second law of thermodynamics, since an information-rich pattern apparently emerges spontaneously from homogeneity. However, they can only operate in open systems, with the provision of an energy substrate. This is made clear by the discovery of several classes of chemical reaction which behave in an analogous manner (Zaikin and Zhabotinsky, 1970; Epstein *et al.*, 1983; Welsh *et al.*, 1983; Epstein, 1984). Such chemical dissipative systems can lead not merely to complex morphologies, but also to complex dynamic changes. For instance, not only may stripes appear in the chemical medium, but "spots" may form which grow until they reach a consistent size, and then divide in two (Pearson *et al.*, 1993; Lee *et al.*, 1994).

Evidence can be found of biological systems that appear to operate in an analogous manner (Meinhardt, 1982; Lechleiter *et al.*, 1991; Kondo and Asal, 1995). It has been suggested that such phenomena may have been important in the evolution of life itself (Eigen and Schuster, 1979; Boerlijst and Hogeweg, 1991; May, 1991). This would neatly resolve the problem of how replication could commence at a time before nucleic acid molecules had error correction mechanisms in place.

Since the initial conceptual breakthrough provided by Turing, other models with similar characteristics have been devised. These often focus on physical forces (Oster *et al.*, 1983; Harris *et al.*, 1984; Oster and Odell, 1984) and can give rise to a variety of "biological" forms and behaviours, such as programmed deformation cylinders, in a manner reminiscent of neurulation. The profoundly important biological process of segmentation can also be modelled mathematically in a way which generates reliable and stable outcomes and which appears to correspond to the results of experiments on embryos (Saunders and Ho, 1995).

One particularly interesting class of self-generating models has been proposed by Murray and his co-workers (J.D. Murray, 1988), where cell migration and adhesion serve as equivalents of diffusion and inhibition. Such models can be shown, at least in principle, to be capable of generating prepatterns which can be interpreted in a number of ways for instance, by striping and spotting morphologies (J.D. Murray *et al.*, 1990). These patterns form striking analogies to pigmentation patterns in animal coats, and make certain predictions which appear to be consistent with observed biological patterns (Cruywagen and Murray, 1992; Cohen and Stewart, 1993). Such mechanisms could also underlie the formation of fingerprints, and since all such models in principle show a degree of indeterminacy, this could explain the intriguing observations that identical twins have different fingerprints (Cummins and Midlo, 1943), and that whorls can form spontaneously in fibroblast cultures (Elsdale and Wasoff, 1976).

Perhaps pigmentation patterns are the best-known examples of such phenomena merely because we can see them so clearly. Many other developmental phenomena may operate by such means, but lack a visible expression that we can recognise. Even spots and stripes may represent a rather trivial by-product of an earlier profound patterning step, rather than the endpoint. This would suggest that they might have no adaptive value in the adult, although adaptive just-so stories can always be told about them. Of course, once these structures have arisen, they may then be selected for another purpose, such as species recognition, and as part of sexual selection. This acquisition of a secondary role for the pattern may have operated in zebras, for instance, as a means of speciation.

Such mechanisms need not operate without gene involvement. Genes could provide the chemical constituents and the means to supply the energy substrate - the chemicals would then provide the pattern. The biological pattern would be parasitic on the chemical information stored in the chemical reaction, just as biology might parasitise energy implicit in minimum energy considerations.

10.6 Parental Specification

An individual's phenotype may be influenced by factors derived from the genes of its parents, rather than by its own. This can operate by maternal effects, or by genomic imprinting.

The actions of maternal effect genes and mutants are well known (Humphrey, 1966; Brothers, 1976; Freeman and Lundelius, 1982; K.V. Anderson and Nusslein-Vollard, 1984; Mohler and Weischaus, 1986). The construction of the egg is the prerogative of the mother, and hence all those phenomena which spring from regionalisation of egg components arise from the maternal genome. More recently discovered is the phenomenon of genomic imprinting (Surani, 1994). Formerly, it was believed that there was no consistent difference between an allele inherited from the mother or father. However, it is now known that a variety of genes show genomic imprinting. For instance, in post-implantation mouse embryos, the gene for insulin-like growth factor 2 (IGF-2), located on chromosome 7, is found to be expressed only in the paternally derived copy. Conversely, the receptor for this growth factor (located on chromosome 17) is only expressed in the late fetus when inherited from the mother. A normal individual with one paternal IGF-2 gene and one maternal receptor gene will have the appropriate pattern of expression of, and response to, this growth factor. An individual with two copies of either gene coming from one parent (resulting from a chromosome exchange accident) will either not express the growth factor properly, or not express the receptor properly, and hence have an abnormal pattern of growth. Here it is not the genes that the individual possesses that control development, but rather the environmental history of these genes.

It might be argued that, since maternal and paternal effects are ultimately specified by the genome (even though not by the genome of the individual), they are not appropriate to this argument. However, there is an important difference between the genome of the individual and the genome of the individual parents. Parental genes may be under different selection pressures and they may accumulate different mutation rates over the two generations across which they act; this may have important implications for biology.

10.7 Tissue Self-assembly

Tissue morphogenesis is often assumed to arise from high-level controls, i.e. mechanisms operating at the level of the tissue itself. However, Steinberg (1970), Steinberg and Poole (1981) have argued persuasively that differential adhesivity of particular cell types can lead to "sorting out" in ways which mimic the deployment of tissues in the body. This indicates directly how low-level properties of cells could lead to higher levels of organisation.

I have argued elsewhere (McLachlan, 1986) that there is a distinct form of cellular self-assembly, representing a direct cellular analogy to self-assembly of subcellular components such as microtubules and viral coat particles. In this mode of development, the individual cells must possess regional characters over their surface. The operation of minimum energy considerations then leads them to produce particular morphologies which are not themselves *directly* specified in the genome. For instance, in the extreme case the genome could specify the shape of a cell as a torus, with adhesive sites on its upper and lower surface. This would then result in the formation of tubes as the minimum-energy conformation.

Such cellular self-assembly may be seen in the formation of kidney tubules (Okada, 1965). These appear during the course of kidney development, and if tubule cells are labelled, and the kidney rudiment is disassembled, then the former tubule cells reform tubules once more. Tubules appear as the differentiation of the tubule cells proceeds, in the form of regional surface specialisation within cells; as the cells lose this surface regionalisation during prolonged culture, so they progressively lose the ability to form tubules.

Cellular self-assembly can in principle explain the form of a number of structures, including blood vessels, elongated muscles, lenses and neural structures. It differs from Steinberg's adhesion model in that (i) it relies on regional differences over the cell surface, and (ii) it can be observed in pure populations of cells (though it need not be confined to them).

In the extreme case it would be possible to argue that such regionalisation of individual cell surfaces could arise entirely from non-genetic mechanisms, as described above for *Tetrahymena*, *Paramecium*, and prion-mediated pathologies. Albrecht-Beuhler (1978) has suggested that even cell behaviour could arise in this way. It is possible to argue that it may be pointless to look for tissue-level controls unless the idea of cell-level control mechanisms have been ruled out. Furthermore, it seems likely that phenomena such as differential adhesivity and cellular surface regionalisation were present at very early levels of evolutionary organisation, and hence it is possible that some of the higher-level morphologies that we view as the consequence of adaptation are in fact the consequence of primitive cell properties.

10.8 Tissue Interaction

A number of tissues can function only when they are in particular spatial relationships with other tissues. However, these relationships need not be specified separately by the genome. Rather, the genome may specify a general starting distribution and the nature of the interaction. The starting condition and the rules will then in turn generate a relationship. Experimentally or environmentally induced changes in the early relationships will lead to new solutions of the relationship equation - the pattern itself is not specified in advance.

A possible mechanism of this kind is where one tissue is specified by the genome, and then issues signals to organise others. For instance, during its development, striated skeletal muscle must be supplied with nerves and blood vessels. However, the difficulty of separately specifying blood vessel, nerve and muscle patterns would be very great. In muscle/nerve development, there operates an interactive mechanism which does not depend directly on the final function. If "muscle-less" limbs are created (Lewis *et al.*, 1981) by irradiating the somite row which is the source of the striated muscle cells, the major pattern of nerve tracts is entirely normal, showing it to be specified by a mechanism other than the muscles. However, the branches which normally supply individual muscles are absent. One plausible explanation is that muscles secrete local factors which attract nerve branches from the major nerve tracts. In vitro studies have revealed that skeletal muscle does indeed release a nerve growth factor which promotes muscle sprouting (Hsu *et al.*, 1982; Gurney, 1984). It can therefore be envisaged that muscle is programmed to produce nerve attractants, which then act on the nerves to produce a relative deployment appropriate to the final function. As far as blood vessels are concerned, it is now clear that there is an extensive family of angiogenic factors, identified first from tumour cells (Folkman, 1985a, b); and muscles have been demonstrated to produce a blood vessel attractant (Morris *et al.*, 1996). It is therefore plausible that muscle becomes vascularised in the same way that it becomes innervated, by attracting blood vessel branches towards it via production of a locally acting growth factor. (Alternatively, muscle vascularisation may be related to use; see next section).

Cartilage also acts on blood vessels, but in an inverse way by producing factors which repel blood vessels (Crum *et al.*, 1985). Blood vessel patterns may therefore be envisaged as arising from a combination of positive and negative influences, rather than by specification in advance.

Such interactive mechanisms are essential in evolutionary terms, since if evolution required a coordinate change in many independent mechanisms, it would proceed only with extreme slowness.

10.9 Specification by Use

Use itself may alter morphology. For instance, muscle mass and diameter increase by hypertrophy on exercise, and fall sharply on immobilisation (Goldspink, 1980). The very persistence of muscle depends on a functioning nerve supply; absence of a functional target leads to nerve atrophy (Bradley, 1970; Bonner and Adams, 1983). Even the differentiation class of the muscle is dependent on the nature of the innervation (Bonner, 1980; Vrbrova, 1980). Similarly, both muscle and tendon require a functional connection for their survival, even though they appear independently (Goldspink, 1974; Shellswell and Wolpert, 1977; Kieny and Chevallier, 1979), and vascularisation is essential to muscle survival (Diwan and Milburn, 1986), though not to its appearance.

Muscle attachment via tendons modifies bone morphology, leading to the formation of rugosities and even prominences which markedly alter the shape of the bone. Bone is in fact markedly susceptible to shape change under mechanical pressure, since stressed bone cells divide in such a way as to alleviate the imposed stresses. The trabeculae in bone thus adopt lines of maximal stress resistance, as described by D'Arcy Thompson, no matter what those stresses are.

Joints rely on mechanical stresses to develop their normal morphology. If mammalian embryos are paralysed so that they cannot show the normal spontaneous movements during the course of the pregnancy, then the joints show a characteristic fused morphology, in which an excess of cartilage develops in the region of the joint (Drachman and Sokoloff, 1966). This shows that the form of the joint itself is not specified directly by the genome, although the knee, for instance, would normally be considered as a moderately complex morphology. Instead the potential for the knee to form is specified, as is a propensity to spontaneous muscle firing leading to random movement. The consequence of the operation of both these programmes is a knee joint, unless they are modified in turn by particular features of the environment. It has been suggested that some classes of abnormality arise from restriction of fetal movement.

Blood vessels also show use responsiveness in terms of their morphology. In the early embryos, some few major blood vessels are predetermined, presumably by the genome. For the remainder, a loose random network develops (Sabin, 1920; Clark and Clark, 1939). When blood begins to flow through this network, those tubes in which the flow is fastest begin to differentiate into substantial vessels. This response includes proliferation and differentiation of the surrounding mesoderm cells as well as the endothelial cells lining the vessel (C.D. Murray, 1926). In consequence, blood vessel patterns are highly variable, particularly when not constrained by factors such as those described in Section 10.8. If a major blood vessel is subsequently occluded, the flow will divert through the surviving minor branches, and the route through these branches with the highest flow rates will subsequently differentiate into a new major vessel, showing that lability persists even into the adult. It has also been argued (Adair *et al.*, 1990) that metabolic by-products induce blood vessel growth towards the site of metabolic activity. This would be a particularly striking example of control of morphology by use, if this hypothesis were confirmed.

Protective use responses may also occur. Callus pads ill-develop in skin exposed to hard wear, and interestingly, common wear sites may have a genetic reinforcement, with preformed callus pads providing a sharp distinction between the genetic programme and the environmental response. In all of these examples the mechanism is specified by the genome, but not the outcome.

10.10 Conclusions

Morphology is often tacitly considered as a kind of Platonic ideal, passed on from generation to generation. In reality, of course, each individual is different from all other individuals in part because of the accidents of their development and subsequent life, and in part because of the operation of the factors described here. In consequence, the ideal itself should be thought of as incomplete. It has areas of inexact specification, occupied sometimes by ranges of possibilities, sometimes by complete alternative forms, which will be defined in practice by the contingent operation of many factors. This is more than the interaction of the environment with the genotype, or reaction norm. Such variation comes from the nature of the developmental programme, which includes mechanisms of high determinacy, but also some with a high degree of indeterminacy. In consequence, development of the individual is stochastic and contingent rather than deterministic, and the operation of identical environmental factors on identical developmental programmes may lead to different morphologies. The norm of reaction is not a property of the individual (Weiss and Gorman 1990), but this variation is.

The indeterminate nature of mechanisms within the developmental programme may help to explain a phenomenon familiar to those who study the structure of one species intensively, as occurs in human anatomy departments. This is the variability of individual members of a species from the morphologies described in the textbook. Mechanisms (6) to (8) may, for instance, lead to individual morphologies which are markedly non-identical but which are functionally equivalent. It is quite possible that the pattern of muscles and nerves is unique to each individual, in the way that fingerprints are, yet is functionally equivalent in each individual, again in the way that fingerprints almost always are.

It is also possible that morphologies may be established by the simultaneous operation of several different mechanisms, to add to the security of development, thus permitting the occasional failure of any particular mechanism.

This may in turn contribute to the understanding of major deviations from the norm, both in terms of the generation of variation, and of birth abnormalities. Abnormalities are often assumed to be either genetic or environmental in origin. However, only about one-third of all known defects can be assigned with certainty to one of these two causes (Whittle and Hanratty, 1986). If several different mechanisms contribute to a particular aspect of morphology, each of which has a small but real probability of going wrong, we would expect that from time to time (the product of the respective probabilities) the operation of chance alone will result in a major abnormality in an organ system. These could be considered as arising stochastically, rather in a deterministic way from genetic specification or from the influence of the environment on the genome.

The operation of mechanisms such as those described here may obviate the need for a great deal of complexity in development. Not everything need be spelled out in detail; in many instances some basic rules may be specified, and development then arises from the contingent operation of these rules in association with other factors external to the developing individual. It could be argued that information is present in the environment, and that the developing organism exploits that information - is perhaps even in a sense parasitic on that information - to complement its own information systems.

Knowledge of the interactions described here may be valuable in evolutionary speculation. These factors provide constraints on possible outcomes that are different in kind to any internal genomic constraints that may exist. They may favour certain kinds of outcome, and may be reinforced by the evolution of independent genetic mechanisms tending to the same outcomes, as Newman (1992) has so persuasively argued. Particular kinds of morphology may initially arise, not because they are themselves adaptive, but because they are high-level consequences of the mechanisms described here; in other words, they are not initially selected because they are adaptive in themselves, but because they are the consequences of another lower-level phenomenon which is adaptive. Once acquired, however, they may then be employed for adaptive purposes. An example of this would be those colour patterns which develop by self-organisation (5), but which may then be employed for other purposes such as species recognition.

A rigid archetype, on the other hand, would be very much slower to respond to environmental changes. The interaction of the environment with the developmental programme to produce variation means that not all variation is genetic, and hence may act to speed up rates of evolution.

Finally, conclusions may be drawn about the adaptive value of morphologies. These may of course have adaptive value in the conventional sense. But it is an implication of the arguments presented here that they need not have adaptive value. They may also have acquired a secondary adaptive use, after appearing for another purpose. Or they may be a visible remnant of an adaptive purpose present earlier in development. Or they may be adaptively neutral. Indeed, if they have been adopted for sexual selection purposes, they may even in themselves be deleterious to some extent.

D'Arcy Thompson believed that he could "rule heredity out of our present account, however true, however important, however indispensable in another setting of the story, such a concept might be" (Thompson, 1942). That is hardly possible now. But perhaps we should take care not to rule out from the modern account mechanisms which do not rely on direct, complex and deterministic specification of morphologies by genes contained within that individual.

Acknowledgements
I gratefully acknowledge Pat Willmer and Mike Ritchie in helping me to formulate this article, even though they disagree with almost everything in it.

REFERENCES

Adair, T.H., Gay, W. and Montani, J. (1990) Growth regulation of the vascular system: evidence for a metabolic hypothesis. *Am. J. Physiol.* 259: 393–404.
Anderson, K.V. and Nusslein-Vollard, C. (1984) Information for the dorsal-ventral pattern of the Drosophila embryo is stored as maternal mRNA. *Nature* 311: 223–227.
Anderson, R.M. (1993) Sorting out natural stone stripes. *Nature* 361: 117.
Albrecht-Beuhler, G. (1978) Daughter 3T3 cells - are they mirror images of each other? *J. Cell. Biol.* 72: 595–603.
Bateson, W. (1894) *Materials for the study of variation.* Macmillan, London.
Beisson, J. and Sonneborn, T.M. (1965) Cytoplasmic inheritance of the organisation of the cell cortex in Paramecium aurelia. *Proc. Nat. Acad. Sci. USA* 53: 275–282.
Boerlijst, M.C. and Hogeweg, P. (1991) Spiral wave structure in prebiotic evolution: hypercycles stable against parasites. *Physica D* 48: 17–28.
Bonner, P.H. (1980) Differentiation of chick embryo myoblasts is transiently sensitive to functional innervation. *Dev. Biol.* 76: 70–86.
Bonner, P.H., and Adams, T.R. (1983) The involvement of nerves in chick myoblast differentiation. In *Limb Development and Regeneration, Progr. in Clin. Biol. Res.*, Vol. 110B, (R.O. Kelly, P.F. Goetinck, and J.A. MacCabe, eds.) pp. 359–368. Alan Liss, New York.
Bradley, S.J. (1970) An analysis of self-differentiation of chick limb buds in chorio-allantoic grafts. *J. Anat.* 107: 479–490.
Brothers, A.J. (1976) Stable nuclear activation dependent on a protein synthesised during oogenesis. *Nature* 260: 112–115.
Clark, E.R. and Clark, E.L. (1939) Microscopic observations on the growth of blood capillaries in the living mammal. *Am. J. Anat.* 64: 251–301.
Clausen, J.C., Keck, D.D. and Heisey, W.M. (1948) Heredity of geographically and ecologically isolated races. *Am. Nat.* 81: 114–133.
Cohen, J. and Stewart, I. (1993) Let T equal tiger... *New Scientist* 140: 40–44.
Cook, T.A. (1914) *The Curves of Life.* Reprinted 1979, Dover Publications, New York.
Crum, R., Szabo, S. and Folkman, J. (1985) A new class of steroids inhibits angiogenesis in the presence of heparin as a heparin fragment. *Science* 230: 1375–1378.
Cruywagen, G.C. and Murray, J.D. (1992) On a tissue interaction model for skin pattern formation. *J. Nonlinear Sci.* 2: 217–240.
Cummins, H. and Midlo, C. (1943) *Fingerprints, palms and soles: an introduction to dermatoglyphics.* Blakiston Co., Philadelphia.
Dent, J.N. (1968) Survey of amphibian metamorphosis. In *Metamorphosis* (W. Etkin and L.

I. Gilbert, eds.) pp. 271–311. New York, Appleton Century Crofts.

Dingle, H. (1985) Migration. In *Comprehensive Insect Physiology, Biochemistry and Pharmacology*, Vol. 9. *Behaviour* (G.A. Kerkut and C. I. Gilbert, eds.). Pergamon Press, Oxford.

Diwan, F.H. and Milburn, A. (1986) The effects of temporary ischaemia on rat muscle spindles. *J. Embryol. Exp. Morphol.* 92: 223–254.

Drachman, D. and Sokoloff, L. (1966) The role of movement in embryonic joint development. *Experimental Biol.* 14: 401–420.

Dodson, S.I. (1988) Cyclomorphosis in *Daphnia Galatea* Mendotas Birge and *D. Retrocurva* Forbes as a predator induced response. *Freshwater Biol.* 19: 109–114.

Douasy, S. and Couder, Y. (1992) Phyllotaxis as a physical self-organised growth process. *Phys. Rev. Lett.* 68: 2098–2101.

Eigen, M. and Schuster, P. (1979) *The Hypercycle: A Principle of Natural Self-Organisation.* Springer, Berlin.

Elsdale, T. and Wasoff, F. (1976) Fibroblast cultures and dermatoglyphics: the topology of two planar patterns. *Wilhelm Roux Archives* 180: 121–147.

Epstein, I.R. (1984) Chemically generated spatial structures. *Nature* 307: 692.

Epstein, I.R., Kustin, K., De Kipper, P. and Orban, M. (1983) Oscillating chemical reactions. *Sci. Am.* 248: 96–108.

Folkman, J. (1985a) Towards an understanding of angiogenesis: search and discovery. *Perspectives in Biology and Medicine* 29: 10–36.

Folkman, J. (1985b) Tumour angiogenesis. *Adv. Cancer Res.* 43: 175–203.

Freeman, G. and Lundelius, J.W. (1982) The developmental genetics of dextrality and sinistrality in the gastropod *Lymnea peregra. Wilhelm Roux Archives* 191: 69–83.

Galloway, J. (1988) Nature's second favourite structure. *New Scientist* 117: 37–39.

Gerhart, J., Ubbels, G., Black, S., Hara, K. and Kirschner, M. (1981) A re-investigation of the role of the grey crescent in axis formation in *Xenopus laevis. Nature* 292: 511–516.

Goldspink, G. (1974) Development of muscle. In *Differentiation and Growth of Cells in Vertebrate Tissues* (G. Goldspink, ed.) pp. 69–99. Chapman & Hall, London.

Goldspink, G. (1980) Growth of muscle. In *Development and Specialisation of Skeletal Muscle* (D.F. Goldspink, ed.), pp. 19–35. Cambridge University Press, Cambridge.

Green, E. (1989) A diet-induced developmental polymorphism in a caterpillar. *Science* 243: 643–646.

Gregg, B.C. and Snow, M.H.L. (1983) Axial abnormalities following disturbed growth in Mitomycin-C treated mouse embryos. *J. Embryol. Exp. Morph.* 70: 135–149.

Gurney, M.E. (1984) Suppression of sprouting at the neuromuscular junction by immune sera. *Nature* 307: 546–548.

Harris, A.K., Stopak, D. and Warner, P. (1984) Generation of spatially periodic patterns by a mechanical instability: a mechanical alternative to the Turing model. *J. Embryol. Exp. Morphol.* 80: 1–20.

Hillman, N., Sherman, M.I. and Graham, C.F. (1972) The effect of spatial development on cell determination during mouse development. *J. Embryol. Exp. Morphol.* 28: 263–278.

Hsu, L., Natyzak, D. and Trupin, G.L. (1982) Neurotrophic effects of skeletal muscle fractions on spinal cord differentiation. *J. Embryol. Exp. Morphol.* 71: 83–95.

Humphrey, R.R. (1966) A recessive factor (o, for Ova Deficient) determining a complex of abnormalities in the Mexican axolotl (*Ambystoma mexicanum*). *Del. Biol.* 13: 57–76.

Jaffe, L.F. (1968) Localisation in the developing Fucus egg and the general role of localising currents. *Advances in Morphogenesis* 7: 295–328.

Jenkin, P.M. (1970) *Control of growth and metamorphosis.* Pergamon Press, Oxford.

Jinks, J.L. (1964) *Extra-chromosomal inheritance.* Prentice Hall, Englewood Cliffs NJ.

Kieny, M. and Chevallier, A. (1979) Autonomy of tendon development in the embryonic chick wing. *J. Embryol. Exp. Morphol.* 49: 153–165.

Kondo, S. and Asal, R. (1995) A reaction-diffusion wave on the skin of the marine angelfish

Pomacanthus. *Nature* 376: 765–768.

Lechleiter, J. Girard, S., Peralta, E. and Clapham, D. (1991) Spiral wave propagation and annihilation in *Xenopus laevis* oocytes. *Science* 252: 123–126.

Lee, K.-L., McCormick, W.D., Pearson, J.E. and Swinney, H.L. (1994) Experimental observations of self-replicating spots in a reaction-diffusion system. *Nature* 369: 215–218.

Lewis, J., Chevallier, A., Kieny, M. and Wolpert, L. (1981) Muscle nerve branches do not develop in chick wings devoid of muscle. *J. Embryol. Exp. Morphol.* 64: 211–232.

McLachlan, J.C. (1986) Self-assembly of structures resembling functional organs by pure populations of cells. *Tissue and Cell* 18: 313–320.

May, R.M. (1991) Hypercycles spring to life. *Nature* 353: 607–608.

Meinhardt, H. (1979) The random character of bifurcations and the reproducible processes of embryonic development. *Ann. New York Acad. Sci.* 316: 188–202.

Meinhardt, H. (1982) *Models of Biological Pattern Formation*. Academic Press, New York.

Mohler, J. and Wieschaus, E.F. (1986) Dominant maternal effect mutations of *Drosophila melanogaster* causing the production of double abdomen embryos. *Genetics* 112: 803–822.

Morris, D., Stock, S. and McLachlan J. C. (1996) Embryonic chick muscle produces an FGF-like activity. *Experientia* 52: 757–761.

Murray, C.D. (1926) The physiological principle of minimum work applied to the angle of branching of arteries. *J. Gen. Physiol.* 9: 835–841.

Murray, J.D. (1988) How the leopard gets its spots. *Sci. Am.* 259: 62–69.

Murray, J.D., Deeming D.C. and Ferguson, M.W. (1990) Size dependent pigmentation-pattern formation in embryos of *Alligator mississippiensis*: time of initiation of pattern generation mechanism, *Proc. Roy. Soc. Lond. Ser. B* 239: 279–293.

Neff, A.W., Malacinski, G.M. and Chung, H.-M. (1985) Microgravity simulation as a probe for understanding early *Xenopus* pattern specification. *J. Embryol. Exp. Morphol.* 89: 259–274.

Nelsen, E.M., Frankel, J. and Jenkins, L.M. (1989) Non-genic of cellular handedness. *Development* 105: 447-456.

Newman, S.A. (1992) Generic physical mechanisms of morphogenesis and pattern formation as determinants in the evolution of multicellular organisation. In: *Principles of Organisation in Organisms, SFI Studies in the Sciences of Complexity. Proc.* (J. Mittenthal, and A. Baskin, eds.) pp. 1–27. Addison-Wesley, London.

Okada, T.S. (1965) Immunohistological studies on the reconstruction of nephric tubules from dissociated cells. *J. Embryol. Exp. Morphol.* 13: 299–307.

Oster, G.F. and Odell, G.M. (1984) The mechanochemistry of cytogels. *Physica D* 12: 333–350.

Oster, G.F., Murray, J.D. and Harris, A.K. (1983) Mechanical aspects of mesenchymal morphogenesis. *J. Embryol. Exp. Morphol.* 78: 83–125.

Pearson, J.E. (1993) Complex patterns in a simple system. *Science* 261: 189–197.

Pruisner, S.B. (1991) Molecular biology of prion diseases. *Science* 252: 1515–1522.

Raup, D.M. (1962) Computer as aid in describing form in gastropod shells. *Science* 138: 150–152.

Raup, D.M. (1966) Geometric analysis of shell coiling: general problems. *J. Paleontol.* 40: 1178–1190.

Ritchie, M.G., Butlin, R.K. and Hewitt, G.M. (1987) Causation, fitness effects and morphology of macropterism in *Chorthippus parallelus* (Orthoptera: Acrididae). *Ecol. Entomol.* 12: 209–218.

Sabin, F. (1920) Origin and development of the primitive vessels of the pig and the chick. *Carnegie Contributions to Embryology* 9: 213–264.

Saunders, P.T. and Ho, M.W. (1995) Reliable segmentation by successive bifurcation. *Bull. Math. Biol.* 57: 539–556.

Schmalhausen, I.I. (1949) *Factors of Evolution: The Theory of Stability*. Blakiston, Philadelphia PA.

Shellswell, G.B. and Wolpert, L. (1973) The pattern of muscle and tendon development in the chick wing. In *Vertebrate Limb and Somite Morphogenesis* (D.A. Ede, J.R. Hinchliffe and M. Balls, eds.) pp. 71–86. Cambridge University Press, Cambridge.
Stearns, B.C. (1989) The evolutionary significance of phenotypic plasticity. *Bioscience* 37: 436–445.
Steinberg, M.S. (1970) Does differential adhesion govern self-assembly processes in histogenesis? Equilibrium configurations and the emergence of a hierarchy among populations of embryonic cells, *J. Exp. Zool.* 173: 395–434.
Steinberg, M.S. and Poole, T.J. (1981) Strategies for specifying form and pattern: Adhesion guided multicellular assembly, *Proc. Roy. Soc. Lond. Ser. B* 295: 451–460.
Stevens, P.S. (1974) *Patterns in Nature*. Little, Brown, and Co., Boston.
Surani, M.A. (1994) Genomic imprinting: control of gene expression by epigenetic inheritance. *Current Opinion in Cell Biology* 6: 390–395.
Thompson, D.W. (1942) *On Growth and Form*, 2nd edn. Cambridge University Press, Cambridge.
Thompson, J.D. (1991) Phenotypic plasticity as a component of evolutionary change. *Trends in Ecol. Evol.* 6: 246–249.
Turing, A.M. (1954) The chemical basis of morphogenesis, *Phil. Trans. Roy. Soc. Lond. Ser. B* 237: 37–72.
Vrbrova, G. (1980) Innervation and differentiation of muscle fibres. In *Development and Specialisation of Skeletal Muscle* (D.F. Goldspink, ed.) pp. 37–50. Cambridge University Press, Cambridge.
Wagner, D.L. and Liebherr, J.K. (1992) Flightlessness in insects. *Trends in Ecol. Evol.* 7: 216–220.
Weiss, A.E. and Gorman, W.L. (1990) Measuring selection on reaction norms: an exploration of the Eurosta-Solidago system. *Evolution* 44: 820–831.
Welsh, B.J., Gomatam, J. and Burgess, A.E. (1983) Three-dimensional chemical waves in the Belousov-Zhabotinskii reaction. *Nature* 304: 611–614.
Werner, B.T. and Hallet, B. (1993) Numerical simulation of self-organised stone stripes. *Nature* 361: 142–145.
West, B.J. and Goldberger, A.L. (1987) Physiology in fractal dimensions. *Am. Sci.* 75: 354–364.
Whittle, M.J. and Hanratty, K.P. (1986) Identifying abnormalities, *Br. Med. J.* 293: 1485–1488.
Woodward, D.E. and Murray, J.D. (1993) On the effect of temperature dependent sex determination on sex ratio and survivorship in crocodilians. *Proc. Roy. Soc. Ser. B* 252: 149–155.
Zaikin, A.N. and Zhabotinsky, A.M. (1970) Concentration wave propagation in two dimensional liquid phase self-oscillating system, *Nature* 225: 535–537.

11

The Role of Chemotactic Cell Movement in Dictyostelium Morphogenesis

Cornelis J. Weijer

Department of Anatomy and Physiology, University of Dundee, Dundee DD1 4HN, Scotland, UK

11.1 Introduction

A major goal of developmental biology is to understand the morphogenesis of multicellular organisms. Development of a vertebrate embryo typically involves the generation of hundreds of cell types from a single cell, the zygote, which are laid down into a highly ordered spatio-temporal pattern. This involves spatio-temporal control of processes such as cell division, cell death, cell differentiation and cell movement. Some cell types may arise in situ at the right position at the right time of development, others may differentiate first as a precursor cell type and then move to their final destinations. The coordination of these processes involves extensive cell–cell communication. These cell types arise through the unfolding of different gene expression programs, which are controlled through thousands of signalling molecules. These molecules interact through positive and negative feedback pathways. Due to this enormous complexity it is very difficult to gain an understanding of morphogenesis of complex organisms such as vertebrates, even at the cellular level. This is one of the reasons to investigate the morphogenesis of much simpler organisms that still show some of the essential characteristics of the development of higher organisms. They are generally more accessible to such a description since they contain fewer cell types and can be easier to manipulate experimentally.

On Growth and Form: Spatio-temporal Pattern Formation in Biology.
Editor Mark A.J. Chaplain, G.D. Singh, J.C. McLachlan

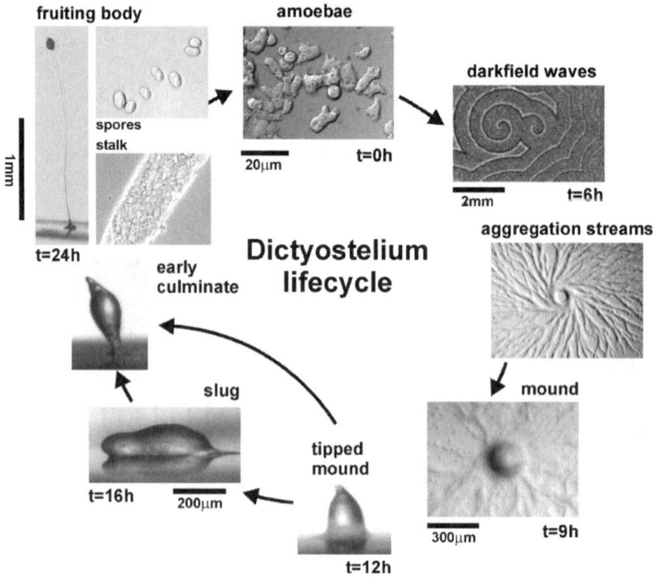

Figure 11.1 The *Dictyostelium discoideum* life cycle. Clockwise from the top, vegetative amoebae, darkfield waves, as observed during aggregation (they reflect the cells in different phases of the movement cycle in response to cAMP waves), aggregation streams, a top view of a mound with incoming streams, a side view of a tipped mound, a side view of a migrating slug and an early culminate and a fruiting body accompanied by high magnification images of the stalk cells and spores. This developmental cycle is starvation induced and takes 24 hours to complete at room temperature.

The social amoeba *Dictyostelium discoideum* (Dd) is one such system. Under normal conditions *Dictyostelium* populations consist of groups of independent, single amoebae. The single amoebae live in the soil where they feed on bacteria and multiply by binary fission. Under unfavourable conditions, when food becomes limiting, these amoebae become "social" and enter a complex multicellular developmental program. This organism therefore stands somewhere between single and multicellular organisms in the evolutionary tree. Upon starvation the cells aggregate chemotactically to form a multicellular structure, the slug, consisting of up to 10^5 amoebae. The slug is motile and, guided by photo- and thermotaxis, it can move away from the place of aggregation to a place suitable to form a fruiting body, normally the soil surface. The fruiting body consists of a stalk, formed by dead vacuolated stalk cells, that supports a mass of spore cells. The spores disperse and each spore can start a new colony under favourable conditions (Figure 11.1). The cells start to differentiate into precursors of the stalk and spore cells during aggregation. A prepattern of prestalk and prespore cells is formed in the slug. The prestalk cells are localised in the anterior 20% of the slug and the prespore cells, which transform later into pores, are localised in the back of the slug.

This organism retains many of the commonly found elements of multicellular development such as differentiation of one cell type, the amoebae, into at least two cells types, the stalk cells and the spores. The proportions of the cell types are highly regulated by extracellular signalling molecules produced by the cells. When a slug is cut into a prestalk and a prespore, both pieces will regulate and form perfectly normal proportioned fruiting bodies. During their multicellular development, the cells do hardly divide (since there is no food), so this regulation involves a transdifferentiation of prespore into prestalk cells and vice versa. Furthermore this implies that morphogenesis, the generation of the form of the organism, results solely from the coordinated movement of individual cells. This coordination of cell movement also requires extensive cell–cell signalling mechanisms.

Some of the features of Dd development are beginning to be understood both at the biological (cellular) and formal level. Here we review the progress made in understanding the cellular basis of Dd morphogenesis.

11.2 Cell Type Differentiation

The signal for entering development is starvation, most likely for essential amino acids. The cells monitor their density and the density of their food source by releasing a glycoprotein called prestarvation factor (PSF), which binds to bacteria and allows the cells to compare their density to that of the bacteria (Clarke and Gomer, 1995). Once they decide to aggregate, they secrete a conditioned medium factor (CMF), which leads to the induction of a number of developmentally regulated genes among which are those important for aggregation (Clarke and Gomer, 1995). Aggregation is the first visible sign of multicellular development. Aggregates may contain up to 10^5 cells. During the late aggregate stage, the cells start to differentiate into a number of different cell types. During late aggregation, cell type specific gene expression is switched on. Prespore cells and some prestalk cell types become distinguishable. These cell types can be recognised by their expression cell type specific genes. They can be visualised by examining the expression of the *lacZ* reporter gene under the control of cell type specific promoters. The cells start to differentiate during aggregation and continue to do so in random positions in the aggregate and mound. They will then sort out to form a relatively simple axial pattern of cell types in the slug. By using specific promoter fragments fused to the *lacZ* reporter gene, it has been possible to further subdivide the prestalk cell population into further subtypes, which finally contribute to different structures in the fruiting body (Figure 11.2). There are the pstA cells, which are located in the tip of the slug and give rise to the stalk of the fruiting body. There is a population of cells pstO which is located just behind the pstA region. These pstO cells will give rise to the upper cup of the fruiting body, which caps the spore mass. There are cells scattered in the back of the slug which express some of the prestalk-specific genes *ecmA* and *ecmB* and can be stained with the dye neutral red. These anteriorlike cells in the back form a slowly exchanging population with the cells in the pstO zone (Abe *et al.*, 1994). They will form the outer part of the basal disc and the lower cup which supports the spore mass from below and is attached to the stalk (A.E. Early *et al.*, 1993; Jermyn and Williams, 1995; Jermyn *et al.*, 1989; Williams and Jermyn, 1991). The differentiation of the cell types is controlled by small molecules such as cAMP and differentiation inducing factor (DIF) (Kay, 1992, 1997; Williams *et al.*, 1995). Furthermore, it seems likely that the initial preference of the cells to differentiate into prespore or prestalk cells is governed by the cell cycle phase

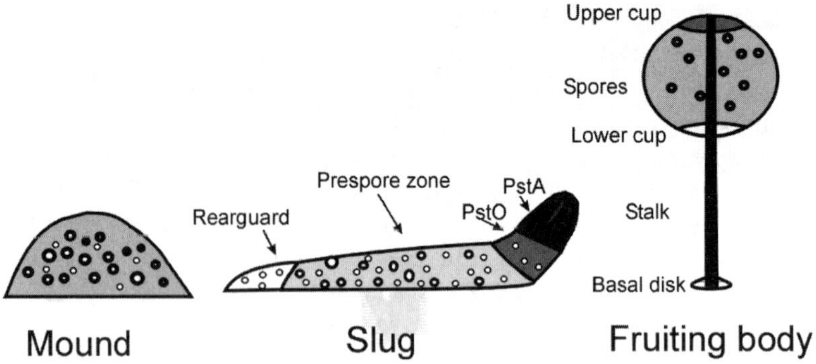

Figure 11.2 Cell type differentiation during *Dictyostelium* development. At the late aggregation stage, the cells start to differentiate into prespore and prestalk cells. They enter the mound in a random pattern and they sort chemotactically to form a simple axial pattern in the slug.

the cells are in at the moment of starvation. It has been shown that the cells which are small (cells in mitosis and early S phase have a high probability of differentiating into stalk cells. Cells which are fat at the moment of starvation, i.e. cells in middle and late G2, will preferentially differentiate into spores (Weeks and Weijer, 1994). In the rest of this chapter I will concentrate on describing the control of cell movement during the different stages of development

11.3 Aggregation

The principles that govern aggregation are now well understood at the cellular level. Aggregation results from three cellular competencies:

1. Autonomous periodic cAMP signalling by cells in the aggregation centre
2. cAMP signal relay by all other cells
3. Chemotaxis towards increasing cAMP concentrations

A few hours after the initiation of starvation, some cells become capable of secreting cAMP. They also become sensitive to cAMP since they start to express cell surface cAMP receptors, which couple to the enzyme that generates cAMP, adenylyl cyclase. The cells can now produce cAMP in response to being stimulated by cAMP, i.e. they become excitable. Some cells start to secrete cAMP in an oscillatory fashion. Groups of these cells presumably form the initial aggregation centres which emit signals in a periodic fashion. It is known that eventually almost all cells develop the capability to form centres (Glazer and Newell, 1981). It is most likely that all the cells will show oscillatory cAMP kinetics at later stages of aggregation.

The cells in the aggregation centre periodically synthesise and secrete cAMP in the extracellular medium. Here it diffuses to neighbouring cells, which detect this signal via highly specific and highly sensitive cAMP receptors. These cells now amplify and pass the signal on to their neighbours, resulting in propagating waves of cAMP. Binding of cAMP to the receptor leads to two competing processes, excitation and adaptation (Parent and Devreotes, 1996). The excitation pathway leads to the activation of the

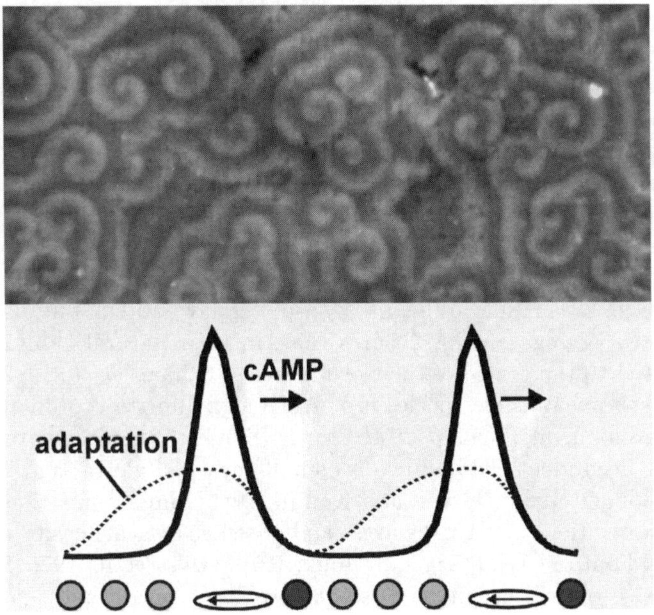

Figure 11.3 Darkfield waves caused by the movement of cells in response to propagating cAMP waves. The top image shows a darkfield image of a monolayer of aggregating cells at low magnification. The cAMP waves propagate as spiral waves. The lower diagram shows two cAMP waves propagating from left to right. The bottom line shows how the cells move in response to these waves. Darkfield waves accurately reflect the underlying cAMP waves and can be used to study their geometry and temporal dynamics.

enzyme adenylyl cyclase, which produces cAMP. This intracellular cAMP is then secreted to the outside via an unknown mechanism. Some of it will bind to the receptors of the same cell, leading to autocatalytic feedback, and some will diffuse away to activate neighbouring cells. The adaptation pathway involves among others a desensitisation of the receptor, which results from phosphorylation of its cytoplasmic tail. Adaptation results in termination of the autocatalytic relay response. The extracellular cAMP is continuously broken down by cAMP phosphodiesterase, causing the extracellular concentration of cAMP to fall. This then results in dephosphorylation and resensitisation of the cAMP receptor. The adaptation process is responsible for the outward propagation of cAMP waves, since cells which have just relayed are refractory to further stimulation by cAMP. The signal transduction pathways leading to cAMP production are well investigated and have recently been reviewed in (Parent and Devreotes, 1996).

The cells also show a chemotactic reaction in the direction of higher cAMP concentrations. The cells move up the gradient, as long as the cAMP concentration is rising. They stop moving as soon as the cAMP wave passes, and the concentration starts to fall since the receptors are now desensitised. This chemotactic response leads to the periodic inward movement of the cells to the aggregation centre, guided by outward-propagating waves of cAMP (Figure 11.3). Superimposed on this cAMP oscillatory system there is a complex feedback of the cAMP oscillations on the expression of various components of the signalling system. The cAMP pulses induce the synthesis of cAMP receptors, adenylyl cyclase and phosphodiesterase, increasing excitability during aggregation (Gerisch, 1987).

It is interesting to note that not all *Dictyostelium* species use a cAMP relay mechanism to aggregate. In some species specialised cells secrete a chemoattractant into the surrounding medium; the chemoattractant is detected but not relayed by surrounding cells. These cells, however, use this signal to move chemotactically towards the source of the signal. This leads to the formation of very small aggregates and fruiting bodies. Other species seem to use a relay system but use different chemoattractants to aggregate. This ensures that, in their natural habitat, different strains are attracted to the centre of their own species (Schaap *et al.*, 1985). There is some evidence that some of these species use cAMP signalling to coordinate the later stages of their development (Schaap *et al.*, 1984). Some of the chemoattractants of other species have been identified and involve small peptides (glorin) and metabolites such as pterins (Bonner, 1983; De Wit and Konijn, 1983; Shimomura *et al.*, 1982).

During early aggregation the cAMP waves can be seen as optical density waves using low-power darkfield optics (Alcantara and Monk, 1974; Gross *et al.*, 1976; Siegert and Weijer, 1989). These optical density waves are correlated with shape changes which cells undergo upon stimulation with cAMP (Figure 11.3). Chemotactically moving cells are elongated and groups of moving cells appear brighter than groups of non-moving cells. This results in the appearance of dark and light bands in fields of aggregating cells. By correlating the cAMP signal via cAMP isotope dilution fluorography with the darkfield waves, it was clearly demonstrated that the optical density waves observed during aggregation represent the propagating cAMP signal (Tomchik and Devreotes, 1981). Most often waves appear as expanding spirals (Figures 11.3 and 11.4) in some strains they also propagate as concentric ring waves. Spiral waves can arise in excitable media and do not need autonomously oscillating cells; but concentric waves clearly need autonomously oscillating cells. I have often observed the development of phase waves, a good indication of autonomous cAMP oscillations. Waves from neighbouring centres collide and annihilate each other, leading to the formation of aggregation territories. The formation of aggregation centres is a developmentally regulated property. When cells that have started to aggregate are washed from the agar plates and replated again at the same density, the number of centres increases dramatically. This is since many more cells have now acquired the ability to oscillate and are able to initiate centres (Gingle and Robertson, 1976).

Quantitative measurements of the optical density waves showed that the period length of the waves decreases during aggregation but the wave propagation speed slows down (Gross *et al.*, 1976; Rietdorf *et al.*, 1996; Siegert and Weijer, 1989). During early aggregation the frequency of the signals may be as low as 6–7 min; later in development the frequency decreases to 2 min. Wave propagation speed decreases simultaneously from 600 μm/min to around 100 μm/min. This means that the chemical wavelength decreases from 3000–4000 μm to 100–200 μm. This is the reason that the spirals can organise big aggregation territories during early aggregation and much smaller structures such as mounds during later development. This change in signalling properties is partly due to the cAMP-dependent expression of components of the oscillatory system, described above, as well as to the dispersive properties of

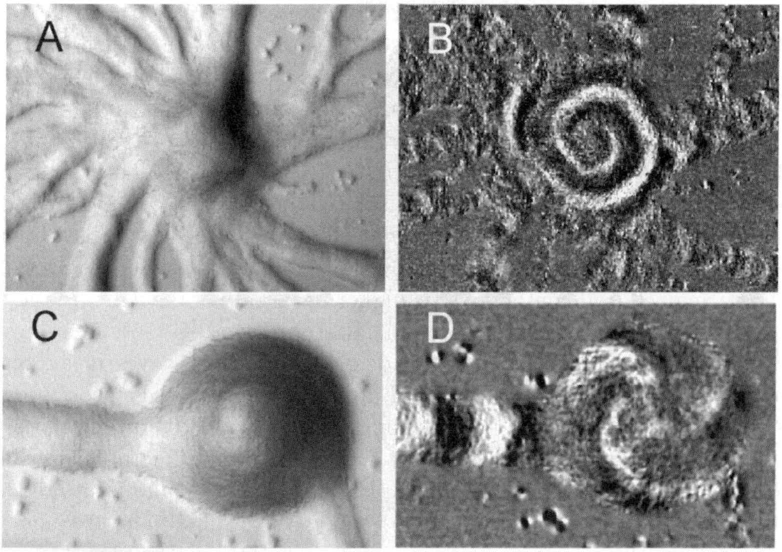

Figure 11.4 Development of optical density waves waves during the aggregation of *Dictyostelium* cells strain DH1. (A,B,C) Original images as recorded under a macroscopic darkfield (Siegert and Weijer, 1989) at three successive times in development: 4,5 and 6 hours. (A, D) show the early darkfield waves. (B,D) show images when the excitation waves are fully developed. (C,F) show images at a time where the cells start to organise themselves into streams and aggregation territories start to develop. The sections shown are 2.5 cm wide.

this excitable medium (Foerster *et al.*, 1990; Gerisch *et al.*, 1987; Siegert and Weijer, 1989).

11.4 Formation of Aggregation Streams

After a number of waves have passed, the cells start to become organised in aggregation streams. These streams radiate from the aggregation centre outward and branch more and more towards the periphery. They do not appear as fractures propagating from a centre outward, but arise rather suddenly more or less everywhere in the aggregation territory at the same time. The theoretical reasons for stream formation are now well understood theoretically and will be discussed in the modelling section.

Cells in streams are highly polarised and elongated. They are thought to make specific end-to-end contact via an EDTA-resistant contact molecule called CSA. Their expression requires the cells being exposed to pulses of cAMP. Mutants which lack the CSA molecules can still aggregate but they lack the clear stream morphology, implying a role for these sites in stream preservation (Harloff *et al.*, 1989; Ponte *et al.*, 1998; Stadler *et al.*, 1989). Another characteristic feature of the cells in aggregation streams is that they generally move faster as isolated aggregation competent cells and their movement is clearly less periodic than that of aggregation stage cells which have not yet made end-to-end contacts. Their movement is more continuous, presumably caused by the strong cell–cell contacts. This higher basal rate of movement is certainly the reason for the increased movement speed (Rietdorf *et al.*, 1996). It is, however, readily possible to observe the propagation of optical density waves in aggregation streams under suitable optical conditions and after appropiate image processing. The wave propagation speed at this stage of development is close to 100 μm/min, i.e. much slower than the 600 μm/min measured during early aggregation (Rietdorf *et al.*, 1996).

Most aggregation centres are organised by spiral waves that break up into individual waves which propagate down the streams as single wavefronts. The frequency and propagation speed of the waves are more or less constant in the streams and the aggregation centre. The shape of the cells changes as they enter the aggregate: in the streams they are highly elongated and polarised but in the body of the aggregate they start to round off and become more amoeboid in shape again, suggesting a change in the adhesion mechanism which operates during later development.

11.5 Mounds

The cells entering the aggregation centre start to pile on top of each other and form a three-dimensional hemispherical structure. I have investigated whether the cell movements in the mound and slug stages of development are still coordinated by propagating waves of a chemoattractant. After suitable image processing I could visualise propagating optical density waves in the later stages of *Dictyostelium* development (Rietdorf *et al.*, 1996; Siegert and Weijer, 1995). Continuous measurements of the optical density waves from late aggregation until tip formation over a period of 3 hours demonstrated that there was a clear evolution in the dynamics of the waves (Rietdorf *et al.*, 1996). Initially the waves propagated fast at low frequency but in the course of aggregation the wave frequency increased, and the wave propagation speed decreased.

There is good evidence that the *Dictyostelium* mound is still organised by cAMP waves (Siegert and Weijer, 1995). It is possible to observe a continuous succession of optical waves from aggregation to the mound stage. Periodic micro-injection of pulses of cAMP into the extracellular space in mounds initiated optical density waves which propagated from the electrode tip outwards. These waves annihilated endogenous waves upon collision, showing a common propagation mechanism. (Rietdorf *et al.*, 1998). It therefore seems very likely that optical density waves in mounds are caused by the same cAMP wave propagation mechanism active during aggregation.

In the mound stage the geometry of the waves can be very variable. The pattern of wave propagation seems to be a characteristic of the strain used. I have found a variety of different wave propagation patterns in mounds. Some strains such as NP55 predominantly show concentric ring waves (Siegert and Weijer, 1995). These waves can originate from one or more centres in the mound. In mounds where target waves arise simultaneously from a number of different centres, the location of these centres may change in time and space during development. These centres are possibly associated with the local accumulation of prestalk cells during the sorting out process The waves generated by these centres compete and interact during the development of the mound and finally one centre seems to become dominant. I also observed simple spiral waves in mounds. In most strains the geometry of the spirals is more complicated and results in the formation of multi-armed spirals (Figure 11.5). I have observed single and multi-armed spirals in the strains Ax2 and Ax3 and DH1. In the case of multi-armed spirals, the waves rotate with high frequency around a common core (Rietdorf *et al.*, 1996; Siegert and Weijer, 1995). I have not yet been able to observe exactly how these multi-armed spiral waves form from the single-armed spirals seen during aggregation. Ax2 and Ax3 produce spiral waves during aggregation, but recently I have also observed multi-armed spiral waves in the strain DH1. This strain produces concentric waves during aggregation and extensive multi-armed spirals at the mound stage of development. Since cell movement is always opposite to the direction of signal propagation (Alcantara and Monk, 1974; Rietdorf *et al.*, 1996; Varnum-Finney *et al.*, 1987), cell movement in mounds organised by concentric waves is slow and directed towards the organising centre. In spiral waves the cell movement is rotational and fast.

It is now known that later in development the cells start to express a number of different cAMP receptors which can all couple to apparently the same cAMP signalling pathways but only differ in their affinity for cAMP (Chen *et al.*, 1996; J.Y. Kim *et al.*, 1998). During early aggregation a high-affinity receptor (car1) is expressed and deletion of this receptor leads to a loss of the ability to aggregate. However, during aggregation the cells also start to express another receptor (car3) of somewhat lower affinity in lower copy number per cells. Deletion of this receptor has no clear morphological consequences. During tip formation a third low-affinity receptor is expressed. This receptor (car2) is mainly expressed by prestalk cells. Its deletion leads to a block at the mound stage of development, showing its importance in tip formation. The prestalk cells also preferentially express the enzymes which will synthesise cAMP adenylyl cyclase and the degrading enzyme cAMP phosphodiesterase (Brown and Rutherford, 1980; Pahlic and Rutherford, 1979; Rutherford *et al.*, 1981). This makes it likely that these cells are seeing higher levels of cAMP and therefore require lower-affinity receptors in order to function well. This switch in expression of receptors could also be responsible for the change in wave propagation observed from concentric or single-

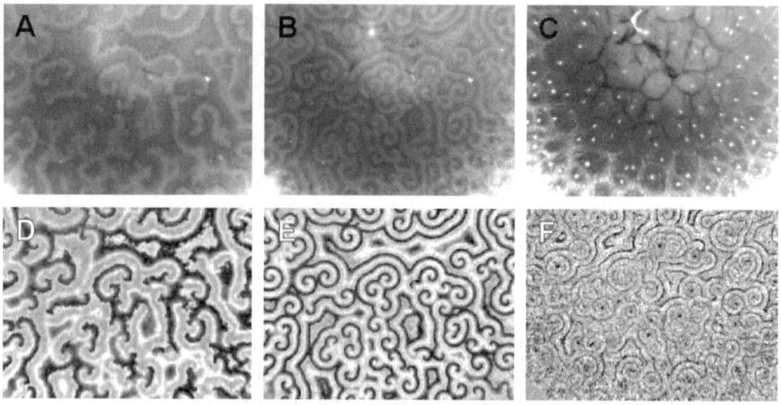

Figure 11.5 Single- and multi-armed spiral in mounds of strain DH1. (A) a DH1 mound with incoming streams. (B) a clear single-armed spiral in the mound shown in (A) obtained after a rolling average subtraction of five images taken 10s apart. (C) Transmitted light image of another DH1 mound. (D) Rolling average subtraction image of the same mound as shown in (C).

armed spirals to multi-armed spirals. This can now be investigated by the study of wave propagation in cAMP receptor mutants, expressing only one receptor of known affinity (J.Y. Kim *et al.*, 1998).

The movement behaviour of cells in mounds has been best analysed in the strain Ax3 (Rietdorf *et al.*, 1996; Sukumaran *et al.*, 1998). This strain is characterised by the propagation of multi-armed spirals at the mound stage of development. As soon as the cells enter the Ax3 mound, their movement speed slows down to 5μm/min. Their movement pattern changes from the straightforward movement seen in aggregation streams to a less oriented movement pattern in the mound; and the shape of the cells changes from elongated to more amoeboid. Then after a while the speed of the cells increases up to 50–60μm/min. This increase in speed temporarily coincides with the time that the cells start to express the cell type specific prespore antigen. The movement of the cells is rather continuous and less periodic, as in aggregation streams their movement is almost fluid-like. Each cell is clearly an individual, but it interacts continuously with all its neighbours. Then after a few hours of high-speed rotation the cells slow down again. Their movement becomes much more periodic and the cells start sorting themselves to form a tip. In strains organised by concentric waves the scenario is very different. In XP55 the cells become almost motionless as soon as they enter the mound; they just sit there and jitter around for a while. Then by some unknown mechanism the prestalk cells also sort towards the top of the aggregate to form the tip.

11.6 Cell Movement in Slugs

During slug formation the hemispherical mound transforms into a long cylinder, the "standing slug". This transformation happens by an intercalation of the cells. The first sign of this intercalation is the formation of a nipple-shaped structure, the tip. The tip forms on top of the mound and seems to direct all subsequent morphogenesis. The tip has some of the classical properties of an organiser. When a tip is transplanted into the side of a slug, it will organise some of the tissue to form a secondary slug. When the tip is cut off a mound or slug, the first thing that happens is that some of the cells will reform a tip. The tip effectively controls the behaviour of cells up to hundreds of cell diameters away, and this clearly requires some form of long-range signalling. However, until now the nature of these signals has not been completely clear. My working hypothesis is that they are propagating waves of cAMP, similar to those being responsible for aggregation and mound development. I will now outline the evidence for and against this hypothesis.

Up to the tipped mound stage it is easy to see optical density waves by looking at the top or the bottom of the mound. However, at the time of tip formation they become invisible. This happens most likely because they change their geometry, and I suspect that they propagate from the tip downward. For this reason I decided to study the patterns of cell movement in the slug. A good description for movement of individual cells in the slug should help me to understand the signals responsible for the control of their movement. The patterns of cell movement should show whether cells move continuously or periodically and whether they move in a coordinated fashion. We have used different methods to visualise cells. I have used the selective staining of prestalk and anteriorlike cells by the vital dye neutral red, as well as the labelling of random cells by loading them with fluorescent dextrans, and more recently by the expression of different spectral variants of the green fluorescent protein GFP under the control of constitutive and cell type specific promoters (Abe *et al.*, 1994; Siegert and Weijer, 1991, 1992). Analysis of movement of neutral red stained and fluorescently labelled cells has shown that the cells in the back of the prespore zone move forwards in the direction of slug migration. Many of the stained cells move in periodic fashion. Analysis of many neutral red stained cells in a small region at the back of a slug showed furthermore that they move periodically in a coordinated fashion. Neighbouring cells move quickly and slowly more or less at the same time suggesting they react to a common periodic signal. It is also clear, however, that they do not keep their neighbours. The cells at the back of the slug move on average with slug speed. Cells in the prestalk zone show a very different pattern of movement. They describe a rotational movement around the long axis of the prestalk zone, slightly slanted to the direction of slug migration (Figure 11.6). This rotational movement is especially strong in the pstO zone when the tip is lifted up from the substrate in the air. Due to their twisted tracks the speed of movement of the individual prestalk cells is greater than the forward movement speed of the slug.

Assuming that the movement of these cells is still controlled by chemotaxis to cAMP, and that the direction of movement is counter to the direction of wave propagation, these measurements allowed me to reconstruct the geometry of the signals. I proposed that the cells in the prestalk zone see a rotating three-dimensional spiral wave of the chemoattractant cAMP (Siegert and Weijer, 1992). The spiral rotates around a core, which is located in the centre of the tip. This induces the rotational movement of the cells. The cells in the back ought to see periodically passing waves (Figure 11.6). One of the big open questions was how this conversion of waveforms could occur, from a rotating wave in the tip to planar waves in the back. I hypothesised it could be the result of differences in the oscillation frequency between cells in the prestalk

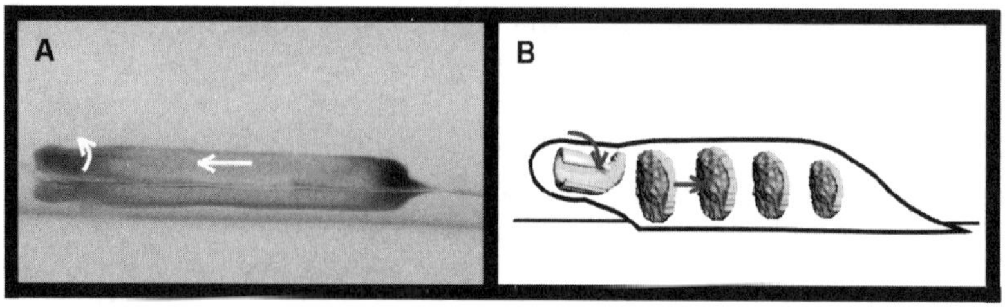

Figure 11.6 Model for wave propagation in slugs. (A) Neutral red stained slug. Cells in the anterior tip (pointing to the left) exhibit rotational movement around the slug's long axis as indicated by the arrow. The cells in the prespore zone move along this axis. (B) Model for the cAMP in the waves in the slug. The prestalk zone is organised by a rotating scroll wave which decomposes into planar waves in the prespore zone.

region of the slug and the prespore region of the slug. This would lead to a change in wave propagation frequency and to a twisting of the wave in the slower oscillating region. Subsequently, using model calculations, we have shown that this difference in excitability between the cells in the prespore zone and in the prestalk zone can give rise to these waveforms (Bretschneider *et al.*, 1995; Steinbock *et al.*, 1993). Depending on the magnitude of difference in excitability between the prestalk and prespore zones, the waves would twist to different degrees. If the difference in excitability between the slug regions was very large, the wave might actually convert into separate planar waves.

The difference in excitability between the prespore and prestalk zones could result from two different mechanisms (1) prestalk cells contain more of the enzymes responsible for the cAMP relay reaction (see above) (2) all the cells in the prestalk zone could relay the signal, but it is relayed by only a few cells, the anteriorlike cells, in the prespore zone of the slug. These cells, on a biochemical level, are very much like prestalk cells but they are arrested at an early stage of prestalk differentiation. They contain all the components necessary for cAMP relay such as cAMP receptors, phosphodiesterase, adenylyl cyclase, etc. They are randomly scattered in the prespore zone and reach a density of around 10–15%. If they were to relay the signal, that would give a good explanation of their existence.

I have also been able to visualise optical density wave propagation in the back end of the prespore zone of *Dictyostelium mucoroides* slugs (Dormann *et al.*, 1997). In these slugs there are clear optical density waves propagating through the slug from the tip to the tail. One of the reasons why I might have been able to visualise the optical density waves here is that the end parts of the slugs are very thin. Furthermore, I found that the motion of the tip is clearly periodic with a periodicity similar to the period of the optical density waves. More recently I have also found propagating planar optical density waves in the prespore zone of *Dictyostelium discoideum* slugs (Dormann and Weijer, unpublished observations). This provides the most direct evidence for wave propagation in the slug.

There is good reason to assume that the cells differ in their relay properties already during the early stages of development. Aggregation stage cells can be separated on density gradients in dense and less dense fractions. Mixing experiments showed that the cells from the dense fractions are going to continue development to cells which sort to the prespore zone of the slug, and differentiate into prespore cells. The cells in the less dense fractions are going to become prestalk cells (Weijer *et al.*, 1984b). The signalling behaviour of these cells has been investigated by following the light scattering oscillations of these cells in a cell suspension. The cells from the dense fraction oscillate at lower frequency when isolated, whereas the less dense fractions oscillate already much faster. When the two cell populations are mixed back together, they will mutually synchronise at some intermediate frequency (Weijer *et al.*, 1984a). This supports the hypothesis of prestalk cells being more excitable than prespore cells.

11.7 Is cAMP the Signal That Controls Cell Movement in the Slug?

A major question still waiting for a definitive answer is whether the signal that controls cell movement in slugs is cAMP. There are many arguments in favour of a role for cAMP in controlling cell movement in slugs. Dissociated slug cells are chemotactically responsive to cAMP. It has also been shown that prestalk A cells when isolated show a more vigorous response to cAMP prestalk O cells or prespore cells (Early *et al.*, 1995). When slugs are placed on agar containing high levels of cAMP, they start to disintegrate and cells leave various parts of the slug in a streamlike fashion (George, 1977). Lower concentrations of cAMP impair the movement of the prespore zone of slugs more than the prestalk zones, resulting in abnormally long slugs. This can now be understood as resulting from a desensitisation of the relatively high-affinity receptors on prespore cells, and the lack of sufficient phosphodiesterase to degrade the cAMP in the local environment of the cells. The prestalk cells, which have lower-affinity receptors and also express more phosphodiesterase, might require higher cAMP concentrations to be inhibited.

Prespore and prestalk cells sort out chemotactically from mounds of cells formed by mixing up slugs. Normally the prestalk cells will go to the top of the mound and form a tip. When the mounds are made on agar containing cAMP, the prestalk cells move preferentially to the outside of the aggregate to form a ring of prestalk cells (Matsukuma and Durston, 1979; Sternfeld and David, 1981) On high concentrations of cAMP, no sorting occurs anymore. Mutants overexpressing the cAMP phosphodiesterase gene under the control of its own prestalk-specific promoter fragment cannot sort out either (Traynor *et al.*, 1992). Finally, I have shown that by periodic micro-injection of pulses of cAMP in the extracellular space of different stages of *Dictyostelium discoideum* development, it is possible to influence optical density propagation and chemotactic cell movement. I could show that, by periodic micro-injection of cAMP pulses into the prespore zone of slugs, it is possible to induce a local aggregation of anteriorlike cells around the electrode (Rietdorf *et al.*, 1998). These results strongly support the hypothesis that the optical density waves observed during early development up to the mound stage, represent propagating cAMP waves.

11.8 Other Possible Mechanisms Controlling Cell Movement in Slugs

There are, however, also a number of arguments against a role for extracellular cAMP in the control of cell movement in mounds and slugs. Recently it has been shown that the development of strains lacking the aggregation stage specific adenylylcyclase gene *ACA* can be made to develop by constitutively overexpressing the catalytic subunit of the cAMP-dependent protein kinase. This mutant can make aggregates and mounds when plated at high density and the mounds will develop to slugs and finally make fruiting bodies (Wang and Kuspa, 1997). Initial experiment have shown that we cannot see any optical density wave propagation in either mounds or slugs of this mutant. Furthermore, there is little evidence for periodic cell movement in slugs of this strain. These new data seem to suggest that, in the absence of periodic cAMP signalling, another mechanism can take over the control of cell movement. This remains to be further investigated. One possibility is that a newly discovered adenylyl cyclase (ACB) takes over the role of the *ACA gene* (H.J. Kim *et al.*, 1998). Maybe cells in slugs can also respond to stationary cAMP gradients possibly generated by this cyclase.

One of the most puzzling observations on the movement of the cells in the later stages of *Dictyostelium* development is that slug stage cells have a clear tendency to stick together and move in groups. Isolated cells are relatively stationary; they extend pseudopodia in all directions but don't make a lot of progress in their movement. Once groups of cells are formed they very often show extensive rotational behaviour. Even little groups of two or three cells can make contact and start to spin around. This is especially obvious after the reaggregation of dissociated slug cells. Initially most of the cells are single but then they rapidly reaggregate and in the process they start to show a pronounced rotational movement. Groups of cells can then reaggregate to form bigger groups of cells, which still rotate. This behaviour is rather similar to the rotational movement observed in some bacterial colonies (BenJacob, 1997). The rotational movement could be caused by the combination of a positive chemoattractant secreted by the cells, which keeps them together, and the drive of slug stage cells to move. The combination of these factors would result in a rotational movement. There is no difficulty in understanding how this mechanism could result in circular cell movement in mounds. In slugs, however, it is much less clear since the movement of all the cells is clearly controlled by the slug tip. Therefore, to do this, the tip must be able to exercise some long-range control over the movement of the cells in the rest of the slug.

11.9 How Do the Cells in the Multicellular Stages of Development Get Their Traction?

One of the problems in understanding the movement of cells in mounds and slugs lies in the fact that it is not immediately obvious how the cells get their traction, which ultimately has to come from the substrate. From the experimental observations it seems clear that the cells do not move on top of each other, since their velocities do not add up. The cells in the higher planes of mound or slug do not move faster than the cells in contact with the substrate. In *Dictyostelium* cell movement is essentially due to the extension (via polymerisation) of the actin network in the leading edge of the cells, which has to be coordinated with a retraction at the back of the cell. Many of the proteins that regulate the polymerisation of actin are also found in the extending pseudopdia. A number of different motor molecules (myosin I isoforms) are found in the extending pseudopodia, whereas the conventional myosin II is localised primarily in the retracting back of the cell. Local stimulation of the cell by a chemoattractant leads to localised actin polymerisation and pseudopod extension, followed by an inhibition of polymerisation and stimulation of depolymerisation at the retracting end of the cell (Maeda *et al.*, 1997). I assume that cells attach to the substrate and other cells via as yet unknown contact molecules which are linked to the internal stationary cytoskeleton. By extending on one site and in a coordinated fashion retracting at the back site, the cell is able to "move". It makes new contact sites continuously near its front and detaches them from the substrate (and/or other cells) at the retracting end. In this manner a cell can translocate and gets its traction through other cells from the substrate. The problem of movement in a densely packed mass of cells then reduces to coordinating the extension and retraction of neighbouring cells. The cells would try to extend in space vacated by retracting cells in front of them. This coordination is achieved by the waves of cAMP. I am now testing this hypothesis in adhesion and cytoskeletal mutants.

11.10 The Mechanism of Cell Sorting

The formation of slugs from a mound involves extensive cell sorting. The cells that form the tip move to the top through all other cells. I have cited evidence that this mechanism involves differential chemotactic cell movement towards cAMP. The difference in effective movement speed towards a cAMP source could be caused by cell type specific differences in the motive force generated by prestalk and prespore cells. Differences in motive force could result from differences in the function of the cytoskeleton, i.e prestalk cells could generate more force in response to a cAMP signal, possibly as a consequence of expressing different cAMP receptors with different kinetic properties. It could also be that all cells develop the same force in response to a signal but differ in their adhesion to the substrate and/or to other cells. There is experimental evidence for both types of mechanisms. Isolated pstA cells, which will sort to the top of the aggregate, are able to move faster to an artificial cAMP source as isolated prespore cells (Early *et al.*, 1995). Furthermore, several mutants with defects in components of the cytoskeleton are arrested at the mound stage and when mixed with non-mutant strains will be excluded from the tip (De Lozanne and Spudich, 1987; Rivero *et al.*, 1996). Using a cold-sensitive myosin mutant it has been shown that there are two stages in development where myosin II is absolutely required for morphogenesis: at the mound stage during tip formation and during culmination (Springer *et al.*, 1994). These mutants support the hypothesis that movement during slug and fruiting body formation requires a high motive force produced by the actin/myosin cytoskeleton. Secondly, in multicellular tissues cell–cell interactions are likely to play an important role in the control of cell movement. Prespore cells are more adhesive than prestalk cells (Siu and Kamboj, 1990), so prestalk cells may move more efficiently in a multicellular aggregate consisting of prespore and prestalk cells. I therefore suspect that cell sorting involves all these mechanisms, i.e. differential chemotaxis towards cAMP, cell type specific differences in the generation of motive force as well as cell type specific differences in cell–cell interactions and cell–substrate interactions.

Cell sorting will provide feedback on the signalling patterns in the tipped mound since prestalk and prespore cells differ in their excitability (see above). Cell sorting should affect the signalling system in the following way: the collection of fast-oscillating prestalk cells in the tip will lead to an increase in excitability in the tip. This will result in a loss of spiral arms to form a simple scroll wave in the tip (Vasiev *et al.*, 1997a, b). The removal of the highly excitable prestalk cells from the body of the mound will produce a decrease in the local excitability of the prespore cell mass and cause the scroll wave in the tip to become a twisted scroll wave in the mound (Bretschneider *et al.*, 1995; Steinbock *et al.*, 1993). This will lead to a twisted rotational cell movement in the base of the mound. As a result, the mound contracts, i.e. the cells intercalate and at the same time the mound elongates and extends up into the air (Vasiev and Weijer, 1999).

11.11 Culmination

Culmination is the last part of the *Dictyostelium* developmental cycle. It also results from complex differential cell movement. After a suitable signal (overhead light) the slug rounds up and the tip migrates to the top of the culminate. Here some of the prestalk cells start to invaginate and form the stalk tube. This process is clearly very complex and requires precisely coordinated cell movement. I investigated cell movement by analysing time-lapse video recordings made during the slug stage and the culmination phase of *Dictyostelium* development. Slugs viewed from the side showed an even, straightforward movement with the tip slightly erected in the air (Dormann *et al.*, 1996). Culmination is initiated by a local aggregation of anteriorlike cells at the base of the slug at the prestalk–prespore boundary, where they form a stationary mass of cells. The mass of cells seems to form by active aggregation of cells from all regions of the slug. This pile seems to contain cells expressing the prestalk cell specific *ecmA* and *ecmB* genes (Jermyn *et al.*, 1996). The other cells seem to follow the tip, move over this pile and are lifted up in the air. As a result, the cells in the pile move to the back of the slug and form a second signalling centre beside the tip. Both centres are characterised by vigorous rotational cell movement. The cells belonging to the basal centre will form the basal disc and the lower cup in the fruiting body. The upper cup will be formed by the prestalk cells rotating most vigorously at the prestalk–prespore boundary. The remaining neutral red stained anteriorlike cells in the prespore zone sort either to the upper or lower organising centre in the fruiting body. It remains to be established what signals the cells in the basal signalling centre use to coordinate their movement.

11.12 Modelling Morphogenesis

Understanding the principles governing the morphogenesis of even a simple organism such as *Dictyostelium* requires understanding many complex cellular interactions. Since these interactions are in general highly non-linear it is very difficult to imagine how they may work, even qualitatively. A population of *Dictyostelium* cells behaves as an excitable biological medium in which the cells communicate by propagating waves with all their inherent pattern-forming properties. On top of that the waves interact with the dynamics of the medium on at least two time-scales. On a short time-scale by inducing a motion of the excitable elements, the cells. On a longer time-scale by interacting with gene expression, which in turn changes the signalling and movement dynamics of the cells. Attempts to model the morphogenesis of *Dictyostelium* will require formalising the essential signalling and cell movement and cell–cell interaction mechanisms and will therefore test our understanding of the principal interactions involved.

11.12.1 Aggregation

There have been several attempts to model the early aggregation process. Two main questions that need to be addressed are: How do the cells produce cAMP waves? And how do they move in response to these waves? Essentially two types of models were developed to mathematically describe the cAMP relay kinetics of *Dictyostelium* amoebae. The first model has been suggested by Martiel and Goldbeter (Goldbeter, 1996). It is based on the assumption that activation/inactivation of the cAMP receptors plays a key role in the response. A second model has been introduced by Tang and Othmer (Tang and Othmer, 1994). In this model the receptor-dependent activation of activating and inhibitory G proteins controls the periodic cAMP production. Both models are able to describe oscillations in the cAMP level in cell suspensions as well as cAMP wave propagation in a dispersed cell population (Tang and Othmer, 1995; Tyson and Murray, 1989). These models basically describe excitable and/or oscillatory media and are very similar to the prototype FitzHugh–Nagumo system:

$$\frac{\partial g}{\partial t} = D\Delta g - \rho[k_g g(g-0.05)(g-1) + k_r r] \tag{11.1}$$

$$\frac{\partial r}{\partial t} = \frac{(g-r)}{\tau} \tag{11.2}$$

The first equation describes the excitation of the medium, defined by the variable (g) over time. This variable is linked to the extracellular cAMP concentration. The second equation defines the recovery process of the medium (r) and could be thought to describe the desensitisation of the cAMP receptors. D is the diffusion coefficient for cAMP; τ is a time scaling factor for the variables r and g; k_g and k_r define the rate of production and hydrolysis of cAMP by one cell. The effect of cell density on the excitability of the medium is accounted for by the factor ρ, which scales the excitability of the medium (rate of production of cAMP). This model can describe wave propagation in excitable and oscillatory media. cAMP waves not only propagate through the cell population but also coordinate cell movement. cAMP orients the direction of otherwise randomly moving cells. There is strong evidence that the cells detect the gradient of cAMP over their length (Caterina and Devreotes, 1991; Gerisch, 1978; Mato *et al.*, 1975). However, there is also evidence that cells use the temporal derivative of cAMP and only move up the gradient as long as the cAMP level is rising (Wessels *et al.*, 1992, 1996). This allows cells to move chemotactically on the wavefront rather than on the waveback. A number of mathematical models have been proposed for chemotactic cell movement. The best-known model is the Keller–Segel model (Keller and Segel, 1970) describing a cell flux $\mathbf{J}$ as a function of cell density ρ and cAMP concentration g:

$$\mathbf{J} = -D(g)\nabla\rho + \chi(g)\rho\nabla g \tag{11.3}$$

The first term on the righthand side describes random cell movement (the velocity

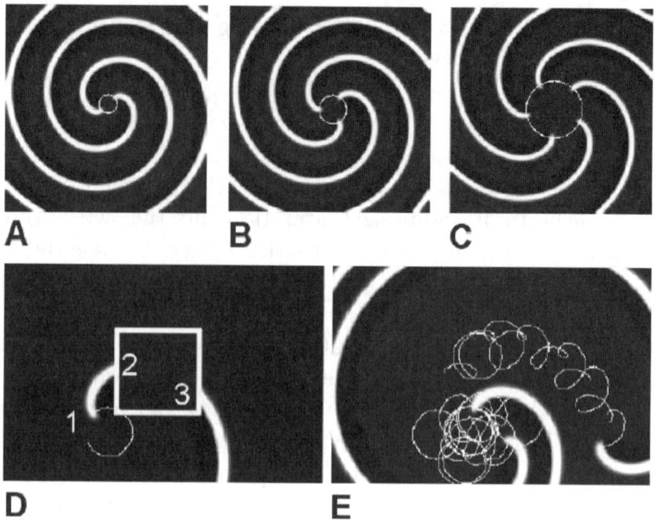

Figure 11.7 Formation of multi-armed spirals in a two-dimensional excitable medium described by the FitzHugh–Nagumo model (11.1). (A,B,C) Two-, three- and five-armed spirals are shown in media with different excitability. The dotted lines in the centre of the patterns indicate the path followed by the spiral tips. (D,E) Breaking of a single spiral can result in the formation of a two-armed spiral. For initial conditions and parameter values see Vasiev *et al.* (1997a).

can depend on the level of cAMP), and the second term describes the directed motion of the cells along the cAMP gradient. Equations (11.1), (11.2) and (11.3) can describe aggregation (Vasiev and Weijer, 1999).

Much effort has been directed towards understanding the physical principles leading to the formation of streaming patterns by amoebae responding chemotactically to propagating cAMP waves. It has been shown analytically and numerically that streams form due to a streaming instability caused by a coupling between the velocity of signal propagation and the density of cells (Höfer and Maini, 1997; Höfer *et al.*, 1995; Levine and Reynolds, 1991; Vasiev *et al.*, 1994). This is presumably related to the angular instability as discussed by Shaffer and Nanjundiah (Nanjundiah, 1973; Shaffer, 1957). Due to the dependence of the rate of cAMP accumulation on cell density, local accumulation of cells will result in a speeding up of the wave propagation. This local deformation of the wavefront will lead to the attraction of even more cells to this region and finally to the formation of bifurcating aggregation streams (Höfer *et al.*, 1995; van Oss *et al.*, 1996; Vasiev *et al.*, 1994), in which the cells move towards the aggregation centre. There they pile on top of each other to form a three-dimensional hemispherical structure, the mound. Multi-armed spirals can then develop in the mound. These waves are only possible if the arms rotate along a relatively large trajectory and the cells behind the wavefront become excitable again long before the original spiral comes around to re-excite them. This way a new arm will fit in between. In theoretical calculations I have shown that multi-armed spirals can form by breaks within the original spiral wave within one wavelength of the wave tip (Figure 11.7). This will generate two new free wave ends. These free ends will develop into spiral wave tips themselves. The new wave end rotating in the opposite direction to the original tip will be expelled from the medium. The wave end rotating in the same direction as the original tip will be attracted by the original tip to form a two-armed spiral. This process, when repeated, results in multi-armed spirals (Vasiev *et al.*, 1997a). This change in signalling pattern is most likely brought about by the change in the expression of the cAMP receptors.

11.12.2 Mound formation

While aggregation patterns were an object of intensive theoretical investigations for almost 30 years, formation of the mound was not considered in any detail until recently. (Savill and Hogeweg, 1997; Vasiev *et al.*, 1997b; Bretschneider *et al.*, 1997; Levine *et al.*, 1997). All of them are very interesting and treat the problem from different points of view. The model by Savill and Hogeweg (1997) is based on the assumption that the main driving force for cell motility is the tendency of the cells to try to decrease their free surface energy by randomly moving around. This energy depends on a cell's adhesive interactions with other cells and with the extracellular matrix. Furthermore, it is strongly affected by the level of cAMP. Using this model the authors were able to simulate all stages of Dd development from single cells to a crawling slug. According to this model, the main force responsible for mound formation is cell–cell adhesion. Another important outcome is that differential cell–cell adhesion could account for cell sorting in the mound.

The model developed by Vasiev *et al.* (1997b) was based on the assumption that the chemotactic movement of the cells can be viewed as fluid flows and described by the Navier–Stokes equations (Figure 11.8). Using this hydrodynamic approach it was possible to simulate the formation of streaming patterns and mound formation. It was found that hemispherical mounds form even in the absence of cell–cell adhesion. Two forces maintain the mound: an inwardly directed force (accumulation by chemotaxis) and a pressure which prevents close compaction of the chemotactically moving cells. A further two papers (Bretschneider *et al.*, 1997; Levine *et al.*, 1997) deal with hybrid models, in which a number of cells were modelled to move according to a number of rules (different rules in each model) in a continuum field of cAMP. The model by Bretschneider *et al.* (1997) reinforces the importance of pressure in the process of mound formation, and another model (Levine *et al.*, 1997) stresses the importance of adhesive forces in maintaining the mound. There is, however, an important advantage in the results obtained using the pressure-based models. In these models the cells continue to move inside the mound (i.e. they rotate along the mound's vertical axis in response to a spiral wave), but in both adhesion-based models the cells stop moving after the formation of the mound, in contradiction with the experimental findings.

11.12.3 Cell sorting and slug formation

Once the cell types have formed, they have to sort themselves to form the axial pattern of cell types found in the slug. Sorting of cells in a mound or slug has also been the subject of extensive theoretical consideration (Meinhardt, 1983; Sekimura *et al.*, 1987; Umeda, 1993). The main problems to be solved are the mechanisms responsible for cell sorting and how the final pattern is stabilised? In order to become sorted, the different

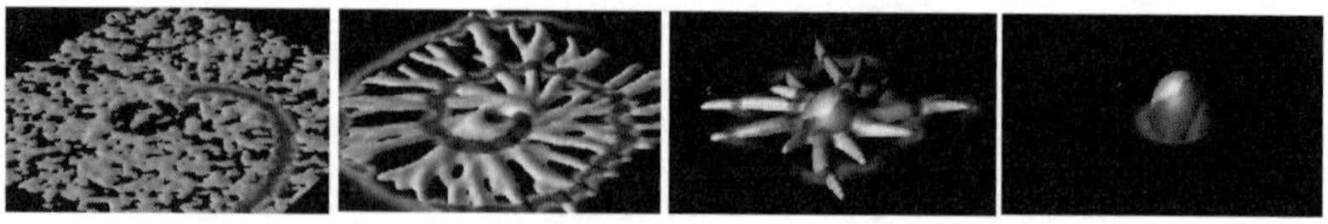

Figure 11.8 Formation of the streaming patterns and the mound in the hydrodynamic model (Vasiev *et al.*, 1997a). Cell density is shown as a white isosurface (= 0.5) and the cAMP concentrations are mapped on this surface from low cAMP (white) to high cAMP (black). The initial density of cells was zero everywhere in 3D space except for the bottom plane. The cell density in each grid of this plane was represented by a random number varying between 0 and 1 so the average density in this plane was equal to 0.5. In response to cAMP, spiral wave cells move and form aggregation streams ($t = 40 - -200$ min) and then mound ($t = 250$ min) which represents a stable solution of the system.

cell types should move differentially inside the mound. This differential motion can be caused by cell type specific differences in the motive force generated by prestalk and prespore cells, caused by differential chemotaxis (different cell types exert different forces due to cell type specific differences in their cytoskeleton) or by differential adhesion (see above). According to several theoretical investigations, the force that drives cell sorting is differential adhesion and the final pattern of cell sorting is defined by the relative adhesion forces between different cell types and between the cells and the substratum (Savill and Hogeweg, 1997). I have shown that cell sorting can be driven only by differential chemotaxis (Vasiev and Weijer, 1999). According to these model calculations, sorting can be achieved if the following conditions are satisfied: (1) a scroll wave of cAMP rotates in the mound and causes counter rotational cell movement; (2) prestalk and prespore cells have different excitabilities, prestalk cells are more excitable than prespore cells; (3) prestalk and prespore cells move chemotactically to cAMP signals as during aggregation, but prestalk cells move faster than prespore cells in response to the cAMP waves; (4) the movement of the cells is treated as a fluid flow.

The model used in this investigation is an expansion of the hydrodynamic model used in Vasiev *et al.* (1997b). Here I consider a mound as consisting of a mixture of two liquids which correspond to prestalk and prespore cells. The motion of each liquid is defined by the momentum balance equation:

$$\frac{\partial(\alpha_i \mathbf{V}_i)}{\partial t} + (\mathbf{V}_i \nabla)(\alpha_i \mathbf{V}_i) = \mathbf{F}_i + \eta \nabla.(\alpha_i \nabla \mathbf{V}_i) - \alpha_i \nabla p + (-1)^i \psi \alpha_1 \alpha_2 (\mathbf{V}_1 - \mathbf{V}_2), \quad (11.4)$$

where $i = 1, 2$ is an index representing prestalk and prespore cells, α_i are the corresponding volume fractions, $\mathbf{V}_i$ the velocities and $\mathbf{F}_i$ the chemotactic forces. The second term on the right-hand side defines the cell–cell interactions between the cells of the same type; the last term defines the interactions between the different cell types. Pressure is defined by the condition of total incompressibility of the mound. I neglect terms containing density derivatives and use the reduced version of equation (11.4):

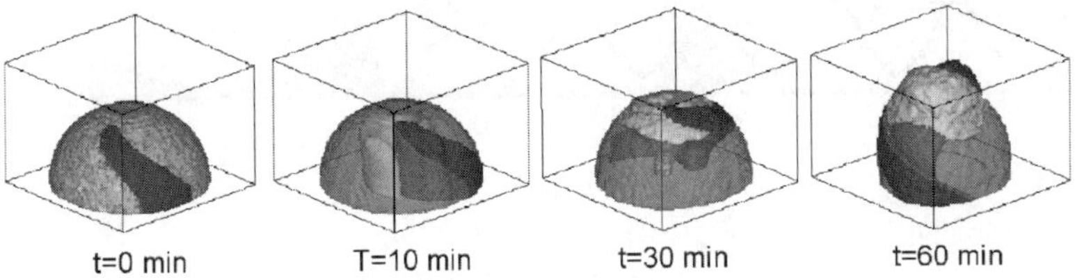

Figure 11.9 Cell sorting in the hydrodynamic model. The mound consists of 20% prestalk cells (light gray) and 80% prespore cells (grey). The cell types differ in their chemotactic response and excitability (Vasiev and Weijer, 1999). The prestalk cells move faster and are more excitable than prespore cells. Initially the mound is a hemisphere in which a cAMP scroll wave (dark grey) rotates clockwise, and both cell types are mixed randomly. Affected by the cAMP waves the cells move and sort, such that the prestalk cells collect at the top of the mound and form a tip.

$$\alpha_i \frac{\partial \mathbf{V}_i}{\partial t} + \alpha_i(\mathbf{V}_i \nabla)\mathbf{V}_i = \mathbf{F}_i + \eta\alpha_i \Delta \mathbf{V}_i - \alpha_i \nabla p + (-1)^i \psi \alpha_1 \alpha_2 (\mathbf{V}_1 - \mathbf{V}_2). \qquad (11.5)$$

I assumed that the force exerted by prestalk cells in response to the chemoattractant cAMP is stronger than the force exerted by prespore cells. The FitzHugh–Nagumo equations describe the cell signalling system, where it is assumed the excitability for prestalk cells k_g in (11.1), is higher than for prespore cells, in agreement with previous experimental data.

My computations show that starting from a random distribution of cell types in the mound, one can obtain a spatially separated pattern of cell types as well as tip formation (Figure 11.9). In response to scroll waves rotating along the vertical axis of the hemispherical mound, the liquid begins to rotate in the opposite direction. The faster-moving fluid (prestalk cells) accumulates in the centre and at the top of the mound (the pressure developing in the centre of the mound drives the prestalk cells to the top of the mound and leads to asymmetry along the vertical axis). Since the faster fluid is more excitable, separation of fluids leads to the mound becoming inhomogeneous with respect to its excitability, so the top of the mound is more excitable than the bottom. This in turn results in a change of the cAMP wave shape, i.e. the scroll becomes twisted and gets a component of velocity directed downwards to the substratum, along the vertical mound's axis. As a consequence, the cells experience a chemotactic force with an upward component which results in the further collection of faster-moving cells on the top. Finally all the faster cells collect at the top of the mound and form a tip, very similar to what happens in real mounds.

According to my numerical simulations, the following mechanisms are responsible for cell sorting and slug formation. (1) The chemotactic cell flows within a mound, arising in response to a scroll wave of cAMP, always transform the hemispherical aggregate into a cylindrical aggregate. (2) The geometry of this cylinder depends on the excitability of the cells; cylinders formed by more excitable cells are thinner. (3) Since the more excitable prestalk cells form the tip, it differs in geometry from the rest of the mound and has a smaller cross-section. (4) Since the twisted scroll originates in the tip and since prestalk cells move to the top of the mound, the mound becomes

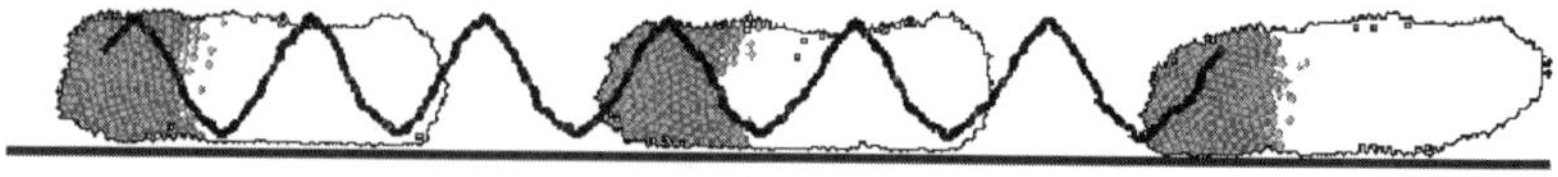

Figure 11.10 Migrating slug in a model system (Bretschneider *et al.*, 1999). Side view of a migrating slug at three successive times. The slug is moving from the right to the left; its tip (formed by prestalk cells) is depicted in grey. It shows the track of one prestalk cell (arbitrarily chosen). The spiral-shaped trajectory indicates that the cell is moving in a rotational fashion. Slug migration is driven by cAMP waves of the shape shown in Figure 11.9.

elongated to form a standing slug.

11.12.4 Slug migration

Probably the most important property of the slug is its ability to crawl. There have been a few attempts to model migration of the slug (Bretschneider *et al.*, 1999; Odell and Bonner, 1986; Umeda, 1989). Slug migration has to result from the coordinated motion of its constitutive cells, and most models assume this motion is chemotactic. Odell and Bonner (1986) were the first to propose that the flow of tissue in a slug could be modelled as the flow of a viscous fluid under the influence of a chemoattractant. One important extra assumption was that all cells produced and destroyed a chemical, which modulated the chemotactic response of all cells, resulting in a gradient from the centre to the outside. This resulted in fountainlike cell flow patterns along the long axis of the slug, which are not in agreement with experimental observations (Siegert and Weijer, 1992). More recent attempts to simulate slug migration using the hydrodynamical model and the discrete pressure-adhesion model are more successful (Bretschneider *et al.*, 1999). They can result in slug movement when the movement of individual cells in the slug is organised by a twisted scroll wave of cAMP (Figure 11.10). They show reasonable patterns of cell movement inside the moving slug. These models assume that the chemotactic force is a body force and that the traction comes from the substrate, via other cells, as discussed above. This now has to be tested experimentally.

11.13 Conclusion and Outlook

Although many of the details are not yet fully understood, *Dictyostelium* will be one of the first organisms where it will be possible to understand the principles that govern its morphogenesis at the cellular level. It is now clearly established that cAMP wave propagation and chemotaxis control the early aggregation stage of development. The formation of aggregation streams can now be understood as a movement-induced instability. The final steady state achieved with one cell type is a hemispherical mound. The behaviour of the system becomes more complex as the cells differentiate into several types with different excitabilities and chemotactic properties. Due to these additional levels of regulation and feedback, complicated waveforms arise, such as multi-armed spirals and twisted scroll waves. The differentiation of these cell types leads to the sorting out of cells with different excitabilities and movement properties.

The spatial separation of these cell types then feeds back on the signal, leading to transformation of the twisted scroll wave in the tip to planar waves in the prespore zone of the slug. This mode of wave propagation can then explain slug movement. It can now be tested by investigating *Dictyostelium* morphogenesis in signalling and motility mutants which should give rise to predictable phenotypes. This is now under way. The culmination phase remains a challenge for the future.

Acknowledgements
I thank my collaborators Dr Dirk Dormann and Dr Bakhtier Vasiev for providing the figures, the experimental data and the model calculations. I thank the BBSRC and the Wellcome Trust for financial support.

REFERENCES

Abe, T., Early, A., Siegert, F., Weijer, C. and Williams, J. (1994) Patterns of cell movement within the *Dictyostelium* slug revealed by cell type-specific, surface labeling of living cells. *Cell* 77: 687–699.

Alcantara, F. and Monk, M. (1974) Signal propagation during aggregation in the slime mould *Dictyostelium discoideum. J. Gen. Microbiol.* 85: 321–334.

Ben Jacob, E. (1997) From snowflake formation to growth of bacterial colonies. 2. Cooperative formation of complex colonial patterns. *Contemporary Physics* 38: 205–241.

Bonner, J.T. (1983) Chemical signals of social amoebae. *Sci. Am.* 248: 114–120.

Bretschneider, T., Siegert, F. and Weijer, C.J. (1995) Three-dimensional scroll waves of cAMP could direct cell movement and gene expression in *Dictyostelium* slugs. *Proc. Natl. Acad. Sci. USA* 92: 4387–4391.

Bretschneider, T., Vasiev, B. and Weijer, C.J. (1997) A model for cell movement during *Dictyostelium* mound formation. *J. Theor. Biol.* 189: 41–52.

Bretschneider, T., Vasiev, B. and Weijer, C.J. (1999) A Model for *Dictyostelium* Slug Movement. *J. Theor. Biol.* 199: 125–136.

Brown, S.S. and Rutherford, C. L. (1980) Localization of cyclic nucleotide phosphodiesterase in the multicellular stages of *Dictyostelium discoideum. Differentiation* 16: 173–183.

Caterina, M.J., and Devreotes, P.N. (1991) Molecular insights into eukaryotic chemotaxis. *FASEB J.* 5: 3078–3085.

Chen, M.Y., Insall, R.H., and Devreotes, P.N. (1996) Signaling through chemoattractant receptors in *Dictyostelium. Trends in Genetics* 12: 52–57.

Clarke, M., and Gomer, R.H. (1995) PSF and CMF, autocrine factors that regulate gene expression during growth and early development of *Dictyostelium. Experientia* 51: 1124–1134.

De Lozanne, A. and Spudich, J.A. (1987) Disruption of the *Dictyostelium* myosin heavy chain gene by homologous recombination. *Science* 236: 1086–1091.

De Wit, R.J.W. and Konijn, T.M. (1983) Pterins and folates as extracellular signals for chemotaxis and differentiation in the cellular slime molds. In: *Biochemical and Clinical Aspects of Pteridines*, H. C. Curtius, W. Pfleiderer and H. Wachter, eds. (New York: Walter de Gruyter), pp. 383–400.

Dormann, D., Siegert, F. and Weijer, C.J. (1996) Analysis of cell movement during the culmination phase of *Dictyostelium* development. *Development* 122: 761–769.

Dormann, D., Weijer, C. and Siegert, F. (1997) Twisted scroll waves organize *Dictyostelium mucoroides* slugs. *J. Cell Sci.* 110: 1831–1837.

Early, A.E., Abe, T. and Williams, J. (1995) Evidence for positional differentiation of prestalk cells and for a morphogenetic gradient in *Dictyostelium. Cell* 83: 91–99.

Early, A.E., Gaskell, M.J., Traynor, D. and Williams, J.G. (1993) Two distinct populations of prestalk cells within the tip of the migratory *Dictyostelium* slug with differing fates at culmination. *Development* 118: 353–362.

Foerster, P., Muller, S. and Hess, B. (1990) Curvature and spiral geometry in aggregation patterns of *Dictyostelium discoideum*. *Development* 109: 11–16.

George, R.P. (1977) Disruption of multicellular organization in the cellular slime molds by cyclic AMP. *Cell Diff.* 5: 293–300.

Gerisch, G. (1978) Cell interactions by cyclic AMP in *Dictyostelium*. *Biol. Cellulaire* 32: 61–68.

Gerisch, G. (1987) Cyclic AMP and other signals controlling cell development and differentiation in *Dictyostelium*. *Ann. Rev. Biochem.* 56: 853–879.

Gerisch, G., Noegel, A., Schleicher, M., Segall, J. and Wallraff, E. (1987) Signal transduction and chemotaxis in *Dictyostelium discoideum*. *Biol. Chem. H-S*, pp. 1045–1046.

Gingle, A.R. and Robertson, A. (1976) The development of the relaying competence in *Dictyostelium discoideum*. *J. Cell Sci.* 20: 21–27.

Glazer, P.M., and Newell, P.C. (1981) Initiation of aggregation by *Dictyostelium discoideum*in mutant populations lacking pulsatile signalling. *J. Gen. Microbiol.* 125: 221–232.

Goldbeter, A. (1996) *Biochemical Oscillations and Cellular Rhythms. The Molecular Bases of Periodic and Chaotic Behaviour.* Cambridge: Cambridge Univ. Press.

Gross, J.D., Peacey, M.J. and Trevan, D.J. (1976) Signal emission and signal propagation during early aggregation in *Dictyostelium discoideum*. *J. Cell Sci.* 22: 645–656.

Harloff, C., Gerisch, G. and Noegel, A.A. (1989) Selective elimination of the contact site A protein of *Dictyostelium discoideum* by gene disruption. *Genes Devl.* 3: 2011–2019.

Höfer, T. and Maini, P.K. (1997) Streaming instability of slime mold amoebae: An analytical model. *Phys. Rev. E* 56: 2074–2080.

Höfer, T., Sherratt, J.A. and Maini, P.K. (1995) *Dictyostelium-discoideum* - cellular self-organization in an excitable biological medium. *Proc. Roy. Soc. Lond. Series B* 259: 249–257.

Jermyn, K.A. and Williams, J.G. (1995) Comparison of the *Dictyostelium Rasd* and *Ecma* genes reveals 2 distinct mechanisms whereby an messenger-RNA may become enriched in prestalk cells. *Differentiation* 58: 261–267.

Jermyn, K.A., Duffy, K.T. and Williams, J.G. (1989) A new anatomy of the prestalk zone in *Dictyostelium*. *Nature* 340: 144–146.

Jermyn, K.A., Traynor, D. and Williams, J.G. (1996) The initiation of basal disc formation in *Dictyostelium-discoideum* is an early event in culmination. *Development* 122: 753–760.

Kay, R.R. (1992) Cell differentiation and patterning in *Dictyostelium*. *Curr. Opin. Cell Biol.* 4: 934–938.

Kay, R.R. (1997) *Dictyostelium* development: Lower STATs. *Current Biology* 7: R723–R725.

Keller, E.F. and Segel, L.A. (1970a) Conflict between positive and negative feedback as an explanation for the initiation of aggregation in slime mould amoebae. *Nature* 227: 1365–1366.

Keller, E.F. and Segel, L.A. (1970b) Initiation of slime mold aggregation viewed as an instability. *J. Theor. Biol.* 26: 399–415.

Kim, H.J., Chang, W.T., Meima, M., Gross, J.D. and Schaap, P. (1998) A novel adenylyl cyclase detected in rapidly developing mutants of *Dictyostelium*. *J. Biol. Chem.* 273: 30859–30862.

Kim, J.Y., Borleis, J.A. and Devreotes, P.N. (1998) Switching of chemoattractant receptors programs development and morphogenesis in *Dictyostelium*: Receptor subtypes activate common responses at different agonist concentrations. *Dev. Biol.* 197: 117–128.

Levine, H. and Reynolds, W. (1991) Streaming instability of aggregating slime mold amoebae. *Phys. Rev. Lett.* 66: 2400–2403.

Levine, H., Tsimring, L. and Kessler, D. (1997) Computational modeling of mound development in *Dictyostelium*. *Physica D* 106: 375–388.

Maeda, Y., Inouye, K., and Takeuchi, I. (1997) *Dictyostelium: A Model System for Cell and*

Developmental Biology. Tokyo: Universal Academy Press.

Mato, J.M., Losada, A., Nanjundiah, V. and Konijn, T.M. (1975) Signal input for a chemotactic response in the cellular slime mold *Dictyostelium discoideum*. *Proc. Natl. Acad. Sci. USA* 72: 4991–4993.

Matsukuma, S. and Durston, A.J. (1979) Chemotactic cell sorting in *Dictyostelium discoideum*. *J. Embryol. Exp. Morphol.* 50: 243–251.

Meinhardt, H. (1983) A model for the prestalk/prespore patterning in the slug of the slime mold *Dictyostelium discoideum*. *Differentiation* 24: 191–202.

Nanjundiah, V. (1973) Chemotaxis, signal relaying and aggregation morphology. *J. Theor. Biol.* 42: 63–105.

Odell, G.M., and Bonner, J.T. (1986) How the *Dictyostelium discoideum* grex crawls. *Phil. Trans. R. Soc. Lond. B* 312: 487–525.

Pahlic, M. and Rutherford, C.L. (1979) Adenylate cyclase activity and cyclic AMP levels during the development of *Dictyostelium discoideum*. *J. Biol. Chem.* 254: 9703–9707.

Parent, C.A. and Devreotes, P.N. (1996) Molecular genetics of signal transduction in *Dictyostelium*. *Ann. Rev. Biochem.* 65: 411–440.

Ponte, E., Bracco, E., Faix, J. and Bozzaro, S. (1998) Detection of subtle phenotypes: The case of the cell adhesion molecule csA in *Dictyostelium*. *Proc. Natl. Acad. Sci. USA* 95: 9360–9365.

Rietdorf, J., Siegert, F. and Weijer, C.J. (1996) Analysis of optical-density wave-propagation and cell-movement during mound formation in *Dictyostelium-discoideum*. *Dev. Biol.* 177: 427–438.

Rietdorf, J., Siegert, F. and Weijer, C.J. (1998) Induction of optical density waves and chemotactic cell movement in *Dictyostelium discoideum* by micro-injection of cAMP pulses. *Dev. Biol.* 204: 525–536.

Rivero, F., Koppel, B., Peracino, B., Bozzaro, S., Siegert, F., Weijer, C.J., Schleicher, M., Albrecht, R. and Noegel, A.A. (1996) The role of the cortical cytoskeleton - F-actin cross-linking proteins protect against osmotic-stress, ensure cell-size, cell-shape and motility, and contribute to phagocytosis and development. *J. Cell Sci.* 109: 2679–2691.

Rutherford, C., Brown, S. and Armant, D. (1981) Enzymatic potential for establishing gradients of cyclic AMP during cell determination in *Dictyostelium discoideum*. *Adv. Cycl. Nucl. Res.* 705.

Savill, N.J. and Hogeweg, P. (1997) Modelling morphogenesis: From single cells to crawling slugs. *J. Theor. Biol.* 184: 229–235.

Schaap, P., Konijn, T.M. and Van Haastert, P.J.M. (1984) cAMP pulses coordinate morphogenetic movement during fruiting body formation of *Dictyostelium minutum*. *Proc. Natl. Acad. Sci. USA* 81: 2122–2126.

Schaap, P., Pinas, J.E. and Wang, M. (1985) Patterns of cell differentiation in several cellular slime mold species. *Dev. Biol.* 111: 51–61.

Sekimura, T., Mimura, M. and Teramoto, E. (1987) A continuous model of pattern formation in *Dictyostelium discoideum*. *Dev. Growth Diff.*, 406.

Shaffer, B.M. (1957) Apects of aggregation in cellular slime moulds. I. Orientation and chemotaxis. *American Naturalist* 91: 19–35.

Shimomura, O., Suthers, H.L.B. and Bonner, J.T. (1982) Chemical identity of the acrasin of the cellular slime mold *Polysphondylium violacium*. *Proc. Natl. Acad. Sci. USA* 79: 7376–7379.

Siegert, F. and Weijer, C.J. (1989) Digital image processing of optical density wave propagation in *Dictyostelium discoideum* and analysis of the effects of caffeine and ammonia. *J. Cell Sci.* 93: 325–335.

Siegert, F. and Weijer, C.J. (1991) Analysis of optical density wave propagation and cell movement in the cellular slime mold *Dictyostelium discoideum*. *Physica D* 49: 224–232.

Siegert, F., and Weijer, C.J. (1992) Three-dimensional scroll waves organize *Dictyostelium* slugs. *Proc. Natl. Acad. Sci. USA* 89: 6433–6437.

Siegert, F. and Weijer, C.J. (1995) Spiral and concentric waves organize multicellular *Dictyostelium* mounds. *Curr. Biol.* 5: 937–943.
Siegert, F. and Weijer, C.J. (1992) Three-dimensional scroll waves organize *Dictyostelium* slugs. *Proc. Natl. Acad. Sci. USA* 89: 6433–6437.
Siu, C.H. and Kamboj, R.K. (1990) Cell-cell adhesion and morphogenesis in *Dictyostelium discoideum*. *Dev. Genet.* 11: 377–387.
Springer, M.L., Patterson, B. and Spudich, J. A. (1994) Stage-specific requirement for myosin II during *Dictyostelium* development. *Development* 120: 2651–2660.
Stadler, J., Keenan, T.W., Bauer, G. and Gerisch, G. (1989) The contact site A glycoprotein of *Dictyostelium discoideum* carries a phospholipid anchor of a novel type. *EMBO J.* 8: 371–377.
Steinbock, O., Siegert, F., Muller, S.C. and Weijer, C.J. (1993) Three-dimensional waves of excitation during *Dictyostelium* morphogenesis. *Proc. Natl. Acad. Sci. USA* 90: 7332-7335.
Sternfeld, J. and David, C.N. (1981) Cell sorting during pattern formation in *Dictyostelium*. *Differentiation* 20: 10–21.
Sukumaran, S., Brown, J.M., Firtel, R.A. and McNally, J.G. (1998) lagC-null and gbf-null cells define key steps in the morphogenesis of *Dictyostelium* mounds. *Dev. Biol.* 200: 16–26.
Tang, Y. H., and Othmer, H. G. (1994) A G protein-based model of adaptation in *Dictyostelium discoideum*. *Math. Biosci.* 120: 25–76.
Tang, Y.H. and Othmer, H.G. (1995) Excitation, oscillations and wave propagation in a G-protein-based model of signal transduction in *Dictyostelium discoideum*. *Phil. Trans. R. Soc. Lond. B* 349: 179–195.
Tomchik, K. J., and Devreotes, P. N. (1981) Adenosine $3',5'$-monophosphate waves in *Dictyostelium discoideum*: A demonstration by isotope dilution-fluorography technique. *Science* 212: 443–446.
Traynor, D., Kessin, R.H. and Williams, J.G. (1992) Chemotactic sorting to cAMP in the multicellular stages of *Dictyostelium* development. *Proc. Natl. Acad. Sci. USA* 89: 8303–8307.
Tyson, J.J. and Murray, J.D. (1989) Cyclic AMP waves during aggregation of *Dictyostelium* amoebae. *Development* 106: 421–426.
Umeda, T.A. (1989) A mathematical model for cell sorting, migration and shape in the slug stage of *Dictyostelium discoideum*. *Bull. Math. Biol.* 51: 485–500.
Umeda, T.A. (1993) A thermodynamical model of cell distributions in the slug of cellular slime mold. *Bull. Math. Biol.* 55: 451–464.
van Oss, C., Panfilov, A.V., Hogeweg, P., Siegert, F. and Weijer, C.J. (1996) Spatial pattern formation during aggregation of the slime mould *Dictyostelium discoideum*. *J. Theor. Biol.* 181: 203–13.
Varnum-Finney, B.J., Voss, E. and Soll, D.R. (1987) Frequency and orientation of pseudopod formation of *Dictyostelium discoideum* amebae chemotaxing in a spatial gradient: Further evidence for a temporal mechanism. *Cell Motil. Cytoskel.* 8: 18–26.
Vasiev, B.N. and Weijer, C.J. (1999) Modeling chemotactic cell sorting during *Dictyostelium discoideum* mound formation. *Biophy. J.* 76: 595–605.
Vasiev, B.N., Hogeweg, P. and Panfilov, A.V. (1994) Simulation of *Dictyostelium-discoideum* aggregation via reaction-diffusion model. *Phys. Rev. Lett.* 73: 3173–3176.
Vasiev, B.N., Siegert F., Weijer C.J. (1997a) Multiarmed spirals in excitable media. *Phys. Rev. Lett.* 78: 2489–2492.
Vasiev, B.N., Siegert, F. and Weijer, C.J. (1997b) A hydrodynamic model for *Dictyostelium discoideum* mound formation. *J. Theor. Biol.* 184: 441–455.
Wang, B. and Kuspa, A. (1997) *Dictyostelium* development in the absence of cAMP. *Science* 277: 251–254.
Weeks, G. and Weijer, C.J. (1994) The *Dictyostelium* cell cycle and its relationship to differentiation. *FEMS Microbiol. Lett.* 124: 123–130.
Weijer, C.J., McDonald, S.A. and Durston, A.J. (1984a) A frequency difference in optical-

density oscillations of early *Dictyostelium discoideum* density classes and its implications for development. *Differentiation* 28: 9–12.

Weijer, C.J., McDonald, S.A., and Durston, A.J. (1984b) Separation of *Dictyostelium discoideum* cells in density classes throughout their development and their relation to the later cell types. *Differentiation* 28: 13–23.

Wessels, D., Murray, J. and Soll, D.R. (1992) Behavior of *Dictyostelium* amoebae is regulated primarily by the temporal dynamics of the natural cAMP wave. *Cell Motil. Cytoskel.* 23: 145–156.

Wessels, D., Shutt, D., Voss, E. and Soll, D.R. (1996) Chemotactic decisions by *Dictyostelium* amoebae in spatial gradients and natural waves of cAMP are made by pseudopods formed primarily off the substratum. *Molecular Biol. Cell* 7: 1349–1349.

Williams, J.G., Abe, T., Nelson, M. and Hopper, N. (1995) Interacting Signaling Pathways Regulating *Dictyostelium* Morphogenesis. *J. Cell. Biochem.* 330–330.

Williams, J.G., and Jermyn, K.A. (1991) Cell sorting and positional differentiation during *Dictyostelium* morphogenesis. In: *Cell-Cell Interactions in Early Development*, J. Gerhart, ed. New York: Wiley-Liss, pp. 261–272.

12

Angiogenesis: Experimental Data Relevant to Theoretical Analysis

Ana M. Schor, Seth L. Schor, Robin Baillie

Oral Disease Group, The Dental School, University of Dundee, Dundee DD1 4HR, Scotland, UK

12.1 Introduction

The formation of new blood vessels during embryonic development occurs by a combination of the progressive association of single endothelial cells and their subsequent morphogenesis (*vasculogenesis* and *intussusception*), as well as by the sprouting or branching of existing vessels (*angiogenesis*) (Noden, 1989; Burri, 1992). Angiogenesis also occurs in the adult, where it contributes to the maintenance of tissue functionality and is an important feature of various pathological processes, such as wound healing and cancer progression (Folkman and Shing, 1992; Carmeliet and Collen, 1998; Hudlicka, 1998). In view of its involvement in such common disease processes, a great effort has been made during the past decade to develop effective means of manipulating angiogenesis (either its inhibition or stimulation) in relevant clinical situations (Pluda, 1997; Kuiper *et al.*, 1998). This endeavour is based on a burgeoning literature documenting the existence of an ever increasing number of chemically disparate angiogenic effector molecules, which either stimulate or inhibit new vessel formation (Cockerill *et al.*, 1995; Norrby, 1997; Carmeliet and Collen, 1998). However, in spite of great advances in our understanding of the manner in which these angiogenic factors function at the cell and molecular level, the basic mechanisms underlying angiogenesis remain unclear.

Angiogenesis is essentially a *morphogenetic* process involving spatial and temporal changes in vascular cell phenotype and cell–cell interactions which result in the formation of new multicellular structures. Further advances in our understanding of these events and the mechanisms by which they are controlled, require a multi-disciplinary approach, including the integration of basic molecular biology at the tissue level of organisation and the testing of experimental hypotheses by theoretical analysis. In this chapter we review various aspects of the biology of angiogenesis with the purpose of providing a conceptual framework relevant to theoretical studies. Emphasis

On Growth and Form: Spatio-temporal Pattern Formation in Biology.
Editor Mark A.J. Chaplain, G.D. Singh, J.C. McLachlan

will be placed on aspects of our own work dealing with the reciprocal interactions between soluble angiogenic factors and insoluble components of the extracellular matrix in the control of angiogenesis. With this objective in mind, data will reviewed regarding (i) the cells involved in angiogenesis, (ii) the sequence of cellular events leading to the formation of new blood vessels in model systems in vivo, (iii) in vitro model systems used to study specific aspects of angiogenesis, and (iv) the interdependent nature of cytokine and matrix control of angiogenesis. Other important aspects of angiogenesis have been reviewed in recent publications (Iruela-Arispe and Dvorak, 1997; Norrby, 1997; A.M. Schor *et al.*, 1997; Bussolino *et al.*, 1997; Sang 1998; Carmeliet and Collen, 1998; Hudlicka, 1998).

12.2 Vascular and Perivascular Cells Involved in Angiogenesis

There is great diversity among blood vessels with respect to size, structure and function. These differences reflect site-specificity, ageing and the influence of environmental factors. *Microvessels* are the smallest type of vessel, ranging in diameter from 5 to 50μm, and include arterioles, venules and capillaries. Venules and capillaries are the simplest structures, being composed of three components: (a) *the basement membrane*, a complex extracellular matrix encircling and supporting the cellular components, (b) *the endothelial cells* lining the lumen and resting on the basement membrane, and (c) *the pericytes* forming a periendothelial cellular network embedded *within* the basement membrane (Palade, 1988; Simionescu and Simionescu, 1988). Although separated by basement membrane material in some areas, direct contact between endothelial cells and pericytes occurs frequently through various types of junctions, such as adhesion plaques, peg-and-socket and gap junctions (Diaz-Flores *et al.*, 1991) The ratio of pericytes to endothelial cells varies according to vessel type, age and pathological state. In the vasculature of skeletal muscle, pericytes have been observed to be randomly positioned in capillaries and concentrated at endothelial cell junctions in venules (Sims, 1991; Sims *et al.*, 1994; A.M. Schor *et al.*, 1997). Larger vessels contain additional concentric layers of cells (e.g. smooth muscle cells) and extracellular matrix. In these vessels the endothelium may also be in anatomical proximity to smooth muscle cells, as these have been reported to send extensions through the internal elastic lamina (Haudenschild, 1980). Certain microvessels contain cells which have been described as intermediate between pericytes and smooth muscle cells (Sims, 1991; Shepro and Morel, 1993; Nehls and Drenckhahn, 1993).

The inner lining of all blood vessels, the *endothelium*, consists of a single layer of endothelial cells presenting a free luminal surface and attached on their abluminal side to the underlying basement membrane (apart from a minimal amount of cell–cell overlap). All endothelial cells can be defined on the basis of shared anatomical and functional characteristics, eg. luminal location and provision of a non-thrombogenic surface. A number of specific cellular markers (e.g. von Willebrand Factor, CD31 and Weibel–Palade bodies) assist in the identification of endothelial cells in vivo and in vitro. In spite of these common characteristics, endothelial cells are heterogeneous with respect to certain morphological and biochemical criteria (e.g. fenestrations and antigen expression) (Simionescu and Simionescu, 1988; Augustin *et al.*, 1994). Both during development and in the adult some of this heterogeneity appears to result from microenvironmental signals. For example, the grafting of quail tissue into chick embryos has demonstrated that the emergence of structural and functional heterogeneity in the graft endothelium is determined not by its site of origin, but rather by the mesenchymal environment of the host tissue; that is, quail mesenteric vessels growing into grafts of chick brain formed capillaries with structural and functional characteristics specific to the brain microvasculature, whereas brain capillaries growing into grafts of mesodermal tissue resembled mesenteric vessels. In a similar way, blood vessels invading renal carcinoma metastases to the brain acquire the typical fenestrated morphology of renal vessels, instead of displaying the continuous morphology characteristic of the parental brain vessels (A.M. Schor *et al.*, 1997). A number of studies have clearly demonstrated that interactions between soluble growth factors and insoluble constituents of the extracellular matrix contribute to the development and maturation of the blood vessels (Risau and Lemmon, 1988; Noden, 1989).

In a mature resting vessel, the functionality and integrity of the endothelium is maintained under steady-state conditions by the biosynthetic activity of the endothelial cells, in conjunction with low levels of cell proliferation and motility. Significant changes in the motility of the endothelial cells, often accompanied by cell proliferation, occur during angiogenesis and in response to vessel injury.

Pericytes extend long filamentous processes which wrap along and around the endothelium. Transmission electron microscopy of these cells indicates the presence of microfilament bundles with focal densities, plasmalemmal vesicles and glycogen deposits (Diaz-Flores *et al.*, 1991; Sims, 1991; Shepro and Morel, 1993; Nehls and Drenckhahn, 1993). Unfortunately, there are no specific biochemical markers for *pericytes*, although the expression of various proteins (including desmin, vimentin, α-smooth muscle actin, non-muscle actin, muscle and non-muscle myosin, high molecular weight melanoma-associated antigen, 3G5 surface ganglioside antigen), in combination with the characteristic location of these cells within the vessel, are used to identify pericytes (Sims, 1991; Nehls and Drenckhahn, 1993; A.M. Schor and Canfield, 1998). Interestingly, cells defined as pericytes solely on the basis of their anatomical location are heterogeneous with respect to expression of these molecules. For example, Nehls and Drenckhahn (1993) found that pericytes in the mid-capillaries of rat mesentery expressed desmin, but not α-smooth muscle actin (α-sma) whereas pericytes in pre- and post-capillaries were positive for α-sma. Other studies have similarly demonstrated heterogeneity among pericytes with respect to staining with antibodies to α-sma, desmin, myosin isoforms and high molecular weight melanoma-associated antigen, HMW-MAA (Nehls and Drenckhahn, 1993; Schlingermann *et al.*, 1991). The functional significance of pericyte heterogeneity remains unknown. Changes in pericyte numbers and phenotype are associated with various pathological conditions, such as diabetes, neoplasia and hypertension (Sims, 1991; A.M. Schor *et al.*, 1997). The use of knockout transgenic mice has indicated that expression of PDGF-B is required for the presence of pericytes in the vasculature (Lindahl *et al.*, 1997).

The non-specific nature of available pericyte biochemical markers must be emphasised. Indeed, some of these markers are also expressed by endothelial and smooth muscle cells, thereby confounding the identification of pericytes. For example, although endothelial cells are generally negative for desmin and α-sma, Fujimoto and Singer (1986) observed that capillary and vein endothelial cells in adult chick tissues may express only vimentin, only desmin or both desmin and vimentin. Similarly, cultured endothelial cells may be induced to express α-sma, and α-sma-positive cells have also been identified in the lumen of atherosclerotic lesions in vivo (Arciniegas *et al.*, 1992). Taken together, these various observations have raised a number of interesting questions regarding the lineage relationship between endothelial cells and pericytes (Section 12.3).

The *fibroblast* is the principal cell type found in the connective tissue stroma surrounding microvessels. In the absence of specific molecular markers, fibroblasts (like pericytes) remain a rather poorly defined cell population whose identity commonly relies on morphological criteria (e.g. location in vivo and spindle-shaped appearance in vitro) and the expression of ubiquitous matrix macromolecules, such as collagen types I and III. Cells classified as fibroblasts on the basis of these criteria are sometimes in close anatomical proximity to the microvasculature, in which position they have been called *extramural* pericytes to distinguish them from the *intramural* pericytes located within the basement membrane (Ashton and DeOliveira, 1966). To avoid confusion, we shall refer to all perivascular spindle-shaped cells as fibroblasts, leaving the term "pericyte" to describe periendothelial cells *embedded within the vascular basement membrane*. Fibroblasts are also a highly heterogeneous cell population, displaying significant developmental, intersite and intrasite variation in the expression of matrix macromolecules, matrix-degrading enzymes and cytokines (S.L. Schor and Schor, 1987, 1997; S.L. Schor *et al.*, 1996a).

12.3 Phenotypic Heterogeneity and Lineage Relationships Amongst Vascular Cells

Heterogeneity in the phenotype within a given cell population results from differences in the pattern of gene expression. Such differences may ultimately be produced by two distinct mechanisms: differentiation and modulation. At the molecular level, we suggest that cell *differentiation* should be defined as the process by which certain genes become *irreversibly* incapable of expression. In this regard, it should be noted that all cells express only a relatively small proportion of the total number of genes they contain. The zygote and a number of its immediate progeny retain the potential of expressing the largest subset of these genes. Differentiation is characterised by the progressive reduction in the potential of gene expression. In contrast, we suggest that *reversible* alterations in the pattern of potentially expressible genes (defined by the state of differentiation) should be known as *modulation*. Accordingly, both differentiation and modulation lead to an alteration in the pattern of gene expression; differentiation is irreversible, whereas modulation is reversible (Figure 12.1). *Cell lineage* defines the sequence of precursor cell populations from which the differentiated cell is derived. In certain well-defined lineages (such as in the haemopoietic system) several distinct lineages may be traced back to a single pluripotential stem cell population (Lajtha, 1983). Both differentiation and modulation may contribute to the existence of phenotypic heterogeneity among vascular cells; i.e. heterogeneity may result from the coexistence of cells at various stages of differentiation along a common cell lineage and/or the modulation of phenotype in cells at a given stage of differentiation in response to microenvironmental stimuli.

Cells believed to belong to a certain lineage may undergo an abrupt alteration in phenotype to one characteristic of a different lineage. This process has been

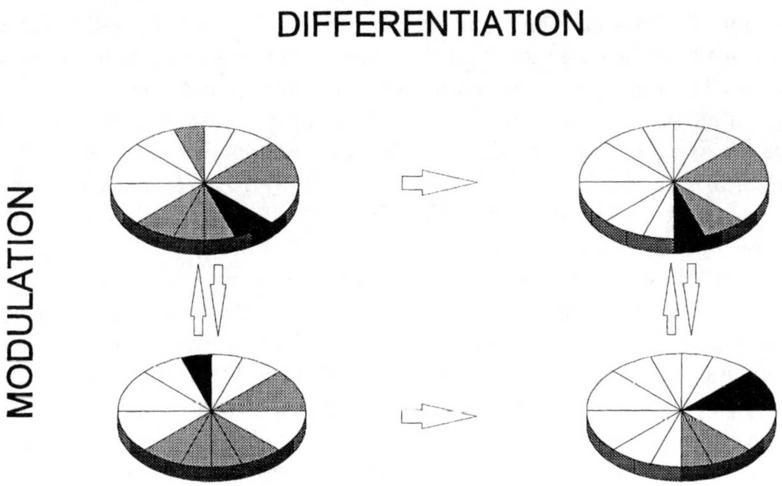

Figure 12.1 Regulation of gene expression. Cells are depicted as moving from left to right along a particular lineage by a process of differentiation; this is characterised by the progressive and irreversible reduction in the set of genes which may be expressed. Modulation, depicted in the vertical direction, is characterised by the reversible alteration in the subset of expressed genes, generally in response to regulatory stimuli. Key: white segment = non-expressible genes; grey = potentially expressible genes; black = subset of expressed genes.

called *transdifferentiation* (Eguchi and Kodama, 1993). The distinction between transdifferentiation and differentiation is difficult to define at the molecular level; accordingly, transdifferentiation should be regarded as an operational definition and implies a digression from the "expected" lineage pathway. As such, it may represent a deficiency in our knowledge of cell lineages, rather than a distinct biological process.

Endothelial stem cells have been reported to be present in the peripheral blood (Asahara *et al.*, 1997), although this observation remains to be confirmed. Similarly, several studies have indicated that either endothelial cells or pericytes function as "undifferentiated" or "primitive" mesenchymal cells in the adult. Such cells have been reported to differentiate (or transdifferentiate) into fibroblasts, osteoblasts and smooth muscle, as well as other cell types (A.M. Schor *et al.*, 1990; Sims, 1991; Shepro and Morel, 1993). For example, Decker *et al.* (1995) implanted tricalciumphosphate ceramic in an artificial defect in the sheep femur and two months later studied the new bone formed within this defect. They found a network of glycogen-rich cells between sprouting capillaries and osteoblasts. Some of these cells were surrounded by the endothelial basement membrane and were therefore classified as pericytes; however, their proximity to the osteoblasts and the presence of glycogen deposits also characterised them as preosteoblasts. Interestingly, Weibel–Palade bodies (a marker of endothelial cells) were occasionally observed in the glycogen-rich cells. Based on the coexistence of Weibel–Palade bodies and glycogen deposits in the pericytes, Decker *et al.* (1995) concluded that endothelial cells serve as stem cells, and pericytes are progenitor cells of the osteogenic cell lineage. Such a lineage relationship involves a movement of cells from a luminal position (e.g. endothelial) to a stromal location.

Using a different in vivo experimental system, Diaz-Flores *et al.* (1992) have also provided evidence that either endothelial cells and/or pericytes may transdifferentiate into osteoblasts. The lineage association between fibroblasts and osteoblasts has been better documented (Owen 1985), and related data have shown that both fibroblasts and pericytes in tissue culture are able to differentiate along the osteogenic pathway (Owen *et al.*, 1987; A.M. Schor *et al.*, 1990, 1995, 1997). As osteogenesis is closely associated with angiogenesis, such examples of "transdifferentiation" among vascular cells may be of relevance to angiogenesis.

Taken together, these various observations indicate that vascular cells and stromal fibroblasts remain poorly defined in terms of their lineage relationships. They are heterogeneous cell populations in which the observed differences in phenotype may result from a combination of differentiation, modulation and transdifferentiation. Such plasticity in cell phenotype is an intrinsic feature of the angiogenic process.

Morphological transitions may also occur at the vessel level. For example, newly formed capillaries have been reported to become arterioles and venules (Clark and Clark, 1939; Simionescu and Simionescu, 1988). In skeletal muscle, pre-existing capillaries can similarly undergo arterialisation by the addition of a smooth muscle layer. Although the term "differentiation" has commonly been employed to describe such structural changes, they may in fact result from cell migration, accretion and proliferation, e.g. the movement of smooth muscle cells from terminal arterioles (Nehls *et al.*, 1992; Price *et al.*, 1994) rather than changes in the pattern of gene expression at the cellular level. In our view, "remodelling" is a more appropriate term to describe changes in vessel type.

12.4 Angiogenesis: A Morphogenetic Process

Angiogenesis is essentially an in vivo phenomenon. Various animal models have accordingly been employed to elucidate the sequence of cellular events which comprise this process. Note that different experimental models may not necessarily yield the same results, as they mimic distinct physiological or pathological situations. (Ausprunk and Folkman, 1977; A.M. Schor and Schor, 1983; Sholley *et al.*, 1984; Ishibashi *et al.*, 1987; Paku and Paweletz, 1991; Hudlicka, 1998). The most commonly used models include the chick chorioallantoic membrane and the rabbit cornea assays. In these models a source of soluble angiogenic stimulus is placed a short distance from the pre-existing microvasculature. Diffusion of the angiogenic factor(s) establishes a gradient in its concentration; when an appropriate concentration reaches the vasculature, it elicits the formation of new blood vessels from the surrounding vasculature. The development of new blood vessels therefore displays a clear directionality, with sprouts growing towards the source of the angiogenic stimulus; in the chick chorioallantoic membrane this results in a characteristic radial pattern of vessel development, whereas the rabbit cornea assay involves a linear movement of vessels from the limbus to the angiogenic stimulus through the normally avascular cornea, thus facilitating the quantification of vessel growth. In these, as well as other in vivo assays, new vessels may originate from several sites along the parent vessel (Sholley *et al.*, 1984). It therefore appears that endothelial cells (and pericytes) capable of adopting the angiogenic phenotype may be present in any location within a blood vessel. New blood vessels most commonly originate from capillaries and venules (the so-called pericytic venules). Nevertheless, it

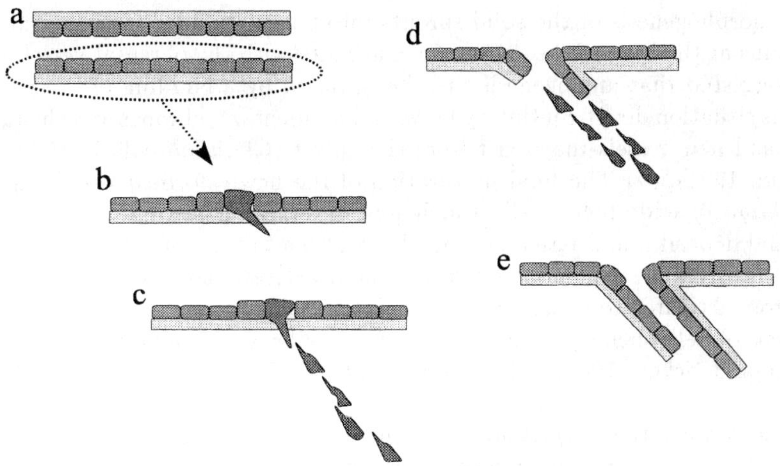

Figure 12.2 Stages of angiogenesis. (a) In a mature microvessel, resting endothelial cells are adherent to the vessel basement membrane and form a monolayer lining the lumen. (b) Activation and migration of the resting endothelial cells. (c) Endothelial cells within the perivascular stroma adopt a sprouting phenotype, aggregate in a head-to-tail fashion to form solid cords or sprouts, and migrate towards the source of angiogenic stimilus. (d) Morphogenesis of solid sprouts into patent vessels in continuity with the parental vessel. (e) Maturation of the newly formed vessel with the synthesis of a new basement membrane and readoption of the resting endothelial phenotype.

is clear that all types of vessels retain angiogenic potential, as new capillaries have also been observed to arise from arterioles and large vessels (Eisenstein, 1991; S.L. Schor and Schor, 1997). The nascent vessels are first observed as solid sprouts, growing parallel to each other. The tips of the sprouts anastomose to form loops and new sprouts emerge from these loops until the angiogenic stimulus is reached.

Using histological and ultrastructural analysis, the events which take place during angiogenesis have been best defined for the endothelial cells. The following sequential, overlapping steps have been described (Figure 12.2). (i) The *activation* of resting endothelial cells lining the lumen of the parental vessel. Cell activation is characterised by blebbing of the endothelial cell surface and loosening of cell junctions (Ausprunk and Folkman, 1977), as well as cell enlargement (Paku and Paweletz, 1991). Alterations to the basement membrane (lysis, fragmentation, reduplication) occur in conjunction with or soon after cell activation. (ii) The movement of the activated cells through the vessel basement membrane into the surrounding connective tissue stroma, where they adopt an elongated sprouting cell phenotype (Ishibashi *et al.*, 1987). Note that this involves the translocation of endothelial cells from a two-dimensional luminal environment into a three-dimensional macromolecular matrix. The composition of the basement membrane, and its susceptibility to hydrolysis is believed to play an important role in this step (Dziadek and Mitrangas, 1989; Carmeliet and Collen, 1998). (iii) The self-aggregation of sprouting cells into solid cords or sprouts and the directional migration of these towards the source of the angiogenic stimulus. (iv) The subsequent morphogenesis of the solid sprouts into patent tubules. The formation of a lumen begins at the base of the sprout, in continuity with the parental vessel. It has also been suggested that angiogenesis may be initiated by a budding of the parental vessel; in this situation direct continuity between basement membranes and the lumens of parental and new vessels may exist from the outset (Clark and Clark, 1939; Paku and Paweletz, 1991). (v) The final maturation of the newly formed vessels involves the *deactivation* of endothelial cells which now revert to a resting phenotype and synthesise and deposit a new basement membrane. Seen in this context, the process of angiogenesis involves the sequential activation and deactivation of vascular cells. As the transition from activated to de-activated state is reversible, it is essentially a reflection of *modulation* of cell phenotype in response to specific microenvironmental stimuli (A.M. Schor and Schor, 1983; A.M. Schor *et al.*, 1997; Iruela-Arispe and Dvorak, 1997).

Investigations into the behaviour of pericytes during angiogenesis have been hampered by the lack of unambiguous markers for these cells. Although it is accepted that the endothelial cells of newly formed vessels originate from the parental vessel, the origin of the pericytes is not clear, as contradictory evidence has been published. For example, it has been shown that pericytes are present at every stage of angiogenesis (Schlingermann *et al.*, 1991; Diaz-Flores *et al.*, 1992); furthermore, they appear to become the leading cells of the newly formed sprouts (Nehls *et al.* 1992; Nehls and Drenckhahn, 1993). Such data suggest that pericytes within the new vessel originate from pericytes present in the parental vessel. However, other experiments have shown that, during angiogenesis, perivascular fibroblasts migrate towards the new vessels and become incorporated therein, suggesting that pericytes originate from these stromal cells (Stein *et al.* 1996). It is possible that both mechanisms may be operative.

Besides their role as a possible source of pericytes, stromal fibroblasts also appear to play an important function in angiogenesis. In this regard, fibroblasts respond to angiogenic factors, and fibroblast migration into granulation tissue has been reported to be temporally correlated with the ingress of endothelial sprouts (Hunt, 1980). Interactions between endothelial cells and stromal fibroblasts have been reported to contribute to the angiogenic process (Javier *et al.*, 1997; Nehls *et al.*, 1998), with the fibroblasts appearing to function as a source of angiogenic factors (Ben-Av *et al.*, 1995). The in vivo relevance of such interactions between fibroblasts was elegantly demonstrated by Fukamura *et al.* (1998) using a transgenic mouse line in which the VEGF promoter drives the expression of GFP (green fluorescent protein). In both wound healing and neoplasia models of angiogenesis, fibroblasts were the most prominent cells showing VEGF promoter activation. Fibroblasts surrounding blood vessels stained positive for GFP, whereas endothelial cells were negative. Fibroblasts recovered from the tumour mass had the same fluorescence profile as fibroblasts propagated in vitro, suggesting that culture in vitro is highly activating for VEGF promoter function.

Both cell proliferation and migration normally contribute to the movement of sprouting cells or vascular buds towards the angiogenic stimulus. Cell proliferation is not equally evident along the extending angiogenic sprout, normally occurring a few cells behind the leading cells (A.M. Schor and Schor, 1983; Nehls *et al.*, 1992). However, angiogenesis, can take place in the absence of cell proliferation (Sholley *et al.*, 1984; Hudlika, 1998); cell migration therefore appears to be of critical relevance to angiogenesis

An ever expanding list of chemically unrelated factors have been reported to function as angiogenic factors; these include various families of peptide *growth factors* (also known as *cytokines*), such as TGFβ, TGFα, EGF, VEGFs, FGFs, scatter factor, TNFα, and angiogenin. Immunohistochemical and biochemical studies have demonstrated that many well-characterised angiogenic factors appear to be ubiquitously present in tissues and body fluids in the absence of demonstrable new blood vessel formation (Berse *et al.*, 1992; Jinno *et al.*, 1998). This discordance between the presence of angiogenic cytokines and overt angiogenesis clearly indicates that the presence of angiogenic factors is not sufficient to induce angiogenesis and further underscores the importance of the microenvironment in the control of vascular phenotype and angiogenesis. In view of the important role played by the extracellular matrix in modulating growth factor bioactivity, these various observations have provided the impetus for us to develop in vitro models of angiogenesis which facilitate the study of cytokine–matrix interactions.

12.5 In Vitro Models to Study Angiogenesis: Interactions between Cytokines and Extracellular Matrix Macromolecules

Endothelial cells, pericytes and fibroblasts from various sources may be isolated and grown in tissue culture with relative ease. Cells maintained in vitro retain many of the characteristics of their in vivo counterparts. For example, endothelial cells in tissue culture express endothelial markers (such as Weibel–Palade bodies, CD31 and vWF) and a non-thrombogenic surface. Endothelial cells derived from both large vessels and microvessels are capable of adopting a sprouting phenotype and forming tubules under appropriate conditions in vitro (A.M. Schor *et al.*, 1983; A.M. Schor and Schor, 1986; Montesano, 1992; Nicosia *et al.*, 1992; Ingber and Folkman, 1989; Vernon *et al.*, 1995). Similarly, cultured pericytes continue to display many of the characteristics of their in vivo counterparts, including ultrastructural features and antigen expression (A.M. Schor *et al.*, 1992; A.M. Schor and Canfield, 1998).

We and others have developed a simple in vitro culture system in which it is possible to induce endothelial cells to express either a resting or sprouting phenotype and thereby mimic specific stages of angiogenesis (Figure 12.3A) (A.M. Schor *et al.*, 1983; A.M. Schor and Schor, 1986; Sutton *et al.*, 1991; Canfield *et al.*, 1986, 1990; Canfield and Schor, 1995; Montesano, 1992; Madri *et al.*, 1989). Accordingly, endothelial cells are plated on a 2D substratum, such as a tissue culture dish or the 2D surface of a 3D collagen gel. Both large-vessel and microvessel endothelial cells cultured on these 2D substrata form a contact-inhibited "cobblestone" monolayer (Figure 12.4A) which exhibits many characteristics of the *resting* endothelium in vivo. This resting phenotype may be maintained indefinitely in the absence of angiogenic factors (e.g. under low serum conditions). Resting cells display a low level of proliferation and a characteristic biosynthetic profile, including the synthesis and deposition of a sub-endothelial matrix (A.M. Schor *et al.*, 1984; A.M. Schor and Schor, 1986; Sutton *et al.*, 1991; Canfield *et al.*, 1986, 1990). The addition of angiogenic factors to these cultures elicits a rapid response characterised by the formation of spindle-shaped *sprouting* cells which ultimately self-associate to form multicellular tandem arrays (Figure 12.4B). Microscopic examination indicates that these sprouting cells are actually below the

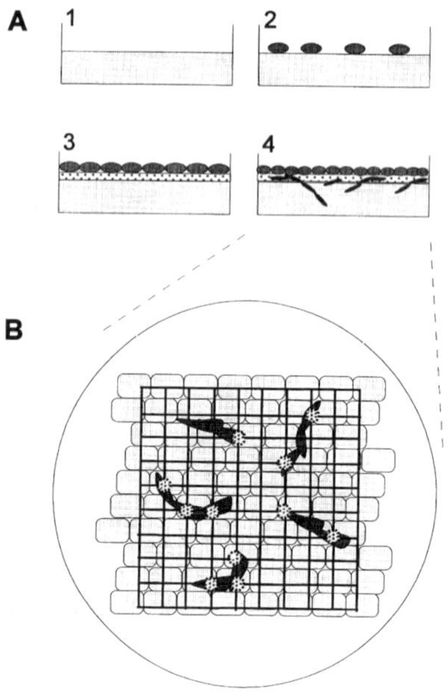

Figure 12.3 Induction and quantification of sprouting cells. (A) The experimental induction of sprouting cells by angiogenic factors in vitro. 1. Collagen gel substratum. 2. Endothelial cells plated on the surface of the collagen gel proliferate and synthesise extracellular matrix components. 3. When confluent, the cells form a "cobblestone" monolayer of resting cells. 4. Following the addition of an angiogenic factor, the resting cells are activated and some of them migrate into the subendothelial matrix and the underlying collagen gel. In this new environment, the cells adopt an elongated "sprouting" phenotype and self-aggregate to form multicellular cords. (B) *En face* view of culture containing sprouting cells beneath a monolayer of resting cells. Quantification of sprouting cells may be achieved with a graticule containing a number of points (defined in this example by the intersections of vertical and horizontal lines). Only those cells touching the intersection points (depicted by circles) are counted and expressed as a percentage of the total number of couting points in the graticule.

Factor	Angiogenesis in vivo	Morphogenesis in vitro
bFGF	stimulates	stimulates
VEGF-A	stimulates	stimulates
PMA	stimulates	stimulates
BB94	inhibits	inhibits
β-xyloside	inhibits	inhibits
pentosan polysulphate	inhibits	inhibits
fumagillin	inhibits	inhibits
suramin	inhibits	inhibits
AGM-1470	inhibits	inhibits
protamine	inhibits	inhibits

Table 12.1 Comparison of in vivo and in vitro effects by stimulators and inhibitors of angiogenesis.

monolayer of cobblestone resting cells (Figure 12.4C), embedded within the three-dimensional subendothelial matrix (Figure 12.4D). If the endothelial cells are cultured on the surface of gel containing type I collagen fibres (Figure 12.3A), the resultant sprouting cells migrate through the subendothelial matrix into the three-dimensional environment of the collagen gel. In this milieu, sprouting cells maintain an elongated and fusiform morphology and tend to associate in a head-to-tail fashion and form complex 3D tubule-like structures (Figure 12.4F). The extent of sprouting may be quantified by various means; we have found the stereological method of point counting to be both quick and reliable (Figure 12.3B).

The induction of sprouting mimics early stages of angiogenesis involving endothelial cell activation and migration in response to stimulation by an angiogenic factor. Consistent with this interpretation, chemically diverse factors which induce angiogenesis in vivo elicit a similar resting-to-sprouting cell transition in vitro. Similarly, known angiogenesis inhibitors prevent the formation of sprouting cells in the presence of angiogenic stimuli (Table 12.1).

As is the case with their in vivo counterparts, expression of a resting or sprouting phenotype is completely reversible by appropriate manipulation of the culture conditions. For example, a cobblestone monolayer of resting endothelial cells may be trypsinised and plated within the three-dimensional matrix of a type I collagen gel as a randomly distributed single cell suspension; under these new culture conditions, the initially round endothelial cells (Figure 12.4E) soon adopt an extended fusiform morphology, actively migrate and eventually aggregate to form multicellular aggregates in the absence of exogenous angiogenic factors (Figure 12.4F).

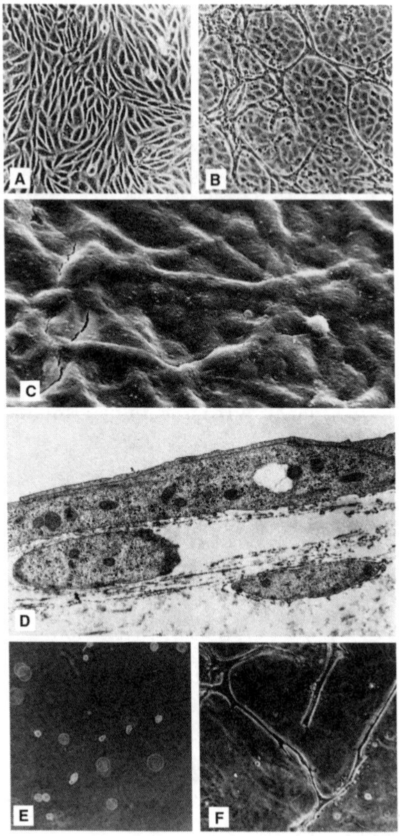

Figure 12.4 Morphology of endothelial cells in culture. (A) Cobblestone monolayer of resting endothelial cells (×200). (B) Sprouting cells (×200). (C) Scanning electron micrograph showing profiles of sprouting endothelial cells beneath the monolayer (×600). (D) Transmission electron micrograph of vertical section through culture containing both cobblestone and sprouting cells (×7000). (E) Cells from a monolayer were brought into suspension and embedded within a 3D collagen gel (×200). (F) Endothelial cells embedded within a collagen gel (as in E) adopt a sprouting morphology within 3 hours after plating and self-associate into complex multicellular sprouts during the subsequent 3–7 days (×200).

Cells cultured under these conditions mimic the stages of angiogenesis involving cell–cell interactions and tubule formation within a three-dimensional environment. The induced transition from resting to spouting phenotype is completely reversible in that isolation of the sprouting cell aggregates and their replating on a two-dimensional substratum results in their readoption of a resting phenotype. Repetition of this cycle is achievable many times during the in vitro lifespan of the cultured cells.

The ability to produce cultures consisting entirely of cells displaying a resting or sprouting phenotype in the absence of exogenous angiogenic factor (i.e. either resting monolayer on the surface of a collagen gel or sprouting cells within a three-dimensional collagen milieu) has made it possible to obtain unambiguous biochemical data regarding the behaviour and biosynthetic activity of these two cell populations. Resting and sprouting cells have accordingly been reported to differ with respect to migratory activity, response to growth factors, proliferation and synthesis of a number of molecules of potential relevance to angiogenesis, including matrix metalloproteases, matrix macromolecules and growth factors (Haas *et al.*, 1998; A.M. Schor and Schor, 1986; Canfield *et al.*, 1986, 1990, 1992; Sutton *et al.*, 1991).

Pericytes also display a sprouting cell phenotype when embedded within a three-dimensional matrix. Sprouting pericytes and sprouting endothelial cells are indistinguishable in terms of morphology and synthesis of certain matrix macromolecules (Schor *et al.*, 1992). In spite of their similarities, these cells display distinctive proliferative behaviour: i.e. sprouting pericytes proliferate within a three-dimensional collagen matrix, whereas endothelial cells do not (A.M. Schor and Schor, 1986; A.M. Schor *et al.*, 1997).

Resting endothelial cells growing on a two-dimensional substratum may also be induced to form "tubules" by other means; these include the deposition of matrix fibrils on top of the resting monolayer, as well as various procedures which reduce the relative strength of cell–substratum adhesion compared to cell–cell adhesion. This latter situation may be achieved by using a poorly adhesive substratum, by exogenous factors and by increasing the contractibility of the cells (Ingber and Folkman, 1989; Vernon *et al.*, 1995). Pericytes produce similar tubular structures on two-dimensional substrata (A.M. Schor *et al.*, 1990, 1991). Interestingly, the formation of endothelial tubules is stimulated by the presence of pericytes and other periendothelial cell types (Watanabe *et al.*, 1997; A.M. Schor *et al.*, 1997). The process of sprouting and tubule formation in vitro has often been called differentiation. However, these processes have been shown to be reversible, so there is no basis for believing that an irreversible alteration in the potential for gene expression has occurred; therefore *modulation* and *morphogenesis* would be more accurate descriptors of this process. Modulation describes changes in cell phenotype and morphogenesis describes the generation of a new multicellular structure.

The implications of these in vitro observations are important for our understanding of angiogenesis. Taken together, they suggest that (i) angiogenic factors function to activate the migration of resting endothelial cells from the cobblestone monolayer into a three-dimensional matrix environment, and (ii) once located within a three-dimensional matrix endothelial cells are induced to express a sprouting phenotype in the absence of angiogenic factor. The mechanisms responsible for this matrix-induced resting-to-sprouting phenotype transition remain unclear. In this context, it is of interest to note that resting endothelial cells adherent to a two-dimensional substratum establish a well defined apical–basal polarity which is lost when they migrate into a three-dimensional matrix and make adhesive interactions over their entire surface. This change in the physical organisation of cell–matrix contact will affect a number of potentially relevant aspects of cytoarchitecture (such as cytoskeletal organisation and receptor clustering) known to influence the pattern of gene expression (Ingber and Folkman, 1989; Vernon *et al.*, 1995).

The precise nature of the extracellular matrix also appears to regulate cellular response to soluble angiogenic factors. Our own results indicate that the macromolecular matrix in contact with the cells may be either *permissive* or *inhibitory* with respect to such factors. For example, ESAF (a low molecular mass angiogenic factor) stimulates the proliferation of microvascular endothelial cells adherent to native type I collagen or fibronectin, but has no effect on cells attached to a denatured type I collagen. The bioactivity of ESAF is cell type specific, as it is without effect on aortic endothelial cells. (Weiss and McLaughlin, 1998; Keegan *et al.*, 1982). More complex matrices (deposited by the endothelial cells themselves under various experimental conditions) may similarly modulate endothelial response to bFGF; in these experiments a confluent (resting) endothelial monolayer grown on a permissive matrix responds to bFGF by the formation of sprouting cells, whereas a resting monolayer adherent to an inhibitory matrix is unaffected by exogenous bFGF. Interestingly, the proliferation of subconfluent cells is stimulated by FGF irrespective of the type of matrix used, thereby indicating that a matrix capable of inhibiting the response of cells to bFGF with respect to one of its functions (e.g. induction of sprouting by confluent cells) is without effect with respect to another function (e.g. stimulation of proliferation by subconfluent cells) (Table 12.2) (Canfield and Schor, 1995).

In vitro assays may also be effectively used to study the induction of directional cell migration by angiogenic factors. This may indeed be the principal function of certain angiogenic factors, for endothelial cells spontaneously adopt a sprouting phenotype when placed within a three-dimensional collagenous matrix. Under these conditions (i.e. in the absence of a concentration gradient of angiogenic factor) the migration of these sprouting cells is essentially random. Angiogenesis also involves the directional migration of perivascular fibroblasts. Interactions between fibroblasts and vascular cells take place through paracrine mechanisms, mediated by both cytokines and matrix molecules. Many peptides that had been originally described as mitogenic and/or motogenic factors for stromal fibroblasts (e.g. FGFs, $TGF\alpha$, $TGF\beta$, PDGF) have been subsequently found to induce angiogenesis when tested in the appropriate in vivo angiogenic assay. A considerable amount of effort has therefore been invested in examining the effects of angiogenic factors on the directional migration of vascular cells and associated stromal fibroblasts.

Cell migration has commonly been studied on two-dimensional substrata, such as plastic tissue culture dishes (wounded monolayer assay) and through the pores of a polycarbonate membrane separating upper and lower medium compartments (Figure 12.5A) (Boyden chamber or transmembrane assay). This assay allows a distinction to be made between *chemokinesis* (defined as random cell movement in response to an isotropic distribution of soluble effector molecule) and *chemotaxis* (defined as directional cell movement in response to an imposed concentration gradient of soluble effector molecule) (Zigmond and Hirsch, 1973). Matrix macromolecules may also affect

	Maximal sprouting (%)		Maximal proliferation (%)	
FGF (ng/ml)	**Substratum** permissive	non-permissive	**Substratum** permissive	non-permissive
0	1.2±2.1	1.2±1.2	2.2±3.4	1.9±6.8
5	4.2±2.9	0.6±0.3	5.3±6.0	5.5±4.4
10	32.1±3.6	0.9±1.2	46.6±5.5	56.1±8.6
50	95.2±9.2	1.9±0.9	91.3±7.9	93.9±6.6
100	100	0.7±0.6	99.7±7.7	100
200	nd	1.5±0.3	nd	nd

Table 12.2 Effects of the extracellular matrix on endothelial sprouting and proliferation induced by bFGF. Permissive and non-permissive matrices were prepared as in Canfield and Schor (1995). Endothelial sprouting and proliferation were assessed in the presence of growth medium containing 1% serum and the indicated concentration of bFGF.

cell migration by *haptotaxis* (defined as directional cell movement in response to an imposed concentration gradient of adsorbed effector molecule). The transmembrane assay has also been used to document the effects of angiogenic factors on endothelial cell self-aggregation into multicellular cords (Garrido *et al.*, 1995).

In view of the important role played by the extracellular matrix in the regulation of cell migration, it is (in our view) essential to develop model systems which will make it possible to study cell movement within a physiologically relevant three-dimensional macromolecular environment in vitro. With this objective in mind, we have devised a migration assay involving the plating of cells on the surface of a collagen gel substratum and quantifying their subsequent migration down into the three-dimensional collagen matrix (Figure 12.5B) (S.L. Schor, 1980). Our data indicate that the motogenic response of fibroblasts and endothelial cells to both cytokines and matrix macromolecules may be quite different in the collagen gel migration assay compared to that obtained in the more commonly employed transmembrane assay. For example, the three mammalian TGFβ isoforms equally stimulate the migration of stromal fibroblasts in the transmembrane assay (Ellis and Schor, 1998; Ellis *et al.*, 1999). In contrast, TGFβ-1 and TGFβ-2 had no effect on the migration of confluent fibroblasts in the collagen gel assay, whereas TGFβ-3 stimulated cell migration; all three isoforms inhibited the migration of subconfluent cells in this assay (Figure 12.6). In a similar vein, we have reported that the gelatin-binding domain of fibronectin is a potent motogenic agent when assessed in the collagen gel assay, but is completely devoid of such activity in the transmembrane assay (S.L. Schor

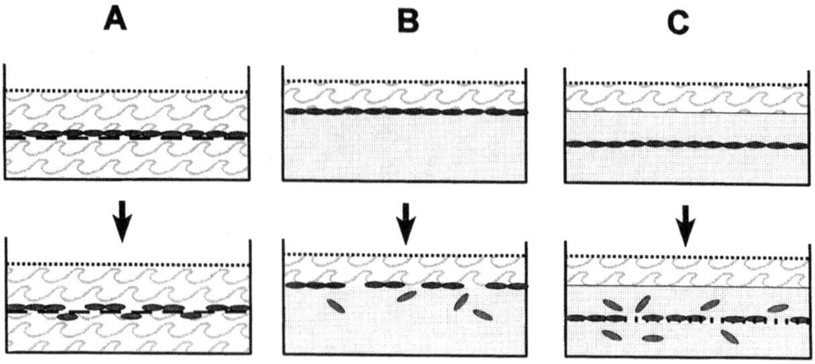

Figure 12.5 In vitro assays for studying cell migration. (A) Transmembrane assay. (B) Collagen gel assay. (C) Sandwich assay.

et al., 1996b). Interestingly, native fibronectin and its cell-binding domain display the reverse activity pattern, inducing cell migration in the transmembrane assay and being without effect in the collagen gel assay. Recent observations further indicate that the gelatin-binding domain of fibronectin also stimulates endothelial cell sprouting in vitro and angiogenesis in vivo. (A.M. Schor, manuscript in preparation). These data are consistent with published reports indicating that the macromolecular matrix to which cells are adherent exerts a profound effect on cellular response to soluble effector molecules, including angiogenic factors (S.L. Schor, 1994; Taipale and Keski-Oja, 1997).

We have recently developed a modified version of the collagen gel assay in which it is possible to distinguish between chemokinetic and chemotactic modes of motogenic response. In the simplest version of this assay (Figure 12.5C), cells are plated on the surface of a native type I collagen gel which is then marked by the application of inert particles. One hour after plating (i.e. sufficient time for the cells to attach to the collagen fibres, but not to spread or induce fibre reorganisation) this substratum is overlaid with a second gel. As a result of this procedure, the plated sheet of cells becomes embedded within an isotropic collagenous matrix. Medium is then applied to the cultures. Angiogenic factors may be added equally to both upper and lower collagen gels and medium overlay (to identify a chemokinetic response) or to only the medium overlay (to identify a chemotactic response to the imposed concentration gradient). The subsequent movement of individual cells (up or down from their marked original location) may then be assessed microscopically. Data presented in Figure 12.7 indicate that PDGF-BB induces both a chemokinetic response and a chemotactic response in target fibroblasts. The advantage of this "sandwich" assay compared to the transmembrane assay is that the effects of soluble factors on directional cell migration may be examined within a physiologically relevant macromolecular matrix. Progressively more complex matrices may be constructed by the stepwise addition of other matrix constituents to the simple collagen gel.

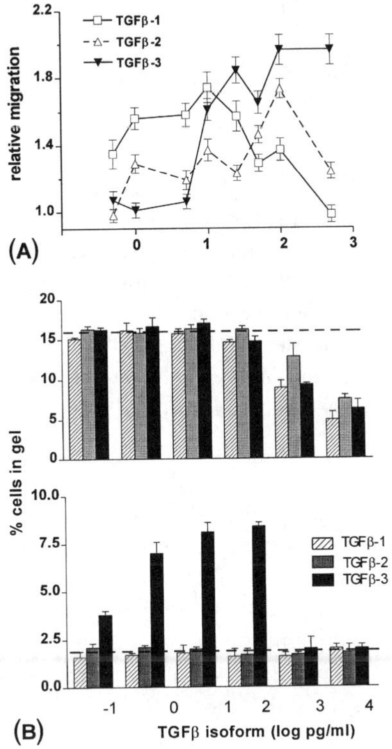

Figure 12.6 Distinct effects of TGFβ isoforms on fibroblast migration in two assays. (A) Effect of TGFβ-1, TGFβ-2 and TGFβ-3 on fibroblast migration in the transmembrane assay. (B, C). Effect of the same TGF isoforms on fibroblast migration in the collagen gel assay.

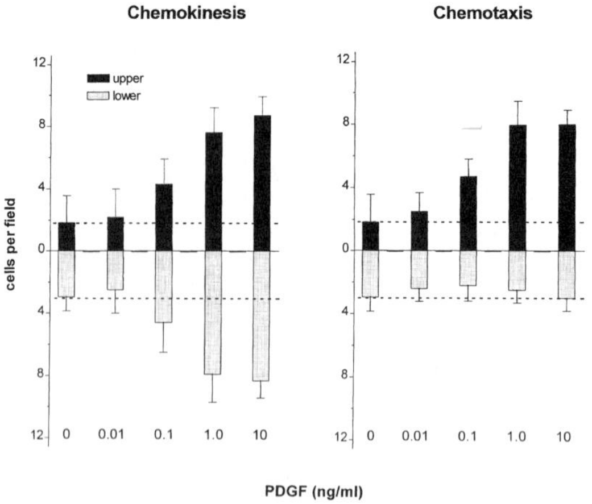

Figure 12.7 The effects of PDGF on fibroblast migration in the sandwich assay. For the assessment of chemokinesis, the indicated concentrations of PDGF were present in the two collagen layers and medium overlay. A chemokinetic response is indicated by the equal stimulation of cell migration into both upper and lower gels. To assess chemotaxis, PDGF was only present in the medium overlay, thereby forming a concentration gradient in the underlying collagen gels. A chemotactic response is indicated by the migration of more cells into the upper gel. The data indicate that PDGF elicits both a chemokinetic response and a chemotactic response.

12.6 Conclusions

Several concepts emerge from the experimental work related to angiogenesis. These are relevant not only in terms of advancing our understanding of the biological processes involved in the control of new blood vessel formation, but further serve to extend the conceptual basis of theoretical modelling of this complex in vivo phenomenon. Key issues raised in this review may be summarised as follows.

(a) Vascular cells and stromal fibroblasts are highly heterogeneous cell populations. Such heterogeneity may arise as a result of several distinct processes operative at the cellular level, including (i) *modulation* of cell phenotype by the microenvironment, (ii) *differentiation* along a specific cell lineage, (iii) *transdifferentiation* from one apparent lineage to another, and (iv) anatomical translocation from one site to another. Mathematical modelling should take into account functional heterogeneity in terms of cellular response to and production of effector molecules (e.g. growth factors and matrix molecules), rather than the tacit assumption that all cells within a given population (e.g. endothelial) behave in a similar manner. Such models may also have to consider the temporal changes in functional heterogeneity which characterise the angiogenic process.

(b) Vascular cells adopt an angiogenic (sprouting) phenotype in response to culture within a three-dimensional macromolecular matrix. This transition from a resting to a sprouting phenotype occurs *in the absence of exogenous angiogenic factors* and is completely reversed when cells are subsequently cultured on the surface of a two-dimensional substratum. This observation raises various possibilities. (i) The prime function of many angiogenic factors involves the stimulation of directional cell migration. This suggestion is consistent with the motogenic activity of angiogenic cytokines (assessed using target vascular cells and stromal fibroblasts) and also with the pivotal role played by modulators of matrix degradation (stimulators and inhibitors). (ii) The expression of *endogenous* angiogenic factors by vascular cells may be regulated by the extracellular matrix.

(c) The extracellular matrix plays an important role in modulating the response of vascular cells and fibroblasts to angiogenic factors. Experimental data clearly indicate that the same angiogenic factor (e.g. TGFβ) may exert quite different effects on cell motility depending on the nature of the macromolecular matrix to which the cells are adherent. The formation of sprouting cells requires the specific combination of stimulatory cytokine(s) and permissive extracellular matrix components. Indeed, even an apparent abundance of stimulatory cytokine is unable to induce sprouting in the presence of non-permissive matrices. This strict dependence of growth factor activity on matrix may contribute to the lack of inappropriate angiogenesis in normal tissues in which immunohistochemical analysis indicates the ubiquitous presence of angiogenic growth factors, such as FGF and VEGF. In view of this modulatory role of the matrix, changes in matrix composition and/or biophysical organisation during the angiogenic process may need to be taken into account in theoretical models. It should also be emphasised that participating cells (e.g. vascular endothelial cells and stromal fibroblasts) are a source of angiogenic growth factors, matrix macromolecules, matrix-degrading enzymes and their inhibitors. Reciprocal paracrine interactions between these cell populations result in the establishment of complex networks of feedback loops whereby neighbouring cells regulate each other's behaviour.

(d) The extracellular matrix contributes to the control of angiogenesis by several mechanisms. Apart from those discussed above, extracellular matrix components may exert direct cell signalling mechanisms, either stimulating or inhibiting angiogenesis. Fragments of hyaluronic acid are a well-characterised example capable of stimulating angiogenesis. The gelatin-binding domain of fibronectin is a more recently documented angiogenic factor. In both cases these angiogenic factors are "cryptic" within the larger (biologically inactive) matrix molecule and are produced when the matrix molecule is hydrolysed. The hydrolysis of matrix macromolecules may also release angiogenic peptides (e.g. FGF, scatter factor) bound to them (Taipale and Keski-Oja, 1997) as well as cryptic angiogenesis inhibitors (angiostatin, endostatin) (O'Reilly, 1997; Sasaki *et al.*, 1998). Therefore, the release of matrix-degrading enzymes by cells participating in the angiogenic process may generate angiogenesis effector molecules and contribute to the modulation of angiogenesis by this means. The generation of such active molecules, including any gradients in their concentration, must also be taken into account in an accurate temporal modelling of angiogenesis.

(e) Considering the well-documented heterogeneity among endothelial cells and the role of the extracellular matrix in the control of cell behaviour, it is clear that the effects of a particular soluble angiogenic factor or inhibitor should be defined in the context of the specific target cell population and its microenvironment. In recognition of these integrative mechanisms operative at the tissue level, we suggest that the control of cell function must ultimately be understood in the context of a *tissue response unit*. At minimum this should consistof the constituent cells and their immediate microenvironment of matrix molecules and cytokines. The considerable advances in our understanding of events occurring at the molecular level should now be integrated within the tissue response unit to understand the interdependent mechanisms operating at the tissue level of organisation.

Acknowledgements
The authors gratefully acknowledge the financial support of the Biotechnology and Biological Sciences Research Council (Grant MMI09008) and the Cancer Research Campaign.

REFERENCES

Arciniegas, E., Sutton, A.B., Allan, T.D. and Schor AM. (1992) Transforming growth factor-β1 promotes the differentiation of endothelial cells into smooth muscle-like cells in vitro. *J. Cell Sci.* 103: 521–529.

Asahara, T., Murohara, T., Sullivan, A., Silver, M., vad der Zee, L.T., Witzenbichler, B., Schatteman, G. and Isner, J.M. (1997) Isolation of putative progenitor endothelial cells for angiogenesis. *Science* 275: 964–967.

Ashton, N. and DeOliveira, F. (1966) Nomenclature of pericytes. Intramural and extramural. *Br. J. Ophthalmol.* 91: 433–457.

Augustin, H.G., Kozian, D.H. and Johnson, R.C. (1994) Differentiation of endothelial cells: Analysis of the constitutive and activated endothelial cell phenotypes. *BioEssays.* 16: 901–906.

Ausprunk, D.H. and Folkman, J. (1977) Migration and proliferation of endothelial cells in preformed and newly formed blood vessels during tumor angiogenesis. *Microvasc. Res.* 14: 53–65.

Ben-Av, P., Crofford, L.J., Wilder, R.L. and Hla, T. (1995) Induction of vascular endothelial growth factor induction by prostaglandin E and interleukin-1: a potential mechanism for inflammatory angiogenesis. *FEBS. Lett.* 372: 83–87.

Berse, B., Brown, L.F., Van de Water, L., Dvorak, H.F. and Senger, D.R (1992) Vascular permeability factor (vascular endothelial growth factor) gene is expressed differentially in normal tissues, macrophages and tumours. *Mol. Biol. Cell* 3: 211–220.

Burri, P.H. (1992) Intussusceptive microvascular growth, a new mechanism of capillary network formation. In: *Angiogenesis: Key Principles, Science, Technology, Medicine.* (R. Steiner, P.B. Weisz, and R. Langer, eds) pp. 32–39. Birkhäuser Verlag, Basel.

Bussolino, F., Mantovani, A. and Persico, G. (1997) Molecular mechanisms of blood vessel formation. *Trends Biochem. Sci.* 22: 251–256.

Canfield, A.E. and Schor, A.M. (1995) Evidence that tenascin and thrombospondin-1 modulate sprouting of endothelial cells. *J. Cell Sci.* 108: 797–809.

Canfield, A.E., Schor, A.M., Schor, S.L. and Grant, M.E. (1986) The biosynthesis of extracellular matrix components by bovine retinal endothelial cells displaying distinctive morphological phenotypes. *Biochem. J.* 235: 375–383.

Canfield, A.E., Boot-Handford, R.P. and Schor, A.M. (1990) Thrombospondin gene expression by endothelial cells in culture is modulated by cell proliferation, cell shape and the substratum. *Biochem. J.* 268: 225–230.

Canfield, A.E., Wren, F.E., Schor, S.L., Grant, M.E. and Schor, A.M. (1992) Aortic endothelial cell heterogeneity in vitro. *J. Cell Sci.* 102: 807–814.

Carmeliet, P. and Collen, D. (1998) Vascular development and disorders: Molecular analysis and pathogenic insights. *Kidney Int.* 53: 1519–1549.

Clark, E.R. and Clark, E.L. (1939) Microscopic observations on growth of blood capillaries in living animals. *Am. J. Anat.* 64: 251–301.

Cockerill, G.W., Gamble, J.R., and Vadas, M.A. (1995) Angiogenesis: models and modulators. *Int. Rev. Cytol.* 159: 113–160.

Decker, B., Bartels, H., and Decker, S. (1995) Relationships between endothelial cells, pericytes and osteoblasts during bone formation in the sheep femur following implantation of tricalciumphosphate ceramic. *Anat. Rec.* 242: 310–320.

Diaz-Flores, L., Gutierrez, R., Varela, H., Rancel, N. and Valladares, F. (1991) Microvascular pericytes: a review of their morphological and functional characteristics. *Hist. Histopath.* 6: 269–286.

Diaz-Flores, L., Gutierrez, R., Lopez-Alonso, A., Gonzalez, R. and Varela, H. (1992) Pericytes as a supplementary source of osteoblast in periosteal osteogenesis. *Clin. Orthopaed. Rel. Res.* 275: 280–286.

Dziadek, M. and Mitrangas, K. (1989) Differences in the solubility and susceptibility to proteolytic degradation of basement membrane components in adult and embryonic tissue. *J. Anat.* 184: 298–310.

Eguchi, G. and Kodama R. (1993) Transdifferentiation. *Curr. Opin. Cell Biol.* 5: 1023–1028.

Eisenstein, R. (1991) Angiogenesis in arteries: review. *Pharmacol. Ther.* 49: 1–19.

Ellis, I.R. and Schor, S.L. (1998) Differential motogenic and biosynthetic response of fetal and adult skin fibroblasts to TGF isoforms. *Cytokine* 10: 281–289.

Ellis, I.R., Banyard, J., and Schor, S.L. (1999) Motogenic and biosynthetic response of adult skin fibroblasts to TGFβ isoforms (- 1, - 2 and - 3) determined by "tissue response unit": Role of cell density and substratum. *Cell Biol. Int.* in press.

Folkman, J. and Shing, Y. (1992) Angiogenesis. *J. Biol. Chem.* 267: 10931–10934.

Fujimoto, T. and Singer, S.J. (1986) Immunocytochemical studies of endothelial cells in vitro. I. The presence of desmin only, desmin plus vimentin, or vimentin only in endothelial cells of different capillaries of the chick. *J. Cell Biol.* 103: 2775–2786.

Fukumura, D., Xavier, R., Sugiura, T., Chen, Y., Park, E.U., Lu, N., Selig, M., Nielsen, G., Taksir, T., Jain, R.K. and Seed, B. (1998) Tumour induction of VEGF promoter activity in stromal cells. *Cell* 94: 715–725.

Garrido, T., Riese, H.H., Aracil, M. and Perez-Aranda, A. (1995) Endothelial cell differentiation into capillary-like structures in response to tumour cell conditioned medium: a modified chemotaxis chamber assay. *Br. J. Cancer* 71: 770–5.

Haas, T.L., Davis, S.J. and Madri, J.A. (1998) Three-dimensional type I collagen lattices induce coordinate expression of matrix metalloproteinases MT1-MMP and MMP-2 in microvascular endothelial cells. *J. Biol. Chem.* 273: 3604–3610.

Haudenschild, C.C. (1980) Growth control of endothelial cells in atherogenesis and tumor angiogenesis. *Adv. Microcirc.* 9: 226–251.

Hudlicka, O. (1998) Is physiological angiogenesis in skeletal muscle regulated by changes in microcirculation? *Microcirc.* 5: 5–23.

Hunt, T.K. (1980) *Wound Healing and Infection: Theory and Surgical Practice.* Appleton-Century-Crofts, New York.

Ingber, D.E. and Folkman, J. (1989) Mechanochemical switching between growth and differentiation during fibroblast growth factor-stimulated angiogenesis in vitro: Role of extracellular matrix. *J. Cell Biol.* 109: 317–330.

Iruela-Arispe, M.L. and Dvorak, H.F. (1997) Angiogenesis: a dynamic balance of stimulators and inhibitors. *Thromb. Haemost.* 78: 672–677.

Ishibashi, T., Miller, H., Orr, G., Sorgente, N. and Ryan, S.J. (1987) Morphologic observation on experimental subretinal neovascularization in the monkey. *Invest. Ophthalmol. Vis. Sci.* 28: 1116–1130.

Javier, R., Sourla, A, Koutsilieris, M. and Doillon, C.J. (1997) Stromal fibroblasts are required for PC-3 prostate cancer cells to produce capillary-like formation of endothelial cells in a three-dimensional co-culture system. *Anticancer Res.* 17: 1551–1557.

Jinno, K., Tanimuzu, M., Hyodo, I., Nishikawa, Y., Hosokawa, Y., Doi, T., Endo, H., Yamashita, T. and Okada, Y. (1998) Circulating vascular endothelial growth factor (VEGF) is a possible tumor marker for metastasis in human hepatocellular carcinoma. *J. Gastroent.* 33: 376–382.

Keegan, A., Hill, C., Kumar, S., Phillips, P., Schor, A. and Weiss, J. (1982) Purified tumour angiogenesis factor enhances proliferation of capillary, but not aortic, endothelial cells in vitro. *J. Cell Sci.* 55: 261–276.

Kuiper, R.A., Schellens, J.H., Blijham, G.H., Beijnen, J.H. and Voest, E.E. (1998) Clinical research on antiangiogenic therapy. *Pharmacol. Res.* 37:.1–16.

Lajtha, L.G. (1983) Stem cell concepts. In: *Stem Cells. Their Identification and Characterisation* (C.S. Potten, ed) pp. 1–11. Churchill Livingstone, New York.

Lindahl, P., Johansson, B.R., Leveen, P. and Betsholtz, C. (1997) Pericyte loss and microaneurysm formation in PDGF-B-deficient mice. *Science* 277: 242–245.

Madri, J.A., Kocher, O., Merwin, J.R. and Bell, L., Yannariello-Brown, J. (1989) The interactions of vascular cells with solid phase (matrix) and soluble factors. *J. Cardiovascul. Pharmacol.* 14: S70–75.

Montesano, R. (1992) Regulation of angiogenesis in vitro. *Eur. J. Clin. Invest.* 22: 504–515.

Nehls, V. and Drenckhahn, D. (1993) The versatility of microvascular pericytes: from mesenchyme to smooth muscle? *Histochem.* 99: 1–12.

Nehls, V., Denzer, K. and Drenckhahn, D. (1992) Pericyte involvement in capillary sprouting during angiogenesis in situ. *Cell Tiss. Res.* 270: 469–474.

Nehls, V., Herrmann, R., Huhnken, M. and Palmetshofer, A. (1998) Contact-dependent inhibition of angiogenesis by cardiac fibroblasts in three-dimensional fibrin gels in vitro: implications fo microvascular network remodeling and coronary collateral formation. *Cell Tiss. Res.* 293: 479–488.

Nicosia, R.F., Bonanno, E. and Villaschi, S. (1992) Large-vessel endothelium switches to a microvascular phenotype during angiogenesis in collagen gel culture of rat aorta. *Atheroscl.* 95: 191–199.

Noden, D.M. (1989) Embryonic origins and assembly of blood vessels. *Am. Rev. Respir. Dis.* 140: 1097–1103.

Norrby, K. (1997) Angiogenesis: new aspects relating to its initiation and control. *APMIS* 105: 417–437.

O'Reilly, M.S. (1997) Angiostatin: an endogenous inhibitor of angiogenesis and of tumor growth. *EXS* 79: 273–94.

Owen, M.E. (1985) Lineage of osteogenic cells and their relationship to the stromal system. In: *Bone and Mineral Research*, Vol. (W.A. Peck, ed.) pp. 1–25. Elsevier, Amsterdam.

Owen, M.E., Cave, J. and Joyner, C.J. (1987) Clonal analysis in vitro of osteogenic differentiation of marrow CFU-F. *J. Cell Sci.* 87: 731–738.

Paku, S. and Paweletz, N. (1991) First steps of tumor-related angiogenesis. *Lab. Invest.* 65: 334–346.

Palade, G.E. (1988) The microvascular endothelium revisited. In: *Endothelial Cell Biology in Health and Disease*, (N. Simionescu, and M. Simionescu, eds.) pp.3–22. Plenum Press, New York.

Pluda,J.M. (1997) Tumor-associated angiogenesis: Mechanisms, clinical implications, and therapeutic strategies. *Sem. Oncol.* 24: 203–218.

Price, R.J., Owens, G.C. and Skalak, T.C. (1994) Immunohistochemical identification of arteriolar development using markers of smooth muscle differentiation. Evidence that capillary arteriolization proceeds from terminal arterioles. *Circ Res.* 75: 520–527.

Risau, W. and Lemmon, V. (1988) Changes in the vascular extracellular matrix during embryonic vasculogenesis and angiogenesis. *Dev. Biol.* 125: 441–450.

Sang, Q.X. (1998) Complex role of matrix metalloproteinases in angiogenesis. *Cell Res.* 8: 171–7.

Sasaki, T., Fukai, N., Mann. K., Gohring, W., Olsen, B.R. and Timpl, R. (1998) Structure, function and tissue forms of the C-terminal globular domain of collagen XVIII containing the angiogenesis inhibitor endostatin. *EMBO J.* 17: 4249–56.

Schlingermann, R.O., Rietveld, F.J.R., Kwaspen, F., van de Kerkhof, P.C.M., de Waal, R.M.W. and Ruiter, D.J. (1991) Differential expression of markers for endothelial cells, pericytes and basal lamina in the microvasculature of tumors and granulation tissue. *Am.*

J. Pathol. 138: 1335–1347.

Schor, A.M. and Canfield, A.E. (1998) Osteogenic potential of vascular pericytes. In *Marrow Stromal Cell Culture: Practical Animal Cell Biology Series*, (J.N. Beresford, and M.E. Owen, eds.) pp. 128-148. Cambridge University Press, Cambridge.

Schor, A.M. and Schor, S.L. (1983) Tumour angiogenesis: a review. *J. Exp. Pathol.* 141: 385–413.

Schor, A.M. and Schor, S.L. (1986) The isolation and culture of endothelial cells and pericytes from the bovine retinal microvasculature: a comparative study with large vessel vascular cells. *Microvasc. Res.* 32: 21–38.

Schor, A.M., Schor, S.L. and Allen, T.D. (1983) The effects of culture conditions on the proliferation and morphology of bovine aortic endothelial cells in vitro: reversible expression of the sprouting cell phenotype. *J. Cell Sci.* 62: 267–285.

Schor, A.M., Schor, S.L. and Allen, T.D. (1984) The synthesis of subendothelial matrix by bovine aortic endothelial cells in culture. *Tiss. Cell.* 16: 677–691.

Schor, A.M., Allen, T.D., Canfield, A.E., Sloan, P. and Schor, S.L. (1990) Pericytes derived from the retinal microvasculature undergo calcification in vitro. *J. Cell Sci.* 97: 449–461.

Schor, A.M., Canfield, A.E., Sloan, P. and Schor, S.L. (1991) Differentiation of pericytes in culture is accompanied by changes in the extracellular matrix. *In Vitro Cell. Dev. Biol.* 27A: 651–659.

Schor, A.M., Canfield, A.E., Sutton, A.B., Allen, T.D., Sloan, P. and Schor, S.L. (1992) The behaviour of pericytes in vitro: relevance to angiogenesis and differentiation. In: *Angiogenesis: Key Principles, Science, Technology, Medicine*, (R. Steiner, P.B. Weisz, and R. Langer, eds) pp. 167–178. Birkhäuser Verlag, Basel.

Schor, A.M., Canfield, A.E., Sutton, A.B., Arciniegas, E. and Allen, T.D. (1995) Pericyte differentiation. *Clin. Orthopaed. Rel. Res.* 313: 81–91.

Schor, A.M., Schor, S.L. and Arciniegas, E. (1997) Phenotypic diversity and lineage relationships in vascular endothelial cells. In: *Stem Cells* (C.S. Potten, ed.) pp. 119–146. Academic Press, London.

Schor, S.L. (1980) Cell proliferation and migration within three-dimensional collagen gels. *J. Cell Sci.* 41: 159–175.

Schor, S.L. (1994) Cytokine control of cell motility: modulation and mediation by the extracellular matrix. *Prog. Growth Factor Res.* 5: 223–248.

Schor, S.L. and Schor, A.M. (1987) Clonal heterogeneity in fibroblast phenotype: implications for the control of epithelial-mesenchymal interactions. *BioEssays* 7: 200–204.

Schor, S.L. and Schor, A.M. (1997) Stromal acceleration of tumour progression: Role of "fetal-like" fibroblast subpopulations. *Pathol. Update* 4: 75–95.

Schor, S.L., Ellis, I., Banyard, J., Dolman, C., Seneviratne, K., Gilbert, A.D. and Chisholm, D.M. (1996a) Fetal-like phenotypic characteristics of gingival fibroblasts: Potential relevance to wound healing. *Oral Diseases* 2: 155–166.

Schor, S.L., Ellis, I., Dolman, C., Banyard, J., Humphries, M.J., Mosher, D.F., Grey, A.M., Mould, A.P., Sotille, J. and Schor, A.M. (1996b) Substratum-dependent stimulation of fibroblast migration by the gelatin-binding domain of fibronectin. *J. Cell Sci.* 109: 2581–2590.

Shepro, D. and Morel, N.M.L. (1993) Pericyte physiology. *FASEB J.* 7: 1013–1018.

Sholley, M.M., Ferguson, G.P., Seibel, H.R., Montour, J.L. and Wilson, J.D. (1984) Mechanisms of neovascularization. Vascular sprouting can occur without proliferation of endothelial cells. *Lab. Invest.* 51: 624–634.

Simionescu, N. and Simionescu, M. (1988) The cardiovascular system. In: *Cell and Tissue Biology: A Textbook of Cytology*, 6th edn. (L. Weiss, ed.) pp. 355–400. Urban and Schwarzenberg, Inc., Baltimore.

Sims, D.E. (1991) Recent advances in pericyte biology- implications for health and disease. *Can. J. Cardiol.* 7: 431–443.

Sims, D.E., Miller, F.N., Horne, M.M. and Edwards, M.J. (1994) Interleukin-2 alters the positions of capillary and venule pericytes in rat cremaster muscle. *J. Submicro. Cytol. Pathol.* 26: 507–513.

Stein, J., Drenckhahn, D. and Nehls, V. (1996) Development of pericyte-like cells during angiogenesis in quail chick chimeras as detected by combined Feulgen reaction and immunohistochemistry. *Anat. Anz.* 178: 153–8.

Sutton, A.M., Canfield, A.E., Schor, S.L., Grant, M.E. and Schor, A.M. (1991) The response of endothelial cells to TGF-β1 is dependent upon cell shape, proliferative state and the nature of the substratum. *J. Cell Sci.* 99: 777–787.

Taipale, J. and Keski-Oja, J. (1997) Growth factors in the extracellular matrix *FASEB J.* 11: 51–59.

Vernon, R.B., Lara, S.L., Drake, C.J., Iruela-Arispe, L, Angello, J.C., Little, C.D., Wight, T.N. and Sage, E.H. (1995) Organized type I collagen influences endothelial patterns during "spontaneous angiogenesis in vitro": Planar cultures as models of vascular development *In Vitro Cell. Dev. Biol.* 31: 120–131.

Watanabe, S., Morisaki, N., Tezuka, M., Fukuda, K., Ueda, S., Koyama, N., Yokote, K., Kanzaki, T., Yoshida, S. and Saito, Y. (1997) Cultured retinal pericytes stimulate in vitro angiogenesis of endothelial cells through secretion of a fibroblast growth factor-like molecule. *Atheroscler.* 130: 101–107.

Weiss, J.B. and McLaughlin, B. (1998) Endothelial cell stimulating angiogenesis factor. *Int. J. Biochem. Cell Biol.* 30: 423–427.

Zigmond, S.H. and Hirsch, J.G. (1973) Leukocyte locomotion and chemotaxis. *J. Exp. Med.* 137: 387–410.

13

Modelling the Growth and Form of Capillary Networks

Mark A.J. Chaplain and Alexander R.A. Anderson

Department of Mathematics, University of Dundee, Dundee DD1 4HN, Scotland, UK

"Passing to what may seem a very different subject, we may investigate a number of interesting points in connection with the form and structure of the blood-vessels....."

(Thompson, 1942, Ch. XV, p. 948)

13.1 Introduction

In Chapter 12 the Schors and R. Bailey gave an excellent experimental overview of several of the key processes involved in angiogenesis. This chapter moves on to the different subject of mathematics and, using mathematical modelling, we may investigate a number of interesting points in connection with the form and structure of the blood-vessels. We will focus on the mathematical modelling of capillary networks as they arise in tumour-induced angiogenesis, although as we have seen angiogenesis is also an important process in embryogenesis (development) and wound-healing (Pettet *et al.*, 1996).

This next section is a summary of the main events which occur during tumour angiogenesis. A summary of mathematical models for angiogenesis leads into a more detailed description of some recent work of Chaplain and Anderson (1997), Anderson and Chaplain (1998b). This chapter follows on naturally from the previous one since the mathematical model it presents focuses specifically on the so-called *tissue response*

On Growth and Form: Spatio-temporal Pattern Formation in Biology.
Editor Mark A.J. Chaplain, G.D. Singh, J.C. McLachlan

unit, which consists of the constituent cells and their immediate microenvironment of cytokines and matrix molecules. It explores the extent to which a (relatively) simple (minimal) mathematical model can reproduce a growing capillary network (more generally, any branching, connected network) not merely in a phenomenological, qualitative manner but, more importantly, in a quantitative and predictive manner.

13.2 Tumour-Induced Angiogenesis

The first event of tumour-induced angiogenesis involves the cancerous cells of a solid tumour secreting a number of chemicals, collectively known as tumour angiogenic factors, or TAF (Folkman, 1985, 1995; Folkman and Klagsbrun, 1987), into the surrounding tissue. These factors (cytokines) diffuse through the tissue (extracellular matrix), thereby creating chemical gradients in the tissue surrounding the tumour. Once the TAF reaches any neighbouring blood vessels, endothelial cells lining these vessels are stimulated into a well-ordered sequence of events affecting their migration and proliferation which culminates in the tumour being penetrated by a capillary network of blood vessels.

First of all, the endothelial cells are induced to synthesise matrix degradative enzymes (e.g. proteinases, collagenases) and to degrade the parent venule basement membranes and then migrate through the disrupted membrane towards the tumour. The initial migratory response of the endothelial cells to these angiogenic factors is a chemotactic one (Sholley *et al.*, 1984; Terranova *et al.*, 1985; Paweletz and Knierim, 1989; Stokes *et al.*, 1990), and the cells begin to move out of their parent vessel into the surrounding tissue towards the tumour. Following this, small, finger-like capillary sprouts are formed by accumulation of endothelial cells which are recruited from the parent vessel. The sprouts grow in length due to the migration and further recruitment of endothelial cells (Warren, 1966; Ausprunk and Folkman, 1977; Sholley *et al.*, 1984) and continue to move toward the tumour directed by the motion of the leading endothelial cell at the sprout-tip. Further sprout extension occurs when some of the endothelial cells of the sprout-wall begin to proliferate. Cell division is largely confined to a region just behind the cluster of mitotically inactive endothelial cells that constitute the sprout-tip. This process of sprout-tip migration and proliferation of sprout-wall cells forms solid strands of endothelial cells among the extracellular matrix. The cells continue to make their way through the extracellular matrix which consists of interstitial tissue, collagen fibre and fibronectin as well as other components (Liotta *et al.*, 1983; Paweletz and Knierim, 1989). Interactions between the endothelial cells and the extracellular matrix are very important and directly affect cell migration (Madri and Pratt, 1986). Although there are several matrix molecules which interact with the endothelial cells, this chapter focuses in particular on the specific interactions between the endothelial cells and fibronectin, a major component of the extracellular matrix, which has been shown to enhance cell adhesion to the matrix.

Fibronectin is a matrix macromolecule which occurs in two distinctly different forms: (i) as a soluble glycoprotein found in various body fluids (including blood), known as plasma fibronectin; (ii) as an insoluble constituent of the extracellular matrix and basement membranes of cells, known as cellular fibronectin (Hynes, 1990). Cultured endothelial cells are known to synthesise and secrete cellular fibronectin (Jaffee and Mosher, 1978) and the expression of this secreted fibronectin by the endothelial cells in cultures (in vitro) closely reflects the distribution of pre-existing fibronectin observed in matrices in vivo (Hynes, 1990). The fibronectin which is produced and secreted by endothelial cells does not diffuse, but remains bound to the extracellular matrix (Hynes, 1990), its central function being the adhesion of cells to the matrix (Schor *et al.*, 1981; Hynes, 1990). It is a major ligand (both cellular fibronectin and plasma fibronectin) between cells and matrix materials in many situations. Endothelial cells use fibronectin for attachment to the matrix via integrins, a family of cell-surface receptors (Johansson *et al.*, 1987; Hynes, 1990).

It has been verified experimentally that fibronectin stimulates directional migration of a number of cell types (including endothelial cells) in Boyden chamber assays (Albini *et al.*, 1987). These results have demonstrated that fibronectin promotes cell migration up a concentration gradient and the results of Quigley *et al.* (1983), Lacovara *et al.* (1984) and McCarthy and Furcht (1984) have further demonstrated that this is a response of the cells to a gradient of adhesiveness of bound fibronectin, termed haptotaxis (Carter, 1965, 1967). Therefore, in addition to the chemotactic response of the endothelial cells to the TAF, there is a complementary haptotactic response to the fibronectin present within the extracellular matrix (Bowersox and Sorgente, 1982).

Initially, the sprouts arising from the parent vessel grow essentially parallel to each other. It is observed that once the finger-like capillary sprouts have reached a certain distance from the parent vessel, they tend to incline toward each other (Paweletz and Knierim, 1989), leading to numerous tip-to-tip and tip-to-sprout fusions known as anastomoses. Such anastomoses result in the fusing of the finger-like sprouts into a network of poorly perfused loops or arcades. Following this process of anastomosis, the first signs of circulation can be recognised, and new buds and sprouts emerge from the primary loops, repeating the angiogenic sequence of events and providing for the further extension of the new capillary bed. The production of new capillary sprouts from the sprout-tips is often known as sprout branching; as the sprouts approach the tumour, their branching dramatically increases until the tumour is eventually penetrated, resulting in vascularisation (Muthukkaruppan *et al.*, 1982).

This process of repeated steps of endothelial cell migration, sprout extension, cell proliferation and loop formation is necessary for the successful vascularisation of the tumour. However, Sholley *et al.* (1984) demonstrated that, in the absence of endothelial cell proliferation, a restricted capillary network, which stops after a few days and never reaches the tumour, is formed. Therefore, unless endothelial cells undergo mitosis, the capillary sprouts cannot complete vascularisation successfully.

Tumour-induced angiogenesis provides the crucial link between the avascular phase of solid tumour growth and the more harmful vascular phase, wherein the tumour invades the surrounding host tissue and blood system (Chaplain, 1996). Once invasion has occurred, the possibility of the cancer spreading to other parts of the body (metastasis) becomes a reality and is far more difficult to treat clinically. However, these apparently insidious features of tumour-induced angiogenesis are now being used to combat cancer growth, and the clinical importance of angiogenesis as a prognostic tool is now recognised. In developing mathematical models of angiogenesis, we hope to be able to provide a deeper insight into the underlying mechanisms which cause the process (Chapter 12). It is therefore essential that we develop a predictive mathematical model which is capable of producing the precise, quantitative *morphological features* of developing blood vessels - "the growth and form of a capillary network".

13.3 Mathematical Modelling of Tumour-Induced Angiogenesis

In recent years several mathematical models, using different approaches, have been developed to describe some of the key features of tumour-induced angiogenesis. Many of these models have used a continuum, deterministic framework in one space dimension (Zawicki *et al.*, 1981; Balding and McElwain, 1985; Chaplain and Stuart, 1993; Byrne and Chaplain, 1995; Orme and Chaplain, 1996; Anderson and Chaplain, 1998a). Although these models were capable of capturing some features of angiogenesis such as average sprout density and network expansion rates, they were unable to provide more detailed information concerning the actual structure and morphology of the capillary network, hence they were of limited predictive value.

More realistic continuum models of angiogenesis in two space dimensions have been considered by Chaplain (1995) and Orme and Chaplain (1997). The results of these models permit a more detailed qualitative comparison with in vivo observations concerning the spatio-temporal distribution of capillary sprouts within the network. However, even with these models, it is not possible to capture certain important events such as repeated sprout branching and hence the overall dendritic structure of the network. More general two-dimensional continuum branching models, with potential application to a wide number of problems, have been considered by Meinhardt (1976, 1982) and Edelstein-Keshet and Ermentrout (1989).

In contrast to these deterministic models, the model of Stokes and Lauffenburger (1991) used a discrete probabilistic framework in two space dimensions, based on stochastic differential equations. This approach had the advantage of enabling the motion of *individual endothelial cells* to be followed. Realistic capillary network structures were generated by incorporating rules for sprout branching and anastomosis. Since parameters were estimated, as far as possible, from available experimental data, this permitted both qualitative and quantitative comparisons with in vivo networks to be made. Although the model incorporated random motility and chemotaxis as mechanisms for cell migration, no account was taken of the important interactions between the endothelial cells and the extracellular matrix. The model was also unable to reproduce the fact that there is an increased frequency of branching at the edge of the network as the capillary sprouts get closer to the tumour. This observed feature of tumour-induced angiogenesis has been described as the *brush border* effect (Gimbrone *et al.*, 1974; Ausprunk and Folkman, 1977; Zawicki *et al.*, 1981; Muthukkaruppan *et al.*, 1982; Sholley *et al.*, 1984).

The rest of this chapter describes recent advances made in modelling tumour-induced angiogenesis and it presents two types of model, one based on a continuum approach using partial differential equations, the other based on a discrete approach using the technique of biased random walks. The mathematical model will focus on the interactions between three very important variables involved in tumour-induced angiogenesis; namely, endothelial cells, tumour angiogenic factors (TAFs) and fibronectin, each of which has a crucial role to play (cf. the tissue response unit).

The morphological events that are involved in new blood-vessel formation have been defined by studies of in vivo systems such as the chick chorioallantoic membrane (CAM), animal corneal models and in vitro examination of endothelial cell migration and proliferation (Folkman and Haudenschild, 1980). The initial mathematical model is based on the implant of a solid tumour in the cornea of a test animal (Gimbrone *et al.*, 1974; Muthukkaruppan *et al.*, 1982). Parameter values used in the model will be based, as far as possible, on estimates obtained from experimental observations.

Once the basic model has been developed and tested (i.e. compared with results from the cornea experiments), theoretical simulations generating a fully three-dimensional capillary network will then be presented in the penultimate section. It is possible to analyse these computed structures for quantitative information such as network expansion rates, loop formation, sprout length and overall network architecture (e.g. vessel area and volume).

13.4 The Continuous Mathematical Model

Denote the endothelial cell density per unit area by n, the TAF concentration by c and the fibronectin concentration by f. As explained in the previous section, let us focus attention on the *tissue response unit* (Chapter 12) and assume that the motion of the endothelial cells (at or near a capillary sprout-tip) is influenced by three factors: random motility (analogous to molecular diffusion), chemotaxis in response to TAF gradients in the surrounding connective tissue stroma (Terranova *et al.*, 1985; Stokes *et al.*, 1990; Stokes *et al.*, 1991) and haptotaxis in response to fibronectin gradients, also present in the surrounding tissue (Schor *et al.*, 1981; Bowersox and Sorgente, 1982; Quigley *et al.*, 1983; Lacovara *et al.*, 1984; McCarthy and Furcht 1984). To derive the partial differential equation governing endothelial cell motion, we firstly consider the total cell flux and then use the conservation equation for cell density. The three contributions to the endothelial cell flux $\mathbf{J}_n$, are given by

$$\mathbf{J}_n = \mathbf{J}_{random} + \mathbf{J}_{chemo} + \mathbf{J}_{hapto}.$$

To describe the random motility of the endothelial cells at or near the sprout-tips, we assume a flux of the form $\mathbf{J}_{random} = -D_n \nabla n$, where D_n is a positive constant, the cell random motility coefficient. We take the chemotactic flux to be $\mathbf{J}_{chemo} = \chi(c) n \nabla c$, where $\chi(c)$ is a chemotactic function. We choose a receptor-kinetic law of the form

$$\chi(c) = \chi_0 \frac{k_1}{k_1 + c}, \qquad (13.1)$$

reflecting the assumption that chemotactic sensitivity decreases with increased TAF concentration, where χ_0, the chemotactic coefficient, and k_1 are positive constants (Lapidus and Schiller, 1976; Lauffenburger *et al.*, 1984; Woodward *et al.*, 1995; Olsen *et al.*, 1997). The influence of fibronectin on the endothelial cells is modelled by the haptotactic flux, $\mathbf{J}_{hapto} = \rho_0 n \nabla f$, where $\rho_0 > 0$ is the (constant) haptotactic coefficient.

Since we are focusing attention on the endothelial cells at the sprout-tips (where there is no proliferation) and given that endothelial cells have a long half-life, on the order of months (Paweletz and Knierim, 1989), we omit any birth and death terms associated with the endothelial cells (endothelial cell proliferation is considered in the section devoted to the discrete model). This assumption models the experiments of Sholley *et al.* (1984), where the cells were irradiated to stop proliferation but the network formed to a certain extent.

The conservation equation for the endothelial cell density n is therefore given by

$$\frac{\partial n}{\partial t} + \nabla \cdot \mathbf{J}_n = 0,$$

and hence the partial differential equation governing endothelial cell motion (in the absence of cell proliferation) is

$$\frac{\partial n}{\partial t} = D_n \nabla^2 n - \nabla \cdot (\chi(c) n \nabla c) - \nabla \cdot (\rho_0 n \nabla f). \tag{13.2}$$

To derive the TAF equation, we first of all consider the initial event of tumour-induced angiogenesis which is the secretion of TAF by the tumour cells. Once secreted, TAF diffuses into the surrounding corneal tissue and extracellular matrix and sets up a concentration gradient between the tumour and any pre-existing vasculature such as the nearby limbal vessels. During this initial stage, where the TAF diffuses into the surrounding tissue (with some natural decay), we assume that the TAF concentration c satisfies an equation of the form

$$\frac{\partial c}{\partial t} = D_c \nabla^2 c - \theta c \tag{13.3}$$

where D_c is the TAF diffusion coefficient and θ is the decay rate. We will assume that the steady state of this equation establishes the TAF gradient between the tumour and the nearby vessels and provides us with the initial conditions for the TAF concentration profile. As the endothelial cells migrate through the extracellular matrix in response to this steady-state gradient (Stokes and Lauffenburger, 1991), there is some uptake and binding of TAF by the cells (Ausprunk and Folkman, 1977; Hanahan, 1997). We model this process by a simple uptake function, resulting in the following equation for the TAF concentration:

$$\frac{\partial c}{\partial t} = -\lambda n c, \tag{13.4}$$

where λ is a positive constant, and the initial TAF concentration profile is obtained from the steady state of (13.3).

Fibronectin is known to be present in most mammalian tissue and has been identified as a component of the tissue of the cornea (Kohno *et al.*, 1987). In addition to this pre-existing fibronectin, it is known that the endothelial cells themselves produce and secrete fibronectin (Jaffee and Mosher, 1978) which then becomes bound to the extracellular matrix and does not diffuse (Hynes, 1990). Therefore the equation for fibronectin contains no diffusion term. There is also some uptake and binding of fibronectin to the endothelial cells as they migrate toward the tumour (Hynes, 1990). These production and uptake processes are modelled by the following equation:

$$\frac{\partial f}{\partial t} = \omega n - \mu n f, \tag{13.5}$$

where ω and μ are positive constants.

Hence the complete system of equations describing the interactions of the endothelial cells, TAF and fibronectin is given by

$$
\begin{aligned}
\frac{\partial n}{\partial t} &= \overbrace{D_n \nabla^2 n}^{random\ motility} - \overbrace{\nabla \cdot \left(\frac{\chi_0 k_1}{k_1 + c} n \nabla c \right)}^{chemotaxis} - \overbrace{\nabla \cdot (\rho_0 n \nabla f)}^{haptotaxis}, \\
\frac{\partial f}{\partial t} &= \overbrace{\omega n}^{production} - \overbrace{\mu n f}^{uptake}, \\
\frac{\partial c}{\partial t} &= - \overbrace{\lambda n c}^{uptake}.
\end{aligned}
\tag{13.6}
$$

This system is considered to hold on a square spatial domain of side L (representing a square of corneal tissue) with the parent vessel (e.g. limbal vessel) located along one edge and the tumour located on the opposite edge. We assume that the cells, and consequently the capillary sprouts, remain within the domain of tissue under consideration and therefore no-flux boundary conditions of the form

$$\underline{\zeta} \cdot (-D_n \nabla n + n (\chi(c) \nabla c + \rho_0 \nabla f)) = 0, \tag{13.7}$$

are imposed on the boundaries of the square, where $\underline{\zeta}$ is an appropriate outward unit normal vector.

We now non-dimensionalise (13.6) by rescaling distance with the parent vessel to tumour distance L, time with $\tau = L^2/D_c$ (where D_c is the TAF diffusion coefficient), endothelial cell density with n_0, and TAF and fibronectin concentration with c_0 and f_0 respectively (where n_0, c_0, f_0 are appropriate reference variables). Therefore setting

$$\tilde{c} = \frac{c}{c_0}, \ \tilde{f} = \frac{f}{f_0}, \ \tilde{n} = \frac{n}{n_0}, \ \tilde{t} = \frac{t}{\tau}$$

and dropping the tildes for clarity, we obtain the non-dimensional system

$$
\begin{aligned}
\frac{\partial n}{\partial t} &= D \nabla^2 n - \nabla \cdot \left(\frac{\chi}{1 + \alpha c} n \nabla c \right) - \nabla \cdot (\rho n \nabla f), \\
\frac{\partial f}{\partial t} &= \beta n - \gamma n f, \\
\frac{\partial c}{\partial t} &= -\eta n c,
\end{aligned}
\tag{13.8}
$$

where

$$D = \frac{D_n}{D_c}, \chi = \frac{\chi_0 c_0}{D_c}, \alpha = \frac{c_0}{k_1}, \rho = \frac{\rho_0 f_0}{D_c},$$

$$\beta = \frac{\omega L^2 n_0}{f_0 D_c}, \gamma = \frac{\mu L^2 n_0}{D_c}, \eta = \frac{\lambda L^2 n_0}{D_c},$$

and subject to the no-flux conditions

$$\underline{\zeta} \cdot \left(-D \nabla n + n \left(\frac{\chi}{1 + \alpha c} \nabla c + \rho \nabla f \right) \right) = 0 \tag{13.9}$$

on the boundaries of the unit square.

Wherever possible, parameter values have been estimated from available experimental data. An average distance from a tumour implant to parent vessels in the cornea is between 1 and 2.5mm (Gimbrone *et al.*, 1974; Muthukkaruppan *et al.*, 1982) and the length scale is taken as $L = 2$mm (cf. Stokes and Lauffenburger, 1991). The experiments of Stokes *et al.* (1990) and Rupnick *et al.* (1988) on endothelial cell migration in response to angiogenic factors provide us with estimates for D_n, χ_0 and c_0. Endothelial cell random motility coefficients in the range 2×10^{-9} to $10^{-8}\text{cm}^2\text{s}^{-1}$ were obtained. However, this was under the assumption that the cells moved independently. Rupnick *et al.* (1988) showed that when the movement of individual endothelial cells is constrained by surrounding cells, their estimate of the random motility coefficient is too large and does not agree with the experimental results. The formation of capillary sprouts causes a dependence on neighbouring cells due to the formation of vessel walls - the endothelial cells which line the sprout-wall are contiguous with one another. Therefore one might expect a smaller random motility coefficient than the value observed by Rupnick *et al.* (1988), so take $D_n = 10^{-10}\text{cm}^2\text{s}^{-1}$ (Bray, 1992). Other in vivo experimental results also indicate that there appears to be very little random motility of endothelial cells at the capillary sprout-tips (Paweletz and Knierim, 1989; Paku and Paweletz, 1991). Stokes *et al.* (1990) measured the chemotactic coefficient of migrating endothelial cells in gradients of acidic fibroblast growth factor (aFGF). The maximum chemotactic response was measured in concentrations of aFGF around 10^{-10}M, giving a chemotactic coefficient of $2600\text{cm}^2\text{s}^{-1}\text{M}^{-1}$, so take $\chi_0 = 2600\text{cm}^2\text{s}^{-1}\text{M}^{-1}, c_0 \approx 10^{-10}$M. In the absence of any available data for the haptotactic coefficient ρ_0, it can be assumed that this is of the same magnitude as χ_0. Since the endothelial cells generally migrate towards the tumour, even in the absence of proliferation, it is reasonable to assume that $\chi_0 > \rho_0$. Terranova *et al.* (1985) found that endothelial cells responded in a haptotactic manner to fibronectin in concentrations of fibronectin around 10^{-10}M and we take this as an estimate for f_0. Estimates for the diffusion coefficient of TAF are in the range 5×10^{-7} to $5.9 \times 10^{-6}\text{cm}^2\text{s}^{-1}$ (Bray, 1992) and for our simulations we take $D_c = 2.9 \times 10^{-7}\text{cm}^2\text{s}^{-1}$. Estimates of the parameters ω, μ, λ were not available since they are very difficult to obtain experimentally (Hynes, 1990).

These parameter values now give non-dimensional values of $D = 0.00035$, $\chi = 0.38$ and we use a non-dimensional value of $\rho = 0.34$. The estimates for L and D_c now give the timescale $\tau = L^2/D_c$ as 1.5 days.

The first event of tumour-induced angiogenesis is the secretion of TAF by the tumour cells. The TAF then diffuses into the extracellular matrix and a concentration gradient is established between the tumour and parent vessel. If we consider the tumour as approximately circular and model TAF diffusion using (13.3) then an approximation of the steady-state solution for such an equation (Chaplain, 1995, 1996) has a concentration field of the form

$$c(x, y, 0) = \begin{cases} 1, & 0 \leq r \leq 0.1, \\ \dfrac{(\nu - r)^2}{\nu - 0.1}, & 0.1 \leq r \leq 1, \end{cases} \tag{13.10}$$

where ν is a positive constant and r is given by

$$r = \sqrt{(x-1)^2 + \left(y - \frac{1}{2}\right)^2}, \tag{13.11}$$

assuming that the tumour is centred on $(1, \frac{1}{2})$, with a radius of 0.1. Taking (13.10) as the initial conditions for the TAF concentration profile might then be a reasonable description of the actual concentration field arising from a small circular tumour implant. To approximate a row of tumour cells (or a larger circular implant) we also consider an initial TAF concentration field of the form

$$c(x, y, 0) = e^{-(1-x)^2/\epsilon_1}, \quad (x, y) \in [0, 1] \times [0, 1], \tag{13.12}$$

where ϵ_1 is a positive constant. In the simulations of the following sections, using initial condition (13.12) will serve to emphasise that any observed lateral motion of the endothelial cells from the simulations is due to the interactions with TAF and fibronectin and is not dependent on the underlying geometry of the system. In all the simulations in the following sections, ϵ_1 was taken to be 0.45.

The initial condition used for the fibronectin assumes that initially the fibronectin is distributed throughout the tissue in a spatially uniform manner, i.e.

$$f(x, y, 0) = f_0 \text{ (constant)}, \ (x, y) \in [0, 1] \times [0, 1]. \tag{13.13}$$

After the TAF has reached the parent blood vessel, the endothelial cells within the vessel form into a few cell clusters (Muthukkaruppan *et al.*, 1982; Orme and Chaplain, 1996) which eventually become sprouts. For simplicity, we will assume that initially three clusters form along the y-axis at $x \approx 0$, with the tumour located at $x = 1$ and the parent vessel of the endothelial cells at $x = 0$. The initial data is given by a distribution which has three discrete peaks of the form $e^{-x^2/\epsilon_2} \sin^2(6\pi y)$, with the positive parameter $\epsilon_2 = 0.001$.

13.5 Model Results

All of the results presented in this section were obtained from numerical solutions using a finite difference approximation of the system (13.8) with boundary and initial conditions (13.9) to (13.13).

Using the initial data (13.12), (13.13) and the boundary condition (13.9), the system (13.8) was solved numerically with the parameter values $D = 0.00035$, $\alpha = 0.6$, $\chi = 0.38$, $\rho = 0.34$, $\beta = 0.05$, $\gamma = 0.1$ and $\eta = 0.1$.

Figure 13.1 shows the spatio-temporal evolution of three initially separate endothelial cell density clusters representing three regions of sprouting. The parent vessel is located along the left-hand edge of the domain ($x = 0$) with the tumour cells distributed along the right-hand edge of the domain ($x = 1$). It can be seen from the figures that as the clusters make their way across the extracellular matrix, each one changes from an initially circular shape to a crescent-like shape. This is due to the lateral migration of the endothelial cell clusters towards one another under the influences of chemotaxis and haptotaxis. By $t = 6$ days the three initially separate clusters have joined to form a continuous band of cell density. At $t = 10.5$ days there

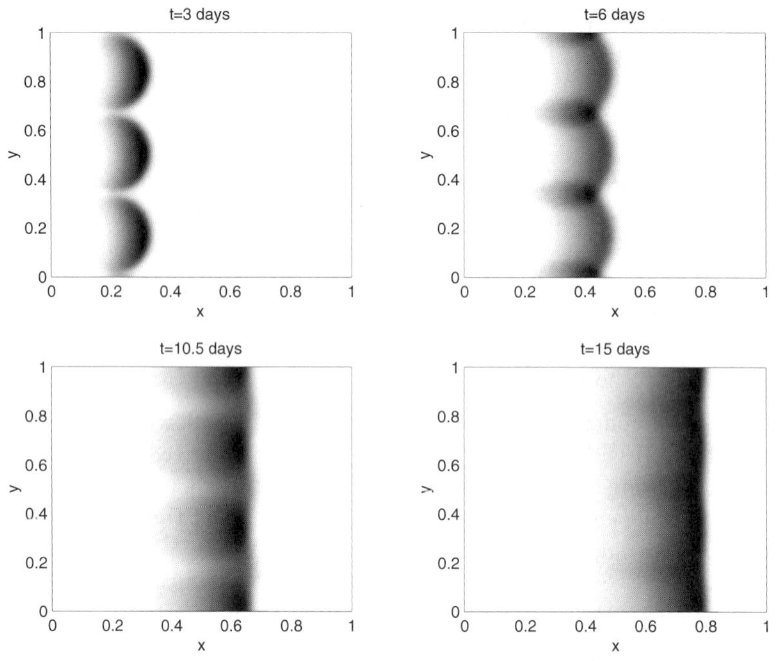

Figure 13.1 Spatio-temporal evolution of endothelial cell density from a numerical simulation of system (13.8) representing endothelial cell migration from the parent vessel towards a line source of tumour cells with both chemotaxis and haptotaxis (see text for parameter values). The initial three regions of high cell density are drawn towards each other and first of all form a looped structure and then a band of high cell density located at the leading edge. Colour graduation is directly proportional to cell density, i.e. black is high density and white low density.

are now four clusters of high endothelial cell density within the band due to the lateral motion and overlapping of the initial clusters. And at $t = 15$ days the endothelial cell density distribution has now formed into a band which is slowly advancing toward the tumour, with the highest cell density at the leading edge.

We now use our model to focus more precisely on the experimental results of Muthukkaruppan *et al.* (1982), where a single spherical tumour (or tumour fragment) was implanted into the cornea of a mouse and the resulting angiogenic response was observed. This involves solving (13.8) numerically with an initial TAF concentration field given by (13.10) (to represent the TAF profile generated by a circular tumour implant) along with the fibronectin initial profile (13.13). In the simulation below we choose a value of $\nu = (\sqrt{5}-0.1)/(\sqrt{5}-1)$. The tumour implant is centred at the point $(1, 0.5)$ and the value of ν was selected to ensure continuity of TAF concentration at $r = 0.1$ and a minimum concentration of TAF in the domain at $x = 0$ approximately equal to its minimum value in equation (13.12).

Figure 13.2 shows the plots of the endothelial cell density profiles for $t = 1.5$ to 22.5 days. The profile at time $t = 1.5$ days immediately shows the different effect of a circular tumour implant on the response of the cells, with the outer clusters of

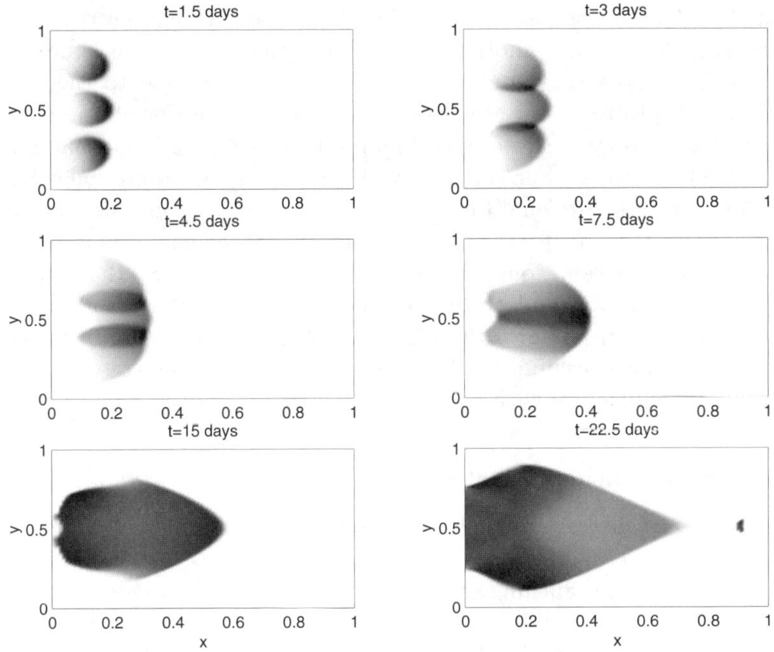

Figure 13.2 Spatio-temporal evolution of endothelial cell density from a numerical simulation of system (13.8) representing endothelial cell migration from the parent vessel toward a circular tumour implant (see text for parameter values). The initial three areas of high cell density are drawn towards the centre of the domain, where they coalesce and eventually form into one central area with the region of highest cell density located at the leading edge. Colour graduation is directly proportional to cell density, i.e. black is high density and white low density as in Figure 13.1.

endothelial cells initially seen to be moving towards the central cluster. Comparing $t = 3$ days in Figure 13.2 with Figure 13.1, we see that endothelial cell migration is somewhat slower in Figure 13.2. This is due to the fact that the gradient of TAF (13.10) is not as steep as before (13.12). As t increases, the outer clusters move laterally and two main clusters are then formed ($t = 4.5$ days). These two clusters subsequently come together to form one large central cluster of high cell density ($t = 7.5$ days). Due to the slower migration of endothelial cells in Figure 13.2 and the rich spatio-temporal dynamic behaviour that is observed, plots of the cell density distribution were generated for later times, $t = 15$ to 22.5 days. At $t = 15$ days we can see that there has been some lateral spread as well as both forward and backward migration of cells. There is a region of high cell density at the leading edge and also, interestingly, near the parent vessel at $x = 0$. From time $t = 18$ days (not shown) to time $t = 22.5$ days, regions of high cell density appear near the parent vessel and migrate forward and also laterally. By a time of $t = 22.5$ days, a small cluster of cells has reached the tumour. At this stage interactions between the endothelial cells and the tumour cells now become important and our model is no longer valid.

The above results show how both the geometry of the TAF gradient and the concentration of TAF are important. In changing the initial TAF concentration profile, we have been able to produce motion of cells directed towards the tumour and also directed back towards the parent vessel. The results are due to the interplay of chemotaxis and haptotaxis and are therefore related to the distributions of both fibronectin and TAF. From the results of Figures 13.1 and 13.2 it is clear that our simple model (13.8) is capable of producing a wide range of spatio-temporal behaviour.

Although these results are qualitatively realistic, since we are dealing with a continuum model, the dependent variable in the endothelial cell equation is endothelial cell *density*. The next section considers a discrete form of (13.8) that enables us to follow the path of a *single* endothelial cell located at a capillary sprout-tip, and hence allows us to simulate the growth of an *individual capillary sprout* (Stokes and Lauffenburger, 1991). This technique will indeed enable us to investigate a number of interesting points in connection with the form and structure of the blood-vessels.

13.6 The Discrete Mathematical Model

Many different types of discrete models, such as coupled map lattice models, fractal models, diffusion-limited aggregation models and L-systems, have already been developed to model general branching processes (including capillary network formation and angiogenesis) in a qualitative and phenomenological way (Bell, 1986; Gottlieb, 1990, 1991a, b; Prusinkiewicz and Lindenmayer, 1990; Kiani and Hudetz, 1991; Landini and Misson, 1993; Nekka *et al.*, 1996).

This section describes a recent discrete mathematical model of tumour-induced angiogenesis due to Anderson and Chaplain (1998b) which enables not only a qualitative but also a quantitative comparison with in vivo experimental results. The model is based around the assumption that the motion of an individual endothelial cell located at the tip of a capillary sprout governs the motion of the whole sprout. This is not unreasonable since the remaining endothelial cells lining the sprout-wall are contiguous (Paweletz and Knierim, 1989; Stokes and Lauffenburger, 1991). The particular technique which we will use to follow the path of an individual endothelial cell at a sprout-tip is that of a biased random walk and is a development of the method used by Anderson *et al.* (1997). Full details of the method as applied to modelling endothelial cell movement can be found in Anderson and Chaplain (1998b).

The technique first of all involves discretising (using standard finite-difference methods) the partial differential equation governing the rate of change of endothelial cell density (13.8). We then use the resulting coefficients of the five-point finite-difference stencil to generate the *probabilities of movement* of an individual cell in response to its local milieu (microenvironment). There is therefore an element of stochasticity (randomness) in our model in the movement rules for the cells. In effect, we derive a biased random walk governing the motion of a single endothelial cell based on the system of partial differential equations (13.8) of the previous section. In this sense, our discrete model is probably most similar in formulation to the reinforced random walk models of Othmer and Stevens (1997), where movement of individuals is modelled in response to a chemical stimulus by considering an equation (discrete in space and continuous in time) governing the probability that a cell is at a given position at time t. This equation is a function of the transition probabilities for one-step jumps to to the orthogonal neighbours.

We now set about formulating the discrete model and deriving the movement probabilities for an individual endothelial cell located at a sprout-tip. Rules for processes such as the generation of new capillary sprouts (branching) and the fusion of two sprouts (anastomosis) will be described later. We first discretise (13.8) using the standard, explicit Euler finite-difference approximation. This involves approximating the continuous two-dimensional domain $[0,1] \times [0,1]$ in the usual way as a grid of discrete points (mesh size h), and time (t) by discrete increments (magnitude k). The full discretised system is given in Anderson and Chaplain (1998b) and for clarity we only consider a simplified form of the endothelial cell equation,

$$n_{l,m}^{q+1} = n_{l,m}^{q} P_0 + n_{l+1,m}^{q} P_1 + n_{l-1,m}^{q} P_2 + n_{l,m+1}^{q} P_3 + n_{l,m-1}^{q} P_4, \tag{13.14}$$

where the subscripts specify the location on the grid and the superscripts the time steps. That is, $x = lh$, $y = mh$ and $t = qk$ where l, m, k, q and h are positive parameters. The precise functional forms of P_i are given in Anderson and Chaplain (1998b).

In a numerical simulation of the continuous model (13.8), the purpose of the discrete equation (13.14) is to determine the endothelial cell density at grid position (l, m), and time $q+1$, by averaging the density of the four surrounding neighbours at the previous time step q. For our discrete model, we will use the five coefficients P_0 to P_4 from (13.14) to generate the motion of an individual endothelial cell. These coefficients can be thought of as being proportional to the probabilities of the endothelial cell being stationary (P_0) or moving left (P_1), right (P_2), up (P_3) or down (P_4).

Each of the coefficients P_1 to P_4 consists of three components,

$$P_n = \text{Random movement} + \text{Chemotaxis} + \text{Haptotaxis}, \tag{13.15}$$

thus showing how the discrete endothelial cell equation is linked to the continuous endothelial cell equation of system (13.8). The coefficient P_0 has a similar form.

The precise forms of P_0 to P_4 are functions of the fibronectin and TAF concentrations at the neighbouring spatial points of an individual endothelial cell (Anderson and Chaplain, 1998b) and hence one can see that this discrete model truly permits one to consider the interactions between cells and their microenvironment. An inspection of the functional forms of P_0 to P_4 allows one to note that if there were no gradients of fibronectin or TAF, then the values of P_1 to P_4 would be equal, with P_0 being smaller or larger, depending on the precise values chosen for the space and time steps i.e. there is no migrational bias in any one direction and the movement of an endothelial cell approximates an *unbiased* random walk. However, if there is a TAF gradient and no fibronectin gradient, chemotaxis dominates and the coefficients P_0 to P_4 will become biased towards the tumour source at $x = 1$. The reverse is true if there is only a fibronectin gradient and no TAF gradient i.e. haptotaxis dominates. When both gradients exist, the complex interplay between chemotaxis and haptotaxis will bias the coefficients accordingly. The motion of an individual cell at the sprout-tip is therefore governed by its interactions with angiogenic factors and matrix macromolecules in its local microenvironment.

Before proceeding to the simulations, we consider how we explicitly incorporate the processes of branching, anastomosis and cell proliferation into the discrete model.

13.6.1 Branching and anastomosis

Although there is a good deal of information regarding the actual events of the generation of new sprouts (sprout branching) and the formation of loops (anastomosis), there is no explanation as to the precise mechanisms which cause them (Paweletz and Knierim, 1989). We will assume that the generation of new sprouts (branching) occurs only from existing sprout-tips. It is also reasonable to assume that the newly formed sprouts are unlikely to branch immediately and that there must be a sufficient number of endothelial cells, near the sprout-tip, for new sprouts to form. We will assume that the density of endothelial cells required for branching is inversely proportional to the concentration of TAF, since the new sprouts become much shorter as the tumour is approached, i.e. as the TAF concentration increases (Muthukkaruppan *et al.*, 1982). From these assumptions we obtain the following three conditions, which must be satisfied before a capillary sprout can branch at its tip and generate a new sprout:

(i) The age of the current sprout is greater than some threshold branching age ψ, i.e. new sprouts must mature for a length of time at least equal to ψ before being able to branch.
(ii) There is sufficient space locally for a new sprout to form, i.e. branching into a space occupied by another sprout is not possible.
(iii) The endothelial cell density is greater than some prescribed threshold level n_b, where $n_b \propto 1/c_{l,m}$.

Given that each of these three conditions is satisfied, we assume that each sprout-tip has a probability P_b of generating a new sprout (branching) and that this probability is dependent on the local TAF concentration. We therefore adopt a simple *positional information* approach with the rule that as the TAF concentration increases, the probability of generating new sprouts (branching) increases. The branching probabilities are chosen on a qualitative basis, i.e. very little branching occurs initially (near the parent vessel, e.g. limbus), but as the endothelial cells migrate closer to the tumour (at $x = 1$) the number of new sprouts slowly increases. A short distance from the tumour, the frequency of branching dramatically increases, creating the "brush border" effect.

Anastomosis, the formation of loops by capillary sprouts, is another very important feature of angiogenesis which can be captured explicitly by the discrete model. As the sprouts progress towards the tumour, driven by the movement probabilities of (13.14), at each time step of the simulation, the endothelial cells at the sprout-tips can move to any of the four orthogonal neighbours on the discrete grid. If upon one of these moves another sprout is encountered, then anastomosis can occur. Experiments have shown that the initial formation of anastomoses occurs at a well-defined distance from the parent vessel (Paweletz and Knierim, 1989). For simplicity we assume that, as a result of the anastomosis, only one of the original sprouts continues to grow (the choice is purely random).

13.6.2 Cell proliferation

As discussed in the introduction, during angiogenesis there is initially no endothelial cell proliferation. Cells are recruited from the parent vessel and migrate toward the tumour. Approximately 36-48 hours into the process, cell mitosis is observed (Paweletz and Knierim, 1989; Sholley *et al.*, 1984) and is confined to a region just behind the sprout-tip. Endothelial cell doubling time has been estimated at 18 hours (Williams, 1987) and we model the process of cell division in the discrete model by assuming that some of the cells behind the sprout-tip divide (into two daughter cells) every 18 hours. We assume that this has the effect of increasing the length of a sprout by approximately one cell length every 18 hours. In terms of the non-dimensional discrete model, this is equivalent to the sprout length being increased by an amount h every half time unit. Owing to the inherent randomness of the discrete model, proliferation will occur asynchronously in separate sprouts, as is observed experimentally (Paweletz and Knierim, 1989). This feature is captured by the discrete model since the age of each cell at a sprout-tip is known and this determines when mitosis occurs.

13.6.3 Simulation process for the discrete model

Each time step of the simulation process involves solving the discrete form of the system (13.8) numerically to generate the five coefficients P_0 to P_4 used in (13.14). Probability ranges are then computed by summing the coefficients to produce five ranges, $R_0 = 0$ to P_0 and $R_j = \sum_{i=0}^{j-1} P_i$ to $\sum_{i=0}^{j} P_i$, where $j = 1$ to 4. We then generate a random number between 0 and 1, and depending on the range in which this number falls, the current individual endothelial cell under consideration will remain stationary (R_0) or move one grid point left (R_1), right (R_2), up (R_3) or down (R_4). The larger a particular range, the greater the probability that the corresponding coefficient will be selected. Each endothelial cell is therefore restricted to move to one of its four orthogonal neighbouring grid points or remain stationary at each time step.

All the simulations of the discrete model were carried out on a 200×200 grid, which is a discretisation of a the unit square, $[0, 1] \times [0, 1]$, with a space step of $h = 0.005$. Given that our unit of length is 2mm, this means that h is equivalent to a dimensional length of 10μm, i.e. approximately the length of one or two endothelial cells (Paku and Paweletz, 1991). A discrete form of the no-flux boundary condition (13.9) was imposed on the square grid, restricting the endothelial cells to within the grid. The initial conditions in all simulations (unless otherwise stated) are given by discrete forms of the fibronectin (13.13) and TAF (13.12) equations. We assume that there are five capillary sprouts initiated by five endothelial cells located at the sprout-tips. Three of them are located at the maximum density positions of the continuous endothelial cell initial data (Figure 13.3) and the other two are placed between them. Therefore we have endothelial cells starting at $y = 0.17,\ 0.3,\ 0.5,\ 0.65,\ 0.84$ all at $x = 0$.

The parameter values used in the following simulations are the same as those used in the previous two-dimensional continuous simulations (unless otherwise stated), i.e. $D = 0.00035$, $\alpha = 0.6$, $\chi_0 = 0.38$, $\rho = 0.34$, $\beta = 0.05$, $\gamma = 0.1$ and $\eta = 0.1$. Through trial and error it was found that a threshold branching age of $\psi = 0.5$ (equivalent to a dimensional time of 0.75 days) produced simulated networks which were qualitatively similar in morphology to those networks observed in vivo.

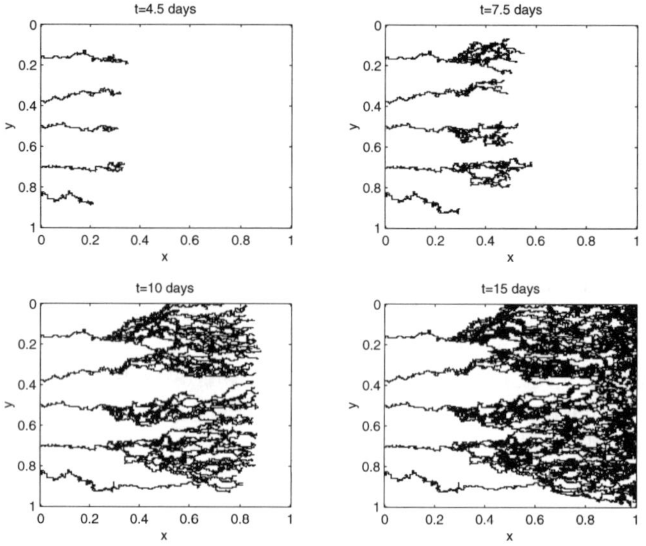

Figure 13.3 Spatio-temporal evolution of a capillary network from a numerical simulation of the discrete model. The figure shows the migration of endothelial cells at the sprout-tips (with both chemotaxis and haptotaxis) towards a line source of tumour cells with endothelial cell proliferation. The network connects with the tumour at between $t = 10$ days and $t = 15$ days, and vascularisation is achieved.

13.6.4 Discrete model simulation results

Figure 13.3 shows the results of simulating the discrete model as described above, incorporating endothelial cell proliferation, sprout branching and anastomosis with a linear source of tumour cells located at $x = 1$. Cell proliferation is assumed to begin 48 hours into the process, i.e. at a (non-dimensional) time of $t = 1.3$, and cell doubling is assumed to occur every 18 hours ($t = 0.5$). By $t = 4.5$ days sprouts one and two (from the top) have almost formed a loop. By $t = 7.5$ days we see that a well-developed branching network is forming, with proliferation providing additional extension of the sprouts. By $t = 10$ there are many connections of sprouts and by $t = 15$ days the sprouts have already connected with the tumour.

Figure 13.4 shows the results of simulating the discrete model (as described above), but this time with a circular tumour implant (shown as the black semicircle). At $t = 4.5$ days we can see that there is already a loop formed by fusion of the bottom two sprouts, and all sprouts are migrating towards the centre of the domain. By $t = 7.5$ days all sprouts have connected in the middle of the domain and further branching and connections occur as the network develops ($t = 10$ days). By $t = 15$ days the sprouts have connected with the tumour. The results of the model are in excellent quantitative agreement with networks which form in the animal cornea experiments (Gimbrone *et al.*, 1974; Muthukkaruppan *et al.*, 1982).

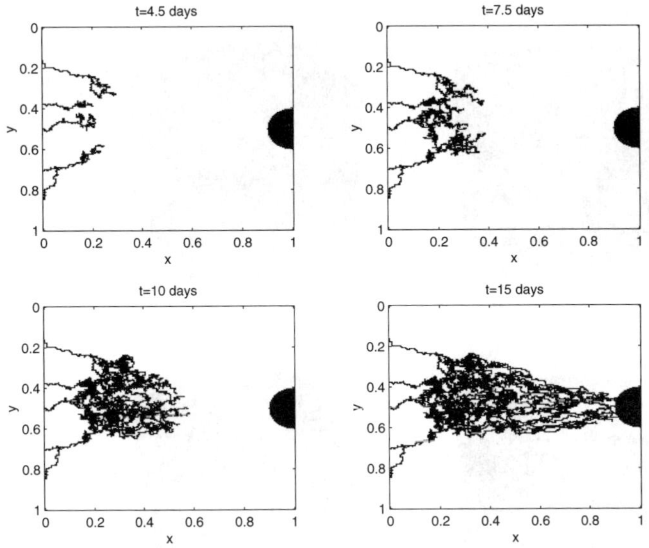

Figure 13.4 Spatio-temporal evolution of a capillary network from a numerical simulation of the discrete model. The figure shows the migration of endothelial cells at the sprout-tips towards a circular source of tumour cells with endothelial cell proliferation. Once again the network connects with the tumour at around $t = 15$ days and vascularisation is achieved.

13.7 Three-Dimensional Simulations

Although the results of the previous section are in excellent quantitative agreement with the experimental results carried out in animal corneal models (Gimbrone *et al.*, 1974; Muthukkaruppan *et al.*, 1982), in vivo angiogenesis is generally a fully three-dimensional (3D) process. Hence we now finally consider extending the model to three spatial dimensions in order to more faithfully represent the actual in vivo situation (see the 3D collagen gel assay in Chapter 12). To extend the discrete model to three spatial dimensions, we simply discretise the spatial terms of the continuous model (13.8) in three dimensions (i.e. we use a seven-point stencil instead of a five-point stencil) which then yields a system of three discrete algebraic equations. Here is the general form of the discrete equation governing the migration of the endothelial cells:

$$\begin{aligned} n^{q+1}_{l,m,w} = n^q_{l,m,w} P_0 \quad &+ \quad n^q_{l+1,m,w} P_1 + n^q_{l-1,m,w} P_2 + n^q_{l,m+1,w} P_3 \\ &+ \quad n^q_{l,m-1,w} P_4 + n^q_{l,m,w+1} P_5 + n^q_{l,m,w-1} P_6, \end{aligned} \tag{13.16}$$

where the subscripts specify the location of an individual cell on the 3D grid and the superscripts denote the time steps. That is $x = lh$, $y = mh$, $z = wh$ and $t = qk$ where l, m, w, k, q and h are positive parameters.

As with the discrete model in two dimensions, we will use the seven coefficients P_0 to P_6 from (13.16) to generate the motion of an individual endothelial cell in response to its 3D microenvironment. Once again these coefficients can be thought of as being proportional to the probabilities of the endothelial cell being stationary (P_0)

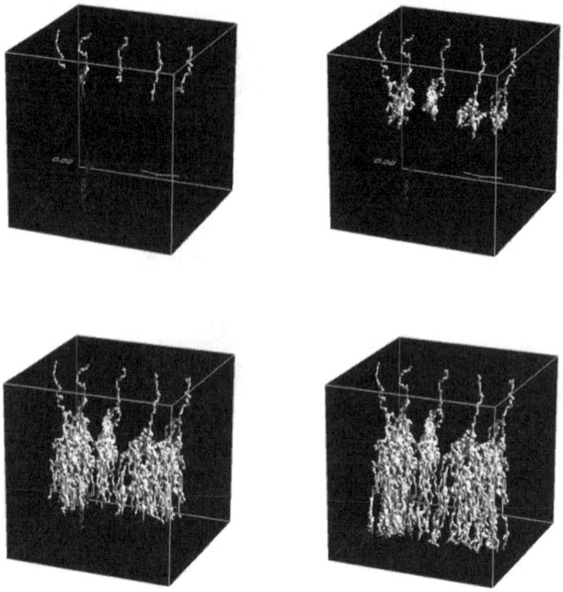

Figure 13.5 Spatio-temporal evolution of 3D capillary network towards a single layer (plane) of tumour cells positioned on the bottom face of the cube. $t = 1.5, 3, 4.5, 6$ days.

or moving west (P_1), east (P_2), north (P_3), south (P_4), up (P_5) or down (P_6). Each of the coefficients P_1 to P_6 consists of the same three components as in (13.15), i.e. random, chemotactic and haptotactic movements, and the coefficient P_0 has a similar form as previously. The exact forms of P_0 to P_6 involve functions of the fibronectin and TAF concentrations near an individual endothelial cell.

All the simulations of the 3D discrete model were carried out on a $100 \times 100 \times 100$ grid, which is a discretisation of a the unit cube, $[0, 1] \times [0, 1] \times [0, 1]$, with a spatial grid size of $h = 0.01$. In dimensional terms, the cube has sides of length 2mm, and as before, one (dimensionless) time unit t is equivalent to 1.5 days. Once again we impose no-flux boundary conditions.

We shall consider the (theoretical) response of capillaries to both linear (more precisely planar) and spherical tumour sources (representing either a large or small solid tumour implant). The linear source is a direct extension of the two-dimensional discrete results. The tumour cells are distributed homogeneously on the lower face of the cube; this provides a TAF gradient from the lower face of the cube to the upper face. Five initial capillary sprouts are initiated at the upper face of the cube and these respond to the TAF gradient and the fibronectin (initially distributed homogeneously throughout the domain). The spherical tumour source is positioned at the centre of the unit cube (representing a small spherical tumour or multicell spheroid) and the resulting TAF concentration field is assumed to be radially symmetric. On five out of the six faces of the cube we randomly place 3 initial sprouts, giving a total of 15

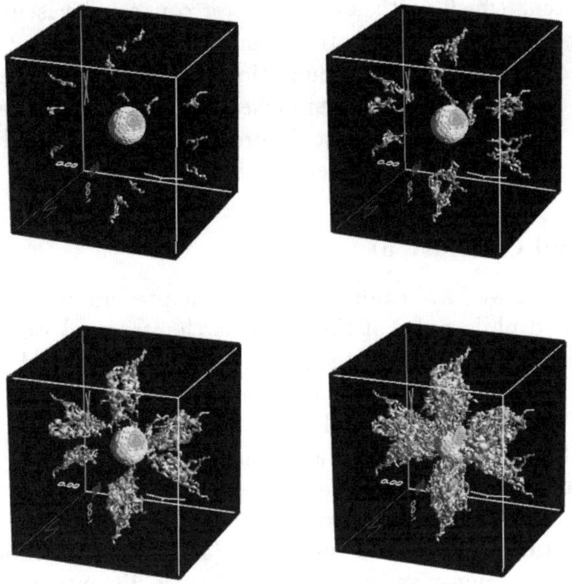

Figure 13.6 Spatio-temporal evolution of 3D capillary network towards a central spherical source of tumour cells (cf. avascular tumour or multicell spheroid). $t = 1.5, 3, 4.5, 6$ days.

capillary sprouts to start the process. Once again the initial fibronectin concentration is assumed to be constant throughout the cube.

13.7.1 3D Simulation Results

Figure 13.5 shows four snapshots in time of the developing capillary network in response to the planar source of tumour cells on the bottom face of the cube. As can be seen, the sprouts grow, branch and anastomose, eventually forming a dense, 3D branched, connected network.

Figure 13.6 shows four snapshots in time of the capillary netowrk formation in response to the spherical source of tumour cells. At $t = 1.5$ days we see that the initial sprouts have grown a short distance towards the tumour, but have only branched a little. As time progresses, the sprouts begin to branch more and migrate further into the matrix towards the tumour. By $t = 3$ days there are sufficient branches for some anastomsis to occur which will lead to a better connected vasculature. In fact, as the sprouts get closer to the tumour, the chance of anastomosis occuring rapidly increases. Finally, at $t = 6$ days we have connection with the tumour and completion of the angiogenic process. The number of branches rapidly increases between $t = 4.5$ days and $t = 6$ days, creating a dense vasculature which almost engulfs the tumour and could be considered as the 3D equivalent of the "brush border".

into the matrix towards the tumour. By $t = 3$ days there are sufficient branches for some anastomsis to occur which will lead to a better connected vasculature. In fact, as the sprouts get closer to the tumour, the chance of anastomosis occuring rapidly increases. Finally, at $t = 6$ days we have connection with the tumour and completion of the angiogenic process. The number of branches rapidly increases between $t = 4.5$ days and $t = 6$ days, creating a dense vasculature which almost engulfs the tumour and could be considered as the 3D equivalent of the "brush border".

13.8 Discussion and Conclusions

The mathematical models we have examined in this chapter are closely linked to the experimental results and philosophy of the previous chapter, and we have explicitly considered the *tissue response unit* as their basis. Migration of endothelial cells is determined by their response to the local milieu/microenvironment consisting of cytokines (e.g. VEGF) and matrix molecules (e.g. fibronectin).

The work presented here has developed a mathematical model for tumour-induced angiogenesis using a novel blend of continuum, deterministic modelling and discrete, stochastic modelling in both two and three spatial dimensions. The model results not only replicate existing experimental protocols (animal cornea models) but also provide a new alternative, non-invasive, predictive three-dimensional "assay". The model results are not only in good qualitative agreement with experimental data but are also in quantitative agreement since parameter values for the models (in particular the length scale L, the cell random motility coefficient D_n, the TAF diffusion constant D_c, the chemotactic coefficient χ_0 and the endothelial cell proliferation rate) have been estimated, as far as possible, from independent experimental measurements, thus grounding the results of the model in a realistic framework.

The results from the continuum model simulations demonstrate two important aspects of capillary network formation. Firstly, in agreement with previous models (e.g. Stokes and Lauffenburger, 1991; Chaplain and Stuart, 1993), a sufficiently strong chemotactic response is necessary for the initial outgrowth of the capillary network. Secondly, the model demonstrates the importance of interactions between endothelial cells and the extracellular matrix. The inclusion of uptake terms for the TAF and fibronectin by the cells enables the creation of local gradients around the areas of high cell density which permits lateral migration. Without such interactions, the model predicts that the cells move directly to the tumour without any significant lateral movement (Anderson and Chaplain, 1998b). These results show that the important large-scale features of angiogenesis can be captured qualitatively using a continuum model. However, important processes on a smaller scale, such as sprout branching, are not captured. This requires the development of a discrete model applicable at the level of a single endothelial cell.

The discrete model presented here was derived from a discretised form of the partial differential equations of the continuum model, and permits the tracking of individual endothelial cells located at the capillary sprout-tip. This, in turn, enables the path of the complete capillary sprout to be followed. Since the parameter values used in the discrete model are directly related to those of the continuum model, the results of the simulations of the discrete model are therefore also firmly grounded in a realistic framework. In addition to this, the discrete model enables us to explicitly incorporate rules for the production of new sprouts (branching), the fusion of sprouts to form loops (anastomosis) and sprout extension through endothelial cell proliferation. The results from the discrete model simulations confirm the predictions of the continuum model that both chemotaxis and haptotaxis are necessary for the formation of a capillary network on a large scale.

On a finer scale, the discrete model simulations produce capillary networks with a very realistic structure and morphology, capturing the early formation of loops (anastomosis), the essential dendritic structure of a capillary network and the formation of the experimentally observed "brush border". The discrete model also incorporates a realistic method of modelling mitosis and its effect on the sprouts, i.e. actual extension in the length of the sprout due to cell division. The theoretical three-dimensional capillary networks generated in the previous section show the real predictive power of this modelling approach. For the first time it is possible to analyse these computed structures for quantitative information such as network expansion rates, loop formation, sprout length and overall network architecture (e.g. vessel area and volume).

The technique of using partial differential equations as the basis for discrete models is therefore clearly very useful, with the ability to generate movements of individual cells based on a continuum model of a population of cells. Indeed, this technique provides a powerful means of linking microscale events to macroscale events, individual behaviour to population behaviour, with potential application to a wide range of problems in mathematical biology. Another advantage of using this technique is the possibility of manipulating the underlying spatial environment. For example, with respect to angiogenesis, it would be easy to incorporate an obstruction to block endothelial cell migration; or to incorporate an underlying spatially heterogeneous medium which more faithfully reflects the make-up of extracellular matrix. Such theoretical experiments are already in progress. The discrete model also offers a useful alternative to stochastic partial differential equations (Stokes and Lauffenburger, 1991), combining, in an elegant and efficient way the strengths of continuum models with the element of randomness.

Although the model has incorporated several key mechanisms involved in the angiogenic process, there are still possible ways to extend it. For example, one might consider explicitly modelling more than one angiogenic factor (it is known that some factors induce a mitogenic response whereas others induce a migratory response). The effect of angiogenic inhibitors may also be considered, as well as more than one matrix macromolecule (e.g. laminin, collagen). However, it seems that these possible extensions are unlikely to provide much more information than is already present (rather they may be considered as fine tuning).

Perhaps a more fruitful idea to develop the modelling (and an idea which echoes the end of Chapter 12) is to turn attention to events at the molecular level and to integrate them within the tissue response unit.

Several angiogenic factors have now been isolated – vascular endothelial growth factor (VEGF), acidic and basic fibroblast growth factor (aFGF, bFGF), angiogenin and others (Folkman and Klagsbrun, 1987) – and *endothelial cell receptors* for these proteins have been discovered (Millauer *et al.*, 1993; Fong *et al.*, 1995; Hewett and Murray, 1996; Duh *et al.*, 1997; Hanahan, 1997). Indeed, there is now clear experimental evidence that *disrupting these receptors* has a direct effect on the final *structure* of the capillary network (Dumont *et al.*, 1994; Fong *et al.*, 1995; Sato *et al.*, 1995; Hanahan, 1997).

There is now clear experimental evidence that disrupting the transmembrane receptor tyrosine kinases (RTKs) has a direct effect on the structural morphogenesis of the capillary network. Results show that disrupting the RTK Tie2 receptor of endothelial cells leads to a poorly formed capillary network which lacks a full branching structure (Dumont *et al.*, 1994; Sato *et al.*, 1995; Hanahan, 1997). Therefore, the number of active receptors on the cell surface can have a direct influence on the outcome of network structure. One way to extend the model further would therefore be to consider the chemotactic function $\chi(c)$ as an explicit function of the number of active TAF receptors, i.e. we could include an extra equation for the receptor kinetics and couple this directly into the chemotactic function (Sherratt, 1994; Höfer *et al.*, 1995). Experimental studies have also shown that angiogenic cytokines such as VEGF, aFGF, bFGF and angiogenin have a marked affinity for heparin and that these cytokines are likely to exist in bound form, as well as soluble form, in the extracellular matrix. Taking these observations into account would involve modelling the distribution of heparin within the matrix and the subsequent binding of TAF to the heparin.

Other important aspects of angiogenesis which can be added to the model in the future include incorporating blood flow through the capillary network (Muthukkaruppan *et al.*, 1982), with potential application to drug delivery and optimisation of chemotherapy regimes; the role of oxygen gradients and oxygen concentration and the role of macrophages. These aspects are also very important to angiogenesis in wound healing.

Finally, it is worth noting that anti-angiogenesis such as the preferential killing of endothelial cells, the inhibition of endothelial cell proliferation via chemicals such as angiostatin and endostatin (O'Reilly *et al.*, 1994), the development of antichemotactic drugs and the development of antihaptotactic drugs (Yamada and Olden, 1978) are now clinically recognised as having enormous potential and promise in the treatment of patients with cancer (Harris, 1997). In particular, their use as an adjuvant chemotherapy is being recognised as a very effective way to treat secondary tumours (metastases), and mathematical models of angiogenesis may have an increasingly important role to play in the development and testing of these therapies.

Acknowledgements
This work was supported by EPSRC Grant GR/K92641 and BBSRC Grant MMI09008.

REFERENCES

Albini, A., Allavena, G., Melchiori, A., Giancotti, F., Richter, H., Comoglio, P.M., Parodi, S., Martin G.R., and Tarone, G. (1987) Chemotaxis of 3T3 and SV3T3 cells to fibronectin is mediated through the cell-attachment site in fibronectin and fibronectin cell surface receptor. *J. Cell Biol.* 105: 1867–1872.

Anderson, A.R.A., Sleeman, B.D.S., Young, I.M. and Griffiths, B.S. (1997) Nematode movement along a chemical gradient in a structurally heterogeneous environment: II. Theory. *Fundam. Appl. Nematol.* 20: 165–172.

Anderson, A.R.A. and Chaplain, M.A.J. (1998a) A mathematical model for capillary network formation in the absence of endothelial cell proliferation. *Appl. Math. Lett.* 11: 109–114.

Anderson, A.R.A. and Chaplain, M.A.J. (1998b) Continuous and discrete mathematical models of tumor-induced angiogenesis. *Bull. Math. Biol.* 60: 857–899.

Ausprunk, D.H. and Folkman, J. (1977) Migration and proliferation of endothelial cells in preformed and newly formed blood vessels during tumour angiogenesis. *Microvasc. Res.* 14: 53–65.

Balding, D. and McElwain, D.L.S. (1985) A mathematical model of tumour-induced capillary growth. *J. Theor. Biol.* 114: 53–73.

Bell, A.D. (1986) The simulation of branching patterns in modular organisms. *Phil. Trans. R. Soc. Lond.* B313: 143–159.

Bowersox, J.C. and Sorgente, N. (1982) Chemotaxis of aortic endothelial cells in response to fibronectin. *Cancer Res.* 42: 2547–2551.

Bray, D. (1992) *Cell Movements.* New York: Garland Publishing.

Byrne, H.M. and Chaplain, M.A.J. (1995) Mathematical models for tumour angiogenesis: numerical simulations and nonlinear wave solutions. *Bull. Math. Biol.* 57: 461–486.

Carter, S.B. (1965) Principles of cell motility: The direction of cell movement and cancer invasion. *Nature* 208: 1183–1187.

Carter, S.B. (1967) Haptotaxis and the mechanism of cell motility. *Nature* 213: 256–260.

Chaplain, M.A.J. (1995) The mathematical modelling of tumour angiogenesis and invasion *Acta Biotheor.* 43: 387–402.

Chaplain, M.A.J. (1996) Avascular growth, angiogenesis and vascular growth in solid tumours: the mathematical modelling of the stages of tumour development. *Mathl. Comput. Modelling* 23: 47–87.

Chaplain, M.A.J. and Anderson, A.R.A. (1997) The mathematical modelling, simulation and prediction of tumour-induced angiogenesis. *Invas. Metast.* 16: 222–234.

Chaplain, M.A.J. and Stuart, A.M. (1993) A model mechanism for the chemotactic response of endothelial cells to tumour angiogenesis factor. *IMA J. Math. Appl. Med. Biol.* 10: 149–168.

Duh, E.J., King, G.L. and Aiello, L.P. (1997) Identification of a VEGF receptor (KDR/FLK) promoter element which binds an endothelial cell-specific protein conferring endothelial selective expression. *Invest. Opthamol. Vis. Sci.* 38: 1124–1125.

Dumont, D.J., Gradwohl, G., Fong, G.H., Puri, M.C., Gertsenstein, M., Auerbach, A. and Breitman, M.L. (1994) Dominant-negative and targeted null mutations in the endothelial receptor tyrosine kinase, TEK, reveal a critical role in vasculogenesis of the embryo. *Genes Dev.* 8: 1897–1909.

Edelstein-Keshet, L. and Ermentrout, G.B. (1989) Models for branching networks in two dimensions. *SIAM J. Appl. Math.* 49: 1136–1157.

Folkman, J. (1985) Tumor angiogenesis. *Adv. Cancer Res.* 43: 175–203.

Folkman, J. (1995) Angiogenesis in cancer, vascular, rheumatoid and other disease. *Nature Medicine* 1: 21–31.

Folkman, J. and Haudenschild, C. (1980) Angiogenesis in vitro. *Nature* 288: 551–556.

Folkman, J. and Klagsbrun, M. (1987) Angiogenic factors. *Science* 235: 442–447.

Fong, G.H., Rossant, J., Gertsenstein, M. and Breitman, M.L. (1995) Role of the FLT-1 receptor tyrosine kinase in regulating the assembly of vascular endothelium. *Nature* 376: 66–70.

Gimbrone, M.A., Cotran, R.S., Leapman, S.B. and Folkman, J. (1974) Tumor growth and neovascularization: An experimental model using the rabbit cornea. *J. Nat. Cancer Inst.* 52: 413–427.

Gottlieb, M.E. (1990) Modelling blood vessels: a deterministic method with fractal structure based on physiological rules. *Proceedings of the 12th International Conference of IEEE EMBS*, pp. 1386–1387. New York: IEEE Press.

Gottlieb, M.E. (1991a) The VT model: a deterministic model of angiogenesis and biofractals based on physiological rules. *Proceedings of the IEEE 17th Annual Northeast Bioengineering Conference*, pp. 38–39. New York: IEEE Press.

Gottlieb, M.E. (1991b) Vascular networks: fractal anatomies from non-linear physiologies. *IEEE Eng. Med. Bio. Mag.* 13: 2196–2197.

Hanahan, D. (1997) Signaling vascular morphogenesis and maintenance. *Science* 227: 48–50.

Harris, A.L. (1997) Antiangiogenesis for cancer therapy. *The Lancet* 349 (suppl. II): 13–15.

Hewett, P.W., Murray, J.C. (1996) Coexpression of FLT-1, FLT-4 and KDR in freshly isolated and cultured human endothelial-cells. *Biochem. Biophys. Res. Commun.* 221: 697–702.

Höfer, T., Sherratt, J.A. and Maini, P.K. (1995) Cellular pattern formation during *Dictyostelium* aggregation. *Physica D* 85: 425–444.

Hynes, R.O. (1990) *Fibronectins.* New York: Springer-Verlag.

Jaffee, E.A. and Mosher, D.F. (1978) Synthesis of fibronectin by cultured endothelial cells. *J. Exp. Med.* 147: 1779–1791.

Johansson, S., Gustafson, S. and Pertoft, H. (1987) Identification of a fibronectin receptor specific for rat liver endothelial cells. *Exp. Cell Res.* 172: 425–431.

Kiani, M. and Hudetz, A. (1991) Computer simulation of growth of anastomosing microvascular networks. *J. Theor. Biol.* 150: 547–560.

Kohno, T., Sorgente, N., Ishibashi, T., Goodnight, R. and Ryan, S.J. (1987) Immunofluorescent studies of fibronectin and laminin in the human eye. *Invest. Opthamol. Vis. Sci.* 28: 506–514.

Lacovara, J., Cramer, E.B. and Quigley, J.P. (1984) Fibronectin enhancement of directed migration of B16 melanoma cells. *Cancer Res.* 44: 1657–1663.

Landini, G. and Misson, G. (1993) Simulation of corneal neo-vascularization by inverted diffusion limited aggregation. *Invest. Opthamol. Visual Sci.* 34: 1872–1875.

Lapidus, I.R. and Schiller, R. (1976) Model for the chemotactic response of a bacterial population. *Biophys. J.* 16: 779–789.

Lauffenburger, D., Aris, R. and Kennedy, C.R. (1984) Travelling bands of chemotactic bacteria in the context of population growth. *Bull. Math. Biol.* 46: 19–40.

Liotta, L.A., Rao, C.N. and Barsky, S.H. (1983) Tumour invasion and the extracellular matrix. *Lab. Invest.* 49: 636–649.

McCarthy, J.B. and Furcht, L.T. (1984) Laminin and fibronectin promote the directed migration of B16 melanoma cells in vitro. *J. Cell Biol.* 98: 1474–1480.

Madri, J.A. and Pratt, B.M. (1986) Endothelial cell-matrix interactions: in vitro models of angiogenesis. *J. Histochem. Cytochem.* 34: 85–91.

Meinhardt, H. (1976) Morphogenesis of lines and nets. *Differentiation* 6: 117–123.

Meinhardt, H. (1982) *Models of biological pattern formation.* London: Academic Press.

Millauer, B., Wizigman-Voos, H., Schnürch, H., Martinez, R., Müller, N.P.H., Risau W. and Ullrich, A. (1993) High-affinity VEGF binding and developmental expression suggest FLK-1 as a major regulator of vasculogenesis and angiogenesis. *Cell* 72: 835–846.

Muthukkaruppan, V.R., Kubai, L. and Auerbach, R. (1982) Tumor-induced neovascularization in the mouse eye. *J. Natn. Cancer Inst.* 69: 699–705.

Nekka, F., Kyriacos, S., Kerrigan, C. and Cartilier, L. (1996) A model of growing vascular structures. *Bull. Math. Biol.* 58: 409–424.

Olsen, L., Sherratt, J.A., Maini, P.K. and Arnold, F. (1997) A mathematical model for the capillary endothelial cell-extracellular matrix interactions in wound-healing angiogenesis. *IMA J. Math. Appl. Med. Biol.*

O'Reilly, M.S., Holmgren, L., Shing, Y., Chen, C., Rosenthal, R.A., Moses, M., Lane, W.S., Cao, Y., Sage, E.H. and Folkman, J. (1994) Angiostatin: A novel angiogenesis inhibitor that mediates the suppression of metastases by a Lewis lung carcinoma. *Cell* 79: 315–328.

Orme, M.E. and Chaplain, M.A.J. (1996) A mathematical model of the first steps of tumour-related angiogenesis: capillary sprout formation and secondary branching. *IMA J. Math. Appl. Med. Biol.* 13: 73–98.

Orme, M.E. and Chaplain, M.A.J. (1997) Two-dimensional models of tumour angiogenesis and anti-angiogenesis strategies. *IMA J. Math. Appl. Med. Biol.* 14: 189–205.

Othmer, H. and Stevens, A. (1997) Aggregation, blowup and collapse: The ABCs of taxis and reinforced random walks. *SIAM J. Appl. Math.* 57: 1044–1081.

Paku, S. and Paweletz, N. (1991) First steps of tumor-related angiogenesis. *Lab. Invest.* 65: 334–346.

Paweletz, N. and Knierim, M. (1989) Tumor-related angiogenesis. *Crit. Rev. Oncol. Hematol.* 9: 197–242.

Pettet, G., Chaplain, M.A.J., McElwain, D.L.S. and Byrne, H.M. (1996) On the role of angiogenesis in wound healing. *Proc. Roy. Soc. Lond. B* 263: 1487–1493.

Prusinkiewicz, P. and Lindenmayer, A. (1990) *The Algorithmic Beauty of Plants*, New York: Springer-Verlag.

Quigley, J.P., Lacovara, J. and Cramer, E.B. (1983) The directed migration of B-16 melanoma-cells in response to a haptotactic chemotactic gradient of fibronectin. *J. Cell Biol.* 97: A450–451.

Rupnick, M.A., Stokes, C.L., Williams, S.K. and Lauffenburger, D.A. (1988) Quantitative analysis of human microvessel endothelial cells using a linear under-agarose assay. *Lab. Invest.* 59: 363–372.

Sato, T.N., Tozawa, Y., Deutsch, U., Wolburgbuchholz, K., Fujiwara, Y., Gendronmaguire, M., Gridley, T., Wolburg, H., Risau, W. and Qin, Y. (1995) Distinct roles of the receptor tyrosine kinases TIE-1 and TIE-2 in blood-vessel formation. *Nature* 376: 70–74.

Schor, S.L., Schor, A.M. and Brazill, G.W. (1981) The effects of fibronectin on the migration of human foreskin fibroblasts and Syrian hamster melanoma cells into three-dimensional gels of lattice collagen fibres. *J. Cell Sci.* 48: 301–314.

Sherratt, J.A. (1994) Chemotaxis and chemokinesis in eukaryotic cells: The Keller-Segel equations as an approximation to a detailed model. *Bull. Math. Biol.* 56: 129–146.

Sholley, M.M., Ferguson, G.P., Seibel, H.R., Montour, J.L. and Wilson, J.D. (1984) Mechanisms of neovascularization. Vascular sprouting can occur without proliferation of endothelial cells. *Lab. Invest.* 51: 624–634.

Stokes, C.L. and Lauffenburger, D.A. (1991) Analysis of the roles of microvessel endothelial cell random motility and chemotaxis in angiogenesis. *J. Theor. Biol.* 152: 377–403.

Stokes, C.L., Rupnick, M.A., Williams, S.K. and Lauffenburger, D.A. (1990) Chemotaxis of human microvessel endothelial cells in response to acidic fibroblast growth factor. *Lab. Invest.* 63: 657–668.

Stokes, C.L., Lauffenburger, D.A. and Williams, S.K. (1991) Migration of individual microvessel endothelial cells: stochastic model and parameter measurement *J. Cell Sci.* 99: 419–430.

Terranova, V. P., Diflorio, R., Lyall, R.M., Hic, S., Friesel R. and Maciag, T. (1985) Human endothelial cells are chemotactic to endothelial cell growth factor and heparin. *J. Cell Biol.* 101: 2330–2334.

Thompson, D.W. (1942) *On Growth and Form.* Cambridge University Press: Cambridge.

Warren, B. A. (1966) The growth of the blood supply to melanoma transplants in the hamster cheek pouch. *Lab. Invest.* 15: 464–473.

Williams, S.K. (1987) Isolation and culture of microvessel and large-vessel endothelial cells; their use in transport and clinical studies. In: *Microvascular Perfusion and Transport in Health and Disease* (ed. P. McDonagh) pp. 204–245. Basel: Karger.

Woodward, D.E., Tyson, R., Myerscough, M.R., Murray, J.D., Budrene, E.O. and Berg, H.C. (1995) Spatio-temporal patterns generated by *Salmonella typhimurium*. *Biophys. J.* 68: 2181–2189.

Yamada, K.M. and Olden, K. (1978) Fibronectin-adhesive glycoproteins of cell surface and blood. *Nature* 275: 179–184.

Zawicki, D.F., Jain, R.K., Schmid-Schoenbein, G.W. and Chien, S. (1981) Dynamics of neovascularization in normal tissue. *Microvasc. Res.* 21: 27–47.

14

On the Mechanochemical Theory of Biological Pattern Formation with Applications to Wound Healing and Angiogenesis

James D. Murray and Kristin R. Swanson

Department of Applied Mathematics, Box 352420, University of Washington, Seattle, Washington 98195-2420, USA

14.1 Introduction and Biological Background

The elucidation of the underlying mechanisms of morphogenesis is of fundamental importance in developmental biology. From the homogeneous mass of dividing cells of the early blastula, emerge the rich spectrum of pattern and structure observed in animals. Little is known about the mechanisms involved in such development. Pattern formation is of course crucial to almost all aspects of nature, whether it is wound healing or angiogenesis, two topics we discuss in detail in this review, or patterns on animal coats and butterfly wings, and so on. Although a great deal of experimental work has focused on the molecular biology of development, genetics can say little about actual mechanisms for the creation of form. Mathematical models for biological pattern generation can suggest possible mechanisms and have played an important role in the study of pattern formation in a variety of cases and have significantly influenced the thinking of experimentalists. Analysis of morphogenetic models can give biological insight into which elements are essential for pattern development and can be useful in predicting the effect of altering a given parameter (and often experimental conditions) on the resultant pattern. Mathematical modelling can also suggest experiments to further elucidate the pattern formation process. This article briefly reviews the basic framework of Murray–Oster mechanical theory and the effective application of the mechanochemical approach to the generation of specific biological patterns and forms. In particular, we discuss the creation of the network structure observed in angiogenesis, a subject of considerable topical importance in view of the recent experimental work on anti-angiogenetic drugs for the control of tumours first suggested by Folkman in

On Growth and Form: Spatio-temporal Pattern Formation in Biology.
Editor Mark A.J. Chaplain, G.D. Singh, J.C. McLachlan

the 1970s, but almost totally ignored until the actual drugs were isolated in 1998 (Folkman, 1972). We also briefly consider a model for wound healing with a aim of furthering our understanding of scar formation, a common feature of full-depth dermal wounds. The brief review of Cook and Murray (1995) describes a variety of models including neural models which are not discussed here.

14.1.1 Theories of morphogenesis

Morphogenesis refers to the development of biological pattern and form during embryogenesis and is responsible for a vast range of patterns and structures. During development, apparently identical cells can follow different developmental pathways depending on their location. Although extensive experimental and theoretical work has elucidated some basic principles necessary for the generation of spatial form, there is no accepted mechanism for generating specific biological patterns. Essentially two distinct theoretical approaches to biological pattern formation have dominated the field over the past 15 years: the Turing reaction-diffusion prepattern theory of morphogenesis and the mechanochemical theory of Murray and Oster.

Turing theory. The Turing theory of morphogenesis (Turing, 1952) is based on the assumption that specialized chemicals, known as morphogens, can react and diffuse in such a way to create a heterogeneous steady-state spatial pattern of chemical concentrations. His mathematical system is the now well-known reaction-diffusion mechanism. He suggested that morphogenesis is the result of cells reacting according to the morphogen concentration in which they find themselves. This concept was exploited experimentally by Wolpert (see, for example, Wolpert 1969, 1971, 1981) with his "positional information" concept. In this framework, cells must be programmed to interpret chemical concentrations and differentiate accordingly. The existence of morphogens is, in most cases, still a controversial issue and how cells interpret a prepattern cannot really begin to be addressed until they are identified. Under the Turing mechanism, development is essentially separated into two stages: (i) the chemical (morphogen) prepattern is set up, (ii) cells respond to the patterned morphogen concentration by differentiating, migrating or changing shape. That is, once the morphogen prepattern is defined, morphogenesis is a slave process. Reaction mechanisms are of open-loop type. Once the prepattern is laid down, the eventual morphogenetic structure is fixed – there can be no feedback from the environment to modify the pattern. Such open-loop systems are unstable and seem an unreliable mechanism for many developmental processes which are, in the main, robust and self-correcting.

Reaction-diffusion models have been applied to a variety of biological problems, numerous examples of which are given, for example, in the books by Meinhardt (1982, 1998) and Murray (1993a). Although it is clear that chemicals are critical in development, the identification of the specialized morphogens is generally still lacking, so it is difficult to suggest practical forms for the reaction diffusion systems. If the final structure is not crucial then a case can be made for a reaction-diffusion approach as in the situation, to a certain extent, of animal coat patterns (Murray, 1981).

Important related mechanisms involve cells and chemotaxis, the aggregation effect whereby cells move up concentration gradients of chemoattractants. The mechanisms generally involve both diffusion and chemotaxis (without some form of dispersal, like diffusion, the solutions of the equations can blow up). Cell movement results from diffusion and chemotaxis with the chemoattractant also diffusing. Simplest models involve cells which produce their own chemoattractant. In this patterning scenario, it is intuitively clear how patterns can be formed (Keller and Segel, 1971). If cells produce their own chemoattractant then a small perturbation in an initially homogeneous distribution of cells gives rise to a gradient in the attractant. The gradient of attractant induces more cells to move towards the higher concentration of cells and thus the heterogeneity is enhanced. But both the cells and the chemoattractant can diffuse, a process which tries to eliminate such heterogeneities. There are thus dispersive and aggregative effects which compete. It is clear that if the aggregative effects (chemotaxis) are greater than the dispersive effects, spatial patterns will be generated. The quantification of these effects, particularly the chemotaxis terms in the model systems, can essentially only be determined by experiment. Using bifurcation theory, a meaningful robustness comparison can be made between reaction-diffusion mechanisms and chemotactic systems (Zhu and Murray, 1995).

Numerous chemotaxis models for specific experimental situations have been investigated with considerable success. Recent dramatic examples include the highly complex bacterial patterns observed with *Salmonella* and *E. coli* where experiment and theory are intimately connected: see, for example, Budrene and Berg (1991), Woodward *et al.* (1995), Ben-Jacob *et al.* (1995), Budrene and Berg (1995), Tyson *et al.* (1999) and other references in these papers. In Tyson *et al.* (1999) a minimal system is presented with, importantly, all of the functional forms in the model equations and all the parameters estimated from experiment.

Mechanochemical mechanisms of generating pattern and form. The mechanochemical modelling framework, developed by Murray and Oster and their colleagues in the early 1980s (Oster *et al.*, 1983; Murray *et al.*, 1983; Murray and Oster, 1984a, b; Murray and Maini, 1986), is based on the observation (Wolpert, 1971) that "relatively simple cellular forces can give rise to complex changes in form". This concept was clearly demonstrated by experiments of Harris and his coworkers (Harris *et al.*, 1980) in which fibroblast cells (discussed further below) are placed on an elastic substratum. The cells spread and migrate, concurrently generating strong traction forces which deform the substratum (Harris *et al.*, 1980) creating tension patterns on the elastic sheet. These experiments, coupled with the elusive nature of morphogens, provided the major impetus for the initial development of the Murray–Oster mechanochemical modelling approach (Murray *et al.*, 1983; Oster *et al.*, 1983).

In embryonic development, major patterning processes take place in two distinct types of tissue: mesenchyme and epithelia (Figure 14.1). The epithelium is a sheet of closely packed cells characterized by folding, growth and movement during development: epithelial cells do not generate traction forces. Mesenchymal cells, on which we focus in this review, are more loosely spread within a three-dimensional fibrous extracellular matrix (ECM) and unlike epithelial cells, generate large traction forces through their finger-like filopodia. As mesenchymal cells migrate within the matrix, they can secrete (and degrade) matrix as well as exert traction forces on the ECM environment. That is, these cells are capable of significantly altering their environment: the wrinkling of elastic substratum by the strong traction forces

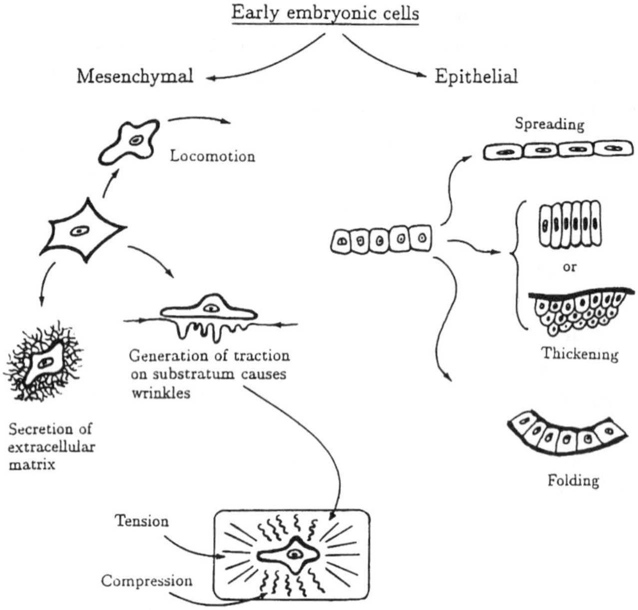

Figure 14.1 Early embryonic cells and their behavior. Although this basic scenario refers essentially to embryonic cells, the cellular behaviors are common throughout cell biology.

generated by mesenchymal cells mentioned above. So, within this simple framework, the cells can create spatial heterogeneity and generate form simultaneously. But perhaps most crucial is that, with such a mechanism, the variables involved and the consequences of their interaction can be observed and measured experimentally. Furthermore, predictions from such models can be tested experimentally with direct relevance to observable quantities. The specific applications we discuss below are but two examples where the theory and experiment are sufficiently close to be able to suggest some of the crucial elements in the biological mechanism which create these specific biological patterns and final structure.

Embryogenesis (and wound healing and angiogenesis discussed here) depends on a series of sequential processes which generate specific patterns at each stage in development. Embryonic development is a highly stable process which must allow for small perturbations during the patterning process. In the mechanochemical modelling approach, pattern formation and morphogenesis occur simultaneously, essentially defining a single process. Since mechanochemical theory subsumes pattern formation and since morphogenesis occurs concurrently, these models allow for self-correction during the morphogenetic process. This makes for a more robust mechanism for morphogenesis than the Turing prepattern approach. In the latter approach, many model systems are very sensitive to parameter variation. It seems that cell–chemotaxis systems, however, are more robust in general.

Robustness of pattern to small random variations in initial and boundary conditions and parameter values is clearly of great importance. In the case of animal coat patterns there is considerable leeway in that the distribution of spots on a cheetah is not important but it is important that they be spots. Zebras, whose stripe pattern is indistinct, for example where the overall pattern impression is black (the default colour), are observed to be quasi-outcasts and tend to remain on the edges of herds. When pattern formation mechanisms are coupled, such as one for epithelial patterning and one for the mesenchyme, the patterning process is much more robust with a more restricted set of stable patterns than each mechanism on its own (Murray, 1993; Cruywagen and Murray, 1993a, b; Murray *et al.*, 1994). Such tissue interaction is ubiquitous in development.

The power of mathematical modelling lies in its ability to make predictions and suggest experiments. Mechanochemical theory is particularly wellsuited to connecting theory with experiment since the model structure is firmly based on empirically measurable quantities (e.g. cell density, matrix density, cell traction forces and known chemicals) and the consequence of varying such quantities can be clearly observed.

The ability of a model to mimic patterns observed in a developmental process is only the first, but necessary, step to its candidacy as a possible mechanism. A model must be able to make testable predictions. From this viewpoint many predictions of reaction-diffusion theory are hard to verify because they are in terms of properties of morphogens yet to be identified, whereas those of mechanochemical theory are based on parameters which are measurable. The processes that occur in development are clearly highly complex. Mathematical models are necessarily based on gross simplification. However, they do tend to focus on particular elements of the process and help to determine the role played by such elements in the overall process. The role of theory is to provide a conceptual framework for the experimenter and to make predictions which encourage or provoke experimentation. From this point of view, mathematical models for development are fulfilling their role and will certainly play a significant part in the understanding of some of the major scientific problems of all time. Several reviews (Murray *et al.*, 1998a; Cook and Murray, 1995) highlight some of the successes. The book edited by Othmer *et al.* (1993) and this volume provide a very good view of the current state of pattern formation modelling in several different areas.

In this chapter we give only a brief introduction to the mechanochemical theory of morphogenesis and, by way of example, focus on two recent successful applications firmly backed up by experiment. An extensive review of the mechanochemical theory is available in the book by Murray (1993a). We first discuss how the models are constructed and how to suggest an appropriate form for the various terms. We then discuss two recent successful applications of the modelling framework: wound healing and vascular network formation. The last section is devoted to a brief discussion of the present usefulness and applicability of mechanochemical modelling and some general comments on the use of mathematics in biology.

14.2 Mechanochemical Theory Framework

The basic mechanical model hinges on two key experimentally determined properties of mesenchymal cells in vitro: (i) cells migrate within a tissue substratum made up of fibrous ECM and other cells; (ii) cells can generate large traction forces. That such traction forces exist in vivo is clear, for example, from the gross scarring that is a feature of the healing of many full-depth (dermal) wounds. The basic framework models the mechanical interaction between the motile cells and the viscoelastic substratum within which they move.

Mesenchymal cells move by exerting forces on their surroundings, consisting of the viscoelastic fibrous ECM and the surface of other cells. They use their cellular protrusions, the filopodia or lamellipodia, which stretch out from the cell in all directions, to grip whatever is available and pull. As the cells move through the ECM, they deform it by virtue of their traction forces. These deformations in the ECM induce anisotropy effects which in turn affect the cell motion. Analysis of models incorporating these various effects shows that coordination of such effects results in spatially organized cell aggregations.

The basic model, a continuum one, consists of three equations governing (i) the cell population density, (ii) the mechanical balance of the forces between the cells and the ECM and (iii) the conservation law for the ECM.

We denote the cell density by $c(\mathbf{x}, t)$ and the matrix (the ECM) density by $\rho(\mathbf{x}, t)$ where $\mathbf{x}$ denotes position in space and t is time. The dynamics are captured mathematically by defining conservation equations for the transport of cells and matrix in and out of a control volume. As we have described above, cells are capable of significantly altering the matrix environment, and vice versa. Cell shape and orientation can also affect the dynamic interaction but we do not include them in this review. Cell orientation can play a significant role in pattern formation (see, for example, Edelstein-Keshet and Ermentrout, 1990). The matrix and cells mechanically interact in such a way as to preserve the conservation of momentum within the tissue, which in view of the slow dynamics reduces to setting the various forces to equilibrium.

14.2.1 Cell density conservation

Several factors, with varying levels of importance, affect the movement of embryonic cells within a tissue substratum which is made up of fibrous ECM. For example, some of the more obvious are:

- Chemotaxis, by which cell motion is directed by a chemical gradient of a soluble chemical.
- Contact guidance, whereby the substratum defines a preferred direction of cell motion.
- Contact inhibition, whereby high cell densities inhibit motion and reduce individual cell traction forces.
- Convection, which is transport of cells and ECM due to local deformation in the matrix. The forces exerted on the ECM by other cells induce passive movement of cells on the ECM.
- Diffusion, which is the usual random dispersal of cells with general motion down a cell density gradient.
- Haptotaxis, which is the process by which ECM gradients (or chemicals *bound* to the ECM) affect cell movement. Cell tractions generate gradients of matrix density, which are in turn associated with the density of adhesive sites for cell lamellipodia to bind to. Since cells can get a stronger grip on denser matrix, cells tend to move up an adhesive/matrix gradient.

- Galvanotaxis, where movement from the field generated by electric potentials, which are known to exist in embryos, provides a preferred direction of motion.

These effects are all well documented experimentally. The model field equations encapsulate the key features which affect cell movement within its extracellular environment. We shall not include all of the effects just mentioned but it will be clear how they can be incorporated and their effect quantified. For simplicity, we include only convection, diffusion and haptotaxis in the sample model we present below; see Murray (1993a) for a fuller discussion.

Within a fixed control volume, cell density is conserved according to

rate of change of cellular density =
convection + diffusion + haptotaxis + mitosis − apoptosis

where apoptosis refers to programmed cell death. In mathematical terms, we write

$$\frac{\partial c}{\partial t} = -\nabla \cdot \mathbf{J}_{\text{cells}} + M \tag{14.1}$$

where M represents the net growth (mitosis − apoptosis) of the cell population. A logistic growth term is commonly used as a plausible metaphor for the dynamics,

$$M = rc(1 - \frac{c}{K}) \tag{14.2}$$

where r refers to the difference between the linear cell mitotic and apoptotic rates and K is the carrying capacity of the cellular environment. $\mathbf{J}_{\text{cells}}$ is the total flux of cells across a unit area in a unit time. In the case we consider here, the flux consists of convection, diffusion and haptotaxis terms:

$$\mathbf{J}_{\text{cells}} = \mathbf{J}_{\text{convection}} + \mathbf{J}_{\text{diffusion}} + \mathbf{J}_{\text{haptotaxis}} \tag{14.3}$$

Defining $\mathbf{u}(\mathbf{x}, t)$ as the displacement of the cell–matrix continuum from its original position, the convective flux can be approximated by

$$\mathbf{J}_{\text{convection}} = c\frac{\partial \mathbf{u}}{\partial t}. \tag{14.4}$$

If cells respond to local variations in the cell density by tending to move down a density gradient, a classical diffusive flux

$$\mathbf{J}_{\text{diffusion}} = -D\nabla c \tag{14.5}$$

contributes to the total cellular flux. Since some cells extend long filopodia, nonlocal effects may also be included in this diffusive term, these give rise to a higher-order term involving $\nabla^2 c$ in the flux (Murray, 1993a). With increasing matrix density, the density of adhesion sites for the lamellipodia to bind to increases, and so an adhesive gradient develops. The cell movement up this gradient is taken to be

$$\mathbf{J}_{\text{haptotaxis}} = ac\nabla\rho \tag{14.6}$$

where a is the haptotactic constant. As with diffusion, long-range effects can also be included in the expression for the haptotactic flux. Note the different signs in the diffusive and haptotactic flux; the former is dispersive while the latter is aggregative.

Contact guidance is crucially important in the formation of cell–matrix networks discussed below. Such guidance can be modelled in different ways. One possibility is to consider the diffusion as directed-diffusion. In this case, the diffusion coefficient $D(\boldsymbol{\varepsilon})$ is a function of the elastic strain tensor $\boldsymbol{\varepsilon}$ defined in Section 14.2.3. This dependence implies that the strain of the matrix can grossly affect the direction of cell movement as we shall see in the section on network formation.

14.2.2 Conservation of matrix density

The ECM has a very complicated fibrous structure that can vary significantly from tissue to tissue. Matrix mechanical behavior is dominated by the fibrous collagen component of the ECM. Local disturbances in the matrix density can induce convection of the matrix. Also, as (mesenchymal) cells move through the matrix environment, they can secrete and/or degrade the ECM: this occurs in wound healing and in angiogenesis.

The conservation equation for the ECM density is then

$$\text{rate of change of matrix density} = \\ \text{convection} + \text{secretion} - \text{degradation}.$$

which in mathematical terms is taken to be

$$\frac{\partial \rho}{\partial t} = -\nabla \cdot \mathbf{J}_{\text{matrix}} + S \tag{14.7}$$

where S represents the secretion of the matrix. The matrix flux is assumed to be predominantly due to convection so $\mathbf{J}_{\text{matrix}} = \rho \partial \mathbf{u}/\partial t$. In many applications, the secretion term S is often negligible on the time scale of the pattern formation. This is the case in the experiments associated with the matrix network related to angiogenesis discussed in later sections.

14.2.3 Mechanical interaction of cell and matrix

Since inertial terms are small in development, the governing equation for the mechanical interactions between the cells and the matrix is simply that the sum of the various forces is in equilibrium within the tissue. We thus have

$$\text{forces on the body} = \\ \text{cell traction forces} + \text{viscoelastic restoring forces of the matrix}$$

This force equilibrium equation can be translated into mathematical terms as

$$\rho \mathbf{F} = -\nabla \cdot \boldsymbol{\sigma} \tag{14.8}$$

where $\mathbf{F}$ is the external force acting on the matrix per unit matrix and $\boldsymbol{\sigma}(\mathbf{x}, t)$ is the stress tensor composed of the cell and ECM components:

$$\boldsymbol{\sigma} = \boldsymbol{\sigma}_{\text{cell}} + \boldsymbol{\sigma}_{\text{matrix}}. \tag{14.9}$$

As the cells and matrix interact, the cells exert traction forces on the matrix and the matrix fibers resist the pulling of the cells: the tensors involve both the matrix and cell density.

Cell traction on matrix. To define the stress tensor $\boldsymbol{\sigma}_{\text{cell}}$ we express the cell traction forces per unit mass of matrix. We expect traction to increase with increasing cell density c, but at high densities there is some cell–cell inhibition and at high enough densities we would expect the traction forces to tend to zero. Mathematically, we might express the traction per unit mass matrix T as a function of the cellular density, c, qualitatively as

$$T(c) = \frac{\tau c}{1 + \lambda c^2} \tag{14.10}$$

where τ is the traction force generated per cell at low cellular densities and λ is a measure of the force inhibition as a result of increasing cell density.

The quantification and form of the cell traction forces is clearly important since these forces play such a crucial role in pattern formation in development. It has proved experimentally difficult to quantify these cellular forces because of the complexity of the cell–matrix interactions and the difficulty of separating out the various mechanical effects in real biological tissue. The seminal paper by Tracqui, Ferrenq and their colleagues (Ferrenq *et al.*, 1997) describes a new experimental technique and general approach for quantifying the forces generated by endothelial cells on ECM. They first developed a mathematical mechanical model, based on the mechanochemical theory described in this article, in which different forms for the cell-generated stress were proposed. They then used it as the basis for a novel experimental device in which cells are seeded on a biogel of fibrin (the matrix) held between two holders, one of which can move and is attached to a force-sensing device. By comparing the displacement of the gel calculated from the model expressions with the experimental data recorded from the moving holder, they were able to justify specific expressions for the cell traction stress: they did this for various experimental setups. They were then able to compare different plausible analytical expressions for the cell traction stress with the corresponding force quantification and to compare the results with experiment and other reported measurements for different kinds of cells by similar and different experimental devices. They show how experimentally justifiable forms for the cell–gel traction stress can be derived and give estimates for each of the parameters involved. Their work thus also provides estimates for some of the key parameters in the mechanical theory of morphogenesis, prime among which is the cell traction per cell. This paper is an excellent example of genuine interdisciplinary mathematical biology research, with theory and experiment each playing an important role in the outcome. For the purpose of this chapter, we retain the simpler form above for pedagogical purposes.

Viscoelastic matrix material. The stress–strain relation in such a complex tissue matrix is almost certainly non-linear and plastic. There is still no experimental evidence which would allow us to postulate the actual form of the relation. Many patterns result from relatively small deformations and cell density differences (one example occurs in the formation of papillae, the dermal condensations of cells which presage feather and scale formation). We make the assumption that for small strains, cell–matrix material forces can be modelled by a linear viscoelastic response (Fung, 1993). The matrix stress tensor then consists of a viscous component and an elastic component:

$$\boldsymbol{\sigma}_{\text{matrix}} = \boldsymbol{\sigma}_{\text{viscous}} + \boldsymbol{\sigma}_{\text{elastic}}. \tag{14.11}$$

with the viscous component of the stress proportional to the rate of change of the strain ε in the matrix

$$\boldsymbol{\sigma}_{\text{viscous}} = \mu_1 \varepsilon_t + \mu_2 \theta_t \mathbf{I} \tag{14.12}$$

where the elastic strain tensor of the matrix is defined by

$$\varepsilon = \frac{1}{2}\left(\nabla \cdot \mathbf{u} + \nabla \cdot \mathbf{u}^T\right). \tag{14.13}$$

and $\theta = \nabla \cdot \mathbf{u}$ is the dilation of matrix. Here $\mathbf{I}$ is the unit tensor and the subscript t refers to differentiation with respect to time. The parameters μ_1 and μ_2 represent the bulk and shear viscosities of the matrix, respectively.

The elastic component of the stress is given by

$$\boldsymbol{\sigma}_{\text{elastic}} = \frac{E}{1+\nu}\left(\varepsilon + \frac{\nu}{1-2\nu}\theta \mathbf{I}\right) \tag{14.14}$$

where the Young's modulus E defines the stiffness of the matrix and the Poisson ratio ν measures the contraction of the matrix in one direction when it is stretched in a transverse direction.

The derivation of the form (14.12), (14.13) and (14.14) for the matrix linear viscoelastic stress tensor is given, for example, in the book by Landau and Lifschitz (1970).

External body forces. There are often external forces acting on the tissue body we are attempting to model. These forces may be due to attachment of the matrix to a substrate or, in the case of network formation we discuss below, to the base of the experimental vessel. The body forces holding the tissue in various places can reasonably be modelled as proportional to the density of matrix and the displacement of the matrix: $\mathbf{F} = -s\mathbf{u}$ where s defines the elastic attachment to the substrate. Alternatively in the matrix network situation discussed later it is more appropriate to relate the attachment force to the viscous drag: we come back to this below.

When this approach was formulated by Murray *et al.* (1983) and Oster *et al.* (1983), the mechanical properties of ECM and cell–matrix interactions had not been quantified, but recent experimental advances have focused on these mechanical aspects of tissue interactions. As a result, many of the parameters described in the model formulation can be estimated. As well as the work by Ferrenq *et al.* (1997) mentioned above, Moore *et al.* (1995) recently quantified the viscoelastic properties of mesodermal cells during early development. In the cell–matrix experiments described below in connection with angiogenesis, parameter estimations have also been obtained. The case of chemical control of eukaryotic cell movement (Sherratt *et al.*, 1993) has also yielded parameter estimates.

In many developmental situations (wound healing is one) chemicals can play a significant role in the pattern formation process and therefore should be included in the model. In principle chemical dynamics can be incorporated relatively simply, but only if we have some idea of how they interact with the cells and matrix. At this stage, these interactions are generally still unknown.

14.2.4 Model analysis

The mathematical investigation of these model equations involves linear analysis about the steady states to obtain the dispersion relation and hence the pattern formation potential. The analysis (see, for example, Murray, 1993a, for a full discussion) results in a remarkably wide variety of dispersion relations which indicate a rich spectrum of non-linear patterns. The non-linear spatial patterns are generally obtained by numerical simulation or singular perturbation methods near bifurcations. As expected, the type of patterns possible depend on which terms are included in the model mechanism. Another aspect of these mechanical models is their robustness to parameter variation. It is essential to know, at least approximately, the parameter domain for specific patterns because this is one of the optimal ways to assess model robustness. Several non-linear analyses have been carried out (for example, Cruywagen *et al.*, 1997) to determine the stable patterns which can be obtained. This non-linear bifurcation type of analysis has been applied to a variety of problems such as sequential patterning in skin morphogenesis (Cruywagen *et al.*, 1993, 1994), to the formation of spots and stripes (Zhu and Murray, 1995) with others reported in these articles. In the remainder of this review we discuss two recent applications of mechanochemical theory, namely wound healing and vascular network formation.

14.3 Wound Healing

Interest in epidermal and dermal wound healing is, of course, of ancient origin. The importance of wound healing is clear with more than 2 million people sustaining burn injuries in the US alone each year with a higher percentage in developing countries where, in rural areas, open fires are the norm. Wound healing has far wider implications; it is involved in the formation of stretch marks during pregnancy, corneal radial keratotomy, hepatic cirrhosis, pulmonary fibrosis, scleroderma and diseases that result in tractional detachment of the retina and so on. However, there is an even broader context in which one can judge the value of wound-healing studies. Wound healing is unique in its position at the meeting point of medicine and developmental biology. Dermal wound healing involves cell differentiation, migration and proliferation, the creation of new tissue and the mechanical deformation of that tissue by the cells within it. Each of these processes has analogies in development and, even if the results of tissue repair were not important to us, the study of wound healing would certainly be worth pursuing for the insights it offers into related developmental processes.

Wound-healing research is particularly active at the present time as a result of advances in quite diverse areas. One of the exciting discoveries, which gave rise to some of the current excitement was in the early 1980s, is that mammalian fetal wound healing is quite different to that in adults – fetuses heal without scars. This gives hope that it may be possible to "turn back the clock" and induce adult wounds to heal without scars. Some modelling of embryonic wound healing has been reported (Sherratt *et al.*, 1992; Sherratt and Lewis, 1993; Sherratt 1993; Dale *et al.*, 1995). Another development is the discovery of the major effects growth factors can have on wound healing, and this has further increased the research effort on growth factors. These chemicals are involved in the control of many of the most important processes of wound healing. Clinical trials involving either promotion or inhibition of growth factors are currently in progress: a preliminary modelling attempt at including such effects is given in Sherratt and Murray (1992). The development of artificial skin for use in grafting also holds great promise, particularly for severe burn wounds.

Wound healing is now a very large field. The theoretical literature is now quite lengthy while the experimental and medical literature is enormous. Rather than try to present even a brief overview of the field, we give some recent references, from which others can be found. We discuss the particular application of the mechanochemical theory in a little more detail as it has been applied to dermal wound healing.

14.3.1 Biological background and motivation

Dermal wounds are deep wounds to the skin, which include serious burns, and frequently result in severe scarring: the process is thus one of repair rather than regeneration. Wound healing often involves significant contraction, or reduction in the area of the wound, which can accentuate the disfiguring nature of the scar and even cause loss of functionality. These aspects of wound healing, contraction and scar formation, have been a major focus of mathematical modelling (see, for example, Murray *et al.*, 1988; Tranquillo and Murray, 1992, 1993; Tracqui *et al.*, 1995b; Cook, 1995; Tranquillo and Barocas, 1999; Murray *et al.*, 1998; Dale *et al.*, 1997; Olsen *et al.*, 1995, 1997a, b, 1998; Barocas and Tranquillo, 1997).

There have also been interesting models proposed to try to understand abnormal wound healing (see, for example, Olsen *et al.*, 1996) and how to manipulate wounds to improve healing or scarring. The most extensive review, including much original work with many unpublished results, of mathematical modelling of wound healing is given by Cook (1995): a small part of this work is presented in Cook *et al.* (1994), Tracqui *et al.* (1995b) and Murray *et al.* (1998a).

The skin is a complex organ but at its basic level consists of two layers of tissue, the epidermis and dermis. Densely packed proliferating cells comprise the outer epidermal layer. Epidermal cells generally correspond, in behavior, to early embryonic epithelial cells shown in Figure 14.1. These cells can spread, thicken or fold tissue. Epidermal wounds heal by a very different process and they will not be discussed here. There have, however, been some successful modelling endeavors (Sherratt and Murray, 1990, 1992; Sherratt *et al.*, 1992; Dale *et al.*, 1995; Gaffney *et al.*, 1999 and references therein).

The main features and gross time course of dermal wound healing is well known (Clark, 1991). Underlying the epidermal layer, the dermis consists of ECM, capillaries and specialized cells called fibroblasts. The fibroblast cells correspond to the schematic representation in Figure 14.1 for early embryonic mesenchymal cells in that they can move, secrete and degrade matrix and generate traction forces. Immediately after wounding, the wound edges retract due to natural skin tension and there follows an inflammatory response. This induces the fibroblasts to proliferate, invade the wound and lay down a matrix, the extracellular matrix (ECM), on which they exert a contractile force, which fills and closes the wound. Angiogenesis (see below) occurs at the same time, and the bright red tissue filling the wound is referred to as granulation tissue. The wound is gradually covered by an invading epithelium but the dermal and epidermal processes are essentially independent. The ECM is continually remodelled, rapidly at first and at a lower rate for many months thereafter. In most cases the result is a scar which differs in appearance from the surrounding skin.

Wound healing has been quantified in several ways, including measuring wound area and contraction rate. Scarring reflects the organization of the collagen in the ECM. Although we are rapidly learning more and more about wound healing, it is still difficult to relate it to the clinically important macroscopic properties of the scar, which is why we believe mathematical modelling has much to offer.

Wound area. Many authors have measured the area and perimeter of the wound over time or have measured the maximum extent of wound retraction and contraction (McGrath and Simon, 1983; Cook, 1995 and references therein). Typically, directly after wounding the wound edges retract due to skin tension. After a latent period of 5 to 10 days the wound area reduces rapidly ($5 - 10\%$ per day) as a result of contractile activity within the wound. The contraction curve eventually bottoms out and the area then changes very little over a long period. Although it varies with the wound, the final contracted area may be around one-fifth of the original.

Wound shape. There is some controversy about the effect of wound shape on contraction rates (McGrath and Simon, 1983; Cook, 1995). The change of shape that accompanies wound contraction is highly dependent on the initial shape, the orientation of the wound to the skin tension lines, and to a lesser extent, on the laxity of the surrounding tissue; Cook (1995) gives a full discussion and relevant references. Circles heal by decreasing their perimeter (taking on an elliptical or oval shape) whereas polygonal wounds tend to have little change in perimeter length; the edges collapse inwards forming star-shaped scars.

Tension. Skin tension affects the extent and course of contraction as well as retraction, and it has also been implicated in the formation of pathological scars. Excessive tension results in wide scars but application of modest tension can improve scar strength (Timmenga *et al.*, 1991).

Mechanical properties. A wound becomes stiffer, stronger and more extensible throughout the wound-healing process. The shape of the stress–strain curve for the tissue changes over time, the reasons for which are not yet known, although various correlations with matrix organization (collagen in particular) have been observed.

Theories. Cook (1995) gives a very thorough review of most of the various theories which exist for wound contraction and scar formation. There are clear differences in organization (including fiber orientation) between scar tissue and normal skin but the causes of pathological scar formation remain unclear (Cook, 1995; Olsen *et al.*, 1996). Pathological scars are characterized by dense nodes of aligned matrix fibers. Modelling has suggested that these scars could be the result of mechanical instabilities in the matrix due to excessive cell tractions.

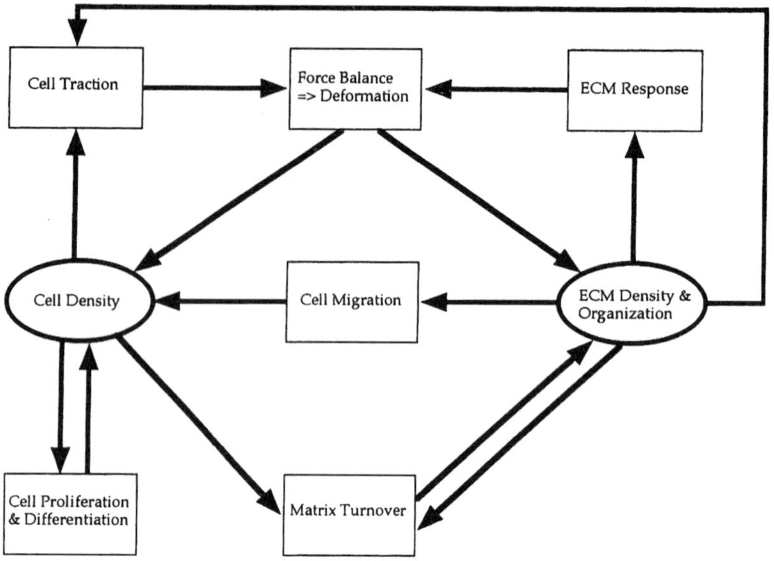

Figure 14.2 Key variables and processes in dermal wound healing. This structure underlies the mathematical model: each variable requires an evolution equation which incorporates terms corresponding to the various processes. The force balance box involves both a variable (momentum) and a conservation equation. This constraint determines another variable, the deformation. ECM refers to extracellular matrix.

14.3.2 Mechanochemical modeling of wound healing

The first mathematical model for dermal wound contraction was that of Murray and Tranquillo (Murray *et al.*, 1988; Tranquillo and Murray, 1992, 1993). It was based on the Murray–Oster continuum-mechanical theory discussed above. This theory extended quite naturally to wound healing since in both instances a key role is played by the mechanical interaction between cells and the viscoelastic ECM which they inhabit and through which they move. Variants of the basic model allow for cell movement (diffusion, chemotaxis, haptotaxis, haptokinesis), ECM turnover, cell proliferation and death, osmotic forces, and mechanical interaction between tissue layers.

One of the achievements of mathematical modelling to date has been the isolation of the critical component processes of wound contraction and scar formation. An outline of the logic of wound contraction and scar formation is as follows (Figure 14.2). Specialized cells, called fibroblasts, invade the wound and are capable of deforming (contracting) the ECM through which they move. The invasion may be enhanced by chemotaxis, haptotaxis or contact guidance. The fibroblasts also play an important role in remodelling (creating and degrading) the ECM. Subdermal attachments and the ECM itself resist the contractile stresses set up by the fibroblasts (or specialized phenotypic variants called myofibroblasts).

The models are complicated, necessarily so if we want to capture as many of the key processes as possible, so we first give the equations in word form:

rate of change of cell density =
passive movement by ECM deformation
+ active movement (diffusion) relative to ECM
+ proliferation/differentiation

rate of change of ECM density = movement due to ECM deformation
+ secretion + degradation

body forces = divergence of [cell traction stress + ECM stress]

The core equations for the initial Murray–Tranquillo model (Murray *et al.*, 1988; Tranquillo and Murray, 1992) are a specific example of the type of model system that has to be analysed with the mechanochemical theory of pattern formation. Many of the later models have a similar structure. The equations comprising the base model are derived respectively from conservation of cells, ECM and momentum of the combined continuum with respect to fixed (Eulerian) coordinates, and they correspond to the above word equations. Cells are convected, diffuse and multiply. ECM is convected. The continuum is subject to active traction forces and also body forces due to elastic tethering to the subdermal layer. The ECM response is linear viscoelastic, one of the grosser assumptions which we briefly discuss below. In view of the timescales involved, inertial effects are ignored. As with the above discussion of the general mechanochemical theory, the form of the various terms is crucial. The most basic system proposed by Murray and Tranquillo is

$$\frac{\partial c}{\partial t} = -\nabla \cdot \left(c\frac{\partial u}{\partial t}\right) + \nabla \cdot (D\nabla c) + rc\left(1 - \frac{c}{K}\right) \tag{14.15}$$

$$\frac{\partial \rho}{\partial t} = -\nabla \cdot \left(\rho\frac{\partial \mathbf{u}}{\partial t}\right) \tag{14.16}$$

$$\chi\rho\mathbf{u} = -\nabla \cdot \tau_o\frac{\rho c}{1+\lambda c^2}\mathbf{I} + \mu_1\frac{\partial \varepsilon}{\partial t} + \mu_2\frac{\partial \theta}{\partial t}\mathbf{I} + \left(\frac{E}{1+\nu}\right)\left(\varepsilon + \left(\frac{\nu}{1-2\nu}\right)\theta\mathbf{I}\right) \tag{14.17}$$

where c is cell density, ρ the matrix density, $\mathbf{u}$ the matrix displacement, $\varepsilon = \frac{1}{2}\left(\nabla \cdot \mathbf{u} + \nabla \cdot \mathbf{u}^T\right)$ the strain tensor, $\theta = \nabla \cdot \mathbf{u}$ the dilation, $\mathbf{I}$ the unit tensor, E the Young's modulus, ν the Poisson ratio, μ_1 and μ_2 the shear and bulk viscosities, D the cell motility (diffusion) coefficient, r the maximum mitotic rate, χ the tethering elasticity coefficient, and λ is a parameter governing the shape of the cell traction function as cell density increases. The first term on the right of the third equation is the active cell traction term, and on the left is the body force due to subdermal attachments. No turnover of ECM is included.

In some versions of the Murray–Tranquillo model (Tranquillo and Murray, 1992) there was a reaction-diffusion equation for a chemical which acted as a chemoattractant. They showed that, with one-dimensional (linear) wounds, experimental contraction profiles (wound area against time; McGrath and Simon, 1983) could be explained by inflammation-mediated mechanisms such as chemotactic accumulation of fibroblasts and growth factor induced traction enhancement.

Subsequent developments. Since the first wound-healing models were published mathematical modelling has increased considerably; see the references given above. More complicated versions of the above model include multiple cell types (with differentiation between types) and multiple types or phases of ECM (which interact and possess different mechanical properties and have different effects on cells). Other models include additional equations for chemicals (such as growth factors) which modulate cell proliferative, motile and contractile behavior. When the mechanical properties of the evolving ECM are modelled with greater realism, further equations must be introduced.

Several extensions of the Murray–Tranquillo model have been made by Maini, Sharratt and their coworkers. One extension (Olsen *et al.*, 1995) includes explicit interconversion between fibroblasts and traction-enhancing myofibroblasts and more complex inflammatory mediator dynamics. Their model captures the experimentally observed temporal changes in the densities of myofibroblasts and growth factors. A wound remains transiently contracted while the slow ECM remodelling is in progress. Another study (Olsen *et al.*, 1996) further extended their model and related pathological scarring to the existence of alternative steady states using a mechanical model and a simpler caricature. A high rate of growth factor production, for example, could cause a spread of the pathological state across the wound and into surrounding tissue. Complementary to this study, they used Dale *et al.* (1996), a non-mechanical model, to investigate the ratio of collagen I to collagen III in scar tissue (this ratio is known to be related to fiber thickness and scar quality) focusing on the regulation by growth factor, TGF isoforms (the latter can be topically applied).

Tracqui and his colleagues (Tracqui *et al.*, 1995b) extended the Murray–Oster formulation to include two types of ECM, an early provisional matrix being replaced by the collagenous matrix characteristic of mature scars. A major feature of this model is that ECM turnover results in plastic behavior whereby the wound remains permanently contracted after the inflammatory response has subsided. Barocas and Tranquillo (1994, 1995) accounted separately for the interstitial fluid and the fibrous network making up the ECM. They were able to account for matrix anisotropy via an orientation tensor which is based on the strain tensor. The fiber orientation tensor governs cell movement (contact guidance) and traction (greatest in the direction of greatest fibre alignment). In circular wounds, contact guidance can reduce the extent of wound contraction as cells align (and pull) circumferentially (Barocas and Tranquillo, 1995).

Biomechanics of evolving tissues. In classical elasticity theory a strain field characterizes the mapping between material points in a reference state and locations in a deformed state. The mechanical properties of the material are incorporated in a constitutive equation which relates stresses to strains. Finally, applied body and boundary forces determine the global deformation via the equations of equilibrium which constrain the stresses in the material. Although finite strains and non-linear constitutive equations provide complications, applying classic viscoelasticity theory to biological materials is no different from applying it to non-living materials. Once experiments have been devised to determine the appropriate constitutive equations for each mechanically distinct component (a nontrivial task), a whole range of biomechanical problems can be solved; Fung (1993) gives many examples.

There are, however, two qualities of living tissues which require classic continuum mechanics to be extended and these have particular bearing on wound healing. Even when all external forces are removed from biomaterials, stresses can still exist within the tissue. The existence of residual stresses violates the assumption of classical elasticity theory that the reference state is globally stress-free. The second problem is that living tissues can change their form, their structure and even their material properties, either as part of the natural developmental process (growth is an example) or in response to other signals (such as injury or applied stress). Classical mechanics, therefore, must be extended to accommodate tissue growth and tissue remodelling.

There have been several attempts to address these problems such as that by Cook (1995) in the particular case of dermal wound healing. Some discussion of this and of other attempts are briefly discussed in Murray *et al.* (1998a).

A major extension of the Murray–Tranquillo model (Murray *et al.*, 1988; Tranquillo and Murray, 1992) was made by Cook (1995). He modelled more realistically the mechanics and structure of an evolving, anisotropic ECM (effective strain is introduced and a fiber orientation tensor is related to that). In addition to providing a measure of scar quality, fiber orientation feeds back to affect (i) cell movement (Cook, 1995 showed that the flux of cells due to contact guidance depends on effective strain; his derivation was based on a biased random walk) and (ii) cell traction (greatest contractile force in the direction of fiber orientation). He also allowed for orthotropic skin tension (and wound orientation relative to skin tension lines). In contrast to previous models, contractile forces were assumed to be zero outside the wound and late in scar formation. This was the first study to attempt numerical solution of the equations in two dimensions.

Cook (1995) also included (i) a study of changes in wound shape, contraction rates, and fiber alignment (wound orientation proves important); (ii) true plasticity: wounds remain contracted (and fiber alignment is anisotropic) in the absence of inflammatory mediators or any other externally imposed differences between skin and mature scar; (iii) strong contact guidance can give rise to instabilities (pattern formation) suggestive of nodules of aligned collagen fibers in pathological scars. Finally, a highly simplified model based on linear elasticity suggested that circular should contract most slowly, which is in agreement with experiments. The role of wound shape is yet to be addressed completely in the full model.

Most of the models beginning with that of Tranquillo and Murray (e.g. Tranquillo and Murray, 1992) involved several gross simplifying assumptions (not surprising in view of the complexity of the processes involved), namely (i) small-strain, linear viscoelasticity based on the Voigt model (Landau and Lifschitz, 1970), which can not be plastically deformed, (ii) isotropic material properties for the ECM, (iii) isotropic contractile stresses, (iv) no attempt to relate ECM turnover and remodelling to changes in mechanical properties (including zero-stress states and residual stresses and strains). By accounting for residual strains and fiber orientation and considering finite deformations, Cook (1995) was the first to address these limitations; see Murray *et al.* (1998a) for a brief description of this work. The model system of equations is more complex since, among other things, it includes an equation for the residual strain and the diffusion coefficient as a function of the strain.

The basic concepts of the new model are the following. Associated with each point in space of the ECM there is a residual strain and fiber density orientation which characterize the scar, determine the mechanical properties of the ECM (using a derived constitutive law), the anisotropic contractile stress tensor and the cell motion. The model involves evolution equations for the ECM density, orientation and residual strain which account for the crucial remodelling and growth. With the simplification that the orientation distribution is uniform in the zero stress state (it is not; there are preferred tension lines in the steady-state skin of mammals) the extra equation is then only for the residual strain. This model system is given in Murray *et al.* (1998a).

The important extensions to the Murray–Tranquillo framework (Tranquillo and Murray, 1992; Tranquillo and Murray, 1993) are that the ECM stress, $\boldsymbol{\sigma}_{\text{matrix}}$, the diffusion matrix, D_{ij}, and the active contractile stress, $\boldsymbol{\sigma}_{\text{cell}}$, all depend upon the effective strain, $\boldsymbol{\varepsilon}$, which in small-strain theory is simply the actual Cauchy strain minus the residual strain. The residual strain represents the correction to the effective strain due to changes in the zero-stress state. The evolution equation for it involves a term which depends upon ECM turnover and a convection term (since we are in an Eulerian frame). The justification for the form of the ECM turnover term is that only some fraction of new ECM is assumed to be added to the existing fibrous network in the unstrained state (so that skin tension can be maintained even with continual ECM turnover).

The diffusion term used for active cell movement represents a form of contact guidance so it is rather “diffusion-like”. Here D_{ij} depends on the strain components and was derived from a random walk argument in which the direction of movement depends upon the orientation of fibers (which depends upon the effective strain). The specific (two-dimensional) form used in the numerical solution given in Figure 14.3 is $D_{ij} = D[2\varepsilon_{ij} + (2 - \varepsilon_{xx})\delta_{ij}]$. Since this “diffusion matrix” involves both derivatives it has a convective as well as a diffusive component arising from the divergence of the flux. The same tensor appears in the cell contractile stress term, $\boldsymbol{\sigma}_{\text{cell}}$, since just as cells move in the direction of fiber alignment, they exert a greater force in that direction. In the following section on angiogenesis, fiber alignment plays an important role in the final pattern.

In his thesis, Cook (1995) studied this model numerically and simulations in two dimensions showed how the fiber orientation within the scar would be expected to evolve over time with different wound orientation to the direction of skin tension. To illustrate the type of results obtained with this new model, we considered an elliptical wound; Figure 14.3 shows how fiber orientation varies across such a wound. The parameters used included as many as were available from experiment. Notice that retraction of the wound along its long axis has realigned the skin outside the wound and increased the ECM density near the wound margins, especially on the lateral margins, the white regions in the figure.

A major deficiency of all of these models for dermal wound healing is that they use small-strain approximations for a phenomenon which involves finite strains. Cook (1995) outlined a finite-strain approach to wound healing which deals more completely with the problem of material anisotropy, so crucial to scar structure and formation. In the more general theory, orientation and effective strain are decoupled. Although a vast amount of work remains to be done in this area, the modelling would achieve a great deal of generality and flexibility if Cook's work (1995) were to be combined with the finite strain (but isotropic growth) theory of Rodriguez *et al.* (1994) and the fabric tensor theory of Cowin (1992). The Rodriguez *et al.* (1994) description of

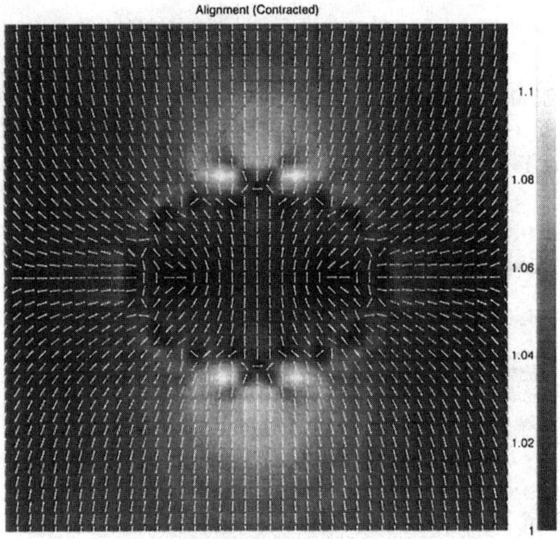

Figure 14.3 Elliptical wound perpendicular to skin tension. With the parameters chosen this represents day 25 post-wounding. Arrows represent the predominant direction of fiber alignment. Light regions are more strongly aligned. See Murray *et al.*, (1998a) for the parameters and further details. The domain is 10cm square with a wound 4.8cm × 3.6cm and initial conditions consisting of no cells and 80% loss of ECM in the wound.

growth and deformation under finite strains has far-ranging implications for wound healing and many other mechanically constrained morphogenetic processes: a brief discussion of this is given in Murray *et al.* (1998a). This article also discusses some open problems which we feel would greatly enhance our understanding of real full-depth dermal wounds.

With regard to wound-healing models (and other modelling endeavors) we feel it is essential to keep certain things in mind, some of which are noted in Murray *et al.* (1998a):

An important point arising from the models to date is that a scar contains its own history. Consider an engineering analogy of our role in managing the wound healing process. It is one thing to suggest that a bridge requires a thousand tons of steel, that any less will result in too weak a structure, and any more will result in excessive rigidity. It is quite another matter to instruct the workers on how best to put the pieces together. It is conceivable that the cells involved in tissue repair have enough "expertise" that given the right set of ingredients and initial instructions they could be persuaded to repair a deep skin wound with a result that looked more like skin than scar tissue. This is perhaps the hope of those who are searching for the optimal cocktail of growth factors to apply topically to a wound. However, it seems to us very likely that the global effect of all this sophisticated cellular activity would be critically sensitive to the sequence of events occurring during tissue repair. As managers we should concern ourselves with how to take advantage of the limited opportunities we have for communicating with the workforce so as to direct the wound healing process towards an acceptable end-product. This may sound rather philosophical, but even a cursory look at the theories of scar formation in the literature reveals a fixation on simplistic explanations. For example, it has variously been suggested that pathological scars result from defects in collagen synthesis or in collagen degradation. We suggest that there need be no misfunctioning of any particular subprocess to explain scar formation. Tissue repair is, after all, fundamentally different from development. The body has only a limited set of "rules" with which it must confront an unlimited number of possible repair problems.

14.4 Vascular Network Formation: Angiogenesis

14.4.1 Biological background and motivation

Angiogenesis refers to the development of vascular structure and is of fundamental importance to wound healing, sustaining tumor growth, morphogenesis, and so on. In the early 1970s Folkman (1971, 1972, 1974) suggested that tumour growth could be arrested if capillary proliferation in the tumour could be inhibited. He even presented experimental evidence that prevention of angiogenesis in the tumour inhibited its growth. Nevertheless, this simple, yet highly revolutionary idea, arguably the most important new concept in the treatment of solid tumour cancers, was almost totally ignored, even ridiculed. Folkman (1995) reviews the clinical implications of angiogenesis. His ideas were to a large extent ignored, in fact, until late 1997 when key chemical factors – Folkman's angiogenesis inhibitors or antiangiogenetic factors – angiostatin and endostatin were discovered (Boehm *et al.*, 1997; O'Reilly *et al.*, 1996, 1997), and experimental results were presented with mice on the remarkable success of not only controlling but eradicating tumours were presented. Other chemicals have an impact on angiogenesis; see the brief review article by Sage (1997) and further references there. The discovery has dramatically affected the cancer research field and has initiated intense experimental interest in vasculature control with a frenetic endeavor to try and discover if it is applicable to human cancers. Amid the media hype, the sanest voice was Folkman's; when asked about the future ability of these chemicals to cure human cancers, he said "If you're a mouse with cancer we can take good care of you."

One of the authors (JDM) had read about Folkman's early work some years ago and initiated research into trying to model the biological processes involved in the network pattern observed in angiogenesis. Even without the cancer component, it is an important and challenging question. In this section we briefly describe the application of the mechanochemical theory (although it is strictly just mechanical) to the patterning process. Although motivated by Folkman's 1970s ideas, this application is directly related to experiments carried out by Sage and Vernon and their co-workers (Vernon *et al.* 1995; Vernon and Sage, 1995). These experiments were also used for comparison between the theory and experiment, and were the basis for most of the parameter estimates which are so essential in any real application to a specific biological problem.

14.4.2 Cell-extracellular matrix interactions for vasculogenesis

The development of in vitro angiogenesis systems is a controlled means of studying vessel formation. Such systems have shown the important mechanical role that the ECM plays in angiogenesis. It provides, among other things, a scaffold necessary for cell migration and morphogenesis (Vernon *et al.*, 1992). Not only can cells produce or degrade their ECM, they can also alter its structure by applying mechanical forces. Through matrix production and degradation, cells can influence the mechanical properties of their ECM and, through their mechanical forces, they can reorganize its fibrous components into matrix lines that the cells use as migratory pathways and movement cues. Such a mechanical scenario, in which cell–ECM interactions can be orchestrated to form complex spatial patterns in development seems to make it a very good candidate for the application of the Murray–Oster mechanical theory. The ability of the cell–matrix interaction to effect an alignment of the matrix, thus influencing cell movement, is quite common in development: some discussion of this is given in Manoussaki *et al.* (1996), Manoussaki (1996) and Murray *et al.* (1998b) but the key reference is the recent book by Little *et al.* (1998) on vascular morphogenesis (appropriately with an introduction by Folkman).

Of particular relevance to this section is the in vitro work on vascular network formation by Vernon, Sage and their coworkers (Vernon *et al.*, 1992). They showed that networks are not specific to endothelial cells since a variety of traction-exerting cells can form networks when cultured on gelled basement membrane (Matrigel). Recent studies by Coger *et al.* (1997) have analyzed this type of aggregation by hepatocyte cells. Moreover, cells cultured on different substrates (Matrigel, type I collagen gels), also form networks provided the matrix is malleable enough. These studies suggest that the mechanism lacks cell or matrix specificity and that a mechanical mechanism could provide a possible explanation for the generality of the network-forming process.

Here we briefly describe a mechanical model mechanism, based on the mechanochemical modelling framework (Manoussaki *et al.*, 1996; Manoussaki, 1996; Murray *et al.*, 1998b), which attempts to capture the key interactions between the mechanical forces generated by the cells and their ECM. We show that a purely mechanical version of the mechanochemical theory could be responsible for the observed pattern and how they are actually formed in development. Such a mechanical model is, in effect, based on very simple mechanical concepts and does not specify the types of cells and matrix involved but rather only considers the possible mechanical interactions between the various components.

14.4.3 Mechanical theory of in vitro vascular network formation

From the experiments of Vernon *et al.* (1992, 1995), the process for all cell lines studied is similar: the cells adhere to the matrix, and the pulling by the cells on the matrix and the resulting cell–ECM interaction eventually form cell aggregates with significant amounts of matrix accumulated underneath the aggregates. Cell traction tension lines appear around the clusters; these lines are reminiscent of lines observed by Harris *et al.* (1980) as a result of traction forces that fibroblasts exert on silicone rubber. In time the matrix components appear to form fibrous lines between neighboring clusters and along the tension lines. Once the lines form, the cells become actively motile and migrate onto the matrix pathways, thus forming cellular cords. Eventually the culture plane is tessellated with polygons with sides defined by the cellular cords (Vernon *et al.*, 1992).

Collagen (an ingredient of tissue matrix) content influences the formation of networks, presumably by altering matrix stiffness and thereby the effect of the cell-exerted traction onto the matrix. Thickness of the matrix layer also influences the size of the networks formed: thin layers of matrix resulted in small or no network formation. On a ramp of increasing matrix thickness, larger cellular networks form on the thicker areas of matrix.

These results suggest that mechanical interactions are of primary importance for the development of pattern. Mathematical descriptions of the interactions, which we quantify below in the mathematical model, confirm the crucial role that mechanical forces play in the network pattern formation and provide a tool for assessing which mechanical properties of the ECM control the patterning process as well as how the parameters may modulate pattern size. It also provides various experimental scenarios to highlight the effect of experimentally variable parameters, such as gel thickness, density of cells and so on. Because the thickness of the experimental gel on which the cells are embedded is very thin, and in particular, relative to the cell–matrix spatial domain, a two-dimensional model is sufficient at this stage.

14.4.4 Mathematical model

The two-dimensional mathematical model quantifies the basic experimental mechanical scenario described above. As before we denote the average local cell density by c (cells/mm^2). We do not include cell proliferation; on the Matrigel cultures the first cell aggregates and lines of tension appear after 4 hours and the networks are complete within about 24 hours. The time between two subsequent cell mitoses is about 17 hours for endothelial cells, so we assume that no significant changes in the cell populations occur which could influence the pattern-forming process. It is easy to include cell and matrix proliferation and such studies are currently under investigation.

Local changes in the cell density are a combination of mainly two movements, the convective flux and an anisotropic strain-dependent random motion tensor. We thus consider that cell movement can be approximated by a random walk that is biased along areas of matrix alignment. The resulting conservation equation for the cell density is

$$\frac{\partial c}{\partial t} = -\nabla \cdot \left(c \frac{\partial \mathbf{u}}{\partial t} \right) + \nabla \cdot \nabla \cdot (D(\boldsymbol{\varepsilon})c) \tag{14.18}$$

Here $D(\boldsymbol{\varepsilon})$ is the random motion tensor, dependent on the matrix strain $\boldsymbol{\varepsilon}$, whose particular form depends on the specific assumption about how cells perform their random movements. Its specific form for small strain was derived under the assumption that the movement bias increases in expansion along one direction and/or compression in the perpendicular direction (Cook, 1995) and is given by

$$D(\varepsilon) = D_0 \begin{pmatrix} 1 + \dfrac{\varepsilon_{xx} - \varepsilon_{yy}}{2} & \dfrac{\varepsilon_{xy} + \varepsilon_{yx}}{2} \\ \dfrac{\varepsilon_{xy} + \varepsilon_{yx}}{2} & 1 - \dfrac{\varepsilon_{xx} - \varepsilon_{yy}}{2} \end{pmatrix} \tag{14.19}$$

Here $\varepsilon_{xx}, \varepsilon_{yy}, \varepsilon_{xy}, \varepsilon_{yx}$ represent the two-dimensional components of the strain tensor ε, and D_0 is the motility coefficient when no strain is present. In the initial analysis (Manoussaki *et al.*, 1996) we did not consider strain-dependent active movement so that we could test whether network formation is possible under cell-exerted traction alone.

Forces within the extracellular matrix. Since the width of the matrix sample in the dish is much larger than the thickness of the matrix, we approximate the matrix as a two-dimensional material. Moreover, we assume that the traction exerted by the cells remains confined in the plane parallel to the dish (or, approximately, the matrix) surface. In other words, we consider the material to be under a *plane stress* assumption. The movement of the matrix is resisted by attachment of the matrix to the dish. In vivo planar angiogenesis is, by comparison, influenced by the potential attachment and continuation of the fibrous and cellular components of the adjacent tissue and is more accurately described by the three-dimensional model equivalent.

The key forces are of course those generated by the cells. The others are generated as a direct consequence. So forces that are present in the tissue are (i) the cell-exerted traction; (ii) the resistance due to the matrix–dish contact; (iii) the viscoelastic forces of the matrix material which are resisting the deformation. Since the time required for generation of the pattern is relatively long and the size of the pattern is small in absolute terms, inertia effects are negligible and the forces at any given point are again considered to be in equilibrium.

In response to a force, the matrix is displaced slowly due to its viscous properties. If strains developed in collagen gels are small (less than 10%) then the gels have been found to respond linearly to applied stresses (Barocas *et al.*, 1995). We therefore describe the matrix response as that of a linear viscoelastic body.

The equation reflecting the balance of these forces is then

$$\underset{\substack{\text{cell–exerted}\\ \text{traction}}}{\mathbf{F}_{\text{cells}}} + \underset{\substack{\text{attachment}\\ \text{on dish}}}{\mathbf{F}_{\text{anchoring}}} + \underset{\substack{\text{viscoelastic}\\ \text{restoring forces}}}{\mathbf{F}_{\text{matrix}}} = 0 \tag{14.20}$$

Again we consider the cell traction forces and the matrix response as arising from the corresponding stresses within the matrix medium. So, $\mathbf{F}_{\text{cells}} = \nabla \cdot \boldsymbol{\sigma}_{\text{cells}}$, with $\boldsymbol{\sigma}_{\text{cells}}$ the stress tensor in the matrix due to the cells and $\mathbf{F}_{\text{matrix}} = \nabla \cdot \boldsymbol{\sigma}_{\text{matrix}}$, with $\boldsymbol{\sigma}_{\text{matrix}}$ the viscoelastic stress tensor in the matrix material. $\mathbf{F}_{\text{anchoring}}$ is an external (body) force resisting the matrix displacement.

The cell traction $\mathbf{F}_{\text{cells}}$ depends on the local cell density as described earlier and it decreases with large enough cell densities. We take the cell-exerted stress on the matrix due to the cells as

$$\boldsymbol{\sigma}_{\text{cells}} = \frac{\tau c}{1+\lambda c^2} \tag{14.21}$$

The parameter τ (dynes/cell) represents the stress that one cell exerts on the matrix at low cell densities and λ is the parameter which measures the force inhibition as a result of increasing cell density. For the densities with which we are concerned, the traction force is approximately given with the cell stress tensor $\boldsymbol{\sigma}_{\text{cells}} \approx \tau c$.

The viscoelastic stresses we again consider to be made up of elastic and viscous parts as in the last section. The difference here is associated with the drag of the matrix across the dish. As the cells pull the matrix, it moves across the dish. Experiments indicate that some matrix fibrils remain attached to the dish, while the rest are dragged across the lower parts of the matrix. The net effect on the matrix surface due to the fibril attachments to the dish, is a resistant force. In this two-dimensional model we treat such resistance as a viscous drag:

$$\mathbf{F}_{\text{anchoring}} = -s\frac{\mathbf{u}_t}{h} \tag{14.22}$$

where $h(x, y, t)$ (mm) represents the thickness of the matrix and s is a measure of the strength of the resistance contributed by the attachment. We thereby assume that the resistance is proportional to the velocity $\mathbf{v} = \mathbf{u}_t$ of matrix displacement.

Matrix thickness. Since the vertical stress $\sigma_{zz} = 0$, we get a relation between the strain components in the three directions $\varepsilon_{xx}, \varepsilon_{yy}, \varepsilon_{zz}$. From Hooke's law in three dimensions we have

$$\sigma_{zz} = \frac{E}{1+\nu}\left[(1-\nu)\varepsilon_{zz} + \nu(\varepsilon_{xx}+\varepsilon_{yy})\right] = 0$$

from which we get

$$\varepsilon_{zz} = \frac{-\nu}{1-\nu}(\varepsilon_{xx}+\varepsilon_{yy}) \tag{14.23}$$

When there is a strain ε_{zz} in the z-direction, the thickness is calculated using $h(x, y, t) = h_0(x, y)(1+\varepsilon_{zz})$ from which we derive the relation giving the thickness h as a function of space and time:

$$\begin{aligned} h(x,y,t) &= h_0(x,y)(1+\varepsilon_{zz}) = h_0\left[1 - \frac{\nu}{1-\nu}(\varepsilon_{xx}+\varepsilon_{yy})\right] \\ \Rightarrow h(x,y,t) &= h_0\left(1-\frac{\nu}{1-\nu}\theta\right) \end{aligned} \tag{14.24}$$

where $\theta = \varepsilon_{xx} + \varepsilon_{yy}$ is the dilation.

Boundary conditions for the model equations. Again, being guided by the experiments, the matrix barely moves along the edge of the dish. We incorporate this into the model by assuming zero displacement as a boundary condition. Hence, for a square dish (square rather than circular simply for ease of numerical programming and reduction of computer time) of dimensions $[ab] \times [ab]$, the condition becomes

$$\mathbf{u}(x, y = a, t) = \mathbf{u}(x, y = b, t) = \mathbf{u}(x = a, y, t) = \mathbf{u}(x = b, y, t) = \mathbf{0} \tag{14.25}$$

where $\mathbf{u}$ is again the matrix displacement. Cells remain in the dish, so we have a zero flux boundary condition for the cells:

$$\mathbf{J}_{\text{cells}} = \frac{\partial \mathbf{u}}{\partial t} c - \nabla \cdot (D(\varepsilon) c) = \mathbf{0} \tag{14.26}$$

The equations for cell conservation and balance of forces together constitute the model mechanism. The equations were solved with the above boundary conditions and initial conditions in cell density and matrix distribution in keeping with the experimental setup.

14.4.5 Parameter values

Because of the close relation between this application and in vitro experiments we were able to derive estimates for various parameters describing cell behavior and matrix properties, for example, matrix stiffness (Young's modulus), the Poisson ratio and the cellular random motility coefficient. Some characterization of the material properties can be achieved by employing appropriate creep and relaxation experiments for the matrix (Fung, 1993). Rheological properties of collagen I gels (Barocas *et al.*, 1995) and tissue-derived gels such as the vitreous body (Lee *et al.*, 1993, 1994a, b) have been described by applying slow-strain techniques (Barocas *et al.*, 1995; Lee *et al.*, 1994a, b). We use such results where they refer to tissues of approximately similar structure to the matrices used in the experiments of Vernon *et al.* (1992).

The *stiffness* of the matrix is represented by the elastic modulus of the material (E dynes/cm^2) and is determined primarily by the type, amount and organization of its fibrous components. For example, the anterior porcine vitreous body and the posterior bovine vitreous body were found to have similar collagen contents and gave elastic moduli of 26.93 dyne/cm^2 and 8.01 dyne/cm^2 respectively (Lee *et al.*, 1994a). Since the collagen gels used in the vascular network-forming experiments have collagen content comparable to that of the vitreous bodies mentioned, we assume that these gels have elastic moduli within a similar range.

The zero *shear viscosity* of 2.1 mg/ml collagen I gel has been measured using creep tests to be 7.4×10^6 poise (Barocas *et al.*, 1995) while that of the posterior bovine vitreous body was estimated to be 2.5×10^2 poise (Lee *et al.*, 1994a).

The *traction* per cell (τ dynes/cell) we consider to be comparable to that of human umbilical vein endothelial cells, which is approximately 6.1×10^4 dynes/cm^2 (Kolodney and Wysolmerski, 1992). From these results we estimate the traction force to be in the region of $\tau \approx 0.15$-0.27 dynes/cell.

The *Poisson ratio* (ν cm/cm) can be theoretically measured by extending a material by a certain amount and measured by its compression in the perpendicular direction. The ratio of relative extension to compression in the perpendicular direction gives the magnitude of the parameter. In practice such measurements for soft materials like collagen gels can be very difficult, not least because the value of ν changes in time.

The Poisson ratio for the fiber network of such gels is estimated at $\nu \approx 0.2$ (Scherer *et al.*, 1991).

The *cell motility* for endothelial cells has been calculated using migration assays (Hoying and Williams, 1996). The values reported for endothelial cells (human microvessel endothelial cells) on gelatin were $9.5 \pm 1.2 \times 10^{-9}$ cm^2/sec in the presence of endothelial cell growth factor and $2.6 \pm 0.6 \times 10^{-9}$ cm^2/sec in its absence and $19.3 \pm 4.22 \times 10^{-9}$ cm^2/sec on fibronectin (Hoying and Williams, 1996).

In our investigation we were concerned with the initial stages of network pattern emergence where the strains are still small and we can assume that the parameters stay approximately constant.

14.4.6 Results from the model analysis

The model mechanism system of equations was analyzed analytically and numerically (Manoussaki *et al.*, 1996). In Figure 14.4 we show a selection of results for the analysis. Analysis showed that small variations in cell density could trigger matrix displacements if the following relation between the model (and real biological) parameters obtains:

$$\frac{\tau}{E} > \frac{1}{1-\nu^2}\left(\frac{1}{c_0}\right)\frac{\left(1+\lambda c_0^2\right)^2}{1-\lambda c_0^2} \tag{14.27}$$

with the parameters as before; c_0 is the initial cell seeding density and λ, the parameter quantifying the decrease in total cell traction upon high density confluence, is chosen so that $\lambda C^2 \approx$ 0.1-0.4 where C is the cell density at confluence. Since the seeding density is an order of magnitude less than C, the last term of the inequality is close to 1, and the inequality takes the approximate form

$$c_0\frac{\tau}{E} > \frac{1}{1-\nu^2} \tag{14.28}$$

Of course c_0, the initial seeding density, must be less than the density at confluence, $c_0 < C$. The last relation suggests a simple rule that *might* make patterns possible: patterns are more likely to form if the cell traction τ is sufficiently large, or if the matrix stiffness E is sufficiently low, or if the initial seeding density is relatively large. The results are in broad agreement with experimental findings: when cell traction was inhibited or matrix stiffness greatly increased, no patterns formed (Vernon *et al.*, 1992). On the other hand, increases in the cell-plating densities facilitates the formation of pattern. If few cells are seeded, they have difficulty in deforming the matrix and starting the patterning process.

The above inequalities predict when pattern could form, but do not predict the type of pattern formed. The pattern described by the equations can be determined by solving the governing equations numerically. Numerical simulations show that indeed formation of networks is possible through such a purely mechanical process as in Figure 14.4.

The numerical simulations clearly showed network formation and subsequent remodelling: some polygons grew larger and others closed in, in a sphincter-like manner. There was a direct correspondence between areas of large cell density and areas of high matrix density, just as we would expect (see the experimental image in Figure 14.5).

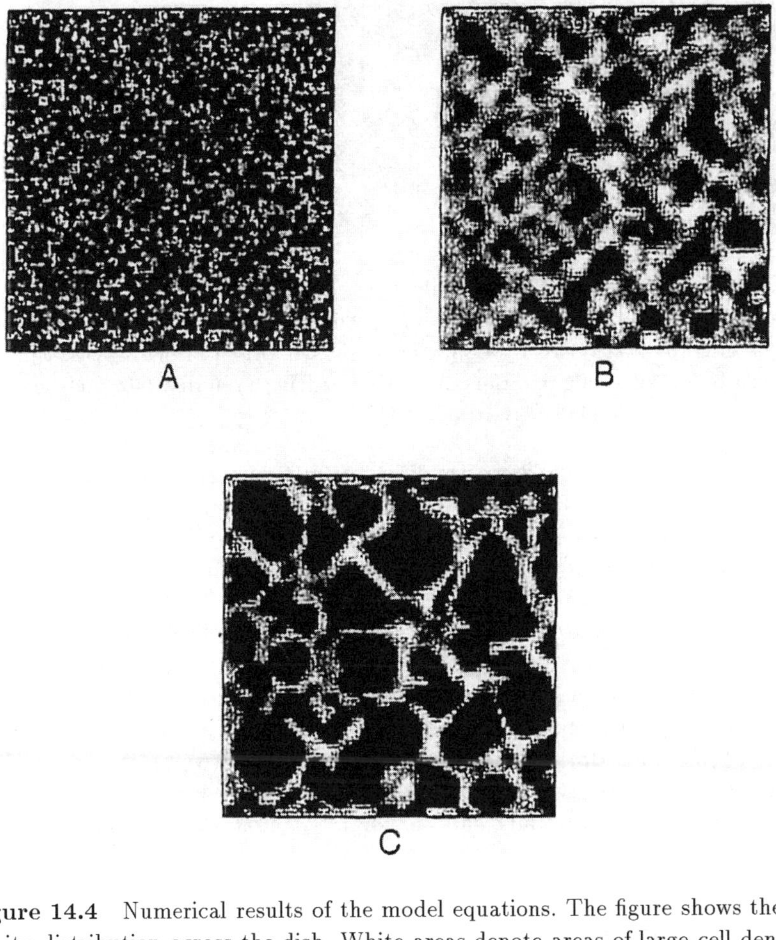

Figure 14.4 Numerical results of the model equations. The figure shows the cell density distribution across the dish. White areas denote areas of large cell densities (up to 6 times the initial seeding densities) and dark areas denote regions of low cell densities. The pattern is shown at (A) 7 h, (B) 20 h and (C) 24 h. The size of one square corresponds to 800μm. The parameter values used in the simulations were $\tau = 0.06$dynes/cell, $E = 20\,$dynes/cm^2, $\mu_1 = 1.5 \times 10^7$poise, $\mu_2 = 0.9 \times 10^7$poise, $D = 0.46 \times 10^{-11}$cm^2/sec, $\nu = 0.2$, $s = 10^{10}$dynes sec/cm^3, $n_0 = 10^5$cells/cm^2, $\lambda = 4 \times 10^{-13}$. The distribution of matrix is similar to that of the cell densities.

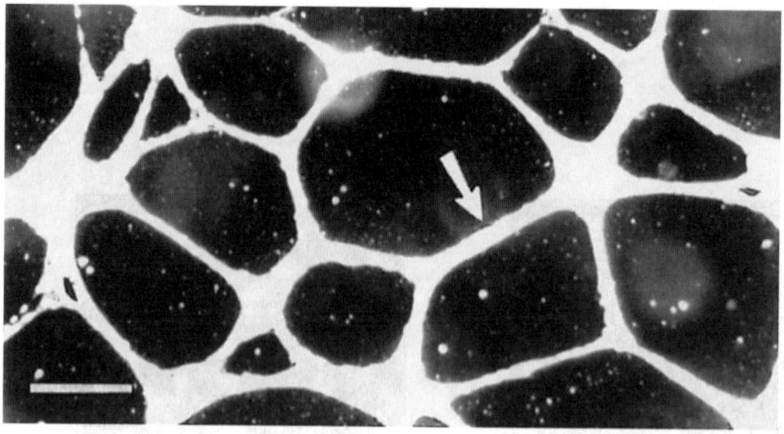

Figure 14.5 Experimental angiogenesis. Bovine aortic endothelial cells, cultured 48 hours in a subconfluent monolayer on top of a layer of Matrigel, organized into a planar network of cords. Each cord (example, arrow) is comprised of many cells. The image is viewed by darkfield illumination. Bar at lower left = 200 μm. (Reproduced with kind permission of Dr C. Little)

The choice of parameters influenced the formation of pattern and the rate at which the network pattern formed. If traction was too low, the plating density too low, or the stiffness too high, no patterns formed. If, on the other hand, stiffness was too low or traction too high, cells and the matrix formed clusters but no matrix cords were observed.

Viscosity of the matrix, as might be expected intuitively, influenced the time it took for the pattern to appear, provided traction and stiffness were within the appropriate range, but it did not determine whether pattern would form or not.

One of the initially surprising and important findings was that random motility of cells was *not* necessary for the formation of pattern. Networks would still form, provided the seeding density was sufficiently large.

Although the model is really only valid for small strains, the simulations are with large strains and for long times.

14.4.7 Results from the model mechanism

Mathematical analysis and simulation of the model mechanism give insight as to which elements are crucial to the development of pattern, help to assess the relative importance of such elements, and they predict how changes in these factors affect pattern growth and shape. In this way, the model is useful in assessing the biological conditions under which a pattern may or may not form.

We have presented a simple, purely mechanical model for network pattern formation in vitro, in which cellular traction forces play a crucial role in forming the planar pattern of cellular aggregates, cords and matrix deformations observed in vitro. In spite of the simplifying assumptions in the model, the results show that in the absence of traction, as in the in vitro system, no pattern forms, supporting our hypothesis that the pattern forms as a result of cellular traction forces. The evolution of the pattern from the model proceeds in the same manner as the in vitro pattern, with irregular polygons both increasing in size and decreasing in number, with the smallest polygons pinching off and vanishing.

As in the in vitro system, patterns form in our system only when the ratio of traction forces and seeding cell density to gel stiffness is above a certain threshold. Our model results confirm that larger polygons will form on thicker gel. It also indicates that isotropic cellular traction is sufficient for pattern formation. We further found that if cellular traction is sufficiently high or the plating densities are sufficiently large, then traction forces alone can lead to formation of networks. Biased cellular migration may be a component of pattern formation in vivo and in vitro, but we have shown it is not a *necessary* feature.

We also found that matrix thickness is an important factor influencing the pattern. Cells spread on a matrix whose thickness increased from one end of the dish to the other, formed polygonal networks whose size increased with increasing thickness of the gel. The numerical simulation results are in agreement with the experimental results of Vernon *et al.* (1992).

The model mechanism is the first mathematical description of cell–matrix interactions for the formation of vascular networks in which all of the component variables are measurable. The theory can be extended to account for the effect of cell-secreted molecules on the matrix material properties, cell proliferation and cell secretion, thus providing an analytical tool for assessing the relevant importance of these factors during angiogenesis. It can also be used to study the effect on the pattern if there is spatial variability in the parameters. We believe that the success of the current model in mimicking experimental observations gives some reason for optimism that it can be used as a guide to predict the outcome of other experiments and, optimally, the consequences of experimentally disturbing the mechanism which obtains in vivo.

14.5 Discussion

Mechanochemical theory has been proven to be important in the study of wound healing, angiogenesis and limb development, to name just a few. One of the significant features of the applicability of this theoretical framework to a wide variety of problems is that the definable parameters are in principle all measurable experimentally. Although a great deal of experimental work has been devoted to the molecular biology of development, few studies have focused on mechanical aspects. With the recent experimental advances and interest in the study of mechanical forces in the development of form, it seems to us that now is an ideal time to work with mechanochemical modelling to study the generation of pattern and form.

Although certain morphogenetic events can be described by chemical and/or mechanical mechanisms, the vast majority of pattern formation events are the result of the interaction of chemical *and* mechanical processeses. For this reason, mechanochemical theory is a broadly applicable, powerful tool in defining mechanisms for numerous developmental events.

Some general concluding comments. Why use mathematics to study something as intrinsically complicated and ill-understood as development, angiogenesis, wound healing and so on? We suggest that mathematics, rather mathematical modelling, must be used if we ever hope to genuinely and realistically convert an understanding of the underlying mechanisms into a predictive science. Mathematics is required to bridge the gap between the level on which most of our knowledge is accumulating (cellular and below) and the macroscopic level of the patterns we see. In wound healing, scar formation itself is a major concern of the surgeon and the patient. A mathematical approach lets us explore the logic of pattern formation. Even if the mechanisms were well understood – and they certainly are not at this stage – mathematics would be required to explore the consequences of manipulating the various parameters associated with any particular scenario. In the case of wound healing, and it will be increasingly so in angiogenesis with the related cancer component, the number of options that are fast becoming available to wound and cancer managers will become overwhelming unless we can find a way to simulate particular treatment protocols before applying them in practice. The latter has already been of use in understanding the efficacy of various treatment scenarios with brain tumours (glioblastomas) (Tracqui *et al.*, 1995a; Cruywagen *et al.*, 1995; Woodward *et al.*, 1996) and new two-step treatment regimes for skin cancer (Jackson *et al.*, 1999).

There is no doubt that we are a long way from being able to reliably simulate actual developmental scenarios, wound healing and angiogenesis. Not only is the active cellular control of the key processes poorly understood, in the case of wounds, for example, they are difficult to reproduce, with similar wounds on different parts of the same body healing at different rates and with different results. Despite these limitations, we argue that exploring the logic of biological processes is worthwhile even in our present state of knowledge. It allows one to take a hypothetical mechanism and examine its consequences in the form of a mathematical model, making predictions and suggesting experiments that would verify or invalidate the model; even the latter casts light on the biology. Indeed, the very process of constructing a mathematical model can be useful in its own right. Not only must one commit to a particular mechanism, but one is also forced to consider what is truly essential to the process, the key players (variables) and the key processes by which they evolve. We are thus involved in constructing frameworks on which we can hang our understanding. The equations, the mathematical analysis and the numerical simulations that follow serve to reveal quantitatively as well as qualitatively the consequences of that logical structure.

In the case of wound healing, none of the individual models we have mentioned, and not even all of them put together, could be considered a complete model of dermal wound healing, even if we think only in terms of wound contraction and scar formation. However, each model has shed light on different aspects of the process and we can now say what must be the most important elements of a complete model. The problem of angiogenesis in the laboratory setting is considerably less complex, although not in vivo, and there is more hope that we shall get a fuller explanation of the processes involved there.

These studies have served to highlight where our knowledge is deficient and to suggest directions in which fruitful experimentation might lead us. A crucial aspect of this research is the interdisciplinary content. A critical test of all these theoretical constructs is in their impact on the experimental community. The field has now achieved some level of maturity and we believe that future dialogue between experimentalists and applied mathematicians will lead us most rapidly towards a fuller understanding, if not a complete one, of the biological processes involved in pattern formation. In the case of wound healing, it could be the goal of scarless wound healing and in angiogenesis, its prevention and elimination in developing tumours.

Acknowledgments
This work was supported in part by grants DMS-9500766 from the US National Science Foundation (JDM), 2P41-RR-01243-17 from the US National Institutes of Health (JDM), and BIR-9256532 from the US National Science Foundation (KRS).

REFERENCES

Boehm, T., Folkman, J., Browder, T. and O'Reilly, M. (1997) Antiangiogenesis therapy of experimental cancer does not induce acquired drug resistance. *Nature* 390: 404–407.

Barocas, V.H. and Tranquillo, R.T. (1994) Biphasic theory and *in vitro* assays of cell-fibril mechanical interactions in tissue-equivalent gels. In: *Cell Mechanics and Cellular Engineering*, V.C. Mow, F. Guilak, R. Tran-Son-Tay and R. M Hochmuth, eds. Springer, New York.

Barocas, V.H. and Tranquillo, R.T. (1995) An anisotropic theory of tissue-equivalent mechanics: the interplay among cell traction, fibril network deformation, and contact guidance. Unpublished manuscript.

Barocas, V.H. and Tranquillo, R.T. (1997) An anisotropic biphasic theory of tissue-equivalent mechanics: the interplay among cell traction, fibrillar network deformation, fibril alignment, and cell contact guidance. *J. Biomech. Eng. Trans. ASME* 119: 137–145.

Barocas, V.H., Moon, A.G. and Tranquillo, R.T. (1995) The fibroblast-populated collagen microsphere assay of cell traction force – Part 2: Measurement of the cell traction parameter. *J. Biomech. Eng.* 117: 161–170.

Ben-Jacob, E., Cohen, E., Shochet, I., Aranson, I., Levine, H. and Tsimring, L. (1995) Complex bacterial patterns. *Nature* 373: 566–567.

Budrene, E.O. and Berg, H.C. (1991) Complex patterns formed by motile cells of *Escherichia coli. Nature* 349: 630–633.

Budrene, E.O. and Berg, H.C. (1995) Dynamics of formation of symmetrical patterns by chemotactic bacteria. *Nature* 376: 49–53.

Clark, R.A.F. (1991) *Cutaneous wound repair* In: *Physiology, Biochemistry, and Molecular Biology of the Skin* ed L.A. Goldsmith,

Coger, R., Toner, M., Moghe, P., Ezzell, R.M. and Yarmush, M.L. (1997) Hepatocyte aggregation and reorganization of EHS matrix gel. *Tissue Eng.* 3: 375–390.

Cook, J. (1995) Mathematical models for dermal wound healing: wound contraction and scar formation. PhD thesis, University of Washington.

Cook, J. and Murray, J.D. (1995) Mechanisms of biological pattern formation. In: *Handbook of Brain Theory and Neural Networks.* pp. 705–710, M.A. Arbib, ed. MIT Press, Cambridge

Cook, J., Tracqui, P., Woodward, D.E., Cruywagen, G.C. and Murray, J.D. (1994) A mechanical model for fibroblast-driven wound healing. In: *Proceedings of the 2nd Congress of the European Society for Mathematical and Theoretical Biology*

Cowin, S.C. (1992) An evolutionary Wolff's law for trabecular architecture. *J. Biomed. Eng.* 114: 129–136.

Cruywagen, G.C. and Murray, J.D. (1993) A new tissue interaction model for epidermal-dermal spatial patterns. In: *Proceedings of the 1st European Conference on the Applications of Mathematics to Medicine and Biology* pp. 423–428, J. Demongeot and V. Capasso, eds. Wuerz Publishing, Winnipeg.

Cruywagen, G.C., Maini, P.K. and Murray, J.D. (1993) Sequential and synchronous skin pattern formation. In: *Experimental and Theoretical Advances in Biological Pattern Formation.* pp. 61–64, H.G. Othmer, P.K. Maini and J.D. Murray, eds. Plenum Press,

New York.

Cruywagen, G.C., Maini, P.K. and Murray, J.D. (1994) Threshold bifurcation in tissue interaction models for spatial pattern generation. *Phil. Trans. Roy. Soc. Lond.* A347: 661–676.

Cruywagen, G.C., Woodward, D.E., Tracqui, P., Bartoo, G.T., Murray, J.D. and Alvord, Jr., E.C. (1995) The modelling of diffusive tumours. *J. Biol. Systems* 3: 937–945.

Cruywagen, G.C., Maini, P.K. and Murray, J.D. (1997) Biological pattern formation on two-dimensional spatial domains: a nonlinear bifurcation analysis. *SIAM J. Appl. Math.* 57: 1485–1509.

Dale, P.D., Sherratt, J.A. and Maini, P.K. (1995) Corneal epithelial wound healing. *J. Biol. Systems* 3: 957–965.

Dale, P.D., Sherratt, J.A. and Maini, P.K. (1996) A mathematical model for collagen fibre formation during foetal and adult dermal wound healing. *Proc. Roy. Soc. Lond.* B263: 653–660.

Dale, P.D., Sherratt, J.A. and Maini, P.K. (1997) Role of fibroblast migration in collagen fiber formation during fetal and adult dermal wound healing. *Bull. Math. Biol.* 59: 1077–1100.

Edelstein-Keshet, L. and Ermentrout, G.B. (1990) Models for contact mediated pattern formation – cells that form parallel arrays. *J. Math. Biol.* 29: 33–58.

Ferrenq, I., Tracqui, L., Vailhe, B., Gumery P.Y. and Tracqui, P. (1997) Modelling biological gel contraction by cells: Mechanocellular formulation and cell traction force quantification. *Acta Biotheor.* 45: 267–293.

Folkman, J. (1971) Anti-angiogenesis: new concept for therapy of solid tumors. *Ann. Surg.* 75: 409–416.

Folkman, J. (1972) Tumor angiogenesis: therapeutic implications. *New Eng. J. Med.* 285: 1182–1186.

Folkman, J. (1974) Tumor angiogenesis factor. *Cancer Res.* 34: 2109–2113.

Folkman, J. (1995) Angiogenesis in cancer, vascular, rheumatoid and other disease. *Nature Medicine* 1: 27–31.

Folkman, J. (1995) Seminars in medicine of the Beth Israel Hospital, Boston: Clinical applications of research on angiogenesis. *New Eng. J. Med.* 333: 1757–1763.

Fung, Y.C. (1993) *Biomechanics: Mechanical Properties of Living Tissues* 2nd edn. Springer-Verlag, New York.

Gaffney, E.A., Maini, P.K., Sherratt, J.A. and Tuft, S. (1999) The mathematical modelling of cell kinetics in corneal epithelial wound healing. *J. Theor. Biol.* 197: 15–40.

Harris, A.K., Ward, P. and Stopak, D. (1980) Silicone rubber substrate: a new wrinkle in the study of cell locomotion. *Science* 208: 177–179.

Hoying, J.B. and Williams, S.K. (1996) Measurement of endothelial cell migration using an improved linear migration assay. *Microcirculation* 3: 167–174.

Jackson, T., Lubkin, S.R., Siemers, N.O., Senter, P.D. and Murray, J.D. (1999) Mathematical and experimental analysis of localization of anti-tumor antibody-enzyme conjugates. *Br. J. Cancer* (to appear)

Keller, E.F. and Segel, L.A. (1971) Travelling bands of chemotactic bacteria: a theoretical analysis. *J. Theor. Biol.* 30: 235–248.

Kolodney, M.S. and Wysolmerski, R.B. (1992) Isometric contraction of fibroblasts and endothelial cells in tissue culture. *J. Cell Biol.* 117: 73–82.

Landau, L. and Lifschitz, E. (1970) *Theory of Elasticity*, Pergamon Press, Oxford.

Lee, B., Mitchell, L. and Buchsbaum, G. (1993) Rheology of the vitreous body. Part 1: Viscoelasticity of human vitreous. *Biorheology* 29: 521–533.

Lee, B., Mitchell, L. and Buchsbaum, G. (1994a) Rheology of the vitreous body. Part 2: Viscoelasticity of bovine and porcine vitreous. *Biorheology* 31: 327–338.

Lee, B., Mitchell, L. and Buchsbaum, G. (1994b) Rheology of the vitreous body. Part 3: Concentrations of electrolytes, collagen and hyaluronic acid. *Biorheology* 31: 339–351.

Little, C., Mironov, V. and Sage, E.H. (eds.) (1998) *Vascular Morphogenesis in vivo, in vitro,*

in mente Birkhauser, Boston.
McGrath, M.H. and Simon, R.H. (1983) Wound geometry and kinetics of wound contraction. *Plastic and Reconstructive Surg.* 69: 975–983.
Manoussaki, D. (1996) Modelling the formation of vascular networks *in vitro*. PhD thesis, University of Washington.
Manoussaki, D., Lubkin, S.R., Vernon, R.B. and Murray, J.D. (1996) A mechanical model for the formation of vascular networks. *Acta Biotheor.* 44: 271–282.
Meinhardt, H. (1982) *Models of Biological Pattern Formation*. Academic Press, London.
Meinhardt, H. (1998) *The Algorithmic Beauty of Sea Shells*. Springer, Heidelberg.
Moore, S.W., Keller, R.E. and Koehl, M.A.R. (1995) The dorsal involuting marginal zone stiffens anisotropically during its convergent extension in the gastrula of *Xenopus laevis*. *Development* 121: 3131–3140.
Murray, J.D. (1981) A pre-pattern formation mechanism for animal coat markings. *J. Theor. Biol.* 88: 161–199.
Murray, J.D. (1993a) *Mathematical Biology*, 2nd edn. Springer-Verlag, Heidelberg.
Murray, J.D. (1993b) General models of pattern formation: some uses, problems and successes. In: *Experimental and Theoretical Advances in Biological Pattern Formation* pp. 1–9. H.G. Othmer, P.K. Maini and J.D. Murray, eds. Plenum Press, New York.
Murray, J.D. and Maini, P.K. (1986) A new approach to the generation of pattern and form in embryology. *Sci. Prog.* 70: 539–553.
Murray, J.D. and Oster, G.F. (1984a) Cell traction models for generating pattern and form in morphogenesis. *J. Math. Biol.* 19: 265–279.
Murray, J.D. and Oster, G.F. (1984b) Generation of biological pattern and form. *IMA J. Math. Appl. Med. Biol.* 1: 51–75.
Murray, J.D., Oster, G.F. and Harris, A.K. (1983) A mechanical model for mesenchymal morphogenesis. *J. Math. Biol.* 17: 125–129.
Murray, J.D., Maini, P.K. and Tranquillo, R.T. (1988) Mechanochemical models for generating biological pattern and form in development. *Phys. Rep.* 171: 59–84.
Murray, J.D., Cruywagen, G.C. and Maini, P.K. (1994) Bifurcating spatial patterns arising from traveling waves in a tissue interaction model. *Appl. Math. Lett.* 7: 63–66.
Murray, J.D., Cook, J., Tyson, R. and Lubkin, S.R. (1998a) Spatial pattern formation in biology: I Dermal wound healing. II Bacterial patterns. *J. Franklin Inst.* 335B: 303–332.
Murray, J.D., Manoussaki, D., Lubkin, S.R. and Vernon, R.B. (1998b) A mechanical theory of *in vitro* vascular network formation. In: *Vascular Morphogenesis in vivo, in vitro, in mente*, C. Little, V. Mironov and E. Helene Sage, eds. Birkhauser, Boston.
Olsen, L., Sherratt, J.A. and Maini, P.K. (1995) A mechanochemical model for adult dermal wound contraction and the permanence of the contracted tissue displacement profile. *J. Theor. Biol.* 177: 113–128.
Olsen, L., Sherratt, J.A. and Maini, P.K. (1996) A mathematical model for fibro-proliferative wound healing disorders. *Bull. Math. Biol.* 58: 787–808.
Olsen, L., Sherratt, J.A., Maini, P.K. and Arnold, F. (1997) A mathematical model for the capillary endothelial cell- extracellular matrix interactions in wound-healing angiogenesis. *IMA J. Math. Appl. Med. Biol.* 14: 261–281.
Olsen, L., Maini, P.K. and Sherratt, J.A. (1997) A mechanochemical model for normal and abnormal dermal wound repair. *Nonlinear Analysis* 30: 3333–3338.
Olsen, L., Maini, P.K. and Sherratt, J.A. (1998) Spatially varying equilibria of mechanical models: Application to dermal wound contraction. *Math. Biosci.* 147: 113–129.
O'Reilly, M.S., Holmgren, L., Chen C. and Folkman, J. (1996) Angiostatin induces and sustains dormancy of human primary tumors in mice. *Nature Medicine* 2: 689–692.
O'Reilly, M., Boehm, T., Shing, Y., Fukai, N., Vasios, G., Lane, W.S., Flynn, E., Birkhead, J.R., Olsen, B.R. and Folkman, J. (1997) Endostatin: An endogenous inhibitor of angiogenesis and tumor growth. *Cell* 88: 277–285.
Oster, G.F., Murray, J.D. and Harris, A.K. (1983) Mechanical aspects of mesenchymal

morphogenesis. *J. Embryol. Exp. Morph.* 78: 83–125.

Othmer, H.G., Maini, P.K. and Murray, J.D. (eds.) (1993) *Experimental and Theoretical Advances in Biological Pattern Formation.* Plenum Press, New York.

Rodriguez, E.K., Hoger, A. and McCulloch, A.D. (1994) Stress dependent finite growth in soft elastic tissues. *J. Biomech.* 27: 455–467.

Scherer, G.W., Hdach H. and Phalippou, J. (1991) Thermal expansion of gels: a novel method for measuring permeability. *J. Non-Crystal. Solids* 130: 157–170.

Sherratt, J.A., (1993) Actin aggregation and embryonic epidermal wound healing. *J. Math. Biol.* 31: 1713–1730.

Sherratt, J.A. and Lewis, J. (1993) Stress-induced alignment of actin filaments and the mechanicsof cytogel. *J. Theor. Biol.* 55: 637–654.

Sherratt, J.A. and Murray, J.D. (1990) Models of epidermal wound healing. *Proc. Roy. Soc. Lond.* B241: 29–36.

Sherratt, J.A. and Murray, J.D. (1992) Epidermal wound healing: the implications of a simple mathematical model. *Cell Transplantation* 1: 365–371.

Sherratt, J.A., Martin, P., Murray, J.D. and Lewis, J. (1992) Mathematical models of wound healing in embryonic and adult epidermis. *IMA J. Math. Appl. Med. Biol.* 9: 177–196.

Sherratt, J.A., Sage, E.H. and Murray, J.D. (1993) Chemical control of eukaryotic cell movement: a new model. *J. Theor. Biol.* 162: 23–40.

Sage, E.H. (1997) Pieces of eight: bioactive fragments of extracellular proteins as regulators of angiogenesis. *Cell Biol.* 7: 182–186.

Timmenga, E.J.F., Andreassen, T.T., Houthoff, H.J. and Klopper, P.J. (1991) The effect of mechanical stress on healing skin wounds. *Br. J. Plastic Surg.* 44: 514–519.

Tracqui, P., Cruywagen, G.C., Woodward, D.E., Bartoo, G.T., Murray, J.D. and Alvord, E.C. Jr. (1995) A mathematical model of glioma growth: the effect of chemotherapy on spatial-temporal growth. *Cell Proliferation* 28: 17–31.

Tracqui, P., Woodward, D.E., Cruywagen, G.C., Cook, J. and Murray, J.D. (1995) A mechanical model for fibroblast-driven wound healing. *J. Biol. Systems* 3: 1075–1085.

Tranquillo, R.T. and Murray, J.D. (1992) Continuum model of fibroblast-driven wound contraction: inflammation-mediation. *J. Theor. Biol.* 158: 135–172.

Tranquillo, R.T. and Murray, J.D. (1993) Mechanistic model of wound contraction. *J. Surg. Res.* 55: 233–247.

Tranquillo, R.T. and Barocas, V.H. (1999) A continuum model for the role of fibroblast contact guidance in wound contraction. (in press)

Turing, A.M. (1952) The chemical basis of morphogenesis. *Phil. Trans. Roy. Soc. Lond.* B237: 37–72.

Tyson, R., Lubkin, S.R. and Murray, J.D. (1999) A minimal mechanism of bacterial pattern formation. *Proc. Roy. Soc. Lond.* B266: 299-304.

Vernon, R.B. and Sage, E.H. (1995) Between molecules and morphology. Extracellular matrix and creation of vascular form. *Am. J. Pathol.* 147: 873–883.

Vernon, R.B., Angello, J.C., Iruela-Arispe, M.L., Lane, T.F. and Sage, E.H. (1992) Reorganisation of basement membrane matrices by cellular traction promotes the formation of cellular networks *in vitro. Lab. Invest.* 66: 536–547.

Vernon, R.B., Lara, S.L., Drake C.J., Iruela-Arispe, M.L., Angello, J.C., Little, C.D., Wright T.N. and Sage, E.H. (1995) Organised type I collagen influences endothelial patterns during 'spontaneous angiogenesis *in vitro*': Planar cultures of models of vascular development. *In Vitro Vasc. Dev. Biol.* 31: 120–131.

Woodward, D.E., Tyson, R., Myerscough, M.R., Murray, J.D., Budrene, E.O. and Berg, H.C. (1995) Spatio-temporal patterns generated by *Salmonella typhimurium. Biophys. J.* 68: 2181–2189.

Woodward, D.E., Cook, J., Tracqui, P., Cruywagen, G.C., Murray, J.D. and Alvord, E.C. Jr. (1996) A mathematical model of glioma growth: the effect of extent of surgical resection. *Cell Proliferation* 29: 269–288.

Wolpert, L. (1969) Positional information and the spatial pattern of cellular differentiation. *J. Theor. Biol.* 25: 1–47.

Wolpert, L. (1971) Positional information and pattern formation. *Current Topics in Developmental Biology* 6: 183–224.

Wolpert, L. (1981) Positional information and pattern formation. *Phil. Trans. Roy. Soc. Lond.* B295: 441–450.

Zhu, M. and Murray, J.D. (1995) Parameter domains for spots and stripes in mechanical models for biological pattern formation. *J. Nonlinear Sci.* 5: 317–336.

15

Statistics and Dynamics of Cellular Shape Changes

Wolfgang Alt

Department of Theoretical Biology, University of Bonn, Kirschallee 1, D-53115, Bonn, Germany

15.1 Introduction

That "description of form" does not only consist of *morphometrics* and *statistics* but also deals with *causality* and *dynamics* was already formulated by D'Arcy Thompson in the introduction to his book *On Growth and Form*: "when we deal with the *subjective conceptions of form*, or movement, or the movement that change of form implies, then *force* is the appropriate term for our conception of the *causes* by which these forms and changes of form are brought about." Thus, in trying to give a survey on the topic statistics and dynamics of cellular shape changes, perhaps I should take up the general topic of the preceeding few chapters – *On the Dynamics of Form* – and combine it with the descriptive level of statistics, very much in the spirit of D'Arcy Thompson. Indeed, in the same introduction, he proclaimed his general plan of a *dynamic morphology* by providing "physico-mathematical descriptions" of biological shapes and shape changes, and he exemplified it by refering mostly to *cells* as standard examples: cell growth, cell division or cell movement. Following him, let me start the survey in the following section with a series of characteristic examples of cellular shape changes:

- Cell growth: tip growth of hyphae
- Cell modulation: deformations of red blood cells
- Cell Division: cleavage of a sand dollar egg
- Cell movement: amoeboid motion

Then, in the third section, to make life simpler and the science clearer, I will restrict myself to the so-called *descriptors of form* in 2D (i.e. two spatial dimensions) only:

1. Geometric shape factors
2. Invariant 2D moments
3. Fourier and Kahunen–Loève expansions

On Growth and Form: Spatio-temporal Pattern Formation in Biology.
Editor Mark A.J. Chaplain, G.D. Singh, J.C. McLachlan

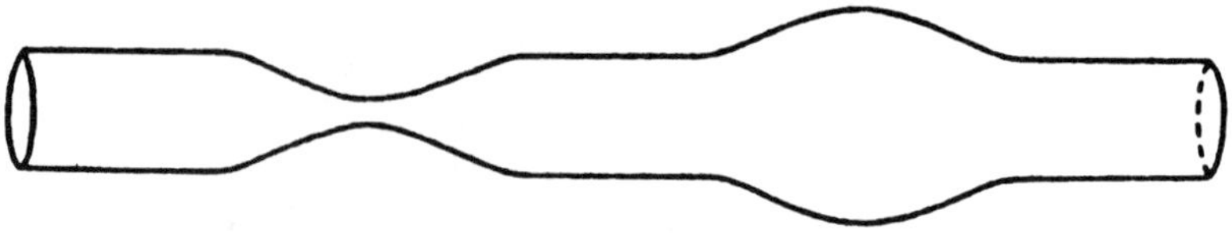

Figure 15.1 Sketch of a heated glass tube as a minimal surface with surface tension raised in region (a) and lowered in region (b). From Thompson (1961, Figure 4)

4. Angular–temporal autocorrelations
5. Protrusion–retraction dynamics

Finally, in the discussion section I'll try to address current and future tasks, including the need for auto- and cross-correlation statistics that relate local dynamic processes and global "morphogenetic" coordination, as well as the search for adequate form descriptors in 3D.

15.2 Examples of Cellular Shape Changes

15.2.1 Tip growth of hyphae

As a first example of cellular shape changes that occur on a very slow timescale, one can investigate the tips of growing fungal hyphae, e.g. of *Streptomyces coelicolor* (Prosser, 1990, Figure 3). The whole cell shape, hemispherical at the apex and unduloidal in the posterior parts of the hypha, can be geometrically described and dynamically understood as a so-called Plateau's surface of revolution. This class of surfaces, which had been widely studied by D'Arcy Thompson in Chapter III "The Forms of Cells" (Thompson, 1961, Figure 9), are characterized as surfaces of minimal area enclosing a fixed volume, and satisfy a pressure balance equation according to Laplace's formula already stated by D'Arcy Thompson:

$$P = p_{ext} + T(\frac{1}{R} + \frac{1}{R'}). \tag{15.1}$$

The last term is the product of surface tension T, and mean curvature equilibrating the pressure difference $P - p_{ext}$ across the surface, which usually vanishes. Thompson himself refers to the analogue example of a heated glass tube (Figure 15.1): here the region under increased (exterior) tension shows an indentation (a), corresponding to a reduced mean curvature mainly due to the appearance of a negative longitudinal curvature, whereas the region under compression bulges out (b), by increasing the mean curvature.

Indeed, this general principle of shape generation due to local modulation of surface tension still seems to be applicable to the case of growing tips in plant or fungal cells. At the apex, new material is continuously incorporated from the cell interior into the outer cell wall, locally softening the surface, so the experimentally measured incorporation rate can be used to estimate the surface tension and predict the nearly hemispherical shape of the apex (Prosser, 1990).

15.2.2 Deformations of red blood cells

The second example reveals a similar principle of shape modulation, namely a pathological change in the shape of red blood cells, the case of elliptocytosis. Whereas normal cells have a soft surface leading to discoidal (biconcave) forms, in this case they are altered into more ellipsoidal forms (Engelhardt and Sackmann, 1988). The more ellipsoidal the cells, the stronger their overall *elastic shear modulus*, a dynamic quantity that can be estimated by exposing the cells to a high frequency electric field and comparing their elongation with corresponding values for a simple model of elastic shells.

This (model-based) description of form relies on the biochemical and biophysical nature of the red cell cortex, which consists of an elastic network of long spectrin filaments interconnected by short actin polymers. Indeed, what happens in patients showing elliptocytosis is a genetically induced modification of the cell cortex; because of the lack of a certain glycoprotein, the spectrin–actin polymer network seems to become more cross-linked, thus increasing the elastic shear modulus and inducing a stronger surface tension.

D'Arcy Thompson was aware of differences between intracellular dynamics determining the shape of cells and fluid dynamics of liquid droplets as a comparable physical analogue, when he claimed that "so long as our purpose is *morphological*, so long as what we seek to explain is regularity and definiteness of form, it matters little if we should happen, here or there, to confuse surface-tension with elasticity, the contractile forces manifested at a liquid surface with those which come into play at the complex internal surfaces of an elastic solid" (Thompson, 1961, p. 87). Nowadays one would not follow him in applying the simple Laplace formula of equation (15.1), in every case, it does happen that simple energy minimization principles, used to characterize two-layered surfaces, do produce similar types of shapes and shape changes, and even a larger variety of forms, including cell budding (Svetina and Žekš, 1991, Figure 3) and cell division (Svetina *et al.*, 1997, Figure 2).

15.2.3 Cleavage of a sand dollar egg

On the other hand, most biological cells are filled with cytoplasm, a highly viscous fluid consisting of various polymers, particularly of flexible actin filaments (F-actin) which, by cross-linking with myosin polymers, make the fluid contractile. If, for example, a sand dollar egg is punctured by a needle, the cytoplasmic fluid is extruded but instantaneously forms a definite rounded bleb (Akkaş, 1990, Figures 4 and 5) which is surrounded by plasma membrane and later contracted. However, even in cells deprived of the plasma membrane surface, the protruded cytoplasm is also retracted, though in a more irregular fashion, so the shape dynamics of these cells resemble the form of solar protuberances (Taylor *et al.*, 1973).

Given this dynamical background, one might better understand the very regular shape changes occuring during the ubiquitous process of cell division. It starts with the appearance of a cleavage furrow around the cell surface, precisely placed in the equatorial plane between the two spindle asters, thus forming the so-called constriction ring. However, if one side of the cleavage furrow is damaged, then it will locally relax while the furrow on the other side will continue to locally indent the cell (Akkaş,

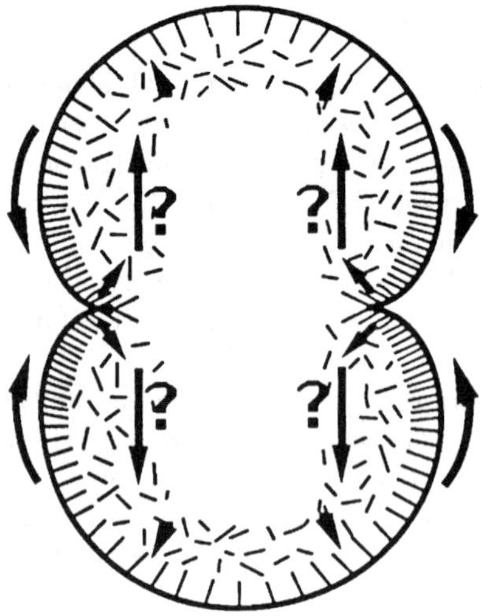

Figure 15.2 A dynamical model of cortical F-actin flow towards the cleavage furrow; see text for details. From Cao and Wang (1990, Figure 6)

1990, Figure 9). This means that surface indentation, this important process of shape change during cell division, is mainly due to the local dynamics of the underlying F-actin cortex, as is well-established nowadays. Within a series of competing model descriptions, the most attractive hypothesis is that cortical flow of F-actin beneath the plasma membrane pulls the cell surface towards the cleavage furrow, where the cytoplasm is actively contracting (Figure 15.2). Within the cell, dissolved actin monomers could then be redistributed by endoplasmic streaming. Indeed, computer simulations based on a corresponding two-phase flow model (Dembo, 1989a) show shape changes that very precisely reproduce the experimentally observed ones (He and Dembo, 1997a; He and Dembo, 1997b, Figure 7).

Although this model-based description of cell division shapes seems to be quite "near to nature" and may become a well-accepted solution of this "century problem", there have been other model descriptions, some of which were wrong in two aspects, e.g. the physical analogue "protocell model" in Schwegler *et al.*, 1985). These models did not reproduce the correct shapes, and they were based on unrealistic assumptions about the dynamic mechanisms that induce the shape changes. Let me take the opportunity to cite D'Arcy Thompson once again, who critically commented on some then fashionable electromagnetic force field models by W.R. Coe, 1899, e.g. shown on the poster by Stuart Pivar during the Dundee conference. Thompson said that though these karyokinetic figures "resemble known physical phenomena, their nature is still the subject of much dubiety and discussion, and neither the *forms produced* nor the *forces at work* can yet be *satisfactorily and simply explained*".

15.2.4 Amoeboid motion

In contrast to this long history of describing cell division, the early descriptions of amoeboid motion were already quite precise and, judged from today's knowlege,

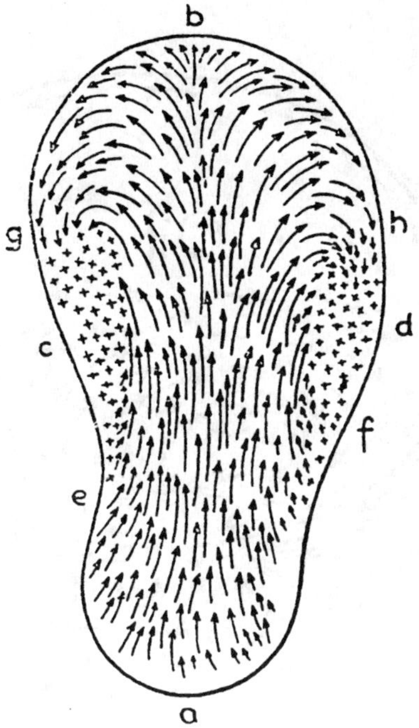

Figure 15.3 Shape and protoplasma current observed in the amoeba *Pelomyxa palustris* by F.E. Schulze in 1875. From de Bruyn (1947, Figure 1)

essentially correct. So, when in 1875 F.E. Schulze depicted the typical shape of a moving amoeba (Figure 15.3), he carefully noticed the fountain streaming of cytoplasm in the front (b), bending to the lateral sides (g,h) and becoming stationary there (c,d). Moreover, he explained this shape-generating dynamic process by supposing contractile forces in the posterior part (e,a,f) and a continuous transfer of contracted material into the central streaming mass, which is then cyclically reincorporated into the lateral parts (compare the similar process in Figure 15.2). Though this dynamic theory was later picked up and reformulated (from 1926 onwards) as posterior gel–sol and anterior sol–gel transformation between G-actin and F-actin, even recently serving as the base for the two-phase flow model descriptions of amoeboid motion (Dembo, 1989b, 1994), there was a time in between when surface tension theories dominated (de Bruyn, 1947). Indeed, at the end of the nineteenth century and during the years when D'Arcy Thompson developed his theoretical views, physical theories of liquid drops and colloid emulsions had a great influence on the biological community. Many (proto)zoologists such as Otto Bütschli with his soap–oil analogue models (Figure 15.4), adhered to the idea that tension gradients along the "cell surface" induce a streaming of the interior as well as the exterior fluid, thereby causing surface protrusion and the typical shape changes of amoeboid motion (Figure 15.5).

Comparing these shape changes with those observed in real amoeboid cell motion (Figure 15.6) shows striking similarities at first. However, a closer look at the dynamical fate of lateral pseudopods, which are continuously retracted towards the posterior part of the cell, reveals eminent differences that are much better explained

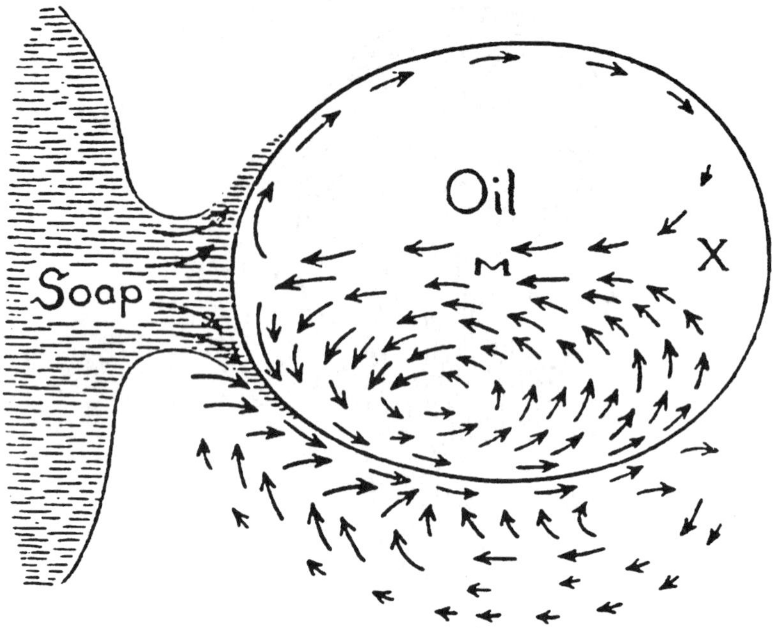

Figure 15.4 Currents in a drop of oil in water and in contact with a soap solution, depicted by Bütschli (1892). From de Bruyn (1947, Figure 3)

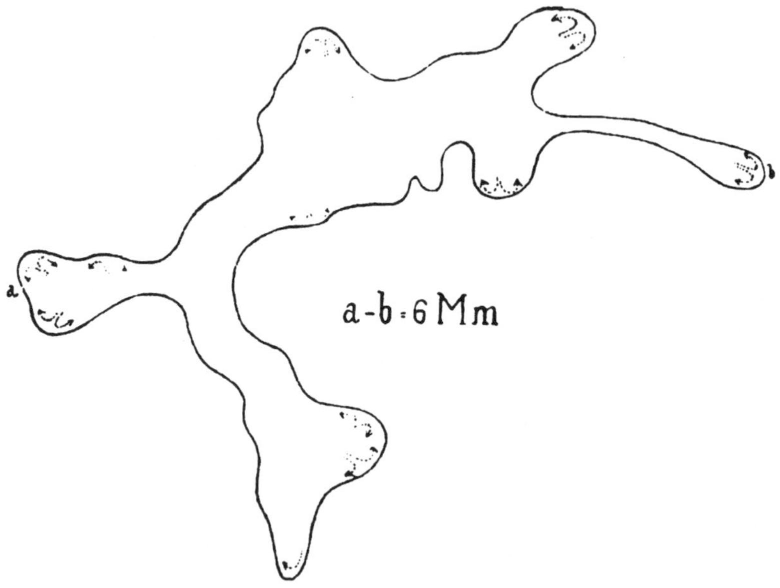

Figure 15.5 Amoeboid motion of soap–oil emulsion drops in experiments by Otto Bütschli. From Bütschli (1892, Figure 6)

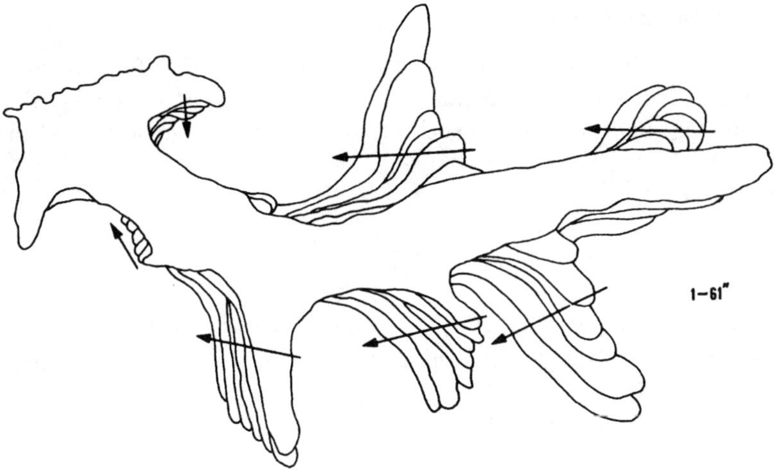

Figure 15.6 Superimposed cell contours of a polytactic amoeba on a water/fluorinert liquid interface at 10-second intervals. From Grebecki (1995, Figure 15)

by the aforementioned contraction theory of the cortical cytoplasm (see above).

These highly dynamic, though relatively slow (in a time scale of minutes), shape changes of the creeping-like amoeboid motion are a ubiquitous phenomenon not only for unicellular organisms such as *Dictyostelium* slime moulds, but also for many tissue and blood cells as fibroblasts, keratinocytes or leukocytes. It seems that typical features of surface protrusions and retractions are very similar in various kinds of cells, suggesting common dynamic mechanisms as candidates for producing these shape changes.

Therefore, during the last twenty years, extensive studies have been performed in order to capture quantitatively the spatial and temporal characteristics of these shape changes, mainly from two-dimensional projections obtained in videomicroscope pictures (Killich *et al.*, 1993, Figure 2), or more recently from three-dimensional reconstructions of optical sections (Soll and Voss 1998, Figures 2.9 and 2.10). Thus, the next part of this chapter considers quantitative descriptors of cell form and its temporal changes.

15.3 Descriptors of Form in 2D

For clarity and simplicity, let me restrict the geometry to two dimensions. I will start with the simplest geometric shape factors, then proceed to the advanced characterization of form by so-called moments. All these are shape descriptors for fixed stationary cells as they appear in snapshot pictures. However, modern videomicroscopy provides longer and mostly regular time series of "movie pictures" which require statistical descriptors of spatio-temporal shape changes. The most important of them will be mentioned and explained using typical examples.

15.3.1 Geometric shape factors

Given a connected two-dimensional domain, one might characterize its shape by simple attributes as elongated, eccentric, rounded, of smooth or wierd boundary. Here are several proposals for quantifications of such attributes which have been used in older literature.

Geometric elongation

$$\frac{L}{W} \geq 1 \qquad \text{or} \qquad 0 \leq (1 - \frac{W}{L}) \leq 1 \, , \tag{15.2}$$

where length L and width W might be determined as the longer and shorter side extension of the best-fit rectangle containing the domain (having minimal area); see Noble and Levine (1986).

Eccentricity (1/circularity): Malinowska factor

$$\frac{P^2}{4\pi A} \geq 1 \qquad \text{or} \qquad (\frac{P}{2\sqrt{\pi A}} - 1) \geq 0 \, , \tag{15.3}$$

with perimeter P and area A satisfying the isoperimetric inequality above. The minimal eccentricity of 1 is attained only if the domain is a circle.

Boundary complexity (standard deviation): Haralick factor

$$\frac{\sqrt{\frac{1}{P} \int_{\mathbf{P}} (r - \bar{r})^2}}{\bar{r}} \geq 0 \, , \tag{15.4}$$

where the so-called polar map r measures the distance from the centroid taken over the boundary $\mathbf{P}$, $\bar{r}$ denoting its mean. This factor just gives the relative standard deviation of boundary distances from the mean; the smoother the boundary, the smaller its value.

Root mean square elongation: (Inverse) Blair and Bliss factor

$$\frac{\sqrt{2\pi \int_{\mathbf{A}} r^2}}{A} \geq 1 \, , \tag{15.5}$$

where the distance r from the centroid is taken over the whole body area $\mathbf{A}$. Again, exactly as for a circular area, we have the minimal value 1 for elongation.

In order to check the value of these geometric factors as shape descriptors, Lewandowska *et al.* have compared the last three factors for a variety of different cell shapes (Lewandowska *et al.*, 1979, Figure 5). In some cases the descriptors were very characteristic, in others not at all. For example, a symmetrically dividing cell and a cell showing an asymmetric protrusion could have the same elongation but also a similar eccentricity, with only their boundary complexity differing. Therefore, these shape descriptors alone do not seem to provide a sufficiently clear and systematic tool for distinguishing relevant cell shapes, as is provided by the method of moment invariants (Section 15.3.2).

An elementary geometric approach has been developed to systematically characterize planar shapes that deviate from a simple circular disc, namely the *skeleton* or *symmetry axis*. This simply defines the centre line of interior points that have maximal distance to the boundary (Blum, 1973) and can be obtained by starting a

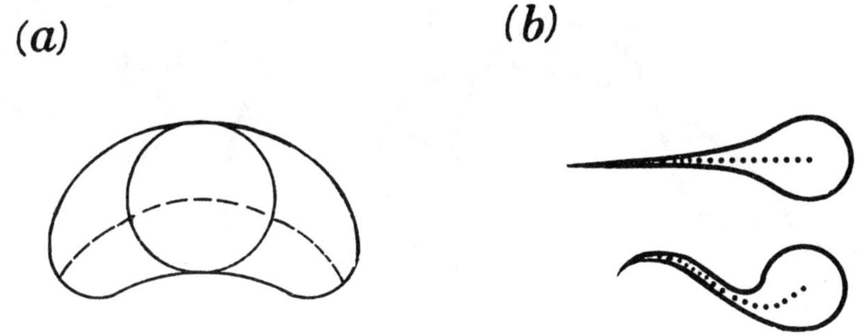

Figure 15.7 Schematic diagram of the skeleton or symmetry axis (a) for a bean-shaped cell (the inscribed circle around a point on the skeleton depicts the one with maximal diameter); (b) for a sperm-like cell, straight and flexed. After Blum (1973, Figure 18)

wave of "grass fire" propagation (or of any other excitable chemical reaction) from the cell boundary, propagating with uniform speed and halting at points where two wavefronts collide: these collision points are the skeleton points. For instance, an elongated cell with one lateral protrusion would generally show a main skeleton line from which a side line bifurcates.

Using the skeleton construction one can give much more appropriate definitions of cell length and cell width. The *cell length* is the maximal length of connected lines in the skeleton graph, and the *cell width* is the maximal diameter of inscribed circles, twice the maximal boundary distance of skeleton points (Figure 15.7a).

The shape of the skeletal boundary can be reconstructed from the skeleton graph and the boundary distance function along it. This is a surprising and powerful property. Thus, any performed flexion of the skeleton keeping the same boundary distance function yields a similar, deformed shape, leaving the following four geometric quantities invariant: area, perimeter, length and width.

However, there are some drawbacks of this geometric descriptor tool: the skeleton as a geometric object is not robust with regard to small perturbations and, related to this, algorithms for numerical computations are sensitive and time-consuming (Noble, 1990).

15.3.2 Invariant 2D moments

A completely systematic and analytic approach to characterize the form of a two-dimensional domain Z uses the so-called physical moments m_{lk} for $l, k \geq 0$, i.e. the mean of the power function $x^l y^k$ over the domain. In particular, the *centroid*, mentioned earlier, is the vector (m_{10}, m_{01}). Then, by shifting the origin of coordinates into the centroid, one defines the following centralized moments of order $l + k \geq 0$:

$$\mu_{lk} := \int_Z (x^l - m_{10})(y^k - m_{01})\, dxdy \quad / \quad \int_Z dxdy\,. \tag{15.6}$$

It is interesting to notice that for a domain Z whose boundary is given as a polygon (the usual situation for digitized cell outlines gained from image processing), the moments

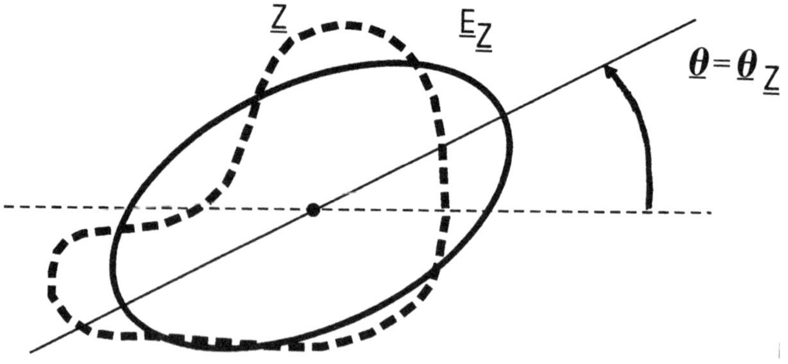

Figure 15.8 Schematic diagram of the momental ellipse E_Z for a boomerang-shaped domain Z. For more details see text.

can be written as explicit sums using only the coordinates of the corners of the polygon (Dunn and Brown, 1990). Of special use is the matrix of second-order moments

$$\begin{pmatrix} \mu_{20} & \mu_{11} \\ \mu_{11} & \mu_{02} \end{pmatrix}$$

which, provided $\mu_{20}\mu_{02} - \mu_{11}^2 > 0$, determines the so-called momental ellipse

$$E_Z := \{(x,y)^T \mid \mu_{02}x^2 - \mu_{11}xy + \mu_{20}y^2 = 4\mu_{20}\mu_{02} - 4\mu_{11}^2\}\,. \tag{15.7}$$

The momental ellipse is the unique ellipse possessing the same centroid and second-order moments as the original domain Z (but not the same area, see below). The angle of its main axis with respect to the x-axis, θ_Z, is the unique angle θ that minimizes the so-called polar curve

$$\mu_{02}(\theta) := \mu_{02}\cos^2\theta - 2\mu_{11}\cos\theta\sin\theta + \mu_{20}\sin^2\theta \tag{15.8}$$

or, equivalently, maximizes

$$\mu_{20}(\theta) := \mu_{02}\sin^2\theta + 2\mu_{11}\sin\theta\cos\theta + \mu_{20}\cos^2\theta\,. \tag{15.9}$$

Moreover, the largest and the smallest radii of the momental ellipse are

$$R_Z^+ = 2\sqrt{\mu_{20}(\theta_Z)}\,, \tag{15.10}$$
$$R_Z^- = 2\sqrt{\mu_{02}(\theta_Z)}\,. \tag{15.11}$$

Since all these quantities can be readily computed from the given momental ellipse, and since the two radii are invariants under translation and rotation, the following shape factors are invariants too, and they can easily be evaluated (Dunn and Brown, 1990):

$$\text{Elongation} := log_2\frac{R_Z^+}{R_Z^-} \geq 0\,, \tag{15.12}$$

$$\text{Dispersion} := log_2\frac{\pi R_Z^+ R_Z^-}{A_Z} \geq 0\,. \tag{15.13}$$

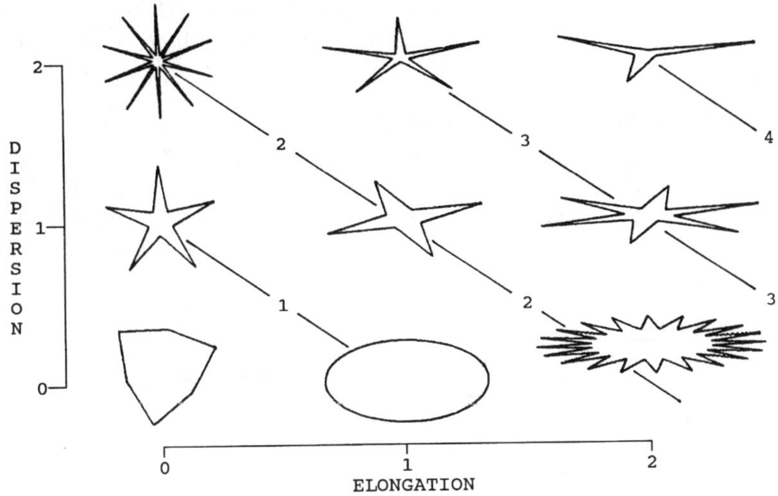

Figure 15.9 Elongation–dispersion diagram: shapes which have the same extension are connected by lines. From Dunn and Brown (1990, Figure 3)

Similar to the geometric shape factors defined in Section 15.3.1, we have the property that both descriptors take on their minimal value 0 if and only if the domain Z has circular shape.

In Dunn and Brown (1990) a third invariant descriptor has been derived from the two others by simply adding them together:

$$\begin{aligned} \text{Extension} &:= \text{Elongation} + \text{Dispersion} \\ &= log_2 \frac{\pi(R_Z^+)^2}{A_Z}\,. \end{aligned} \tag{15.14}$$

The examples in Figure 15.9 display different types of two-dimensional shapes in such a way that shapes having the same extension are connected by lines; obviously this concept coincides with an intuitive classification. Indeed, the two shape factors might be used to discriminate mouse fibroblast cells treated by some chemical (phorbolacetate) from control cells (Dunn and Brown 1990, Figure 5): Whereas *elongation* is not changed by the treatment, *dispersion* is significantly increased, giving rise to the conclusion that these shape changes do not affect the length/width ratio of cells but the type of membrane protrusions (cells with higher dispersion show more irregular boundaries).

In principle a systematic description of form could proceed by successively studying invariant moments of third and higher order. However, this has not been achieved so far, and no applicable list of higher-order shape factors is available. There are only plots of the third-order polar curve $\mu_{30}(\theta)$, which is defined in analogy to equation (15.9). Successfully depicted by Dunn and Brown (1990), these polar curves have the property of triple symmetry for roughly triangular cells, but have extremely large values in those angular directions where the cells have extensive protrusions.

Though this third-order geometric shape descriptor indicates some qualitative features of cell protrusion, the most restrictive drawback is that it (like the other shape factors mentioned so far) does not consider the temporal aspect of cellular shape changes. This important aspect is taken up during the following sections.

15.3.3 Fourier and Kahunen–Loève expansions

Angular Fourier analysis

Given a time series of cell outlines, for instance represented by a time-dependent polar distance function $r = z(\varphi, t)$ of cell boundary points from the centroid, the classical angular Fourier expansion

$$z(\varphi,t) = \frac{1}{2}a_0(t) + \sum_{k=1}^{\infty} a_k(t)\sin(k\varphi) + b_k(t)\cos(k\varphi) \tag{15.15}$$

$$= \frac{1}{2}a_0(t) + \sum_{k=1}^{\infty} c_k(t)\cos(k\varphi - \psi_k(t)) \tag{15.16}$$

provides a decomposition of cell shape into symmetric and periodic k-modal components, the harmonics, whose amplitudes $c_k(t)$ and whose angular positions $\psi_k(t)$ on the periphery, can then be investigated with respect to their temporal evolution. This has been performed for physico-chemical analogues of "amoeboid cell motion" as pulsations of the so-called mercury beating heart (Avnir, 1989) or of evaporating liquified gas drops (Takaki *et al.*, 1989) on certain surfaces. In particular, quite frequently one observes stationary pulsating triodes (with wave mode $k = 3$) or pulsating pentodes (with wave mode $k = 5$) or temporal transitions between them.

This classical Fourier approach is limited to a description of moderate changes in the cell outline (compare its physical application only within the validity of linear elastic deformation theories). Therefore, a more geometric variant of this approach has been developed in the so-called elliptic Fourier analysis by Diaz *et al.* (1990). This might plot a basal ellipse as the first harmonic and then higher-order harmonic cycloids along it. Thus, corresponding Fourier spectra are able to reveal characteristic periodic protrusion patterns around the cell periphery. Specific applications have been made to characterize the "spine-like" structures of *Echinocytes* or *Chondrocytes* (Diaz *et al.*, 1990, Figure 3).

Moreover, for the purpose of pattern recognition or reproduction, a totally different approach has been invented and used – the Fourier expansion of an alternative boundary parameterization, namely the *deviation between angular direction and arc length of boundary points*

$$\Delta\psi(s) := \psi(s) - \frac{2\pi s}{L} \quad 0 \le s \le L\,. \tag{15.17}$$

Using this method, the shapes of number digits are already well reproduced using only the first ten harmonics (Lord and Wilson, 1984, Figure 228).

Lestrel (1997) has recently given an almost complete review of these and other Fourier descriptors plus a valuable survey of many biological applications.

Kahunen–Loève expansion

Another, more adaptive method to analyse a time series of polar distance functions $\{z(\varphi,t), 0 \leq t \leq T\}$ is to consider, for each directional angle φ, the mean extension $D(\varphi) := < z(\varphi,t) >_t$ averaged over time and then construct, for the corresponding temporal deviations

$$d(\varphi,t) := z(\varphi,t) - D(\varphi) \tag{15.18}$$

and suitable angular discretizations, φ_i, $(i = 1,\ldots,N)$, the quadratic covariance matrix

$$R = (r_{ij}) \quad := \quad < d(\varphi_i,t) \cdot d(\varphi_j,t) >_t \tag{15.19}$$

$$= \quad \frac{1}{M} \sum_{m=1}^{M} d(\varphi_i,t_m) \cdot d(\varphi_j,t_m), \tag{15.20}$$

where the second representation is valid for experimentally given discrete time series. R is the symmetric matrix of averaged covariances $(i \neq j)$ and variances $(i = j)$ over time for the polar distance function in various directions. Then by standard principal component analysis one obtains an ordered series of N eigenvectors of R

$$\mathbf{s_k}(\varphi) = (s_k(\varphi_1), \ldots, s_k(\varphi_N)), \tag{15.21}$$

such that for each $k = 1, \ldots, N$ the kth Kahunen–Loève coefficient

$$c_k(t) \quad := \quad \mathbf{d}(t) \cdot \mathbf{s_k}(\varphi) \tag{15.22}$$

$$= \quad \sum_{i=1}^{N} d(\varphi_i,t) \cdot s_k(\varphi_i,t) \tag{15.23}$$

quantifies the contribution of the kth component in the polar deviation function $d(\varphi,t)$ at any time t. Thus $c_1(t)$ describes the temporal development of how much the main component $s_1(\varphi)$ is represented in $d(\varphi,t)$; the same holds for the successively lower components. Suppose the main component were strictly positive, $s_1(\varphi) > 0$, then $c_1(t)$ would eventually reflect a temporal variation in total cell area; if, in addition, the second component $s_2(\varphi)$ were a bimodal periodic function (like $\cos 2\varphi$) and $c_2(t) > 0$, then this would indicate a standing bimodal extension of the cell outline, whereas an oscillating coefficient $c_2(t)$ would indicate a pulsating diode, as an overlay on top of the cell area change mentioned before.

The pulsating diode pattern has been extracted from observations of *Dictyostelium* cells (Killich *et al.*, 1993, Figure 4). The authors also describe the appearance of a "circulating diode" and its reconstruction by a linear superposition of the first two main components, both of them being bimodal in φ but phase-shifted by $\pi/2$ (Vicker and Wei, 1997, Figure 3). They also suggest that observed intracellular calcium concentration waves having a similar rotation speed might cause the periodic shape changes; see Section 15.3.4.

This shape analysis has the ability to describe angular extension patterns that deviate from classical harmonics. It is able to describe repetitive protrusions localized in a narrow region of the cell periphery, e.g. in a leading lamellipod. However, the Kahunen–Loève method also has some disadvantages. Since it performs a correlation analysis by comparing fixed angular directions, in general it does not easily reflect rotations of localized extension patterns. Moreover, since the mean peripheral extension function $D(\varphi)$ depends on the particular time-series sample, hence on each individual cell, a standardized comparison of shape changes between different types of cells is difficult to establish.

These tasks are somewhat better performed via the following method, which provides a correlation analysis by comparing peripheral protrusions not only at different angular positions but also at different times (Brosteanu *et al.*, 1997).

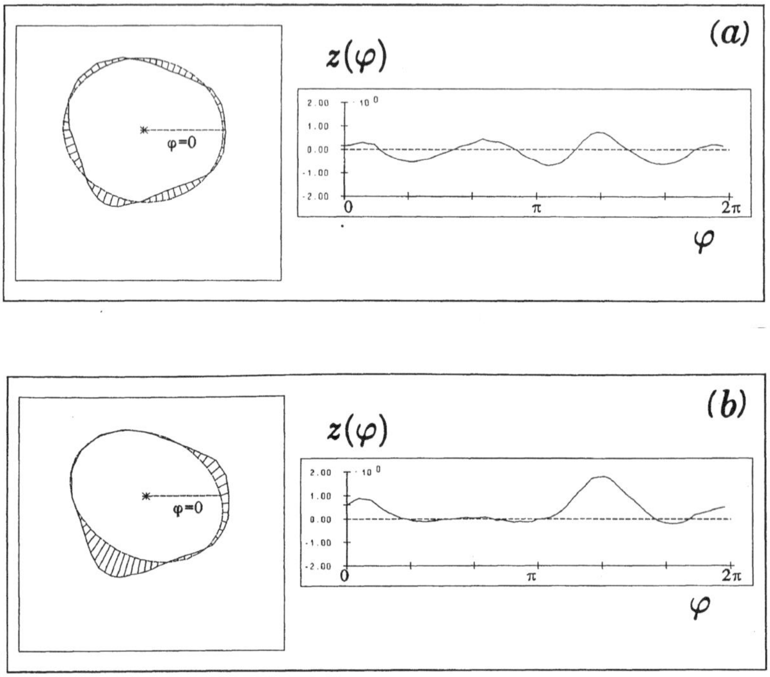

Figure 15.10 Pictures of (a) the momental ellipse and (b) a smaller "weighed" momental ellipse for the cell image of a leukocyte, together with plots of the corresponding normal distance functions $z(\varphi)$ between boundary points and the ellipse. From Brosteanu (1994)

15.3.4 Angular–temporal autocorrelations

Supposing a stationary circular time series $z(\varphi, t)$ of an "extension" function, then for the deviations

$$d(\varphi, t) := z(\varphi, t) - D\,, \tag{15.24}$$

from the total mean $D := < z(\varphi, t) >_{\varphi,t}$ averaged over angle and time, one defines the autocorrelation function

$$C(\alpha, \tau) \quad := \quad < d(\varphi, t) \cdot d(\varphi + \alpha, t + \tau) >_{\varphi,t} \tag{15.25}$$

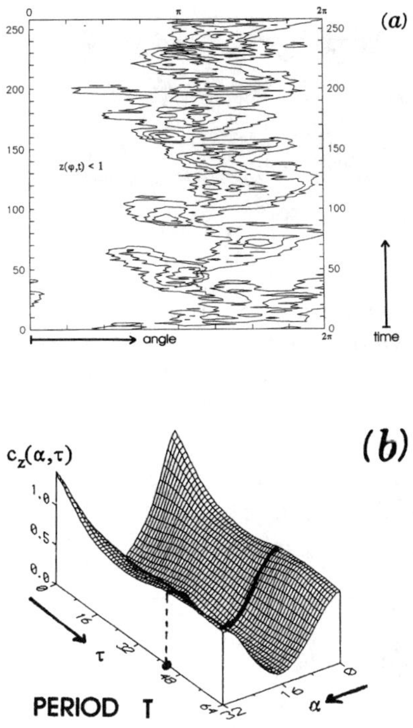

Figure 15.11 (a) Topographic level-line plot of the extension function for a videomicroscope image series of a leukocyte, showing waves of protrusions. (b) Corresponding angular–temporal autocorrelation function with the estimated period T indicated. From Brosteanu (1994)

$$= \frac{1}{N}\frac{1}{M}\sum_{i=1}^{N}\sum_{m=1}^{M} d(\varphi_i, t_m) \cdot d(\varphi_i + \alpha, t_m + \tau)\,, \qquad (15.26)$$

where again the second representation is valid for a discrete version of the time series. $C(\alpha, \tau)$ measures the averaged correlation between extensions z at angular positions separated by a correlation angle α, and at times separated by a correlation time τ. This autocorrelation function is periodic in α and, for $\tau = 0$, it describes the purely angular autocorrelations of extensions, whereas for $\alpha = 0$ the purely temporal autocorrelations.

When this autocorrelation analysis was simply applied to the polar distance function $z(\varphi, t)$ which has been used up until now (Section 15.3.3), it turned out, e.g. for recorded cell outlines of leukocytes, that the results were quite difficult to interpret, because the extension function z is a superposition of global cell shape deformations and local lamellipodial protrusions (Brosteanu ,1994). Therefore, the obvious idea was to obtain a first approximation of cell shape, at any time instant t, by computing the momental ellipse $E_{Z(t)}$, (15.7), and then plotting a new elliptic extension function $z_E(\varphi, t)$ defined as the normal distance to the momental ellipse (Figure 15.10a). However, an even better representation of peripheral protrusions by z_E was achieved when the momental ellipse was adjusted (by computing so-called weighed moments) to mainly describe the slowly changing shape of the "cell body" (Figure 15.10b).

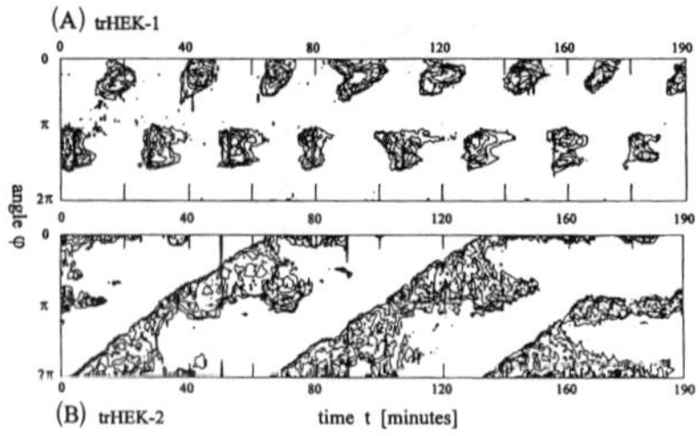

Figure 15.12 Topographic level-line plot (as in Figure 15.11a) for cell outlines of transfected human keratinocytes: (A) pulsating cell deformations, (B) lamellar protrusion rotating around the cell periphery with changing width. From Hinz and Brosteanu (1997)

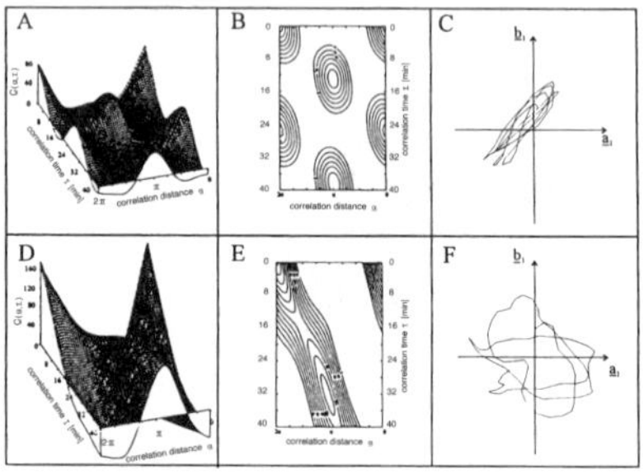

Figure 15.13 Angular–temporal analysis of the two extension time series in Figure 15.12. (A, D): Corresponding autocorrelation functions; (B, E): their topographic line plots; (C, F): time diagrams of the first two Fourier coefficients $c_1(t)$, $c_2(t)$. From Hinz and Brosteanu (1997)

Performing the angular–temporal autocorrelation analysis for the adjusted "weighed" elliptic extension function z_E leads to a proper quantitative characterization of protrusion patterns on the cell periphery of leukocytes (Brosteanu, 1994) and keratinocytes (Alt *et al.*, 1995). For example, a wave of localized protrusions in the leading lamella of leukocytes (Figure 15.11a) induces a typical autocorrelation pattern (Figure 15.11b), from which the angular width as well as the temporal period of protrusion can be estimated.

Human epidermal keratinocytes show similar protrusionpatterns, e.g. pulsating antipodal extensions or rotating travelling waves (Figure 15.12), for which the angular–temporal autocorrelation function (Figure 15.13) is able to extract quantities as period and speed, and eventually detect periodic modulations or asymmetries in the wave patterns; see the figures and comments in Hinz and Brosteanu (1997).

Still in debate are possible reasons for the appearance of such regular protrusion dynamics. We have already mentioned theories about intracellular reaction-diffusion waves of some messenger substance that would stimulate processes near the plasma membrane (e.g. polymerization) and entrain the described membrane protrusion waves. However, alternative explanations based on fluid mechanical models (Alt, 1996) or viscoelasticity models use force and stress balances in the cell cortex in addition to local assembly–disassembly kinetics for the contractile filament system. In particular, a simplified one-dimensional differential equation model for concentration and tangential velocity of the cortical filament system succeeded in reproducing the observed pulsating and rotating wave patterns (Alt and Tranquillo, 1995, 1997). In any case, decisions about the preference of one model over another depend on further progress not only in biochemical and molecular genetic research, but also on detailed observations and statistical image analyses of the spatio-temporal protrusion–retraction dynamics. This is the theme of the last section.

15.3.5 Protrusion–retraction dynamics

One well-established method for measuring cellular shape changes from a time series of two-dimensional cell outlines, is to superimpose two consecutive pictures Z_t and Z_{t+dt}, then compute the area of the difference set $(Z_{t+dt} \setminus Z_t)$ as protrusion area and of $(Z_t \setminus Z_{t+dt})$ as retraction area. Also the peripheral location of protrusion and retraction zones have been represented by vectors from the cell centroid to the centroid of the corresponding zones (Dunn *et al.*, 1997, Figure 1). Then statistical cross-correlation analysis can be performed in order to reveal a dependence between protrusions and retractions. Indeed, for chick heart fibroblasts, Dunn and coworkers (Dunn *et al.*, 1997) found out that the plot of the protrusion area lags behind the retraction area with an average of 1min. Moreover, the autocorrelation function for protrusion area decays much slower than for retraction area. From these results the authors conclude that, on average, retraction is the faster step and is dominant over protrusion.

An even more detailed analysis was performed in Dunn and Zicha (1995). During situations where the cumulative area of cells increased, this was mainly induced by a transient reduction in the amount of retracted area. Similarly, when the cumulative area decreased, the protrusion area would typically remain constant, whereas the retraction area would transiently enlarge. Again, this indicates that retraction might play the leading role in determining the time course of cellular shape changes.

Analogous methods to quantify protrusions and retractions in reconstructed three-dimensional images of *Dictyostelium* cells, for instance, have been started during recent years (Soll and Voss, 1998). However, an appropriate statistical correlation analysis has not yet been achieved (see Section 15.4).

Another attempt to understand the details of shape changes at the periphery of cells that migrate on a two-dimensional substratum, has been developed in Alt *et al.* (1995) and Hinz *et al.* (1999). The illumination profile of phase contrast video images is recorded along a section line transversal to the cell border, from which the position of the cell edge (tip of lamellipodial protrusion) and the motion of dark structures (filament-enriched membrane ruffles) can be extracted. It turns out that not only the cell edge performs quite regular protrusions and retractions with a period of the order of 2min, but that the ruffles are periodically retracted toward the cell body in phase with the retraction of the lamellipodium and with uniform speed of about $1-2\mu$m/min (Hinz and Brosteanu, 1997, Figure 1). These findings suggest that the dynamics of cell surface retraction is mainly determined by the action of the contractile actin–myosin filament systems, whereas the dynamics of protrusion could be a combined effect of several actions: hydrostactic or osmotic pressure (eventually acting against the contractile forces) could disrupt connections between intracellular filaments and proteins in the plasma membrane; this pressure and/or filament polymerization behind the membrane could push the cell boundary outwards. Some of these hypothetical mechanism of the protrusion–retraction dynamics have been implemented in a one-dimensional mathematical model, whose simulations show qualitative agreement with the observed spatio-temporal patterns of cell surface dynamics. For more details see Alt and Dembo (1998).

15.4 Discussion: Current and Future Tasks

This survey of two-dimensional shape descriptors and adequate spatio-temporal statistics might have shown how different aspects within the general task of a "Description of Form" lead to different methodological approaches and quantitative characterizations of shape and shape changes.

The most elementary aspect is to define global shape properties' of a given domain, circumscribed by the cell outline, leading to a quantification by various shape factors. One such task was to guarantee that different factors are complementary to each other (Section 15.3.2), this aspect being most clearly realized in the expansion methods providing a complete (infinite) set of shape factors (Section 15.3.3). A further task has been identified, namely to define descriptors that are invariant under certain transformations of space (Sections 15.3.1 and 15.3.2). It would be desirable to study the influence of symmetries on shape factors more systematically. What should be investigated too are potential relations between various mathematical expressions, e.g. between the higher-order moments (Section 15.3.2) and the higher-order Fourier coefficients (Section 15.3.3).

The next important aspect is to register temporal shape changes. Here the approach was to plot the values of instantaneous shape factors over time, and describe phenomena like sudden or slow transitions between different forms or periodicities in certain shape properties. More intrinsic approaches, still to be developed further, try to record the changes of shape directly by comparing two or more successive images, perhaps by measuring the protrusion and retraction zones (Section 15.3.5) or by several variations of autocorrelation analysis (Section 15.3.4). In many cases these methods yield distinguishable characterizations of shape changes, of their temporal course and their spatial or angular extension, which are applicable to discriminate different types of cells and their motile behaviour.

However, the necessary process of averaging over time and space often prohibits a more detailed description of local events as, for example, transient protrusions of small lamellipods or of so-called blebs. Therefore, an important future task for a dynamic morphology would be to describe and distinguish relations between local processes of shape deformations and their global coordination over the cell body or cell periphery. This would require more intricate statistics able to extract information from a wider range of temporal and spatial scales.

For example, local and fast dynamics of lamellipods should be recorded and related to more global and slow shape changes of the cell border; potential methods might be linear (but also non-linear) cross-correlation analyses as well as suitable local shape parameterizations over temporally varying "skeletons" or "polygons" representing the global cell shape.

Moreover, remembering D'Arcy Thompson's view of a dynamic morphology having the task to reveal the causes which induce shapes to change, further statistical methods should be inspired by and based on appropriate dynamic models of the possible mechanisms for shape deformation and, also, for shape maintenance.

These models should, on one hand, include the local molecular kinetics and energetics of membrane proteins, filament polymerization, their attachment and cross-linking as well as their local chemical regulation by enzymes and by diffusible second messengers. On the other hand, hypothetical candidates for global regulation and coordination of these local processes should be considered as there are the electric membrane potential, ion currents, pressure gradients, tension and stress fields.

Finally, most of what has been performed so far, envisaged or required for descriptors of two-dimensional shape changes, suitable for the analysis of cells adhering and spreading on flat substrata, is only in the early developmental stage for corresponding 3D descriptors of cellular shape changes; only the very first and elementary steps have been undertaken to define analogous shape factors or generalized extension–retraction statistics, e.g. for reconstructed 3D pictures of cells migrating on (or floating within) a homogeneous medium.

Even more difficult, but no less important, would be a study of the statistics and dynamics of single cells moving in a heterogeneous environment as a filamenteous extracellular matrix or a cellular tissue. A further valuable task would be to describe cooperative shape changes of connected cell pairs or of interacting cells forming tissues or even organs. This would have wide applications in developmental biology, e.g. for embryogenesis, and in medicine, e.g. for wound healing or tissue remodelling and for cancer formation.

Acknowledgements

For the opportunity to speak at the D'Arcy Thompson Conference in Dundee and for the stimulating ideas and discussions during the inspiring meeting there, I gratefully thank Mark Chaplain and his young crew.

REFERENCES

Akkaş, N. (1990) Some basic principles of mechanics as applied to cytology. In *Biomechanics of Active Movement and Deformation of Cells*, (Akkaş, N. ed.), pp. 109–154, Springer, Berlin.

Alt, W. (1996) Biomechanics of actomyosin mediated motility of keratinocytes. *Biophysics* 41: 181–188.

Alt, W. and Dembo. M. (1999) Cytoplasm dynamics and cell motion two-phase flow models. *Math. Biosci.* 156: 207–228.

Alt, W. and Tranquillo, R.T. (1995) Basic morphogenetic system modeling shape changes of migrating cells: How to explain fluctuating lamellipodial dynamics. *J. Biol. Systems* 3: 905–916.

Alt, W. and Tranquillo, R.T. (1997) Protrusion-retraction dynamics of an annular lamellipodial seam. In *Dynamics of Cell and Tissue Motion*, (Alt, W., Deutsch, A. and Dunn, G.A. eds.), pp. 73–81, Birkhäuser-Verlag, Basel.

Alt, W., Brosteanu, O., Hinz, B. and Kaiser, H.W. (1995) Patterns of spontaneous motility in videomicrographs of human epidermal keratinocytes (HEK). *Biochemistry and Cell Biology*, 73: 441–459.

Avnir, D. (1989) Chemically induced pulsations of interfaces: The mercury beating heart. *J. Chem. Education* 66: 211–212.

Blum, H. (1973) Biological shape and visual science. *J. Theor. Biol.* 38: 205–287.

Brosteanu, O. (1994) *Methoden zur Analyse der Lamellipodienaktivität von Leukozyten.* PhD thesis, Universität Bonn.

Brosteanu, O., Plath, P.J. and Vicker, M.G. (1997) Mathematical analysis of cell shape. In *Dynamics of Cell and Tissue Motion* (Alt, W., Deutsch, A. and Dunn, G.A. eds.), pp. 29–32, Birkhäuser-Verlag, Basel.

Bütschli, O. (1892) *Untersuchungen über mikroskopische Schäume und das Protoplasma.* W. Engelmann, Leipzig.

Cao, L. and Wang, Y. (1990) Mechanism of the formation of contractile ring in dividing cultured animal cells. II. Cortical movement of microinjected actin filaments. *J. Cell Biol.*, 111: 1905–1911.

de Bruyn, P.P.H. (1947) Theories of amoeboid motion. *Quart. Rev. Biol.* 22: 1–24.

Dembo, M. (1989a) Field theories of the cytoplasma. *Comments on Theoretical Biology* 1: 159–177.

Dembo, M. (1989b) Mechanics and control of the cytoskeleton in *Amoeba proteus*. *Biophys. J.* 55: 1053–1080.

Dembo, M. (1994) On free boundary problems and amoeboid motion. In: *Biomechanics of Active Movement and Division of Cells*, (Akkaş, N. ed.), pp. 231–283, Springer, Berlin.

Diaz, G., Quacci, D. and Dell'Orbo, C. (1990) Recognition of cell surface modulation by elliptic Fourier analysis. *Comp. Meth. Prog. Biomed.* 31: 57–62.

Dunn, G.A. and Brown, A.F. (1990) Quantifying cellular shape using moment invariants. In: *Biological Motion*, Vol. 89, (Alt, W. and Hoffmann, G. eds.), pp. 10–34, Springer, Berlin.

Dunn, G.A. and Zicha, D. (1995) Dynamics of fibroblast spreading. *J. Cell Sci.* 108: 1239–1249.

Dunn, G.A., Weber, I. and Zicha, D. (1997) Protrusion, retraction and the efficiency of cell locomotion. In: *Dynamics of Cell and Tissue Motion*, (Alt, W., Deutsch, A. and Dunn, G.A. eds.) pp. 33–46, Birkhäuser-Verlag, Basel.

Engelhardt, H. and Sackmann, E. (1988). On the measurement of shear elastic moduli and viscosities of erythrocyte plasma membranes by transient deformation in high frequency electric fields. *Biophys. J.* 54: 495–508.

Grebecki, A. (1995) Relative motion in *Amoeba proteus* in respect to adhesion sites II. Ectoplasmic and surface movements in polytactic and heterotactic amoebae. *Protoplasma* 127: 31–45.

Hinz, B. and Brosteanu, O. (1997) Periodicity in shape changes of human epidermal keratinocytes. In: *Dynamics of Cell and Tissue Motion*, (Alt, W., Deutsch, A. and Dunn, G.A. eds.), pp. 15–20, Birkhäuser-Verlag, Basel.

Hinz, B., Alt, W., Johnen, Ch., Herzog, V. and Kaiser, H.-W. (1999) Quantifying lamella dynamics of cultured cells by a new computer-assisted motion analysis: The role of tyrosine kinases for keratinocyte motion. *Exper. Cell Res.* Submitted.

He, X. and Dembo, M. (1997a) A dynamical model of cell division. In: *Dynamics of Cell and Tissue Motion*, (Alt, W., Deutsch, A. and Dunn, G.A. eds.), pp. 55–65, Birkhäuser-Verlag, Basel.

He, X. and Dembo, M. (1997b) On the mechanics of the first cleavage division of the sea urchin egg. *Exp. Cell Res.* 233: 252–273.

Killich, Th., Plath, P.J., Wei, X., Bultmann, H., Rensing, L. and Vicker, M.G. (1993) The locomotion, shape and pseudopodial dynamics of unstimulated *Dictyostelium* cells are not random. *J. Cell Sci.* 106: 1005–1013.

Lestrel, P.E. (1997) *Fourier Descriptors and Their Applications in Biology.* Cambridge Univ. Press, Cambridge.

Lewandowska, K., Doroszewski, J., Hämmerli, G. and Sträuli, P. (1979) An attempt to analyze locomotion of leukemia cells by computer image processing. *Comp. Biol. Med.* 9: 331–344.

Lord, E.A. and Wilson, C.B. (1984) *The Mathematical Description of Shape and Form.* Ellis Horwood, Chichester.

Noble, P.B. (1990) Images of cells changing shape: pseudopods, skeletons and motile behavior. In: *Biological Motion.* (Alt, W. and Hoffmann, G. eds.), Vol. 89, pp. 42–67, Springer, Berlin.

Noble, P.B. and Levine, M.D. (1986) *Computer-assisted analyses of cell locomotion and chemotaxis.* CRC Press, Boca Raton FL.

Prosser, J.A. (1990) Comparison of tip growth in prokaryotic and eukaryotic filamentous microorganisms. In: *Tip Growth in Plant and Fungal Cells.* (Heath, I.B. ed.), pp. 233–259, Academic Press, San Diego CA.

Schwegler, H., Tarumi, K. and Gerstmann, B. (1985) Physicochemical model of a protocell. *J. Math. Biol.* 22: 335–348.

Svetina, S. and Žekš, B. (1991) Mechanical behavior of closed lamellar membranes as a possible mechanism for the establishment of developmental shapes. *Int. J. Dev. Biol.* 35: 359–365.

Svetina, S., Božič, B. and Žekš, B. (1997) Shape behavior of closed layered membranes. In: *Dynamics of Cell and Tissue Motion,* (Alt, W., Deutsch, A. and Dunn, G.A. eds.), pp. 67–71, Birkhäuser-Verlag, Basel.

Soll, D.R. and Voss, E. (1998) Two- and three-dimensional computer systems for analyzing how animal cells crawl. In: *Motion Analysis of Living Cells,* (Soll, D.R. and Wessels, D. eds.), pp. 25–52, Wiley-Liss, New York.

Takaki, R., Katsu, A., Arai, Y. and Adechi, K. (1989) Vibration of a flattened drop. III. Mechanisms of mode transition. *J. Phys. Soc. Japan* 58: 129–139.

Taylor, D.L., Condeelis, J.S., Moore, P.L. and Allen, R.D. (1973) The contractile basis of amoeboid movement I. The chemical control of motility in isolated cytoplasm. *J. Cell Biology* 59: 378–394.

Thompson, D.W. (1961) *On Growth and Form.* Cambridge Univ. Press, Cambridge (abridged and edited by J.T. Bonner).

Vicker, M.G. and Wei, X. (1997) Self-organized F-actin autowaves govern pseudopodium projection and the non-random locomotion of *Dictyostelium* amoebae. In: *Dynamics of Cell and Tissue Motion,* (Alt, W., Deutsch, A. and Dunn, G.A. eds.), pp. 21–28, Birkhäuser-Verlag, Basel.

16

Shape Asymmetry and Developmental Stability

Jean-Christophe Auffray[1], Vincent Debat[1] and Paul Alibert[2]

[1]*Institut des Sciences de l'Evolution, UMR 5554 CNRS, CC064, Université Montpellier 2, 34095 Montpellier cedex 05, France*
[2]*Biogéosciences-Dijon, UMR 5561 CNRS, Bâtiment Gabriel, Université de Bourgogne, 6 Bd Gabriel, 21 000 Dijon, France*

16.1 Introduction

Waddington (1957) defined developmental homeostasis as the mechanisms responsible for ensuring phenotypic constancy in organisms despite the great variability of genetic, environmental and developmental features. Developmental homeostasis is considered as a combination of two components: developmental canalisation and developmental stability (Zhakarov, 1989). Developmental canalisation corresponds to the processes that buffer development to produce consistent phenotypes under a range of environmental and genetic conditions. Developmental stability, on the other hand, is defined as the ability of organisms to withstand genetic or environmental disturbances during development, so as to produce a predetermined phenotype (Zakharov, 1989). In other words, if canalisation ensures the constancy of phenotypes for different genotypes under varying environmental conditions, developmental stability enhances this constancy for a given genotype and environment.

Inasmuch as developmental stability reflects the capacity of organisms to produce an optimum phenotype despite perturbations encountered during development, its appraisal may thus be used to evaluate these stresses, as well as the ability of genotypes to correct them. The stresses can be of environmental or genetic origin. A large number of environmental stresses – food deprivation, temperature, pollution, and so on – have been shown to significantly impair the developmental stability of organisms (Parsons, 1990; Møller and Swaddle, 1997). Although the genetic basis of developmental stability remains mostly unknown, it is presumed to depend on certain genetic states, including genomic coadaptation and heterozygosity (Graham, 1992; Clarke, 1993). Any cause, which impairs one of these two conditions may thus be considered as a genetic stress such as cases (1) of inbreeding, which is expected to be accompanied both by a

On Growth and Form: Spatio-temporal Pattern Formation in Biology.
Editor Mark A.J. Chaplain, G.D. Singh, J.C. McLachlan

reduction of heterozygosity, and a higher fixation rate of deleterious alleles (Clarke, 1993), (2) of incorporation of new alleles (e.g. an insecticide resistant allele) which may disrupt genomic co-adaptation (McKenzie and Clarke, 1988) as well as (3) most of the cases of hybridisation (Graham, 1992). These properties have led developmental stability to be the focus of an increasing number of studies during the last decade in numerous fields of evolutionary biology and ecology.

Developmental stability is most often appraised by the study of fluctuating asymmetry (FA). Given that the two sides of a bilaterally symmetrical trait are produced by the same gene complexes and develop under similar environmental conditions, any deviation from symmetry will express the inability of the organism to correct developmental errors, i.e. its level of developmental stability (Møller and Swaddle, 1997). Therefore, developmental stability is widely estimated by a measure of the variability, within a population, of the differences between the right and left sides of organisms, the so-called FA. For a single metric trait, FA can be assessed by the variance of the left-minus-right $(L-R)$ distribution in a population, as well as by any other expression of this variability (Palmer, 1994).

16.2 Nature of Asymmetries

In bilateral organisms, however, some asymmetries are adaptative and hence thought to be genetically determined, rendering them uninformative for estimating developmental noise. Thus, fluctuating asymmetry needs to be distinguished from two other types of asymmetry: directional asymmetry (DA) and antisymmetry (AA). DA occurs when one side of a bilateral character is systematically larger than the other, so the mean of the $(L-R)$ normal distribution of the population will be different from zero. The AA also corresponds to a systematic deviation from symmetry, but in this case the side that is larger varies at random among individuals. A typical antisymmetric trait leads to a bimodal $(L-R)$ distribution centered on zero.

Numerous illustrative cases of DA are found in bilateral organisms, e.g. internal organs in mammals. On the other hand, typical antisymmetric traits are more rarely observed; the claws of the male fiddler crabs (genus Uca) are the most frequently cited example – one of them is always much larger than the other. In most cases, however, DA and AA may be subtle. A trait can be characterised by a slight but significant DA, and any normal-like distribution of the $(L-R)$ may hide a slight but true AA (Palmer and Strobeck, 1992). A prior assessment of DA and AA is crucial in studies of fluctuating asymmetry, not only for the biological information it provides but also because several indices available for estimating fluctuating asymmetry are potentially biaised by DA, e.g. mean of the $|R-L|$ distributions (Palmer, 1994). If the purely statistical bias is more or less easily resolved (Palmer, 1994; Rowe *et al.*, 1997; Graham *et al.*, 1998), the meaning of DA or AA in terms of developmental stability is far from trivial. Several studies have persuasively suggested that FA, DA and AA are dynamically interrelated (Graham *et al.*, 1993, 1998; Leamy, 1999). In wild house mice, measurements (maximum length and width) of the three lower molars are often used in studies of asymmetry. While five of these six highly related traits do not exhibit any DA, several independent studies have shown that the width of the third lower molar presents significant DA, the right third molar being larger (Alibert *et al.*, 1997; Chatti *et al.*, 1999). However, it should not be concluded that one specific trait will systematically exhibit a constant DA across populations. Studies on mandible mensurations in a house mouse strain (CV1) have demonstrated that several traits displayed significant DA, but not always towards the same sides (Leamy, 1993, 1999). Additionally, while DA is thought to be heritable and FA to be an expression of developmental noise, heritability estimates of DA have been shown to be low, and only slightly higher than those reported for FA (Leamy *et al.*, 1997; Leamy, 1999). These results have led Leamy (1999) to stress the relevance of investigating DA as a marker of developmental stability. Formerly, the emergence of DA in developmentally unstable organisms during selective experiments (Graham *et al.*, 1993) emphasised the potential of DA in expressing developmental instability. In the same manner, several reports of transition from FA to AA have suggested an equivalent potential of AA (McKenzie and Yen, 1995; Leary and Allendorf, 1989). Although, Rowe *et al.* (1997) have demonstrated that several emblematic studies of FA were actually dealing with AA, several studies have more directly reported the relevance of antisymmetric signals in terms of developmental instability (McKenzie and Clarke, 1988; Rowe *et al.*, 1997).

In summary, despite the fact that the nature of DA and AA, as well as their meaning in terms of developmental stability, are increasingly debated, these types of asymmetry are still believed to present a certain genetic component. Therefore, their statistical significance has still to be carefully assessed to avoid confounding effects. The recent studies of Graham *et al.* (1998) and Rowe *et al.* (1997) respectively reconsider the statistical distinction between FA and DA and between FA and AA.

16.3 Nature of Characters

The influence of the number and the nature of characters used to appraise FA has to be considered. As a matter of fact, there is no general rule in the choice of these characters, unless it is guided by the aim of the study, such as secondary sexual characters for sexual selection studies. The influence of the choice of a peculiar set of traits in FA studies strongly depends on the congruence of these characters in revealing developmental perturbations. This question may be adressed at an individual or a population level.

Clarke (1998a) pointed out that when several characters are considered among individuals, a non-random developmental noise among these traits would lead to a significant concordance of individual asymmetry. This would then allow the estimation of an individual asymmetry parameter (IAP) over all characters. However, it is noteworthy that such a parameter is most often undetected (Clarke, 1998a; see table W2.1 at `http://www1.oup.co.uk/MS-asymmetry/`; Møller and Swaddle, 1997). Several reasons could account for this. A first reason for not detecting an IAP, which is currently assessed by correlative procedures between individual asymmetries exhibited by several characters, would be related to statistical bias due to sample sizes and to low repeatability of measurements (Whitlock, 1996). Hence, it has been reported that an individual asymmetry parameter would be conveniently detected if high sample sizes were used (Leamy, 1993).

Another reason would lie with the independence of developmental processes and control of developmental stability of the characters considered. As already suggested in the literature, differences in developmental time, morphogenetic or developmental independence between characters as well as differences in the developmental control systems may also explain the absence of concordance between individual asymmetries (Møller and Swaddle, 1997; Clarke, 1998a). Such hypotheses could be more conveniently tested if one used the concordance of characters among samples to express the levels of developmental instability. In other words, if a population exhibits the highest level of FA for one character, this population is expected to present the highest levels of FA for the other characters. If this concordance of characters is verified, it would allow the establishment of a population asymmetry parameter (PAP, Soulé, 1967), i.e. a synthetic estimation of FA among populations across characters.

Møller and Swaddle (1997) reported that PAP estimates were often significant (for 21 out of 30 studies reported in table 2.1 at the URL cited above). In contrast, Clarke (1998a) has recently shown that, in most of the cases, the characters were not concordant in providing the correct ranking of FA across samples within several invertebrate species. In his review the studies involved three to six characters which were not closely related in terms of development. The biological interpretation of the presence or absence of such a PAP is undoubtely the same as for the IAP. Yet it has been shown that the PAPs were more significant when their estimates, instead of being based on the whole set of characters of a morphological structure, were based on several sets of morphogenetic units (Leamy, 1993). For example, in rodent teeth, i.e length and width of lower and upper molars, this concordance is verified most of the time (Alibert *et al.*, 1994, 1997; Auffray *et al.*, 1999; Chatti *et al.*, 1999). It is likely that teeth share a common and synchronous developmental process and potentially a common mechanism for controlling stability. In contrast, among more independent morphological structures, these processes could differ (Møller and Swaddle, 1997), leading to insignificant PAPs. Moreover, each case in which the concordance of characters was not verified, corresponded to sets of populations which did not exhibit significant differences in FA levels for any of the traits considered (Auffray *et al.*, 1996b; Fontanillas and Auffray, unpublished). Clarke (1998a) reported a similar pattern, i.e. significant PAPs arising only when FA levels were significantly different, trait by trait.

However, the concordance of characters in ranking different populations according to their level of developmental stability and leading to a significant PAP is most often reported in the literature. This suggests a genome-wide process of control of asymmetry, but the respective amounts of FA displayed by the different characters might differ and thus their sensitivity to developmental stability.

16.4 Developmental Stability and Canalisation

Palmer (1994) stressed the fact that inferring differences in developmental stability among samples required the use of developmentally independent characters. However, even developmentally related characters, which are expected to be congruent in providing the rank of FA levels among samples, may considerably differ in their sensitivity to developmental stability. Any researcher in the field of developmental stability has experienced the heterogeneity in the amplitude of response of characters to developmental perturbations. A clear example was provided by the mole rat (*Spalax ehrenbergi*) populations of the Near East (Auffray *et al.*, 1999). Among eight tooth traits (six for molars, two for incisors) which were concordant in providing ranks, the transversal diameter of the incisor clearly exhibited one of the lowest levels of FA whereas the other incisor parameter, the anteroposterior diameter, presented some the highest. In these populations the level of FA was related to the mean size of characters. Whitlock (1996) stressed the necessity to use the coefficient of variation of FA to compare levels of FA among traits. In the case of the mole rat, while the FA estimates strongly differed among traits, their respective coefficients of variation were more homogeneous (between 1.28% and 1.71%) suggesting that the amount of FA partly depended on character size.

Most of the attention should, however, be given to the relationship between FA levels and variances among traits. The phenotypic variance of a character in a population arises from several components: (1) the genetic variance, which is related to differences in genetic constitution, (2) the environmental variance resulting from exposure to different environmental conditions, (3) an interaction term between these two factors and (4) a stochastic part due to the developmental noise. It is well known that all characters are not similarly sensitive to genetic or environmental conditions. The less variable they are, the more they are canalised. The genetic control of canalisation is poorly known, but it has clearly been established that a high level of canalisation for a character can result from selective processes (Scharloo, 1991; Møller and Swaddle, 1997).

Waddington (1957) has suggested that characters for which phenotypic constancy is important in terms of individual fitness will be developmentally less variable (better buffered) than characters for which constancy is less important. He has thus proposed a form of stabilising selection, called canalising selection. Canalising selection acts to eliminate from the population those genotypes that render developmental pathways for a given character sensitive to genetic or environmental variation. This suggests that the degree of developmental homeostasis is character-dependent and is hypothesised to reflect the functional contribution of the character to the fitness of the organism. In the same manner it also implies that the genetic mechanism of developmental homeostasis is character-dependent. Such assumptions, however, still remain to be tested rigorously (Clarke, 1998b).

Waddington (1957) also presumed that the two components of developmental homeostasis, canalisation and developmental stability, have phenotypical and genetic independent properties. Although the genetic control of these two mechanisms is mostly unknown, this idea has been widely accepted. A simple way, however, to assess the level of dependence between canalisation and developmental stability is to appraise the relationship between morphological variability and FA. For example, Clarke (1998b) has recently shown that, in some insect and crustacean species, FA levels exhibited by characters were usually correlated with their variability estimated by the coefficient of variation.

However, in order to study the relative levels of phenotypic variance of characters, and their covariance with FA levels, and to potentially relate these patterns with the functional importance of characters, a high number of characters of presumably different functional importance is required. Given that, in studies of asymmetry, both sides of a single character have to be measured with several replicates on sufficiently high sample sizes, the number of characters studied is rarely very large. The necessity to increase the number of characters would thus undoubtely discourage this kind of approach. What was required was a method which made it possible to appreciate the whole morphology as well as to estimate FA, while preserving reliable information on the covariation among traits.

Such an opportunity was provided by geometric morphometrics based on landmark data, which are suitable to precisely describe and quantify the covariation among components of a morphological structure (Bookstein, 1991; Rohlf and Marcus, 1993). In addition, geometric morphometrics can properly estimate fluctuating asymmetry using morphological structures as a whole.

16.5 Shape Asymmetry

The use of geometric morphometrics to appraise fluctuating asymmetry was initially proposed by Bookstein (1991). Geometric morphometric methods are based on landmark data for the study of shape variation (Rohlf and Marcus, 1993; Dryden and Mardia, 1998). In this case, instead of being represented by several pairs of measurements, the right and the left sides of an individual are represented by their configuration of landmarks in two or three dimensions.

The estimation of individual asymmetry is based on the calculation of Procrustes distances between the configuration of one side and the mirror image of the other. Procrustes methods, first introduced in 1970 by Schanemann and Gower (Bookstein, 1991), are based on the least-squares technique in order to compute the best-fitting superimposition of landmark configurations (Rohlf, 1990). The mathematical and computational procedures of Procrustes least-squares superimposition are detailed in Rohlf (1990) and in Rohlf and Slice (1990). Briefly, least-squares superimposition of two configurations (Figure 16.1), e.g. one side and the mirror image of the other, requires both to be conventionally scaled to unit centroid size, i.e. the root of the sum of the squared distances of all landmarks to the centroid of their own configuration will be equal to unity. Once the configurations are scaled, the best-fit superimposition of the two configurations involves the translation of the configurations in order to superimpose their centroids, and the rotation of one of them into a position of best-fit to the other one. The optimal superimposition is obtained by minimising the sum of squared distances (A^2) between the corresponding landmarks of the two configurations.

Bookstein (1991) as well as Smith *et al.* (1997) depicted the geometric meaning of this estimate of asymmetry: it is proportional to the sum of the areas of circles centred on each landmark, of which the radii are the vectors connecting corresponding landmarks. The square root of A^2, A, is thus a measure of net (or raw) asymmetry between the two sides of an organism. An analogy can be made between this measure of asymmetry and the $|R - L|$ parameter used in the traditional calculation of the FA1 index of Palmer (1994). Both are unsigned and when these individual asymmetry values are averaged over a sample to provide a mean estimate of asymmetry, both can be very biased by DA or AA. Thus, the mean Procrustes distance between the two sides in a sample, as well as $\overline{|R - L|}$ in traditional approches of FA, can be considered as relevant measures of fluctuating asymmetry only if DA or AA is absent (Palmer, 1994).

Here the directional asymmetry is estimated in a sample by the Procrustes distance between the mean right and the mean left configurations after their superimposion. These two mean configurations are computed by a generalised least squares (GLS) superimposition which is a generalisation over a sample of the two-configuration least-

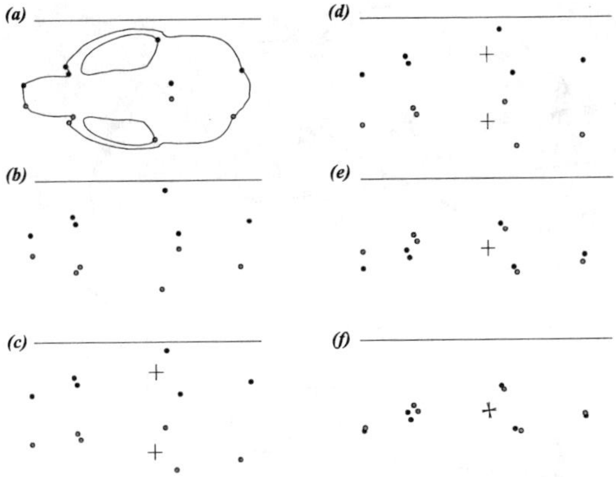

Figure 16.1 Two-dimensional representation of the Procrustes least-squares superimposition of the two sides of a mouse skull: (a) location of digitised landmarks on the two sides of the skull (landmarks on each side have a different color); (b) representation of the raw configurations of landmarks; (c) reflection of one of the sides and location of the centroid for each side (+); (d) scaling configurations to equal centroid size (in practice both configurations are scaled to unit centroid size); (e) translation of configurations in order to superimpose their centroids; (f) rotation of one configuration using the criterion of minimum squared distances between corresponding landmarks on both sides; the rotation angle is shown by the + signs (Bookstein, 1991; Auffray et al., 1996b).

square superimposition. Computational details can be found in Rohlf (1990) or Rohlf and Slice (1990). This procedure minimizes the sum of Procrustes distances from all individuals to the Procrustes mean configuration. GLS superimposition is an iterative procedure which begins by fitting all configurations to one of them used as a reference. Once all configurations are superimposed according to the criterion of minimizing their Procrustes distances to this reference, a mean configuration is calculated onto which all configurations are re-superimposed. A new mean configuration is then calculated and so on. The procedure converges rapidly to a stable mean configuration (Rohlf, 1990; Slice, 1996).

In Bookstein's (1991) studies and Auffray *et al.*'s (1996b) studies of Procrustean asymmetry, GLS superimposition was only used to estimate the mean configuration of each side, and individual asymmetries were computed one by one. Smith *et al.* (1997) simplified the procedure by superimposing both sides of all individuals using GLS superimposition. This procedure, which does not modify the Procrustes distance between sides of individuals, allows one to use the new coordinates of superimposed configurations to estimate individual asymmetry as well as directional asymmetry. At this step, it should be noted that in the statistical techniques reported here, the use of coordinates of superimposed configurations is strictly similar to that of residuals around each mean landmark. Once all sides are superimposed by GLS (Figure 16.2a), the whole variability at each landmark is expressed by the scatter of points around the mean landmark. Within the scatter around a given landmark k, each individual is represented by two points I_{kr} and I_{kl}, corresponding to the kth landmark of the

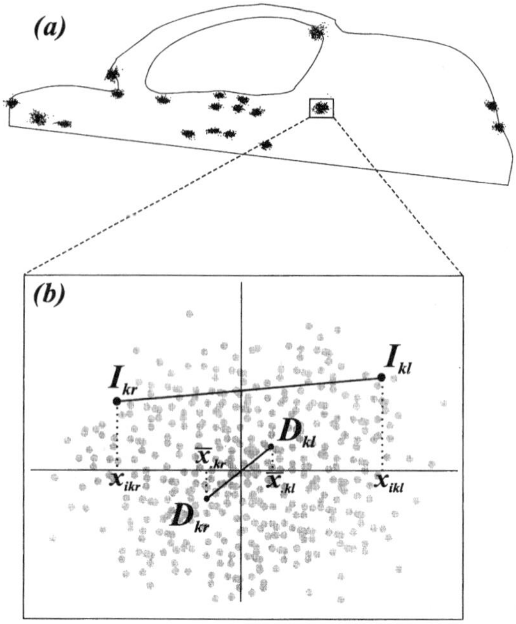

Figure 16.2 Procrustean estimate of shape asymmetry according to the procedure of Smith *et al.* (1997): (a) generalised least squares (GLS) superimposition of configurations for both sides in a sample; (b) basic design of the partition of raw asymmetry into its directional and fluctuating terms shown for one coordinate of a given landmark: I_{kr} and I_{kl} respectively correspond to the kth landmark of the right and left configuration of individual i; D_{kr} and D_{kl} respectively represent the kth mean landmark for the right and left sides in the sample (see text).

right side configuration and of the left one respectively (Figure 16.2b). The mean right point D_{kr} and the mean left one D_{kl} are computed and correspond respectively to the kth landmark of mean right and mean left configurations. If we consider $x_{ikr}, x_{ikl}, \overline{x}_{.kr}$ and $\overline{x}_{.kl}$ as the projections of respectively I_{kr}, I_{kl}, D_{kr} and D_{kl} onto the x-axis, the distances between I_{kr} and I_{kl} projected onto this axis can be expressed as

$$\underset{(1)}{(x_{ikr} - x_{ikl})} = \underset{(2)}{(\overline{x}_{.kr} - \overline{x}_{.kl})} + \underset{(3)}{(x_{ikr} - \overline{x}_{.kr})} + \underset{(4)}{(\overline{x}_{.kl} - x_{ikl})} \tag{16.1}$$

where (2) corresponds to the difference after projection, between the kth landmarks of the right and left mean configurations, and (3) and (4) correspond to the deviation of the landmarks of each side to their respective mean side landmark over the sample. Similar distances can be computed on the y-axis, using the y-coordinates for the same landmarks.

Computing the distances between I_{kr} and I_{kl} in two dimensions requires that terms be squared and summed. Thus, for landmark k, the square distance between I_{kr} and I_{kl}, $d^2(I_{kr}, I_{kl})$ is

$$d^2(I_{kr}, I_{kl}) = d^2(D_{kr}, D_{kl}) + F_{ik} \tag{16.2}$$

where F_{ik} corresponds to the expression of the distance between the kth landmark of both sides around the norm. The norm is distinguished from perfect symmetry in that it integrates the expected mean difference between sides due to DA.

Summing the terms across all landmarks for an individual i corresponds exactly to the calculation of the square Procrustes distance, A_i^2, between the configurations of its two sides. It is expressed by

$$A_i^2 = Ad^2 + F_{i.} \tag{16.3}$$

where Ad^2 corresponds to the Procrustes distances between the mean right and the mean left configurations, in other words the squared DA, and $F_{i.}$ is an expression of the distance between the two sides around the norm.

Summing A_i^2 across the n individuals in a sample and averaging them provides the mean raw asymmetry $\overline{A^2_.}$ of the sample as

$$\overline{A^2_.} = Ad^2 + \overline{F_{..}} \tag{16.4}$$

where $\overline{F_{..}}$ corresponds to the fluctuating component of the mean square raw asymmetry averaged over the sample.

Testing differences of mean raw asymmetry among samples can be done by a one-way ANOVA on the root of the square Procrustes distances, i.e. A_i (Smith *et al.*, 1997; Klingenberg and McIntyre, 1998). As such an index of asymmetry can be very biased by DA, it is necessary to assess the absence of DA in the samples considered. On a single sample, the significance of DA can be appraised by testing the equality of the mean right and mean left shapes. For reasons related to the geometric procedure of Procrustes superimposition, and to the loss of degrees of freedom (Bookstein, 1996, Dryden and Mardia, 1998), the use of general linear models such as MANOVAs on Procrustes residuals is allowed only when variation is small and the number of individuals is high with regard to the number of landmarks. If these conditions are fulfilled, a one-way MANOVA on all coordinates with the factor side can be used to appraise the equality of mean side shapes. If not, non-parametric tests such as permutation tests should be used (Dryden and Mardia, 1998).

Since 1991, when this method was first proposed by Bookstein, it has gathered few followers. So far the studies of Auffray *et al.* (1996b) and Smith *et al.* (1997) are the only published works using this method, and Smith *et al.* (1997) is a re-examination of the study reported by Bookstein in 1991. These studies have yielded relevant biological signals, but as is often the case for newly arising methods, they were unsatisfactory from several points of view. None of them took into account the size asymmetry. However, although each configuration had been scaled to a unit centroid size (see above), the size component of FA could have been assessed from the raw centroid sizes as for any traditional trait (Palmer, 1994). Another problem lay with appraising the measurement error, which was absolutely required since FA results from subtle differences between sides. Auffray *et al.* (1996b) introduced a procedure to control the level of the error of measurement by estimating Procrustes distances between the two replicates of left configurations in a sample. However, this procedure did not yield an

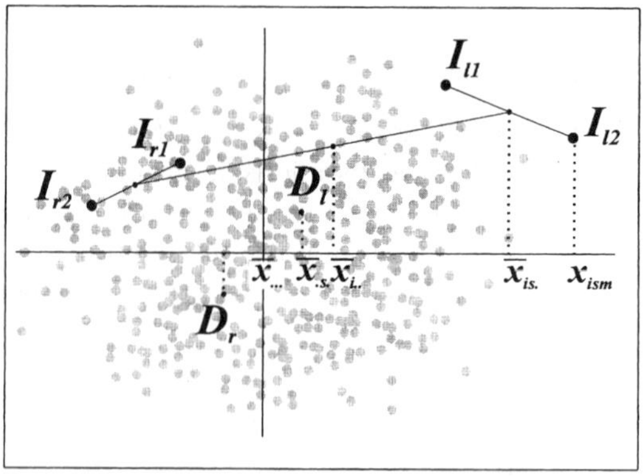

Figure 16.3 Basic design of the partition of whole shape variability according to several effects following the procedure of Klingenberg and McIntyre (1998), shown for one coordinate at a given landmark. $I_{r1}, I_{r2}, I_{l1}, I_{l2}$ respectively correspond to two replicates of the right and left sides of an individual i for the same landmark, D_r and D_l respectively depict the mean landmark for the right and left sides over the sample (see text).

estimate of the fluctuating asymmetry index from which the error effect was removed, just as for the FA10 index of traditional approaches (Palmer, 1994).

Such an estimate of fluctuating asymmetry using geometric morphometrics has been recently proposed by Klingenberg and McIntyre (1998). Coordinate by coordinate, the basic design of this procedure differs from that of Smith *et al.* (1997) in that, instead of partitioning the total distance between the two sides into directional and fluctuating components, it partitions the deviation of each replicate to the grand mean according to several potential sources of variation, i.e. side, genotype (or individual) as well as their interaction and error. By analogy with the two-way mixed-model ANOVA proposed by Palmer and Strobeck (1986) and Palmer (1994) to test the significance of FA relative to the measurement error, this approach can thus involve several replicates of both sides. The generalisation of the basic partition design, by considering the sum of squares at a landmark or at the whole configuration, provides a means to appraise the significance of these effects as sources of variation.

In practice each side of an individual is represented by the number of its digitised replicates. By analogy with Figure 16.2, the basic partition design of the Procrustes residual used in the approach of Klingenberg and McIntyre (1998) is represented in Figure 16.3.

As for any two-way ANOVA, it can be stated that, in terms of the projection on the x-axis of the scatter of points around a given landmark, the deviation, in equation (16.5), of one of the replicates m of side s of individual i, x_{ism}, from the grand mean $\overline{x}_{...}$, can be partitioned as

$$\underbrace{(x_{ism} - \overline{x}_{...})}_{(1)} = \underbrace{(\overline{x}_{.s.} - \overline{x}_{...})}_{(2)} + \underbrace{(\overline{x}_{i..} - \overline{x}_{...})}_{(3)} + \underbrace{(x_{ism} - \overline{x}_{is.})}_{(4)} + \underbrace{F_{ism}}_{(5)} \qquad (16.5)$$

where (2) is the distance between the mean side over replicates and individuals, $\overline{x}_{.s.}$ and the grand mean, (3) the distance between the mean individual over sides and replicates, $\overline{x}_{i..}$ and the grand mean, (4) the residual error between the replicate considered and the mean side s of the individual i over replicates, $\overline{x}_{is.}$, and (5) an interaction term F_{ism} between the two factors such as

$$F_{ism} = (\overline{x}_{is.} - \overline{x}_{i..} - \overline{x}_{.s.} + \overline{x}_{...}) \,. \tag{16.6}$$

For a given coordinate, this expression is squared and summed over all replicates, sides and individuals, leading to a classical ANOVA sum of squares. As stated already, the main interest of this procedure is to partition morphological variability according to several sources of variation: (1) the variation among individuals, (2) the variation related to side, (3) the added random variance component due to non-directional asymmetry, which reflects the variation in the between-sides difference among individuals such as FA, AS or error, and (4) the variance due to the error of measurement. This partition of the whole variability at one landmark coordinate is strictly similar to the one performed in the two-way mixed-model ANOVA used for the estimation of FA10 (Palmer, 1994).

Summing mean squares for each effect accross the x- and y-coordinates for a given landmark could provide a relevant picture of the sensitivity of each landmark to the sources of variation considered. However, applying F-ratios of ANOVAs at this step would yield only scarce information on the significance of the different sources of variation. Landmark by landmark, these residuals are not independent (Bookstein, 1991; Auffray *et al.*, 1996a). If one landmark is more variable than the other ones, i.e. if it represents a very localised variability, the GLS superimposition will dilute this localised variability over all landmarks and lead to its underestimation, as noted by Klingenberg and McIntyre (1998).

To take into account the variability of the whole morphological structure over a sample, sums of squares are added for each of the effects across the x- and y-coordinates of all landmarks. These Procrustes sums of squares are then used to compute mean squares for each effect by dividing them by the relevant degrees of freedom, which are the conventional degrees of freedom (Palmer, 1994) multiplied by the number of coordinates minus 4. Four degrees of freedom are lost during the GLS superimposition, two for translation, one for scaling and one for rotation (Bookstein, 1991). F-tests can then be applied on these mean squares. However, as for any inferences following Procrustes superimposition (see above), non-parametric tests such as permutation tests are recommended for all statistics dealing with Procrustes coordinates or residuals.

In traditional FA as well as in the Procrustes approach, the significance of the third term indicates that the interaction variance is greater than the error of measurement. In other words, it indicates that the differences between sides among individuals vary more than would be expected given the size of the measurement error (Palmer, 1994). If this term is significant, it is possible to extract the measurement error variance from the between-side variance in order to get a fluctuating asymmetry index from which the effect of measurement error has been removed. The shape fluctuating asymmetry index can then be computed in a similar way to the traditional FA10 of Palmer (1994). A corresponding size FA10 index for size can also be computed from the centroid size of each replicate. In this case, the significance of the ANOVA effects can be classically assessed by F-ratios, i.e. using parametric statistics.

More interestingly, Klingenberg and Mcintyre (1998) developed a method to assess and depict the morphological variability related to each effect as well as their relationships. Their procedure involved performing principal components analyses (PCAs) on different covariance matrices computed from the whole Procrustes coordinates or the residual matrix, but corresponding to each potential source of variation. The covariation of landmarks in relation to the inter-individual source of variation was then appraised by a PCA based on a covariance matrix computed from the individual means, FA from individual left–right differences, and the error of measurement from the residual of the replicate measurements over the mean of each side.

Furthermore, the computation of eigenvectors defining the principal components emerging from these analyses allows one to depict, at each landmark of the original mean configuration, the vectors representing the variability expressed at a given principal component for any of the effects considered. At this step, permutation tests are used to establish the independence between eigenvectors, or to appraise the agreement between the morphological expression of developmental instability (FA), the interindividual variability and the error of measurement.

In their study, Klingenberg and McIntyre (1998) provided a detailed example, in which they focused on the similarity of covariation of landmarks for different sources of variation (interindividual variability, FA and error of measurement) rather than on the comparison of FA levels among groups. The significant correlation between these three covariance matrices, as well as the non-independence between emerging PCs and their highly similar pictorial patterns, led the authors to consider that there was no reason to suspect that specific developmental processes were affecting asymmetry or buffering against it, other than those influencing the mean shape across body sides. They also admitted that, since this study was the first one, their conclusions could not be considered as a general rule.

16.6 Towards the Form Asymmetry

Further uses and developments of geometric morphometric methods in the field of developmental stability would undoubtely be extremely informative about the processes governing developmental stability as well as those patterning the emergence of FA. This would be the case in particular for the relationship between size, shape, and their respective levels of asymmetry. But even though examination of the relationships between the centroid size and the shape or size asymmetry has the same interest as in traditional approaches to developmental stability (Palmer and Strobeck, 1986; Palmer, 1994; Rowe *et al.*, 1997; Graham *et al.*, 1998), assumptions about the patterns of the relationship between size and shape asymmetries are far from being stated. One can, however, hypothesise that an independence between these asymmetries as well as a negative relationship would provide interesting clues to numerous unanswered questions such as the relative functional importance of size and shape, or the potential complementing effects between size and shape components in the maintenance of symmetry. Our experience on several independent samples of the house mouse skull has indicated that shape and size asymmetries are positively correlated, both among individuals and among samples (Debat and Auffray, unpublished). This leads us to

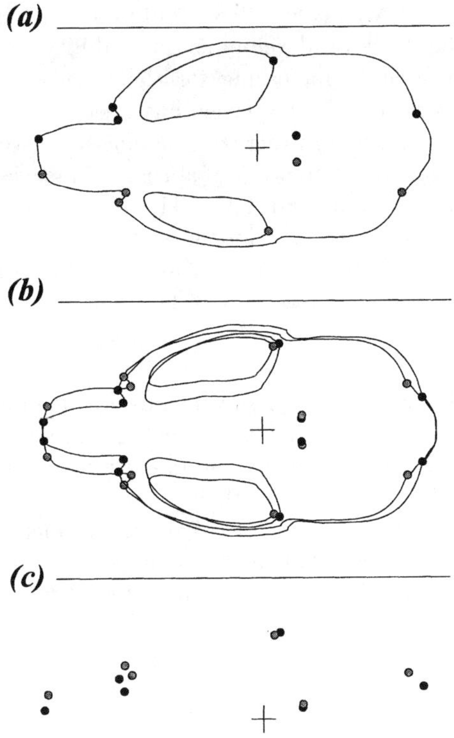

Figure 16.4 Superimposition procedure of a whole symmetric structure in order to estimate form asymmetry: (a) location of landmarks digitised on the two sides (the landmarks on each side have a different color); (b) least-squares superimposition of the whole configuration onto its mirrored image; (c) description on one side of the form distances between sides; since both sides are not scaled to each other, the raw asymmetry shown here corresponds to raw asymmetry of form.

address the question of the potential weakening effect of partitioning of FA signals into size and shape terms on the statistical appraisal of overall FA. As size and shape are classically conceived to be components of form, it would be possible to compute a form asymmetry index combining both size and shape asymmetries. This would, however, require the use of single symmetric morphological structures, i.e. containing a midline (e.g. a skull) instead of two disconnected ones around an axis of symmetry (e.g. wings). Instead of superimposing both sides of an organism, which involves scaling each side to a unit centroid size, the GLS superimposition of whole morphological structures (both connected sides) onto their mirror images would preserve the size of both sides relative to each other (Figure 16.4). This procedure would lead to a perfectly symmetric distribution of residuals (or coordinates) for both sides around the midline, which should not be confused with the axis of symmetry. Consequently, the residuals at each landmark of one side, and eventually those of the expected axis of symmetry, would contain all the information on the covariation of landmarks in terms of form. Except for the fact that in this case size asymmetry does not have to be considered, and that the degrees of freedom should be adequately modified, all the procedures of Klingenberg and McIntyre (1998) would remain fully applicable.

Although studies in the field of developmental stability as well as geometric morphometrics have been steadily increasing for the last two decades, the use of the geometric morphometrics to infer FA levels has been rare until now. This was probably due to the fact that the use of Procrustes distances as estimates of FA had not been expressed within the general framework of statistical assessement of FA, mostly represented by "the primer" of Palmer (1994). The procedure of Klingenberg and McIntyre (1998) has not only conveniently done this, it has also extended the study of covariation of landmarks to much more integrative aspects of morphological variability, e.g. the relationships between the components of developmental homeostasis.

Acknowledgements

We thank Christian P. Klingenberg and Larry Leamy for helpful discussions during the preparation of the manuscript and for having sent preprints of their work in press; Janice Britton-Davidian for comments on an earlier version of the manuscript; Sandrine Peired and Josette Catalan for their help with the bibliography. This study was supported by the Programme CNRS Paléoenvironnement et Evolution des Hominidés: Stress génétiques et environnementaux subis par les populations actuelles et fossiles sous l'effet de l'anthropisation (coordinator Jean-Christophe Auffray). This is publication ISEM No. 99-134.

REFERENCES

Alibert, P., Renaud, S., Dod, B., Bonhomme, F. and Auffray, J.-C. (1994) Fluctuating asymmetry in the Mus musculus hybrid zone: a heterotic effect in disrupted co-adapted genomes. *Proc. Roy. Soc. Lond.* B258: 53–59.

Alibert, P., Fel-Clair, F., Manolakou, K., Britton-Davidian, J. and Auffray, J.-C. (1997) Developmental stability, fitness, and trait size in laboratory hybrids between European subspecies of the house mouse. *Evolution* 51: 1284–1295.

Auffray, J.-C., Alibert, P., Latieule, C. and Dod, B. (1996a) Relative warp analysis of skull shape across the hybrid zone of the house mouse (*Mus musculus*) in Denmark. *J. Zool.* 240: 441–455.

Auffray, J.-C., Alibert, P., Renaud, S., Orth, A. and Bonhomme, F. (1996b) Fluctuating asymmetry in *Mus musculus* subspecific hybridization: Traditional and Procrustes comparative approach. In: *Advances in Morphometrics*, Marcus L.F., Corti M., Loy A., Slice, D. and Naylor, G. eds. pp. 275–283. Plenum press, New York.

Auffray, J.-C., Renaud, S., Alibert, P. and Nevo, E. (1999) Developmental stability and adaptive radiation in *Spalax ehrenberghi* superspecies in the Near East. *J. Evol. Biol.* 12: 207–221.

Bookstein, F. L. (1991) *Morphometric Tools for Landmark Data: Geometry and Biology.* Cambridge University Press, New York.

Bookstein, F. L. (1996) Combining the tools of geometric morphometrics. In: *Advances in Morphometrics*, Marcus L.F., Corti, M., Loy A., Slice, D. and Naylor, G. eds. pp. 113–152. Plenum press, New York.

Chatti, N., Said, K., Catalan, J., Britton-Davidian, J. and Auffray, J.-C. (1999) Developmental instability in wild chromosomal hybrids of the house mouse. *Evolution* (in press).

Clarke, G.M. (1993) The genetic basis of developmental stability. I Relationships between stability, heterozygosity and genomic coadaptation. *Genetica* 89: 15–23.

Clarke, G.M. (1998a) The genetic basis of developmental stability. IV. Individual and population asymmetry parameters. *Heredity* 80: 553–561.

Clarke, G.M. (1998b) The genetic basis of developmental stability. V. Inter- and intra-individual character variation. *Heredity* 80: 562–567.

Dryden, I.L. and Mardia, K.V. (1998) *Statistical Shape Analysis.* John Wiley, Chichester.

Graham, J.H. (1992) Genomic coadaptation and developmental stability in hybrid zones. *Acta Zoologica Fennica* 191: 121–131.

Graham, J.H., Freeman, D.C. and Emlen, J.M. (1993) Developmental stability: a sensitive indicator of populations under stress. In: *Environmental toxicology and risk assessment*, Landis, W.G., Hughes, J.S. and Lewis, M.A., eds., pp. 136–158. American Society for Testing and Materials, Philadelphia PA.

Graham, J.H., Emlen, J.M., Freeman, D.C., Leamy, L.J. and Kieser, J.A. (1998) Directional asymmetry and the measurement of developmental instability. *Biol. J. Linn. Soc.* 64: 1–16.

Klingenberg, C. P. and McIntyre, G.S. (1998) Geometric morphometrics of developmental instability: analyzing patterns of fluctuating asymmetry with Procrustes methods. *Evolution* 52: 1363–1375.

Leamy, L. (1993) Morphological integration of fluctuating asymmetry in the mouse mandible. *Genetica* 89: 139–154.

Leamy, L. (1999) Heritability of directional and fluctuating asymmetry for mandibular characters in random-bred mice. J. Evol. Biol. 12: 146–155.

Leamy, L., Routman, E.J., Cherevud, J.M. (1997) A search for quantitative trait loci affecting asymmetry of mandibular characters in mice. *Evolution* 51: 957–969.

Leary, R.F. and Allendorf, F.W. (1989) Fluctuating asymmetry as an indicator of stress: implication for conservation biology. *Trends in Ecology and Evolution* 47: 214–217.

McKenzie, J.A. and Clarke, G.M. (1988) Diazinon resistance, fluctuating asymmetry and fitness in the Australian sheep blowfly, *Lucilia cuprina. Genetics* 120: 213–220.

McKenzie, J.A. and Yen, J.L. (1995) Genotype, environment and the asymmetry phenotype. Dieldrin-resistance in *Lucilia cuprina* (the Australian sheep blowfly). *Heredity* 75: 213–220.

Møller, A.P. and Swaddle, J.P. (1997) *Asymmetry, Developmental Stability, and Evolution.* Oxford University Press, Oxford.

Palmer, A.R. (1994) Fluctuating asymmetry analysis: a primer. In: *Developmental Instability: Its Origin and Evolutionary Implications*, Markow T.A., ed., pp. 335–364. Kluwer Academic, Dordrecht.

Palmer, R.A. and Strobeck, C. (1986) Fluctuating asymmetry: measurement, analysis, patterns. *Annual Review of Ecology and Systematics* 17: 391–421.

Palmer, R.A. and Strobeck, C. (1992) Fluctuating asymmetry as a measure of developmental stability: implications of non-normal distributions and power of statistical tests. *Acta Zoologica Fennica* 191: 57–72.

Parsons, P.A. (1990) Fluctuating asymmetry: an epigenetic measure of stress. *Biol. Rev.* 63: 131–145.

Rohlf, F.J. (1990) Rotational fit (Procrustes) method. In: *Proceedings of the Michigan Morphometrics Workshop*, Rohlf, F.J. and Bookstein F.L., eds., pp. 227–236. University of Michigan Museum of Zoology, Ann Arbor MI.

Rohlf, F.J. and Marcus, L.F. (1993) A revolution in Morphometrics. *Trends in Ecology and Evolution* 8: 129–132.

Rohlf, F.J. and Slice, D. (1990) Extensions of the Procrustes methods for the optimal superposition of landmarks. *Systematic Zoology* 39: 40–59.

Rowe, L., Repasky, R.R. and Palmer, A.R. (1997) Size-dependent asymmetry: fluctuating asymmetry versus antisymmetry and its relevance to condition-dependent signaling. *Evolution* 51: 1401–1408.

Scharloo, W. (1991) Canalization: genetioc and developmental aspects. *Annual Review of Ecology and Systematics* 22: 65–93.

Slice, D. E. (1996) Three-dimensional generalized resistant fitting and the comparison of least-squares and resistant-fir residuals. In: *Advances in Morphometrics*, Marcus, L.F., Corti, M.,

Loy, A., Slice, D. and Naylor, G. eds., pp. 275–283. Plenum press, New York.
Smith, D.R., Crespi, B.J. and Bookstein, F.L. (1997) Fluctuating asymmetry in the honey bee, *Apis mellifera*: effects of ploidy and hybridization. *J. Evol. Biol.* 10: 551–574.
Soulé, M.E. (1967) Phenetics of natural populations. II Asymmetry and evolution in a lizard. *American Naturalist* 101: 141–160.
Waddington, C.H. (1957) *The Strategy of the Genes.* Macmillan, New York.
Whitlock, M. (1996) The heritability of fluctuating asymmetry and the genetic control of developmental stability. *Proc. Roy. Soc. London* B263: 849–854.
Zakharov, V.M. (1989) Future prospects for population phenogenetics. *Sov. Sci. Rev.* 4: 1–79.

17

Invariance and Morphometrics: A Critical Appraisal of Statistical Techniques for Landmark Data

Subhash Lele

Department of Mathematical Sciences, University of Alberta, Edmonton, Alberta T6G 2G1, Canada

17.1 Introduction

Here may be found homely problems, such as often tax the highest skill of the mathematician and reward his ingenuity. (D'Arcy Thompson, 1942)

One may agree or disagree with particular problems and methods proposed by D'Arcy Thompson; however, one must agree that the issues he raised in his seminal book *On Growth and Form* have kept mathematicians and biologists of the twentieth century extremely busy. I have no doubt the same can be said for the mathematicians and biologists of the twenty-first century.

Since the beginning of the study of biological organisms, biologists have always suspected that form follows function (Cuvier, 1828; Allen *et al.* 1998). When studying forms, one usually asks two key questions: Why do organisms look the way they do? Why do organisms differ from each other? Biological systems exist in the present physical world. In order to survive in this world, they have to adapt to the changing environment and the availability of the ecological niches. Also there are purely biomechanical and physiological constraints to consider. It was pointed out by Thompson himself: "The form, then, of any portion of matter, whether it be living or dead, and the changes of form which are apparent in its movements and in its growth, may in all cases alike be described as due to the action of force." Thompson further

On Growth and Form: Spatio-temporal Pattern Formation in Biology.
Editor Mark A.J. Chaplain, G.D. Singh, J.C. McLachlan

states: "We want to see how, in some cases at least, the forms of living things ... can be explained by physical considerations, and to realize that, in general, no organic form exists, save such as are in conformity with physical and mathematical laws." Finding the mathematical laws underlying the form change is certainly a noble and overarching goal for mathematical biology and morphometrics to strive for. However, in practice, there are several limiting factors that make achieving this goal difficult. This chapter discusses these limiting factors and their implications to morphometrics, the quantitative study of form and form change.

17.2 Limiting Factors: Three Major Classes

Limited information. Information about form can be limited in two ways:

(a) *Sample size*: In practice it may be difficult to get a large number of observations. The reasons for this are diverse, for example, the number of available fossils may be limited or the disease under study may be rare to obtain enough cases.

(b) *Number of landmarks*: The form of a biological object is extremely complex. In order to study it biologists often represent a given form by a set of points called landmarks. One of the important characteristics of landmarks is that they reflect the underlying biological processes. Unfortunately, the number of such points may be limited.

Biological variability. It is a fact of life that no two individuals, no matter how closely related, are exactly the same. Before studying differences between forms of two species or groups, one of the key questions that needs to be answered is What is the normal variability within a group? The knowledge of this innate biological variability is essential to decide whether the difference between two species is statistically significant or not. The knowledge of the innate biological variability is also essential to address questions such as, How much of this variability is due to the genetic factors and how much is due to the environmental factors? The presence of this innate biological variability is what makes a statistical approach essential in morphometrics. Paul Holland (1992) puts it succinctly: "The ubiquity of noise is why statisticians and statistics are useful to science - if there is no noise, no uncertainty, then there is no need for us or it." The presence of biological variability also makes it difficult to uncover mathematical laws.

Lack of a common coordinate system. This limitation is extremely severe and also the most difficult to understand and explain. In my experience, this limitation is unappreciated by both biologists and statisticians in the field of morphometrics. The lack of a common coordinate system has serious implications in morphometrics. It is, therefore important that biologists and statisticians understand these implications thoroughly. The rest of this chapter attempts to explain the implications of the lack of a common coordinate system for the study of morphology.

Remarks. We start with the assumption that the biological landmarks are available and are agreed upon by the researchers. Any consequent limitations are accepted and understood. It is up to the biologists and not the statisticians to argue about such a choice. The loss of a common coordinate system imposes serious limitations on the estimation of biological variability. These limitations are discussed in detail by Lele (1993) and Lele and McCulloch (1998). This chapter concentrates solely on the limitations imposed by the loss of a common coordinate system when comparing two forms.

17.3 Comparison of Related Forms

In a very large part of morphology, our essential task lies in the comparison of related forms rather than in the precise definition of each (D'Arcy Thompson, 1942)

Let us begin by ignoring the biological variability. Suppose we are studying two groups of organisms, say two different species of animals. Suppose a typical, representative form for each of the groups is available. One of the questions posed in practice is, Are these two forms different from one another? Although important in some situations where the differences in the forms are subtle, in most practical situations the differences are substantial and the answer is trivially affirmative. The more important question is, How do the two forms differ from one another? For example, we want to make statements of the type: such and such part of the form is most different, such and such part of the form has changed the least. In addition, we also want to quantify these differences. Such information can be a useful guide for the biologists' intuition about the underlying physiological mechanisms that may be causing the form change. The answer to the second question is what we call "localization of form differences". This, the localization, is the most important issue in the quantitative analysis of forms and shapes.

To understand what such a localization could mean, consider a simple situation where the number of landmarks is 3 and the dimension of the object is 2. Let us conduct a simple experiment with three transparencies. We will take these transparencies to be circular with different diameters. Start with drawing the three coordinates of the first object, the type form of the first group, on one of the transparencies using a red pen. Put the other transparency on top of the first and change the location of each of the red landmarks anyway you like and use a green pen to locate these points. The object on the second transparency is our type object from the second group. Take a third transparency and, *without disturbing* the position of the first two transparencies, draw the vectors from the red landmark 1 to the green landmark 1, the red landmark 2 to the green landmark 2, and so on. These vectors legitimately represent the form change from group 1 to group 2. Because these vectors tell us precisely which landmark was changed and by how much, they "localize" the changes from one form to the other. Now remove the third transparency with the vectors depicting the true form change and keep it aside for future reference. In the following, I refer to the transparency with red markings as the "red transparency", the transparency with the green markings as the "green transparency" and the transparency with the vectors depicting the form change as the "true form change" transparency.

To simulate the effect of the loss of the common coordinate system, in the next stage of the experiment, move the red and green transparencies in an arbitrary fashion by throwing them on the floor and picking them back up. The question is, Is it possible to recreate the vectors that depict the true form change using these two transparencies?

There is a class of approaches in morphometrics called the "superimposition-based approaches" to the study of form change that is presumed to recreate the vectors which depict the true form change using these two transparencies.

17.3.1 Superimposition approaches

The basic idea behind the superimposition approach is quite intuitive. Fix the first object, the red transparency. Translate and rotate the second object, the green transparency, so that the landmarks on the green transparency lie in as close proximity as possible to the corresponding landmarks on the red transparency. Then draw the vectors from the red landmark 1 to the green landmark 1, the red landmark 2 to the green landmark 2, and so on. These vectors are presumed to depict the true form change and are used to infer how the first form has changed into the second form (Siegel and Benson, 1982).

The key choice in superimposing two figures is the measure of proximity. A commonly used measure of proximity is based on the Procrustes distance. In this case, after fixing the red transparency, we translate and rotate the green transparency so that the sum of the squared distances between the corresponding landmarks in the red and green transparency is minimized. See Goodall (1991) or Dryden and Mardia (1998) for details on the Procrustes approach to superimposition. Another commonly used superimposition scheme translates and rotates the green transparency so that instead of minimizing the sum of the squared distances, we minimize the sum of the absolute distances. Siegel and Benson (1982) describe this and related superimposition schemes known as the robust superimposition schemes. There are infinitely many different measures of proximity and the corresponding superimposition schemes. See Lele (1991) for a general discussion of the superimposition approach to comparing forms.

Now take the red and green transparencies and use the Procrustes superimposition scheme to superimpose them on each other. Draw the vectors that join the corresponding landmarks, landmark 1 on the red transparency to landmark 1 on the green transparency, etc. This is the representation of form change using the Procrustes superimposition approach. The question that we now ask is, Are the vectors drawn after Procrustes superimposition *identical* to the vectors drawn on the true form change transparency?

Similarly, given the two transparencies with the red and green markings, we could use the robust superimposition scheme and obtain another representation of the form change. The question again is, Are the vectors drawn after robust superimposition *identical* to the vectors drawn on the true form change transparency?

It is trivial to see that these superimposition schemes do not recreate the true form change vectors. Because if they did recreate the true form change, the answers obtained from the Procrustes superimposition and robust superimposition should be identical to each other as well. Of course they are not. Hence it is clear that superimposition schemes do not recreate the true form change. In fact, the superimposition approach would produce the true form change if and only if *all* possible superimposition schemes were to provide the same answer. Those who are not convinced by this should conduct the experiment and see for themselves that the true form change vectors cannot be reconstructed exactly.

Since the vectors representing the true form change (as given on the third transparency that we kept aside) cannot be reconstructed using superimposition, such superimposition schemes do not truly "localize" the form change. The main reason why it is impossible to recreate the original vectors which correctly describe the form change is that once the original relative positions of the red and green transparencies are disturbed, they cannot be put back together in their original positions. There is no common coordinate system to put the transparencies back together.

Besides superimposition, another commonly used method to study form change is transformation grids or the deformation approach.

17.3.2 Transformation grid or deformation approach

The deformation of a complicated figure may be a phenomenon easy of comprehension (D'Arcy Thompson, 1942)

To understand how the loss of a common coordinate system affects this description of form change, let us conduct another thought experiment. Consider a sheet of pasta dough. Suppose we draw a square grid on it. Let us plot the coordinates for the first object (the red transparency in the previous experiment) on this dough. Let us roll the dough in any manner we like. After rolling, the relative positions of the landmarks will change and so will the look of the grid. This changed grid is an excellent descriptor of the form change. It localizes the form difference. Now take another circular transparency and transfer the changed landmark positions onto this transparency using a green marker. As before call these transparencies the red and the green transparencies. Now keep aside the dough with the grid markings.

To simulate the loss of the common coordinate system, the red and the green transparencies are shuffled arbitrarily as in the previous experiment. The question is, Given these two transparencies, is it possible to recreate the grid that is on the dough? The transformation grids are presumed to recreate this true form change grid. However, there are several reasons why the transformation grids can not recreate the true form change grid.

Choice of the deformation

It is well known that there are many ways in which a given object can be deformed into any other given object. The choice of one of these deformations is arbitrary and may not match the deformation caused by our rolling action. In fact D'Arcy Thompson recognized this: "We might suppose that by the combined action of appropriate forces any material form could be transformed into any other." As a solution, he suggests: "We shall strictly limit ourselves to cases where the transformation ... shall be of a simple kind [because] it is a maxim in physics that an effect ought not to be ascribed to the joint operation of many causes if few are adequate to the production of it."

Unfortunately, even under this maxim of parsimony, the limiting factors of the number of landmarks and the lack of a coordinate system cause problems in finding an appropriate deformation and hence the transformation grid associated with the form change.

Number of landmarks and choice of deformations

Let us start with the limitation imposed by the number of landmarks. Recall that we have only three landmarks on each of the objects. With only three landmarks on each of the objects, the simplest deformation that deforms one object into another is the affine deformation. Goodall and Green (1986) provide an excellent description of affine deformation. Although our original deformation caused by the rolling of the dough may be anything other than affine, because we have only three landmarks, the best we can infer under the maxim of parsimony is that the deformation is affine (see the next paragraph for the mathematical reason). This situation is very similar to the situation in classical regression. If only two data points are available, only a straight line can be fitted and inferred; no matter how non-linear the true relationship might be between the covariate and the response variable. If one wants to get a proper idea about the underlying non-linear relationship, one has to have a substantial number of data points (in principle, infinitely many). Similarly, in the study of the deformation of objects, unless one has a substantial number of landmarks, only certain types of deformation may be inferred. The transformation grids based on a limited number of landmarks may not necessarily give an adequate representation of the true biological processes.

Lack of a common coordinate system

Suppose that the underlying deformation is not highly non-linear and that a few landmarks will in fact provide adequate information. Is such a deformation identifiable given the red and green transparencies which have been disturbed from their original positions? Let us assume that the underlying deformation caused by the rolling of the dough is in fact affine. This affine deformation can be described as follows. Let M_1 and M_2 denote the landmark coordinate matrices corresponding to the two objects in their original positions; that is, the coordinates on the transparencies before they are disturbed or thrown on the floor. These are 3×2 matrices with each row representing the coordinates of the corresponding landmark. Under the assumption that the transformation from M_1 to M_2 is affine, we find this transformation by solving the equation: $M_2 = M_1 A + \underline{1}t$ where A is a 2×2 matrix corresponding to the affine transformation, t is a 1×2 vector and $\underline{1}$ is a 3×1 matrix of 1's. This is a linear system of six equations in six unknowns with a unique solution. The matrix A fully describes the deformation from M_1 to M_2. However, we do not know M_1 and M_2 in their original version. All that we know are their rotated and translated versions; that is, we know M_1^* and M_2^* where $M_1^* = M_1 R_1 + \underline{1}t_1$ and $M_2^* = M_2 R_2 + \underline{1}t_2$ with the rotation parameters (R_1, R_2) (where R_1, R_2 are orthogonal matrices) and translation parameters (t_1, t_2) being unknown. Given this situation, suppose we solve the equation $M_2^* = M_1^* A^* + \underline{1}t^*$. The question is, Does $A^* = A$? That is, can we obtain the full and correct description of the true affine deformation? Simple algebra shows that $A^* \neq A$; only the singular values of the two matrices are equal but the left and right eigenvectors are different. The lack of a common coordinate system precludes us from finding the information about the affine deformation fully; singular values can be obtained unambiguously but not the eigenvectors. Lele and McCulloch (1998) give a detailed discussion of this lack of identifiability for some important features under affine and non-affine deformations.

The moral is simple and clear: the number of landmarks and the lack of a common coordinate system limit the kind of information about the form change we can obtain from the landmark coordinate data. Superimposition does not provide the vectors that depict the true form change vectors. Similarly, the transformation grids do not recreate the true form deformations.

The next section puts these ideas in mathematical terms.

17.3.3 Orbits of equivalent forms

Let M be a $K \times D$ matrix, a landmark coordinate matrix. The form of an object as represented by this collection of landmark coordinates is that characteristic which remains invariant under the group of transformations consisting of rotation, reflection and translation. Consider a collection of all $K \times D$ matrices that can be obtained by rotation, reflection and translation of M. This collection is called an orbit described by M under the group of transformations consisting of rotation, reflection and translation. Notice that, under the definition of "form", all the matrices in the same orbit represent exactly the same form.

Strictly, as an aid to understanding the idea of an orbit, consider a topographical map of a mountainous area. Each contour on the map corresponds to points on the surface which have the same altitude. The orbits in the space of all matrices defined above are similar to the contours on a topographical map.

Now suppose a person is hiking somewhere on the surface described by the topographical map. And suppose this person carries a sensor that sends out a signal telling what altitude he is at. That is, all we know about the location of this person is the contour he is on but not the exact position along that contour. This is the same situation we are in when dealing with the landmark coordinate data. Because the transparencies are disturbed from their original position, all that we can tell with certainty is the orbit in the space of all $K \times D$ matrices to which the form under study belongs but not the exact location on the orbit.

Continuing with the hiker example, suppose we know that the hiker started at altitude 2000 feet and is now at altitude 2560 feet. Thus we know which contour he started on and which contour he is on now; but we have no idea which point he started from and which point he has reached now. This is in perfect analogy with the problem of form comparison. We know which orbit the first form belongs to and which orbit the second form belongs to; but not their exact locations on these orbits.

Now suppose we want to describe the movement of the hiker from one orbit to the other orbit. It should be clear that realistically all that we can describe is that he has gone up 560 feet. Although we would like to describe the exact path that he took to gain this height, we cannot do that. The description of the exact path can be provided if and only if we know exactly the point where the hiker started and every intermediate point that he visited. We simply do not have the information to obtain such a description. We should acknowledge this limitation of the data and try to do something with the information we do have: the initial contour and the final contour. Similarly, when comparing forms, we can use only that information which is available unambiguously: the orbit on which the first form lies and the orbit on which the second form lies. We do not know the exact locations of the forms on these orbits and we should acknowledge and accept this limitation.

In precise mathematical terms, let M_1 be any point on the first orbit and let M_2 be any point on the second orbit in the space of the landmark coordinate matrices. Let $Diff(M_1, M_2)$ denote a description of the difference between the two forms. Because we only know the orbits on which the two forms lie but not the exact location on the orbits, this description should be such that $Diff(M_1, M_2) = Diff(M_1R_1 + \underline{1}t_1, M_2R_2 + \underline{1}t_2)$, that is, any such description should be invariant to the point on each orbit which is chosen to describe the difference (Lele and McCulloch, 1998). Going back to the hiker example, recall the description that the hiker has gained 560 feet elevation is invariant to the choice of the point on the initial contour and the point on the final contour. In the presence of this invariance requirement, some natural questions to ask here are, Is it possible to define such an invariant descriptor of form change? Are such invariant descriptors useful to the biologists to infer about the underlying biological process? Fortunately, the answers to these questions are affirmative. One such answer is described below.

17.3.4 Euclidean distance matrix analysis (EDMA)

The first step in finding such an answer would be to find a representation of the form or shape of an object which is invariant to rotation, reflection and translation. To give an example, suppose we have an object with three landmarks. From high school geometry we know that the form of a triangle can be summarized by its three sides. This representation of the form of a three-landmark object is invariant to rotation, reflection and translation. Such an invariant representation of form can be generalized for objects with any number of landmarks in any number of dimensions. The form of an object with K landmarks in D dimensions can be represented in an invariant fashion using the vector of distances between all possible pairs of landmarks (Lele, 1991, 1993). This vector is called a form matrix.

Now suppose we have two objects with K landmarks. To compare their forms, we first construct the form matrices, i.e. the vector of all possible pairwise distances for each one of the objects. These vectors are of length $K(K-1)/2$. Given these two vectors, there are many ways one can describe the difference between them. One particular description that has been used (Lele and Richtsmeier, 1991, 1992, 1995) is the vector of the ratios of the corresponding distances, the form difference matrix. Thus, for three-landmark objects, the form difference matrix is a 3×1 vector with the first element being $d_{12}(B)/d_{12}(A)$, the ratio of the distance from landmark 1 to landmark 2 in object B and the same distance in object A, and so on.

Notice an important property of this description: it only depends on the orbits to which the two forms belong, not on the exact locations along these orbits. Consider the experiment we conducted earlier with the transparencies. Suppose we calculate the form difference matrix as described above for the two objects before the red and green transparencies were thrown on the floor; that is, before their original positions were disturbed. Now suppose we throw the transparencies on the floor and then calculate the form difference matrix. It can be guaranteed that this will be identical to the original form difference matrix. This invariant description of the change in the form is identifiable even after the original positions of the transparencies are disturbed. The lack of a common coordinate system has no impact on the description of form change that satisfies the invariance requirement. Not only does the description of form change using the form difference matrix satisfy the invariance requirement, it has also proved useful in drawing biological inferences (Richtsmeier *et al.* 1993a, 1993b, 1998). But using distances to study form and form change is not a novel idea. Biologists have traditionally used distances in the quantitative study of forms (Reymant *et al.*,

1984, Mosimann, 1970). Recently Rao and Suryawanshi (1996) and Rao (1998) further developed such invariant representations of forms and shapes based on landmark coordinate data as well as outline data for use in various applications such as pattern recognition and discrimination.

In summary, there are descriptors of form change that satisfy the invariance requirement and hence are identifiable even with the disturbed transparencies. The lack of a common coordinate system does not affect these invariant descriptors. These descriptors have also proved useful for drawing biological inferences. I emphasize strongly, however, that even if a particular descriptor of form change satisfies the invariance requirement, it may not necessarily provide biologically meaningful and useful information. A good descriptor of form change should be both invariant and biologically useful.

17.4 Non-Invariant Descriptors and Scientific Consequences

Any descriptor of form change that does not satisfy the invariance requirement is mathematically non-identifiable. That is, if one computes the form change using the undisturbed transparencies and then tries to reconstruct that form change using the disturbed transparencies, the two answers do not match. What are the consequences of this mathematical non-identifiability to the scientific inferences?

Such mathematical non-identifiability will have less serious scientific consequences if the scientific conclusions drawn from $Diff(M_1, M_2)$ and $Diff(M_1R_1 + \underline{1}t_1, M_2R_2 + \underline{1}t_2)$ are not significantly different for any choice of rotation parameters R_1, R_2 and translation parameters t_1, t_2. Such an invariance may be called scientific invariance. On the other hand, if the scientific conclusions vary substantially depending on the choice of the points on the orbits that are used to describe the form change, then one should be extremely cautious in using a non-invariant descriptor.

To understand what such a scientific invariance would mean, consider the hiker example again. Suppose all possible paths our hiker can take to go from altitude 2000 feet to altitude 2560 feet are only minor variations of each other. Although in a strict mathematical sense the paths are different, because they are similar enough to each other, describing any one of these paths will lead us to practically the same information and conclusions as any other path. Although not invariant in a strict mathematical sense, they are invariant in a practical sense.

Can the non-invariant descriptors of form change based on superimposition and deformation be scientifically invariant? We will look at the superimposition approach in more detail. The deformation approach can be handled in a similar fashion. Let us consider the red and green transparencies again. In this simple example, scientific invariance will hold provided *all* possible superimpositions of the two transparencies give practically the same information about how one form was changed into the other. In other words, scientific invariance will hold provided all possible superimpositions provide the form change vectors that are very similar (but not necessarily identical) to the true form change vectors. Unfortunately such scientific invariance does not necessarily hold in real biological examples, as will be shown below.

The following is a slight modification of an example discussed by Richtsmeier (1987, p. 474). Suppose we are comparing the skull of a normal child with the skull of a child with a genetic disorder that affects skull development. Figure 17.1A shows

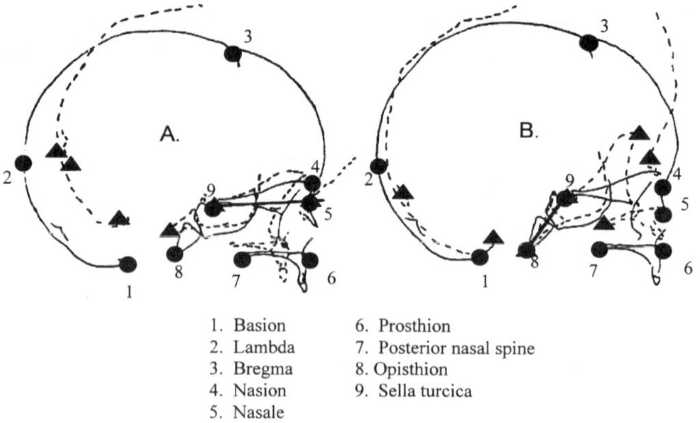

Figure 17.1 An illustration of the scientific non-invariance of the superimposition method. The solid outline depicts the form of the skull of a normal child and the dotted line depicts the form of the skull of a child with a genetic disorder that affects skull development. Part (A) corresponds to one possible superimposition and part (B) corresponds to another possible superimposition. The same biological processes cannot underlie the form changes depicted in both parts.

one possible superimposition of the two forms. If this superimposition is used, a biological conclusion is that posterior neurocranium in the malformed skull is shallower than the normal skull and the superior neurocranium is elongated anterosuperiorly. Orbital morphology is also drawn forward and upward. Finally the palate has shifted posteriorly and superiorly. Figure 17.1B compares the same two forms using another superimposition scheme. In this comparison the posterior and middle neurocranium are nearly identical in both skulls but the anterior cranial base of the deformed skull is elongated superiorly and the palate is displaced superiorly. The same biological processes cannot underlie the form changes depicted in Figure 17.1A. According to Figure 17.1B, there is no difference between skulls local to the posterior neurocranium; whereas according to Figure 17.1A there is a major difference in this region. According to Figure 17.1A, the palate of the deformed individual is displaced posteriorly; whereas according to Figure 17.1B the palate is displace superiorly.

Now let us think of the surgeon reviewing the superimposition and planning the possible corrective surgical procedures. Suppose Figure 17.1A is the *true form change* that has been caused by the genetic mutation. But not knowing this, the surgeon chooses Figure 17.1B to plan the reconstructive surgical procedures. Because, according to Figure 17.1B, the posterior neurocranium is unchanged between the normal and the deformed skull, the surgeon leaves this region untouched. Lack of corrective surgery in this region will result in increased posterior displacement of the posterior neurocranium. Using superimposition Figure 17.1B, the surgeon would plan to lower the palate, when in fact it should have been protracted anteriorly. A surgical procedure based on Figure 17.1B will correct local regions that do not need correcting, leave untouched regions that are truly dysmorphic, and result in a skull that bears little resemblance to a normal skull.

What this example illustrates is that the non-invariance of the superimposition approach is not simply a mathematical curiosity. Ignoring this non-invariance and pretending to be able to obtain the *true form change* using a superimposition approach can have significant implications in practical terms.

17.5 Conclusions and Future Directions

What can we conclude about the methods for the description of form change using landmark coordinate data? The first and foremost mathematical requirement of any descriptor of form change is that it satisfy the invariance condition, namely $Diff(M_1, M_2) = Diff(M_1 R_1 + \underline{1} t_1, M_2 R_2 + \underline{1} t_2)$. However, a particular descriptor of form change that satisfies the invariance requirement may not necessarily provide biologically meaningful and useful information. *A good descriptor of form change should be both invariant and biologically useful.* The descriptors used by EDMA are both invariant and biologically useful. In contrast, the descriptors based on superimposition and deformation ideas are not invariant. This non-invariance also has serious and significant scientific consequences.

Not being a visionary like D'Arcy Thompson, instead of speculating on the future directions, perhaps I should conclude with this quotation: "The reasonings about the wonderful and intricate operations of nature are so full of uncertainty, that, as the wise-man truly observes, hardly do we guess aright at the things that are upon earth, and with labour do we find the things that are before us" Stephen Hales (1727).

Acknowledgments

I am indebted to Joan Richtsmeier and Tim Cole for patiently teaching me the essentials of biology throughout our collaborations. I am also indebted to Charles E. McCulloch for helping me think and present my ideas in morphometrics precisely and clearly. I would like to thank Michael Zumpano for assisting me in writing the description of the biological example and commenting on the manuscript. I would also like to acknowledge the support given by Jim Cheverud, Matt Cartmill and Robert Ehrlich during the initial development of EDMA. This work would not have been possible without these colleagues. I would also like to acknowledge Kanti Mardia and Paul O'Higgins for listening to my arguments and explaining their viewpoint and the organizers of the conference, Mark Chaplain, Dave Singh and John McLachlan for giving me the opportunity to present my work. I, of course, owe a lot to D'Arcy Thompson and would like to thank him for creating such an exciting field of study.

REFERENCES

Allen, C., Bekoff. M. and Lauder, G. (1998) *Nature's purposes: Analysis of Function and Design in Biology.* MIT Press, Cambridge, MA.

Cuvier, G. (1828) *Le Regne Animal Distribue d'apres Son Organisation.* 2nd edn, Fortin, Paris.

Dryden, I. and Mardia, K.V. (1998) *Statistical Analysis of Shapes.* John Wiley, New York.

Goodall, C.R. (1991) Procrustes methods in the statistical analysis of shape (with discussion) *J. R. Stat. Soc., Ser. B*: 285–339.

Goodall, C.R. and Green, P.B. (1986) Quantitative analysis of surface growth. *Bot. Gaz.* 147: 1–15.

Hales, S. (1727) *Vegetable Staticks*, p. 318. (As quoted by D'Arcy Thompson in *On Growth and Form*).

Holland, P. (1992) Discussion of "The impact of sociological methodology on statistical methodology" by C.C. Clogg. *Stat. Sci.* 7: 198.

Lele, S. (1991) Some comments on coordinate free and scale invariant methods in morphometrics. *Am. J. Phys. Anthropol.* 85: 407–418.

Lele, S. (1993) Euclidean distance matrix analysis of landmark data: Estimation of mean form and mean form difference. *Math. Geol.* 25: 573–602.

Lele, S. and McCulloch, C. E. (2000) Invariance and morphometrics. *J. Am. Stat. Assoc.* (to appear).

Lele, S. and Richtsmeier, J.T. (1991) Euclidean distance matrix analysis: A coordinate free approach for comparing biological shapes using landmark data. *Am. J. Phys. Anthropol.* 86: 415–427.

Lele, S.R. and Richtsmeier, J.T. (1992) On comparing biological shapes: Detection of influential landmarks. *Am. J. Phys. Anthropol.* 87: 49–65.

Lele, S.R. and Richtsmeier, J.T. (1995) Euclidean distance matrix analysis: Confidence intervals for form and growth difference. *Am. J. Phys. Anthropol.* 98: 73–86.

Mosimann, J. E. (1970) Size allometry: size and shape variables with characterizations of the lognormal and generalized gamma distributions. *J. Am. Stat. Assoc.* 65: 930–948.

Rao, C.R. (1998) Geometry of circular vectors and pattern recognition of shape of a boundary. *Proc. Natl. Acad. Sci.* 95: 12783–12786.

Rao, C.R. and Suryawanshi, S. (1996) Statistical analysis of shape of objects based on landmark data. *Proc. Natl. Acad. Sci.* 93: 12132–12136.

Reymant, R.A., Blacklith, R.E. and Campbell, N.A. (1984) *Multivariate Morphometrics*, 2nd edn. Academic Press, New York.

Richtsmeier, J.T. (1987) Comparative study of normal, crouzon and apert craniofacial morphology using finite element scaling analysis. *Am. J. Phys. Anthropol.* 74: 473–493.

Richtsmeier, J.T., Cheverud, J., Danahey, S., Corner, B. and Lele, S.R. (1993a) Sexual dimorphism of ontogeny in the crab eating macaque (*Macaca Fascicularis*). *J. Human Evol.*. 25: 1–30.

Richtsmeier, J.T., Corner, B.D., Grausz, H.M., Cheverud, J.M. and Danahey, S. (1993b) The role of postnatal growth pattern in the production of facial morphology. *Systematic Biology* 42(3): 307–330.

Richtsmeier, J.T., Cole, T.M., Krovitz, G., Valeri, C.J. and Lele, S. (1998) Preoperative morphology and development in sagittal synostosis. *J. Craniofac. Genet. Dev. Biol.* 18: 64–78.

Siegel, A.F. and Benson, R.H. (1982) A robust comparison of biological shapes. *Biometrics* 38: 341–350.

Thompson, D.W. (1942) *On Growth and Form*, 2nd edn. Macmillan, New York.

18

Statistical Shape Analysis and its Applications

Kanti V. Mardia

Department of Statistics, University of Leeds, Leeds LS2 9JT, England, UK

18.1 Introduction

Objects are everywhere – natural and man made. It is therefore not surprising that their study has absorbed biologists, anthropologists, engineers, image analysts, computer vision experts, etc. This chapter treats the shape of an object as the geometrical information that is left after location, scale and rotational effects are removed. If scale effects are not removed, then we are led to form (size and shape). In our discussion the underlying theme will be random shapes that is, there is underlying shape variability. We will not consider random form in this context.

D'Arcy Thompson's book, written in 1917, is a classic text on raising problems about the analysis of objects in a biological framework. D'Arcy Thompson looked at the solid shape of objects themselves rather than going through any parsimonious description such as in a set of landmarks. Most recent developments have been through landmark data, where the total shape description of objects is assumed to be contained in a set of landmarks (points of correspondence/homologous points). However, his definition of form is the modern definition of affine shape:

The precise definition of an ellipse introduces us to all the ellipses in the world; the definition of a "conic section" enlarges our concept, and a "curve of higher order" all the more extends our range of freedom. By means of these large limitations, by this controlled and regulated freedom, we reach through mathematical analysis to mathematical synthesis. We discover homologies or identities which were not obvious before. (Thompson, 1942, p. 1027)

On Growth and Form: Spatio-temporal Pattern Formation in Biology.
Editor Mark A.J. Chaplain, G.D. Singh, J.C. McLachlan

His interests have been mainly on deformation, and landmarks seem to appear somewhat indirectly:

A series of points, a, b, c, along a certain line persist as corresponding points a', b', c', however the line connecting them may lengthen or bend; and as with points, so with lines, and so also with areas. Ear, eye and nostril, and all the other great landmarks of cranial anatomy, not only continue to exist but retain their relative order and position throughout all our transformations. (Thompson, 1942, pp. 1085-1086)

There has also been progress in landmark-free shape description, i.e. the object as a whole. The main emphasis in this chapter will be on landmark-based methods but it also describes some procedures which are landmark-free. It describes methods which retain full geometry in a direct way. This is similar to the strategy adopted by D'Arcy Thompson's studies in deformation, e.g. deformation of fish from one species into another (Thompson, 1917; reproduced in Dryden and Mardia, 1998 p. 8, figure 6). Fred Bookstein and David Kendall pioneered developments in the late 1970s and early 1980s and have continued to work on landmark-based methods; see Dryden and Mardia (1998) for further historical comments. For a differential geometric approach to shape, see Kendall (1984, 1989), Le and Kendall (1993) and Small (1996).

18.2 Examples

The simplest example is three labelled vertices of a triangle. Figure 18.1 shows five labelled right-angled triangles A, B, C, D, E and another triangle F. Triangle A is the same shape as B and C and triangle D has the same shape after relabelling. Triangle E is a reflection of triangle D, so in our discussion E and D do not have the same shape. Triangle F is none of these. We will be concentrating on shapes without reflection.

We will be working through an example where the points are taken around the outline of a mouse vertebra (T2) shown in Figure 18.2. It has six mathematical landmarks which are extremal points of curvature; in between each pair of mathematical landmarks are seven points, equally spaced pseudo-landmarks. These key points around the outline of the bone will be taken to describe the bone. An example of a biological landmark is say the tip of the nose on a face. The key features on the face can be taken to describe eyes, nose, chin, etc.

18.3 Bookstein Shape Coordinates

Since location, size and orientation are not relevant for shape, it is convenient to "register" the objects by choosing a standardization. We first describe Bookstein registration, which has the following steps:

- Choose two landmarks labelled 1 and 2; this is the base (Figure 18.3a).
- Translate the object so that landmark 1 lies at the origin (0,0) (Figure 18.3b).
- Rotate the object so that landmark 2 lies on the positive x-axis (Figure 18.3c).
- Scale the object so that landmark 2 lies at (1,0) (Figure 18.3d).

The final coordinates of landmark 3 are shape coordinates and are called Bookstein coordinates. Galton (1907) himself used these coordinates a long time ago, but the statistical developments are due to Bookstein (1986). Mathematically, let us denote

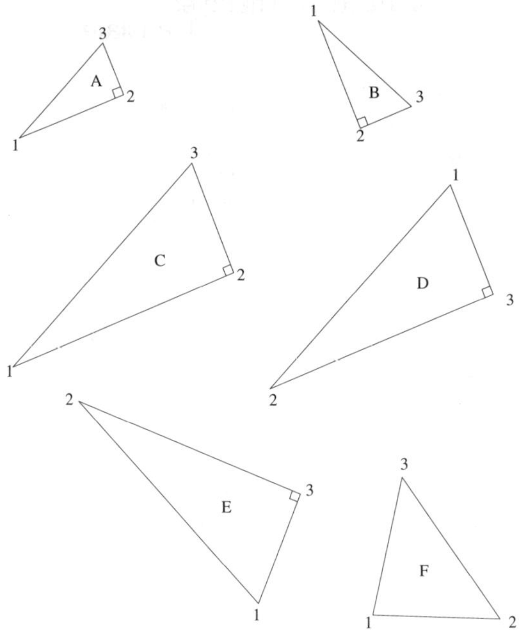

Figure 18.1 Six labelled triangles: A, B and C have the same shape; D has the same shape after relabelling; E can be reflected to D; F has a different shape from the rest. From Dryden and Mardia (1998, p. 6).

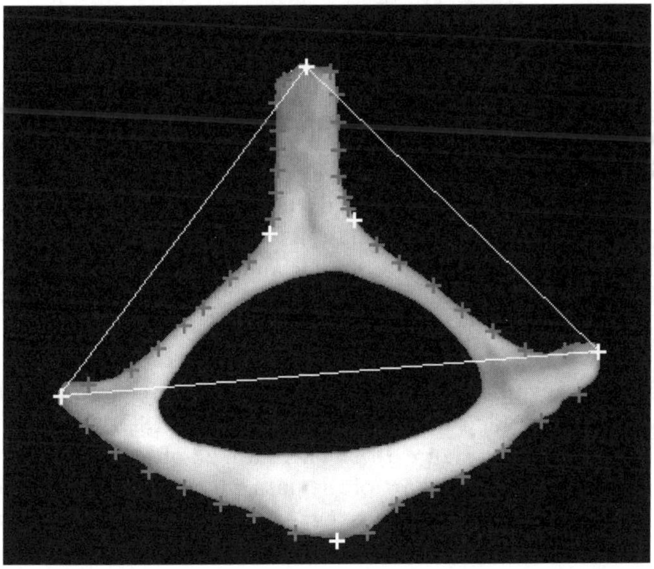

Figure 18.2 Grey level image of T2 mouse vertebra with six mathematical landmarks (dark +) and 42 pseudo-landmarks (light +) and a triangle. Adapted from Dryden and Mardia (1998, p. 4).

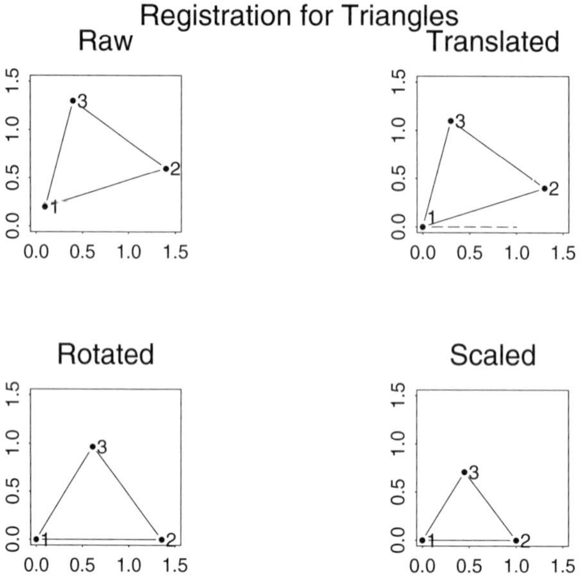

Figure 18.3 The geometrical interpretation of Bookstein coordinates (see text).

these landmarks by $x_1, x_2, ..., x_k \in \mathbb{C}$ in the complex plane so that Bookstein shape coordinates are given by

$$u_j = \frac{x_j - x_1}{x_2 - x_1}, \quad j = 3, ..., k.$$

Note that the use of complex coordinates gives a simple expression.

We can use Bookstein coordinates for the following:

(a) Plot a collection of shape data using this registration for visualization.
(b) Turn shape analysis (which is complicated due to the non-Euclidean shape space) into a branch of multivariate analysis by replacing k landmarks on each object with $2k-4$ Bookstein coordinates and carrying out the standard multivariate techniques using these $2k-4$ variables, e.g. compute average shape, carry out statistical tests under multinormality.

Let us consider as an example the vertebrae bones of two groups of mice, small (S) versus medium (M) (Dryden and Mardia, 1998, pp. 313–316). Take only three landmarks A, B, C as shown in Figure 18.2. There are 23 observations in each group. Figure 18.4 shows a scatter plot of a part of the data with their mean triangles. Notice that the mean shapes are symmetrical as expected, so both are close to isosceles triangles: group S leads to a slightly flatter isosceles triangle than group M. The strengths of this representation are as follows:-

- It is easy to do.
- It gives concrete representations of different shapes.

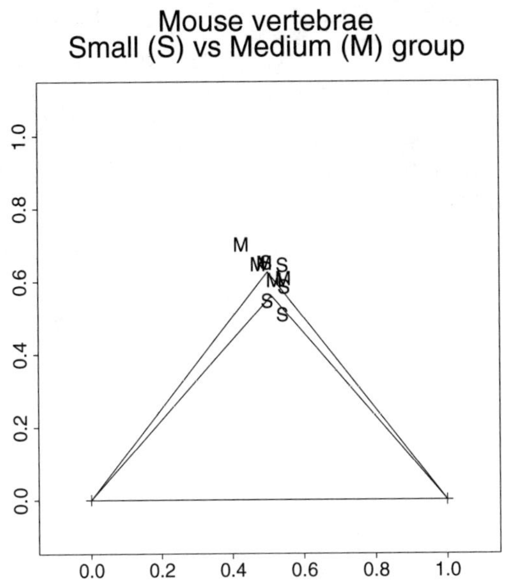

Figure 18.4 Scatter plot of vertebrae data (S = small, M = medium) in Bookstein coordinates.

- For concentrated shape distributions it makes very little difference which two landmarks are used as the base, but the representation does depend on the choice of landmarks used for the base.

and the weaknesses are:

- It does not work if the two landmarks for the base are nearly coincident.
- It does give asymmetrical results due to the preferred role of the chosen base. In particular, one has to be careful in using this system for examining shape variability.

18.4 Kendall's Shape Space and Procrustes Coordinates

We again consider the two-dimensional case with $x_1, ..., x_k$ representing landmarks in complex variable notation. Kendall shape space is defined as follows (with distance on the space as the Procrustes distance):

(i) Remove location

$$v_j = x_j - \bar{x}, \ j = 1, ..., k.$$

(ii) Remove size (or rescale)

$$z_j = \frac{v_j}{\sqrt{\Sigma |v_j|^2}} = \frac{x_j - \bar{x}}{\sqrt{\Sigma \parallel x_j - \bar{x} \parallel^2}}, \ j = 1, ..., k.$$

If $x = a + ib, |x| = \sqrt{(a^2 + b^2)}$.

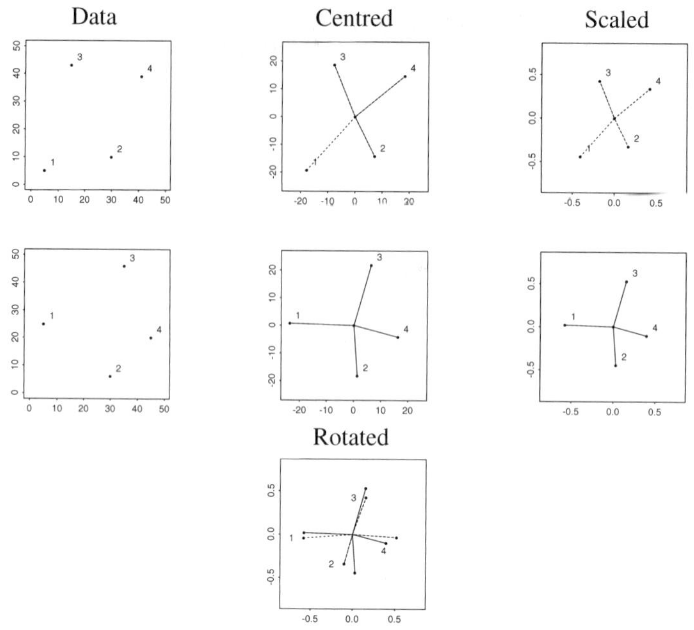

Figure 18.5 The geometry of Procrustes fits in calculating shape distances. Adapted from Bookstein (1996).

(iii) Remove rotation by identifying $\mathbf{z}e^{i\theta}\{0 < \theta \leq 2\pi\}$ and $\mathbf{z}$ to be the same point, and let this equivalence class be denoted by $[\mathbf{z}]$ where $\mathbf{z}^T = (z_1, ..., z_k)$. Then $[\mathbf{z}]$ is the shape of the configuration defined by $x_1, ..., x_k$. Note that (i) implies

$$\mathbf{1}^T\mathbf{z} = 0 \tag{18.1}$$

and (ii) implies

$$\mathbf{z}^*\mathbf{z} = 1 \tag{18.2}$$

where $\mathbf{z} = (z_1, ..., z_k)^T$. This sphere is called the preshape sphere. Thus shape space is complex projective space of k-2 dimensions since the point $\mathbf{z}$ and $\mathbf{z}e^{i\theta}$ are equivalent. The situation is like axial data in directional statistics, where directions $\mathbf{x}$ and $-\mathbf{x}$ are identical for a point $\mathbf{x}$ on a sphere $|\mathbf{x}| = 1$. (Mardia, 1972).

We now discuss Procrustes shape coordinates which are directly related to Kendall's space. First, consider these ideas in matching two configurations $\mathbf{z}_1$ and $\mathbf{z}_2$ (in complex notation):

(i) Centre the two configurations (shown in Figure 18.5; middle top two).
(ii) Scale the two configurations (Figure 18.5 extreme RHS).
(iii) Rotate and rescale $\mathbf{z}_1$ to $\mathbf{z}_2$ by least squares (Figure 18.5 bottom).

Thus, we get this Procrustes matching of shape $\mathbf{z}_1$ to shape $\mathbf{z}_2$. We can extend this procedure to obtain the Procrustes coordinates. Let $\mathbf{z}_i$, $i = 1, ..., n$ be n configurations.

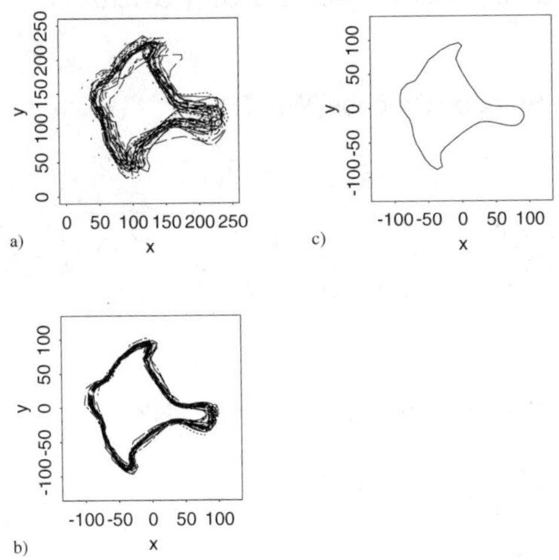

Figure 18.6 (a) Full data for T2 small mouse vertebrae, (b) Procrustes matching and (c) Procrustes mean.

(i) Centre the configurations $1^T \mathbf{z}_i = 0$.
(ii) Scale the configurations $\| \mathbf{z}_i \| = 1$.
(iii) Compute the (full) Procrustes coordinates

$$P_i = (\mathbf{z}_i^* \hat{\boldsymbol{\mu}})\mathbf{z}_i \ , \ i = 1, ..., n,$$

where $\hat{\boldsymbol{\mu}}(\| \hat{\boldsymbol{\mu}} \| = 1)$ is the largest eigenvector of $\sum_{i=1}^{n} \mathbf{z}_i \mathbf{z}_i^*$, and $\mathbf{z}_i^*$ denotes the transpose of the complex conjugate of $\mathbf{z}_i, i = 1, ..., n$. Note that step (ii) indirectly assumes that the size is measured by $\| \mathbf{z}_i \|$.

We can now use these Procrustes coordinates P_i as we have done before for Bookstein coordinates by turning shape analysis into a branch of multivariate analysis using Euclidean approximations (Goodall, 1991; Kent, 1994). Here is an example:

(i) The representation of Procrustes average shape is $\bar{P} = \frac{1}{n} \Sigma P_i$.
(ii) For shape variability use the tangent space

$$\mathbf{u}_i = P_i - \bar{P}, \ i = 1, ..., n.$$

(iii) The Procrustes distance between shapes $\mathbf{z}_i$ and $\mathbf{z}_j$ is approximately equal to the Euclidean distance between $\mathbf{u}_i$ and $\mathbf{u}_j = \| \mathbf{u}_i - \mathbf{u}_j \|$.
(iv) Take $\mathbf{u}_i, \ i = 1, ..., n$ to be a sample from a (singular) multivariate normal population for inference.

Figure 18.6 shows the full vertebra data together with Procrustes matching, and their Procrustes mean. We can also display shape variability as follows:

Given a training set of shapes we can apply principal component analysis on $\mathbf{u}_i, \ i = 1, ..., n$ to find the dominant modes of shape variation. Following Cootes *et al.* (1992), we plot

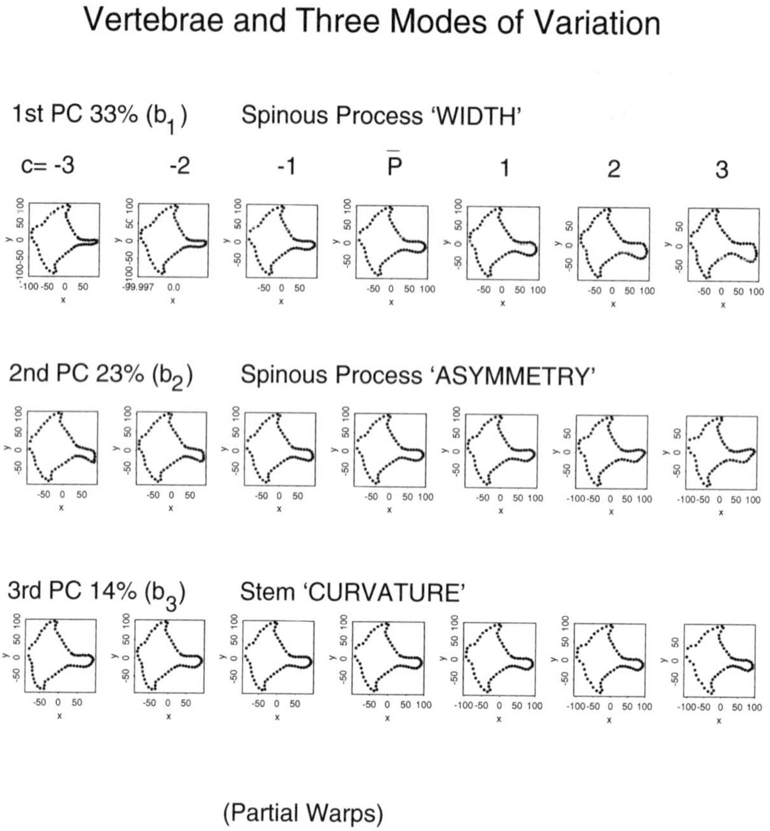

Figure 18.7 Three rows showing a series of T2 vertebra shapes evaluated along the first three PCs.

$$\bar{P} \pm c\lambda_j^{\frac{1}{2}} \mathbf{b}_j, j = 1, 2, 3; c = 0, \pm 1, \pm 2, \pm 3,$$

where $\mathbf{b}_j$ is the jth principal component (PC) of the $(2k \times 2k)$ matrix of the sums of squares and products of

$$[\mathrm{Re}\,(P_j - \bar{P}),\ \mathrm{Im}(P_j - \bar{P})],\ j = 1, ..., n,$$

with eigenvalue λ_j. Note that the rank of this matrix is $2k - 4$.

This step is similar to standard PCA but has the advantage of the graphical display of modes of variation. Note that the PCA in Bookstein registration can lead to spurious correlation even for isotropic models (Dryden and Mardia, 1998, p. 173).

Figure 18.7 shows the variations for the vertebra data. The first PC includes a contribution to the variation in spinal process, the second PC includes variation in the asymmetry of spinal process and the third PC includes variation in stem curvature. The first PC accounts for 33% of the variability, the second for 23% and the third for 14%. The use of principal warps and relative warps of Bookstein (1989, 1991) provides an alternative method of visualization.

18.5 Shape Change and Deformations

We first start with a quotation from D'Arcy Thompson on the importance of deformations:

In a very large part of morphology, our essential task lies in the comparison of related forms rather than in the precise definition of each; and the *deformation* of a complicated figure may be a phenomenon easy of comprehension, though the figure itself have to be left unanalysed and undefined. This process of comparison, of recognising in one form a definite permutation or *deformation* of another, apart altogether from a precise and adequate understanding of the original "type" or standard of comparison, lies within the immediate province of mathematics, and finds its solution in the elementary use of a certain method of the mathematician. This method is the Method of Coordinates, on which is based the Theory of Transformations. (original italics, Thompson, 1942, p. 1032).

To understand shape changes, we consider deformations which can be used to change one shape into another. Let A be the source landmarks in the x-plane with $x = (x_1, x_2)$ and let B be the target landmarks in the y-plane with $y = (y_1, y_2)$. Let the source landmarks be $x_i = (x_{1i}, x_{2i}) \in R^2, \quad i = 1, ..., n$, and let the target landmarks be $y_i = (y_{1i}, y_{2i}) \in R^2, i = 1, ..., n$. The deformation problem reduces to finding a smooth transformation of A to B,

$$y_1 = \phi_1(x), y_2 = \phi_2(x)$$

given that source landmarks are mapped to the target landmarks. One such transformation consists of a pair of thin-plate splines (Bookstein, 1989) which has certain physical properties. Its extension is a pair of kriging predictors for ϕ_1 and ϕ_2 (Mardia *et al.*, 1996b). It is interesting to note that in the following quotation, perhaps D'Arcy Thompson foreshadowed the method.

But as soon as we have marked out a number of points in the gorilla's or chimpanzee's skull, corresponding with those which our coordinate network intersected in the human skull, we find that these corresponding points may be at once linked up by smoothly curved lines of intersection, which form a new system of coordinates and constitute a simple "projection" of our human skull. (Thompson, 1942, p. 1083).

And how these points should have homology is further elucidated by him:

The cuttlefish eye seems as perfect, optically, as our own; but the lack of an invariant relation of position between them, or lack of true homology between them (as we naturalists say), is enough to show that they are unrelated things, and have come into existence independently of one another. (Thompson, 1942, p. 1086).

The simplest deformations are affine (Dryden and Mardia, 1998, pp. 201–202) and a famous example of the affine deformation is the transformation of fish *Argyropelecus* to fish *Sternoptyx* of Thompson (1942, p. 1062, Figures 517, 518; also reproduced in Dryden and Mardia, 1998, p. 202, Figure 86).

Figure 18.8 gives an example with the thin-plate splines deformations. The effect of such deformations can be understood more easily by deforming images (Dryden and Mardia, 1998, Ch. 11). However, it is worthwhile to note the importance of this task from a biological point of view as summarized by Medawar:

The best and most famous chapter of *Growth and Form* is that which embodies D'Arcy Thompson's own most completely original contribution to biology: the Method

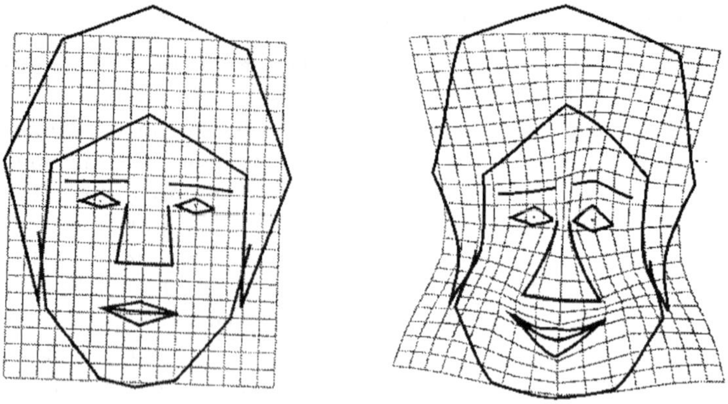

Figure 18.8 (a) A face with 33 landmarks and (b) its deformation (Mardia and Hainsworth, 1993). Reproduced by kind permission of Carfax Publishing Limited.

of Transformations, in the chapter 'On the Comparison of Related Forms'. Consider the shapes of two organisms of related genera. Being closely related, the two shapes will be 'homeomorphic' – that is, roughly speaking, they will be qualitatively the same, so that the one could be changed into the other, as two different faces could be, by some process of plastic remodelling. But although two such shapes may be qualitatively similar, they will in fact differ in a multitude of particular little ways. The anatomist's method of comparing shapes is to do so piecemeal, feature by feature, with perhaps an occasional measurement of proportion. D'Arcy saw that all these particular little differences of disposition, angle, length, and ratio might be simply the topical expressions of some one comprehensive and pervasive change of shape. He grasped the transformation *as a whole*. (Medawar, 1956, pp. 229–230, original italics).

Averaging images

To start with, consider that we have two images that we wish to combine. Direct superimposition will usually be unsuccessful because key features will not be matched.

1. Define landmarks in each image (Figure 18.9a)
2. Average the two shapes together (Figure 18.9b)
3. Deform each image so its landmarks match those of the average shape (Figure 18.9c)
4. Superimpose the two deformed images (Figure 18.9d).

Figure 18.10 gives an average of before and after operation. The eventual aim could be to predict growth from child to adult. Note that the problem of averaging photographs goes back to Galton (1878), who used an "analogue" method rather than the digital method of Mardia (1993) that we have described. For various other applications of averaging objects, see Glasbey and Mardia (1998).

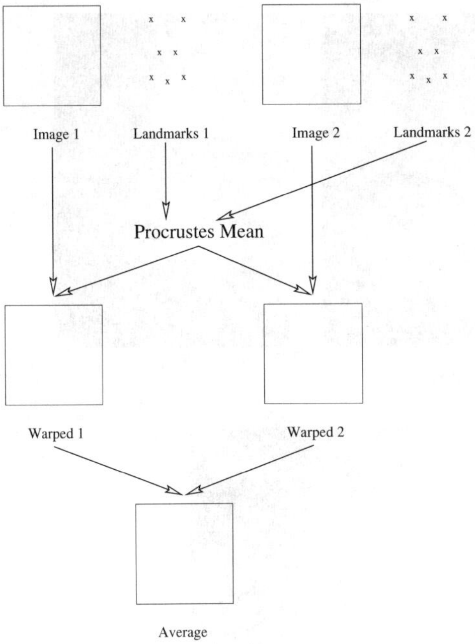

Figure 18.9 A schematic diagram of merging two images (see text).

Landmark-free Approach

Suppose we have to work with grey-level objects. Using certain morphometric measurements obtained from photographs, we may wish to discriminate between two species of fish, e.g. haddock and whiting, not an easy problem (Strachan *et al.*, 1990).

In general we can proceed as follows. Let $I_k(x, y), k = 1, ..., K$ be K images. Suppose that $\{u_k(x, y), v_k(x, y)\}$, is the kth warp (to be estimated) and let warped images be $I'_k(u_k(x, y), v_k(x, y))$. According to Glasbey and Mardia (1998), we can compute

$$\bar{I}(x, y) = \sum_{k=1}^{K} I'_k(u_k(x, y), v_k(x, y))/K$$

where the warps $(u_k, v_k), k = 1, ..., K$, minimize a sum of two terms: mean squared difference between $\bar{I}(x, y)$ and $I'_k(u_k, v_k)$, $k = 1, ..., K$, and a factor λ times the sum of shape roughness penalties for warping $I_k(x, y)$ to $I'_k(u_k, v_k)$. The multiplying factor $\lambda(> 0)$ is a "smoothing" parameter. Glasbey and Mardia (1998) give details of how to discriminate between haddock and whiting given a grey-level image of a fish from these two species. A landmark-free approach for outlines has been introduced by Bookstein (1997).

18.6 Models and Concentrated Data

We discuss the two-dimensional case. Suppose that $x = (x_1, ..., x_k)^T \sim \mathbb{C}N(\mu, \sigma^2 I_k)$ so that the landmark vectors have a complex multivariate normal distribution with mean vector μ and covariance matrix I_k where I_k is the $k \times k$ identity matrix. Hence

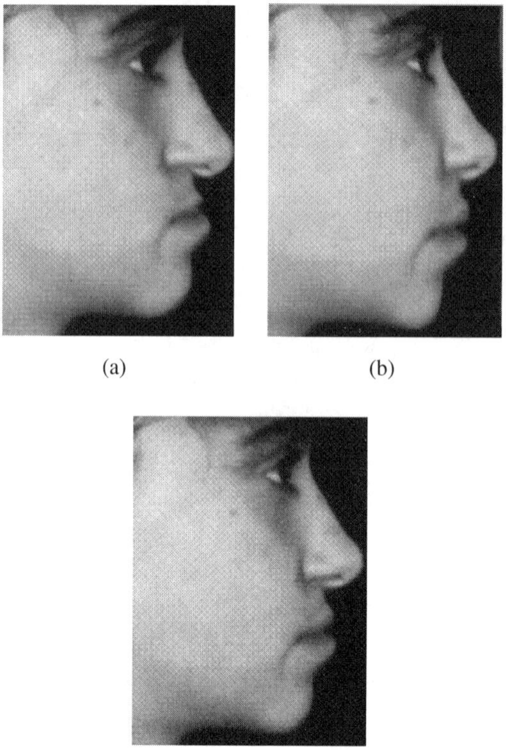

(a) (b) (c)

Figure 18.10 Photographs (a) and (b) show a patient before and after an operation to correct a cleft plate. Photograph (c) is merged between (a) and (b) (reproduced from Mardia, *et al.*, 1995, by permission of Oxford University Press).

it is assumed that the population configuration is perturbed around its mean.

If the means are coincident then we have a uniform shape distribution leading to Kendall's uniform measure (Kendall, 1984). For the case when the means are not coincident and take general values, then Mardia and Dryden (1989a,b) give the exact distribution. These distributions are basically offset normal shape distributions. This work can be extended further for general covariances (Dryden and Mardia, 1991) but the maximum likelihood estimation requires computational care. For higher dimensions, Goodall and Mardia (1991, 1992, 1993) provide various extensions. For the triangle case, one method is to use the relationship between $\mathbb{C}P^1$ with the sphere in three-dimensions and then to use the Fisher–Bingham distribution; Kent's family (Kent, 1982) has the advantage of leading to a bivariate normal density for concentrated data.

Alternatively, one could use a distribution in a preshape space with rotational symmetry so that it can be treated as a shape distribution. The complex Bingham distribution (Kent, 1994) is one example and this case assumes complex symmetry but the distribution is quite tractable. In particular, the maximum likelihood estimate of the mean shape turns out to be the dominant eigenvector γ of sums of squares and

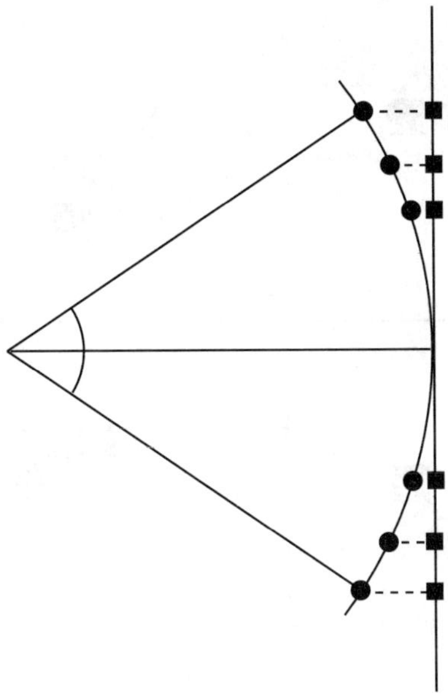

Figure 18.11 Concentrated angular data and their projection on a tangent line.

products of preshapes. A particular case of interest is the complex Watson distribution (Mardia and Dryden, 1998); in spirit it is the same as an offset shape distribution for the isotropic case but it is simpler and has various interesting properties, including that the maximum likelihood estimate of the mean shape is again γ.

So far we have only come across biological data sets which are highly concentrated (O'Higgins, 1998). Thus in practice, we can use the tangent space to the shape space in practice (e.g. we can use the Procrustes method described earlier). To illustrate the idea of concentrated data, consider the angular data given in Figure 18.11. Here we could use a tangent projection of the data, namely projection on a tangent line. Thus the angular data become linear data and classic statistics can be used.

Figure 18.12 gives concentrated configurations after Procrustes matching. We assume that there is a preferred configuration; this could be the Procrustes mean or it could be any other configuration labelled $\mathbf{a}$ in the "neighbourhood" of the observed configurations. The selected configuration forms a rectangle. The offset normal distributions are symmetric functions of the Procrustes distance only, and lead to a multi-normal distribution in the tangent space. The assumption of multinormality in the tangent space is thus tenable.

18.7 Different Approaches

I now give some insight into other methods based on landmark data. There are alternative methods such as using distances and angles. Distances were used by Karl Pearson in shape description as early as the 1930s. The angles were used for shape by Mardia *et al.* (1977) and recently by Rao and Suryawanshi (1998). Lele (1993) and

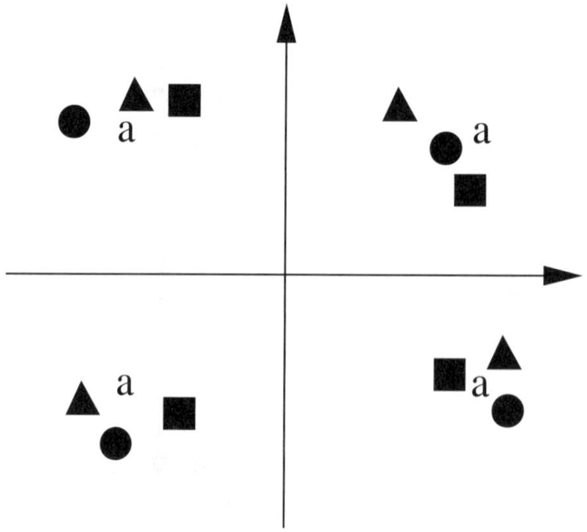

Figure 18.12 Concentrated configurations (▲, ■, •) and a 'neighbourhood' configuration (**a**).

Stoyan and Stoyan (1994) proposed a method called EDMA. Rao and Suryawanshi (1996) and Mardia *et al.* (1996a) proposed the use of log distances in shape space. As pointed out by Kendall, angles for flat triangles suffer from theoretical disadvantages. Distances, of course, are appropriate for reflected shapes and most of these methods are equivalent for calculating the mean shape for highly concentrated data. However, the distance-based methods seem to have limitations for describing deformations.

The distance-based methods (EDMA) have some advantages for analysis of form (Dryden and Mardia, 1998, pp. 282–284). These are computationally straightforward in comparison to the offset normal shape distributions but sensitive to outliers since the fourth-order moments are used in removing bias in distances. Note that under the assumption of a factorized covariance for the normal configuration, we have a *consistent* estimator for the mean and the covariance matrix. See Dryden and Mardia (1998, p. 280) for various comments on the meaning of a consistent estimator and its implications for shape analysis.

Taking logs of distances for shape requires care if two landmarks are too close together or too far apart. Rao and Suryawanshi's (1996) method of stacking vector of log distances followed by a multivariate analysis seems to create highly non-linear space since in two dimensions from $2k$ landmarks we move to $k(k-1)$ distance variables (Chapter 20). Mardia *et al.* (1996a) have introduced a similar method for shape analysis but in their case, the data are genuinely a subset of interpoint distances.

Direct landmark methods based on the Procrustes approach can be justified through isotropic models. A more general approach is to use the offset shape distribution or the complex Bingham distribution; the maximum likelihood estimators for the complex Bingham distribution are comparatively easy to compute. Although computational aspects of general models have not been fully developed, one can use multinormal approximations for analysis of concentrated data, since biological shape data always seem to be concentrated (Chapter 20). There is, of course, the advantage in being able to model deformations of outline/solid shapes, etc., which are difficult to explore using distances alone. There are also various visualization tools for understanding shape variability using direct landmark methods.

In Chapter 17 Subhash Lele has raised various questions about Procrustes methods for matching. Firstly, note that the configurations are labelled. Secondly, Procrustes analysis uses the least squares principle which is statistically acceptable. However, one can use robust methods or other loss functions; hence there is nothing strange in Figure 17.1 (Chapter 17). The question does not seem to be much different than even in the basic statistics of whether to use mean, mode or median, etc., for data in one dimension (Section 18.8). Also using different loss functions in EDMA will lead to different results. The issue of using different loss functions is sometimes misunderstood and exaggerated by the exponents of EDMA. The appropriateness of a loss function must depend on the biological problem to some extent. Furthermore, registration is an important issue, and Procrustes methods give us a mathematical solution. In examples like plastic surgery, it will be necessary to bring external knowledge into account to decide what registration might be the most appropriate.

Deformations have some similarity with curve fitting. For example, fitting a straight line, one can work with 2, 3 or 4 points, and so on. In general, the greater the number of points, the higher the accuracy. Also deforming by thin-plate splines (or other similar deformations), we have a situation like regression of y on x or of x on y. Thus the deformation depends on what is the source and what is the target. The same is true for the principal and partial warps. Also, if distortions are not severe, one expects there will be no folding in deformations. A pair of thin-plate splines stretches a plane most smoothly (in a particular mathematical sense) in deformation. Perhaps this represents Medawar's thinking.

So it might be, for example, with the change of shape produced by distorting a sheet of rubber on which has been drawn a house, or face, or any other figure whatsoever: the shapes of the drawings change in every single particular, but the transformation as a whole might be defined by some quite simple formulae describing the way in which the rubber had been stretched. (Medawar, 1958, p. 230).

This approach seems to give some answer to Medawar's following statements:

The reason why D'Arcy's method has been so little used in practice (only I and one or two others have tried to develop it at all) is because it is analytically unwieldy. (Medawar, 1958, p. 231).

It is clear that the subject is still in its infancy and at least there are no theoretical difficulties when one comes to problems like deformable templates in object recognition. Of course, one has to have a general perspective. In classic statistics there are different measures of locations (mean, mode, median, etc.) and different measures of variability (interquartile range, variance, etc.). i.e. different loss functions provide different strategies. And on the whole, shape analysis is no exception, including EDMA. In factor analysis, there is invariance under rotation (but we can use varimax rotation so that some loading coefficients are zero) and in shape analysis there is invariance under the similarity transformations. Certain techniques in shape analysis are inherently distinct. The problem is similar in extending univariate to multivariate analysis or time series to spatial analysis, where some techniques are inherently multivariate and spatial.

It seems that measures for calculating a shape mean are in general equivalent for concentrated data (Kent, 1994) but the problem of measuring a shape change is a different matter. Some methods will report via deformations, some by pairs of points, some by regions where the transformation is particularly non-linear; these representations have different diagrams, different statistics and different implications. This discussion reminds me of the holistic principle (Anekāntvāda) of Jain logic, which uses the story of six blind men and an elephant. Each blind man touches a different part of the elephant and they come to a differing opinion. Hence if we wish to understand what kind of object the elephant is, we must look at it from all sides (Mardia, 1996), i.e. use Anekāntvāda!

18.8 Future Perspectives

Most of the description here is based on landmark data, but landmark-free methods are quite important as well. In future one will see further developments in the landmark-free approaches. A few pointers have been given in Section 18.5. However, early historical work involved the use of interlandmark distances, measured with calipers on biological specimens. With modern technology we can now measure landmarks directly in both two- and three-dimensions. Thus, development of landmark-based methods has been more recent. Also there are ways of obtaining landmarks automatically, e.g. Mardia *et al.* (1997) who use scale space methods to obtain landmarks from given photographs. A historical point to note is that David Kendall dealt with diffuse shapes whereas Fred Bookstein dealt with concentrated shapes. The models such as complex Bingham, offset normal, etc., also allow us to study non-concentrated shapes.

Stochastic deformable templates based on geometric properties of objects, average shape with textures, and human atlas building for anatomy are other growing areas. Thin-plate splines deformation, finite element methods, fluid mechanical models, and smooth versus discontinuous deformations will continue to be challenging.

I hope to see further developments, especially in non-parametric methods for deformation rather than the parametric approach described earlier. Perhaps D'Arcy Thompson is alluding to non-parametric deformation in the following quotation:

The problem is closely akin to that of the cartographer who transfers identical data to one projection or another; and whose object is to secure (if it be possible) a complete correspondence, *in each small unit of area*, between the one representation and the other. The morphologist will not seek to draw his organic forms in a new and artificial projection; but, in the converse aspect of the problem, he will enquire whether two different but more or less obviously related forms can be so analysed and interpreted that each may be shown to be a transformed representation of the other. (Thompson, 1942, p. 1033, original italics).

The discussion here has focused on simple shape descriptors but one foresees developments in the use of curvature, torsion, ridge curves, etc., in three dimensions. In general, the work in higher manifolds has a lot of scope. Overall, functional analysis (Ramsay and Silverman, 1997) has a lot to offer to shape analysis. Also there is further scope for new methods to analyse shape with covariates such as size, directions, etc. Similarity shape is a particular aspect of large areas of pattern analysis, led by Grenander and his group (Grenander, 1993). The approach emphasizes the compositional nature and how regular structures are formed from "generators" leading to image algebra. The methods also provide a computationally efficient method to derive optimal solutions.

Growth studies is another challenging area. Indeed, D'Arcy Thompson was preparing a third edition of *On Growth and Form* (Archive, University of St Andrews) and the following insertion to his second edition on page 1088 sets a scene for problems in this area while acknowledging the work of Medawar (1944):

It goes almost without saying that this method of comparing one animal with another applies equally well, perhaps more easily still, to the successive phases of growth, or stages of development, in a single individual. An orderly series of transformations, or progression of change, leads from infant to child, to youth and to man; Durer's drawings, or C.H. Jackson's, show this at a glance. Mr. P.B. Medawar has made an analytical study of such a development, treating it as a continuous set of transformations, "maintaining a definite general sense of distortion, alike in space and time." We may learn nothing new by doing so; but we gain a clearer insight into the transformation involved, and the analytical properties which are involved in it.

We have treated only some areas where shape statistics have an impact on morphology. More to the point, morphometrics and images continue to provide new and exciting developments in shape methodology. It will be worthwhile to end with a quote from Thompson (1942, p. 1096):

But it is not for the sake of polemical argument that I have written, and the doctrines which I do not subscribe to I have only spoken of by the way. My task is finished if I have been able to show that a certain mathematical aspect of morphology, to which as yet the morphologist gives little heed, is interwoven with his problems, complementary to his descriptive task, and helpful nay essential, to his proper study and comprehension of Growth and Form.

Acknowledgements

I am grateful to Ian Dryden, John Kent, Richard Morris and Kevin de Souza for their helpful comments. I have been interested in D'Arcy Thompson's work for quite some time and wish to express my thanks to Mark Chaplain and Dave Singh for inviting me to the conference. In particular, the conference gave me a chance to discuss again with Subhash Lele the mathematical aspects of shape analysis.

REFERENCES

Bookstein, F.L. (1986) Size and shape spaces for landmark data in two dimensions (with discussion) *Statistical Science* 1: 181–242.

Bookstein, F.L. (1989) Principal warps: thin-plate splines and the decomposition of deformations. *IEEE Transactions on Pattern Analysis and Machine Intelligence* 11: 567–585.

Bookstein, F.L. (1991) *Morphometric Tools for Landmark Data: Geometry and Biology.* Cambridge University Press, Cambridge.

Bookstein, F.L. (1996) Biometrics, biomathematics and the morphometric synthesis. *Bulletin of Mathematical Biology* 58: 313–365.

Bookstein, F.L. (1996) Applying landmark methods to biological outline data. In: em Image Fusion and Shape Variability Techniques, Mardia, K.V., Gill, C.A. and Dryden, I.L. eds.,

pp. 59–70, Leeds University Press, Leeds.
Cootes, T.F., Taylor, C.J., Cooper, D.H. and Graham, J. (1992) Training models of shape from sets of examples. In: *British Machine Conference*, Hogg, D.C. and Boyle, R.D. eds. pp. 9–18. Springer Verlag, Berlin.
Dryden, I.L. and Mardia, K.V. (1991) General shape distributions in a plane. *Advances in Applied Probability* 23: 259–276.
Dryden, I.L. and Mardia, K.V. (1998) *Statistical Shape Analysis.* Wiley, New York.
Galton, F. (1878) Composite portraits. *Journal of the Anthropological Institute of Great Britain and Ireland* 8: 132–142.
Galton, F. (1907) Classification of portraits. *Journal of the Anthropological Institute of Great Britain and Ireland* 8: 132–142.
Glasbey, C.A. and Mardia, K.V. (1998) Statistical aspects of image warping. Submitted for publication.
Goodall, C.R. (1991) Procrustes methods in the statistical analysis of shape (with discussion) *Journal of the Royal Statistical Society, Series B* 53: 285–339.
Goodall, C.R. and Mardia, K.V. (1991) A geometrical derivation of the shape density. *Advances in Applied Probability* 23: 496–514.
Goodall, C.R. and Mardia, K.V. (1992) The noncentral Bartlett decompositions and shape densities. *Journal of Multivariate Analysis* 40: 94–108.
Goodall, C.R. and Mardia, K.V. (1993) Multivariate aspects of shape theory. *Annals of Statistics* 21: 848–866.
Grenander, U. (1993) *General Pattern Theory.* Clarendon Press. Oxford.
Kendall, D.G. (1984) Shape manifolds, Procrustean metrics and complex projective spaces. *Bulletin of the London Mathematical Society* 16: 81–121.
Kendall, D.G. (1989) A survey of the statistical theory of shape. *Statistical Science* 4: 87–120.
Kent, J.T. (1982) The Fisher-Bingham distribution on the sphere. *Journal of the Royal Statistical Society, Series B* 44: 71–80.
Kent, J.T. (1994) The complex Bingham distribution and shape analysis. *Journal of the Royal Statistical Society, Series B* 56: 285–299.
Le, H.L. and Kendall, D.G. (1993) The Riemannian structure of Euclidean shape spaces: a novel environment for statistics. *Annals of Statistics* 21: 1225–1271.
Lele, S. (1993) Euclidean distance matrix analysis (EDMA): estimation of mean form and mean form difference. *Mathematical Geology* 25: 573–602.
Mardia, K.V. (1972) *Statistics of Directional Data* Academic Press, New York.
Mardia, K.V. (1993) Discussion to papers on "Gibbs sampler and other MCMC methods". *Journal of the Royal Statistical Society, Series B*, 53: 83–84.
Mardia, K.V. (1996) *The Scientific Foundation of Jainism.* 2nd ed. Motilal Banarsidass, Delhi.
Mardia, K.V. and Dryden, I.L. (1989a) The statistical analysis of shape data. *Biometrika* 76: 71–282.
Mardia, K.V. and Dryden, I.L. (1989b) Shape distributions for landmark data. *Advances in Applied Probability* 21: 742–755.
Mardia, K.V. and Dryden, I.L. (1998) The complex Watson distribution and shape analysis. Technical report, Department of Statistics, University of Leeds.
Mardia, K.V. and Hainsworth, T.J. (1993) Image warping and Bayesian reconstruction with grey-level templates. In: *Statistics and Images* Vol.1 pp. 257–280, Mardia K.V. and Kanji, G.K., eds., Carfax, Oxford.
Mardia, K.V., Edwards, R. and Puri, M.L. (1977) Analysis of Central Place Theory. *Bulletin of the International Statistical Institute* 47: 93–110.
Mardia, K.V., Rabe, S. and Kent, J.T. (1995) Statistics, shape and images. In *IMA Proceedings on Complex Stochastic Systems and Engineering Applications* pp.85–103, Clarendon Press, Oxford.
Mardia, K.V., Coombes, A., Kirkbride, J., Linney, A. and Bowie, J.L. (1996a) On statistical

problems with face identification from phototgraphs. *Journal of Applied Statistics* 23: 655–675.

Mardia, K.V., Kent, J.T., Goodall, C.R. and Little, J.L. (1996b) Kriging and splines with derivative informations. *Biometrika* 83: 207–221.

Mardia, K.V., McCulloch, C., Dryden, I.L. and Johnson, V. (1997) Automatic scale-space method of landmark detection. In: *Proceedings of the Leeds Annual Statistics Research Workshop* pp.17–29, Mardia, K.V., Gill, C.A. and Aykroyd, R.G., eds. University of Leeds Press, Leeds.

Medawar, P.B. (1944) The shape of the human being as a function of time. *Proceedings of the Royal Statistical Society, Series B* 132: 133–141.

Medawar, P.B. (1958) Postscript: D'Arcy Thompson and growth and form. In: *D'Arcy Wentworth Thompson* pp. 219–233, by Ruth D'Arcy Thompson. Oxford University Press, Oxford.

Ramsay, J.O. and Silverman, B.W. (1997) *Functional Data Analysis.* Springer-Verlag, New York.

Rao, C.R. and Suryawanshi, S. (1996) Statistical analysis of shape of objects based on landmark data. *Proceedings of the National Academy of Science USA* 93: 12132–12136.

Rao, C.R. and Suryawanshi, S. (1998) Statistical analysis of shape through triangulation of landmarks: A study of sexual dimorphism in hominids. *Proceedings of the National Academy of Science USA* 95: 4121–4125.

Small, C.G. (1996) *The Statistical Theory of Shape.* Springer, New York.

Stoyan, D. and Stoyan, H. (1994) *Fractals, Random Shapes and Point Fields: Methods of Geometric Statistics.* Wiley, Chichester.

Strachan, N.J., Nesvadba, P.P. and Allen, A.R. (1990) Fish species recognition by shape analysis of images. *Pattern Recognition* 23: 539–544.

Thompson, D.W. (1917) *On Growth and Form.* 1st edn. Cambridge University Press, Cambridge.

Thompson, D.W. (1942) *On Growth and Form.* 2nd edn. Cambridge University Press, Cambridge.

19

Sphenoethmoidal Growth, Malgrowth, and Midfacial Profile

Scott Lozanoff

Department of Anatomy and Reproductive Biology, University of Hawai'i School of Medicine, Honolulu, Hawai'i, 96734, USA

19.1 Introduction

The midface is the craniofacial region innervated by the maxillary division of the trigeminal nerve (cranial nerve V). It is a critical portion of the craniofacial region since it houses the major sensory organs of the head, including the nasal capsule, ocular orbit, oral cavity and the auditory apparatus. The midface is a composite of tissues arising from an assortment of embryonic primordia. Stem cell populations of neural crest and cranial mesenchyme contribute to the formation of the midface. These precursor cell populations become locally modified during their differentiation and contribute to diverse developmental processes, including chondrogenesis, endochondral ossification, intramembranous ossification, ultimately achieving a final midfacial form. In addition, functional attributes such as occlusion, mastication, respiration and phonation provide epigenetic stimuli affecting growth outcomes of the midface. Due to these complex interactions, a clear understanding of how the adult midfacial profile is achieved remains obscure.

The cranial base represents one bony structure that undoubtedly contributes to midfacial positioning. This underlying feature forms a strut system upon which the midface develops. The anterior component of the cranial base forms the central core of the cranium extending from the anterior tip of the cartilaginous nasal septum extending posteriorly to the sella turcica (Figure 19.1). It is a composite of developmental structures formed from four major components: the cartilaginous nasal septum; the endochondrally derived sphenoidal body and perpendicular plate of the ethmoid; and the vomer that ossifies intramembranously. The anterior cranial base undergoes extensive growth during the late embryonic and early fetal periods in the human (Diewert, 1983). Due to its dramatic growth activity during the prenatal period, the anterior portion of the cranial base is considered to play an important role in the emergence of normal midfacial form. An understanding of cranial base

On Growth and Form: Spatio-temporal Pattern Formation in Biology.
Editor Mark A.J. Chaplain, G.D. Singh, J.C. McLachlan

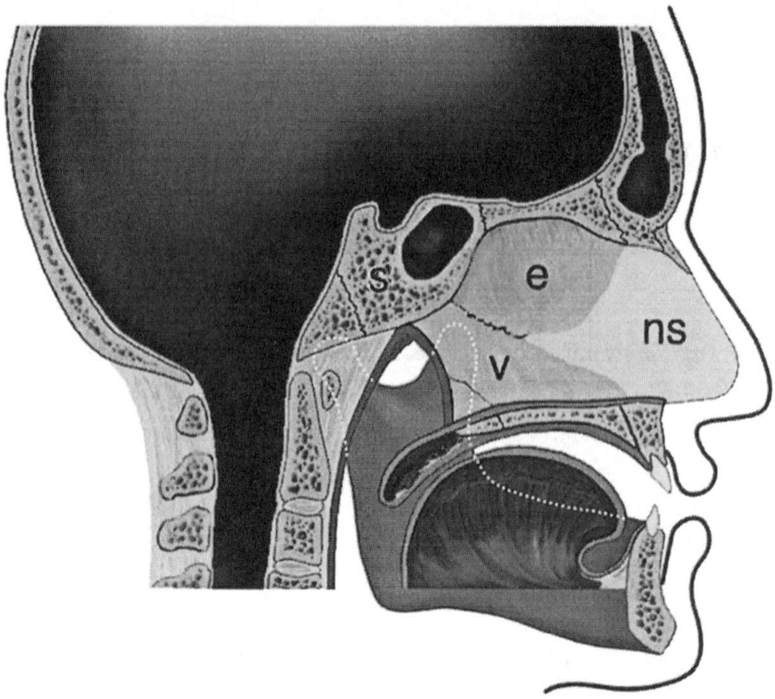

Figure 19.1 Schematic representation of the human cranial base viewed from the sagittal perspective. Nasal septum (ns), ethmoid (e), vomer (v) and sphenoid (s).

malgrowth is equally important so that dysmorphologies and extreme variations resulting from altered growth trajectories of the midface can be understood and corrected clinically to achieve a desired functional outcome.

In order to investigate the contribution of the cranial base to the emergence of midfacial morphology, a suitable animal model must be identified. I have described a mouse mutant, called Brachyrrhine (*Br*) which shows severe midfacial morphological deficiencies (Figure 19.2). Adult heterozygotes display severe midfacial retrognathia and deficiencies in midfacial calcification (Lozanoff, 1993). Midfacial retrognathia is recognizable externally as early as gestational day 15 (Ma and Lozanoff, 1993). In addition, various renal and reproductive system deficiencies are evident (Ma and Lozanoff, 1993). Utilizing the *Br* mouse as a model for midfacial retrognathia, I delineated and compared regional growth trajectories in the cranial base of normal mice and midfacially deficient *Br* mutants. In order to achieve this goal, technical obstacles, including computerized reconstruction and modelling of cranial base growth trajectories, needed to be overcome. This chapter describes a system for modeling the three-dimensional growth of craniofacial components in the prenatal and young postnatal mouse. And based on in vivo and in vitro experiments, it delineates important components of the cranial base contributing to both normal and retrognathic midfaces.

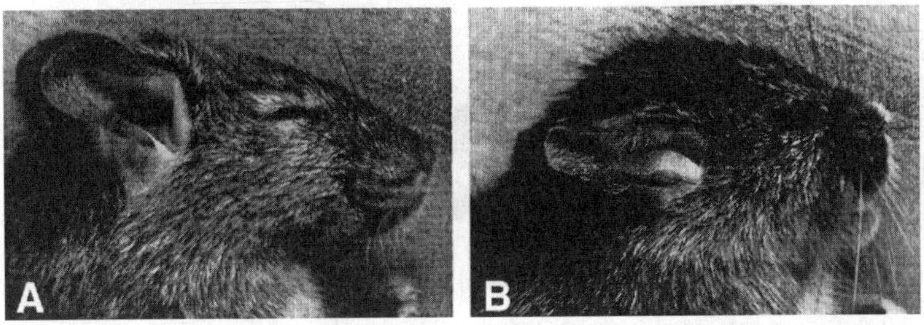

Figure 19.2 Lateral craniofacial profiles of an adult C3H mouse: (A) showing normal midfacial prognathism and an adult *Br* mutant; (B) displaying severe midfacial retrognathia.

19.2 Computerized Reconstruction of the Murine Cranial Base

Computerized modeling of anatomical and histological morphologies has become very useful for visualizing complex three-dimensional forms. Computer models not only provide a means to visualize complex morphology derived from two-dimensional tissue outlines, they also permit mathematical modeling of growth or functional attributes. Numerical simulation and prediction can be used to test functional or morphogenetic hypotheses. Prior to the 1970s, graphical reconstruction represented a primary means to generate 3D models and consisted of tracing outlines of histological structures on opaque paper. Outlines were superimposed providing a three-dimensional model. Solid modeling reconstruction using wax plates that were cut out and assembled provided another popular approach (Gaunt and Gaunt, 1978).

Computerized applications for 3D modeling began during the 1970s. Structural edges were digitized and histological structures were expressed as line reconstructions. Although this was a useful first step, these reconstructions lacked surface information. Solid fills attempted to compensate for this deficiency. Considerable effort was spent during the 1980s to develop algorithms capable of modeling surfaces. Two general approaches were developed: surface modeling and solid modeling. Surface modeling reconstructs the object as a shell. Structures are digitized manually or automatically and tissue edges are represented as streams of coordinates or contours. A viewing space is defined by mapping a window and its world coordinate contents into the viewport through a scaling function. The scaling function consists of the magnification factor and tissue thickness values. Geometric mapping algorithms are used to map coordinates for each tissue contour to the corresponding contours on subsequent sections, providing a wire-mesh model. The mesh model is rotated to a viewing angle and each surface tile is shaded according to the position of the light source and the viewer's eye.

Solid modeling represents the object as a discrete volume. A viewing space is defined as above and data input consists of voxels (pixel volumes). Voxels are stacked upon one another and the intensity is computed and applied to the stack. The solid model is rotated and the intensity values are applied to the image based on the position of the stack to the viewer's eye. An alternate approach attempted to merge these two concepts and was first described by Boissonnant (1988) and subsequently expanded upon by Lozanoff and Deptuch (1991). Using this approach, volumes are

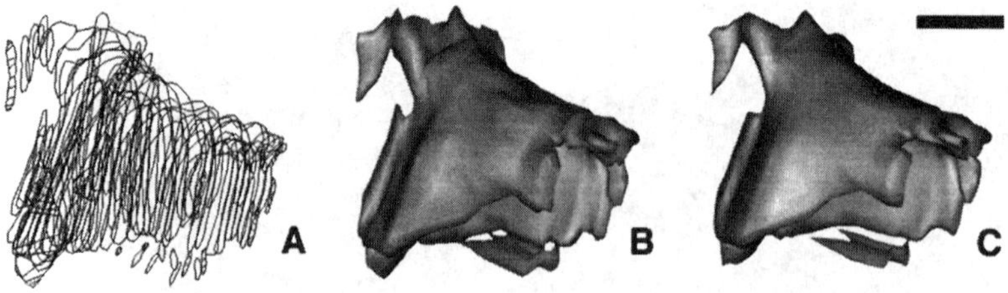

Figure 19.3 Computerized representations of an embryonic nasal capsule generated with SURFdriver: (A) wire frame, (B) rendered and (C) smoothed models viewed from the anterolateral perspective. Bar = 2.0 mm.

constructed around contours using Delaunay and Voronoi diagrams. Intersection planes between contours and volumes are identified and surface tiles are "carved" from these intersections. Tiles are sorted according to depth and the resulting surface is rendered. This approach has been used extensively for morphological analyses of craniofacial growth and form (Lozanoff *et al.*, 1993; Diewert *et al.*, 1993; Diewert and Lozanoff, 1993a, b).

The newest generation of the reconstruction program used in our laboratory is called SURFdriver and it runs on inexpensive Mac and Windows 95 platforms (Moody and Lozanoff, 1998). The software consists of data acquisition and display routines. The data acquisition program enables image loading and the establishment of a scaling function based on units selected by the user. SURFdriver transparently manipulates image size and placement, as well as the vertex coordinate system on the images, to ensure continuity of the anatomical objects as they are reconstructed. Objects on the image surface are segmented and each distinct area of the object within the image is mapped with a separate contour. A variety of tools facilitate the placement of contours, including the Autocontour tool, which selects areas of similar color and places vertices automatically around the object edge using a user-specified density. Once the contours have been assigned to each serial image containing the object, vertices are connected using a surfacing routine based on these earlier versions. The current routine uses a hashing structure to compute and refine a best-fit solution for assembling the vertices into a volumetrically optimized solid. This routine virtually eliminates false triangles caused by inappropriate vertex connections. Surfaced objects are rendered in the SURFdriver viewing window, where a variety of lighting conditions can be applied. Multiple objects created using the same image files can be loaded concurrently with correct alignment. Objects in the viewer can be exported using various 3D-file formats including DXF and IGES, both used by a variety of CAD and 3D software rendering packages. An additional feature of the viewer is the ability to "smooth" objects, or reduce the surface angularities caused by aberrant vertices, using a modified shortest path algorithm (Figure 19.3). This results in cleaner surface definition and better rendering by external 3D software. Recent work has centered on the development of export functions for texture application (Moody and Lozanoff, 1998). Future work on SURFdriver is aimed at integrating additional morphometric procedures for growth modeling facilitating simulation and prediction of growth outcomes (Chapter 20).

We developed a system permitting 3D reconstruction of the murine cranial base (Figure 19.4). Using this system, a murine cranium is glued to a block, mounted in

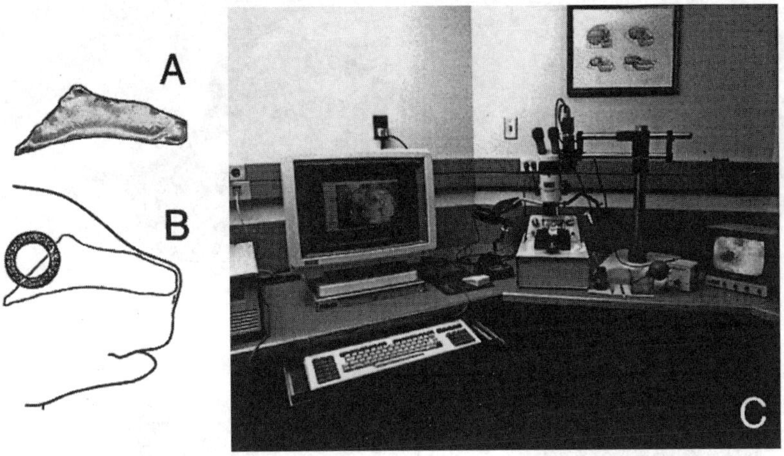

Figure 19.4 The cranial base: (A) computerized model and (B) a corresponding schematic representing its position within the craniofacial region of the mouse. (C) Laboratory setup for generating 3D reconstructions of the murine cranial base.

a vibratome and positioned beneath a standard Leitz dissecting microscope equipped with a video camera. Heads are sectioned with the vibratome and the surface of the block is lightly stained with 10% alcian blue (8GS) following each cut, enhancing the contrast of the anterior cranial base cartilage. The surface of the block is video recorded and this process continues until at least 30 sections are recorded. In this fashion, a standardized reference system is established prior to sectioning, avoiding the need for tissue contour alignment, a problem requently encountered when one attempts to stack tissue sections subsequent to sectioning. Initial trials showed that the morphology of the nasal capsule and anterior cranial base could be successfully modeled using this computerized approach (Lozanoff and Deptuch, 1991). To validate this system, cranial bases from a group of young postnatal mice were collected, treated with dispase and dissected free of surrounding connective tissue. Using a dissecting scope, a set of 11 linear measures was recorded directly from the tissue. A second group of age and gender matched animals were collected and the crania were fixed in either 2% glutaraldehyde (in 0.05 M cacodylate buffer) or 10% neutral buffered formalin. These tissues were reconstructed using the computer system and the same battery of linear measures was recorded directly from the models. Coefficients of error for each group were determined and results showed that measures recorded from the actual tissues displayed variability that was significantly greater than those recorded from the models (Lozanoff, 1992). These results indicated that measures derived from the computer models were highly precise and they were therefore suitable for morphometric analysis. Therefore, it was necessary to develop a system whereby morphological change in the murine cranial base resulting from growth could be captured and measured.

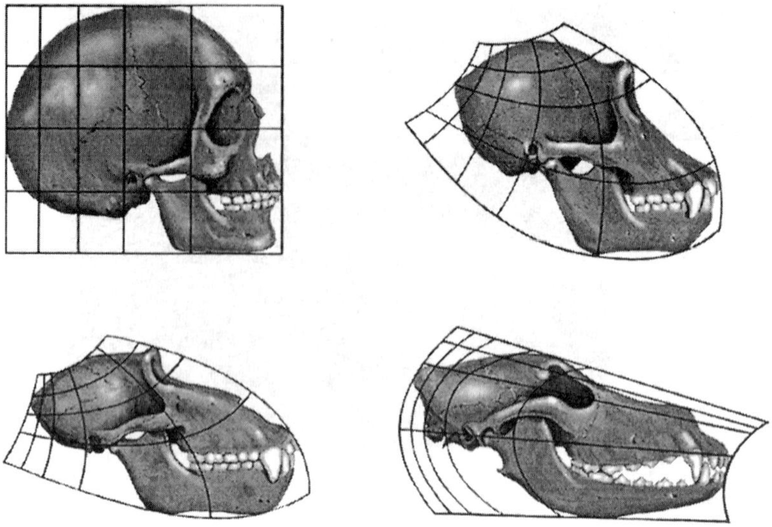

Figure 19.5 Thompsonian grid transformation represents anatomical change as a deformation of a homology map. FEM describes the transformation as a set of strains for each element of the grid that can be decomposed into measures of size or shape change.

19.3 Thompsonian Grid Transformation Analysis Using Finite Element Methods

The scope of methods for measuring anatomical form has broadened dramatically in recent years (Bookstein, 1991; Richtsmeier *et al.*, 1992; Marcus *et al.*, 1996). These new techniques are aimed at describing alterations between configurations of anatomical landmarks, thus capturing and partitioning form alteration into size and shape components. One strategy utilizes concepts of finite element modelling (FEM) and continuum mechanics for the description of anatomical change (Lozanoff and Diewert, 1986, 1989). This approach views alterations in anatomical form as a continuous deformation of an initial pattern into a final configuration. A strain is generated during this transformation and a quantitative expression of the internal force provides a numerical representation of form change. Even though this approach was developed with a mechanical engineering context, it became attractive to morphologists owing to its applicability for measuring form change akin to a Thompsonian transformation grid (Figure 19.5). A grid is constructed so that crosshairs fall on homologous landmarks. The finite element method is used to determine principal extensions at the center of each element. These in turn can be used to generate measures of size and shape. The mathematical calculations for this approach are described in detail elsewhere (Lozanoff and Diewert, 1986, 1989).

To validate this system for use in measuring craniofacial growth change, experiments were conducted by inducing a known change in craniofacial morphology in embryonic rats (Lozanoff and Diewert, 1986). Pregnant Sprague–Dawley rats received a single IP injection of diazo-oxo-norleucine (DON) at 15 days of gestation (E15). This drug is known to reduce glycosaminoglycan synthesis. Since Meckel's cartilage undergoes an extensive growth spurt at E16, the mandible in DON-treated embryos was expected to remain retrognathic compared to normal control animals. Treated and control embryonic rats were collected at E17 and craniofacial configurations were compared (Figure 19.6). Results showed that principal extensions reflected mandibular retrognathia in an expected fashion. That is, the nasal septal regions were roughly similar between treated and control groups. However, oral and mandibular regions were significantly affected in the DON-treated animals compared to the control sample, with the principal extensions reflecting the primary difference in the position of the rostral process. Results from this experiment indicated that principal extension reflected an induced anatomic form change in an expected fashion.

19.4 Localizing Anatomical Form Change in the Cranial Base of Normal and *Br* mice

The cranial base is a likely anatomical candidate to be involved in midfacial malgrowth resulting in retrognathia since it forms the central core of the cranium, providing a major support structure on which the midface develops (Burdi, 1969; Diewert, 1983; Dibbets, 1996). But specific cranial base units responsible for midfacial retrognathia have not been delineated. Traditionally, the study of midfacial retrognathia has been performed using cephalometric analysis in human groups and experimental studies have used extirpation approaches in animals in an attempt to identify craniofacial regions which direct midfacial advancement (Sarnat, 1997). Based on extirpation experiments, theories of midfacial advancement have been proposed, including a nasal septal mechanism which states the nasal septum pulls the midface forward through the septovomeronasal ligament (Scott, 1953; Latham, 1970; Copray, 1986; Mooney *et al.*, 1989). Therefore, it was expected that the anterior regions of the cranial base (cartilaginous nasal septum) would be affected in the *Br* mouse. We used computerized reconstruction and FEM in an attempt to localize dysmorphic regions in the cranial base associated with midfacial retrognathia in *Br* mice (Lozanoff *et al.*, 1994; Ma and Lozanoff, 1996). Crania from groups of young postnatal mice, age and gender matched, were subjected to computerized three-dimensional reconstruction.

Ten landmarks were identified on each anterior cranial base model. As a test for reliability, landmarks were digitized repeatedly (five times) on two models from each of three age groups. Models were scaled to equivalent sizes using Procrustes analysis. Residuals were compared and error among landmarks was significantly greater between individuals within an age than between replicates. Therefore, the configurations were deemed reliable. In order to localize differences, average configurations for each age group were generated and these averages were compared using Procrustes analysis (Figure 19.7). Results showed that *Br* cranial base models were significantly different compared to internal and external controls (3H1 +/+ and C3H/HeJ samples, respectively). In order to generate maps of form change and thus localize morphological differences, average models were compared using FEM. Three principal stretches were determined at 500 randomly selected points. Size values were determined at each point (product of the three principal extensions) and these were subjected to k-means clustering and the clusters were compared using a ratio criterion (Lozanoff *et al.*, 1994). The mapping procedure showed that the cranial base displayed two distinct growth trajectories (Figure 19.8); one located posteriorly (presumptive sphenoethmoidal

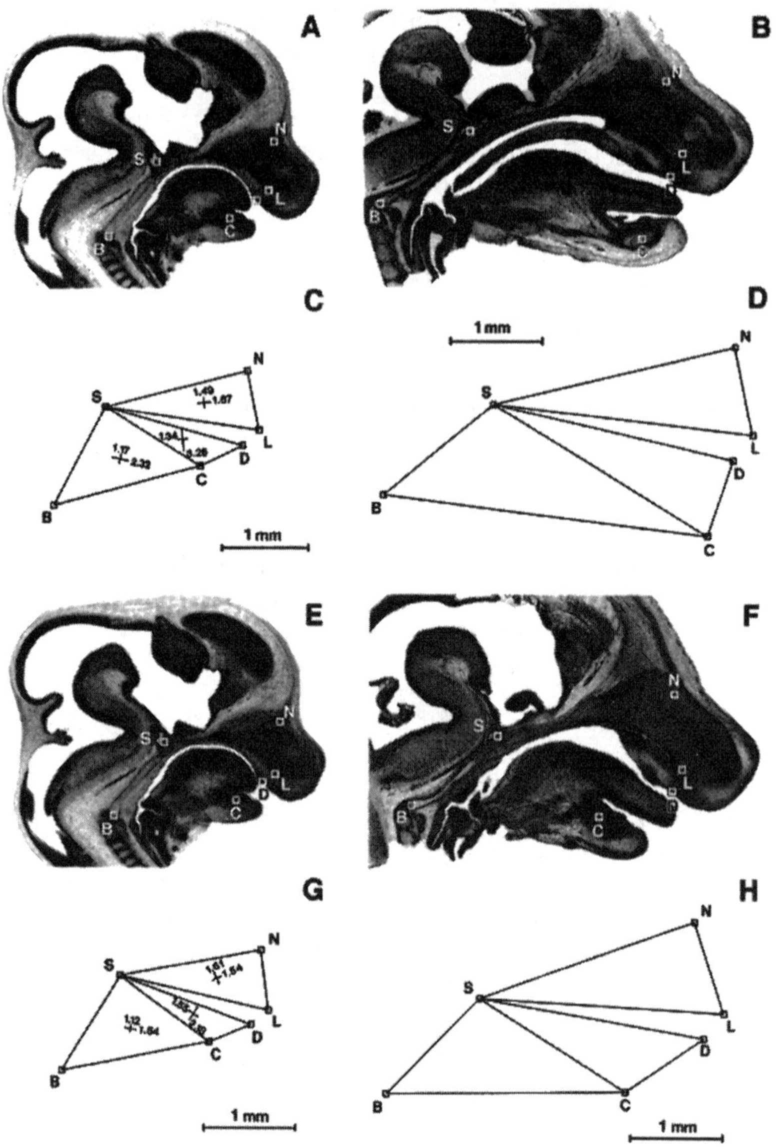

Figure 19.6 The craniofacial region of the prenatal rat can be divided into nasal (SNL), oral (SDC) and mandibular (SBC) elements. Normal growth shows a large anterior and inferior trajectory through the rostral process (C) reflected by a large principal extension (326%) when a normal E15 embryo (A, C) is compared to a normal E17 rat (B, D). However, this trajectory is altered dramatically with the administration of DON when the normal E15 rat (E, G) is compared to the E17 treated animal (F, H). For details see Lozanoff and Diewert (1986, 1989).

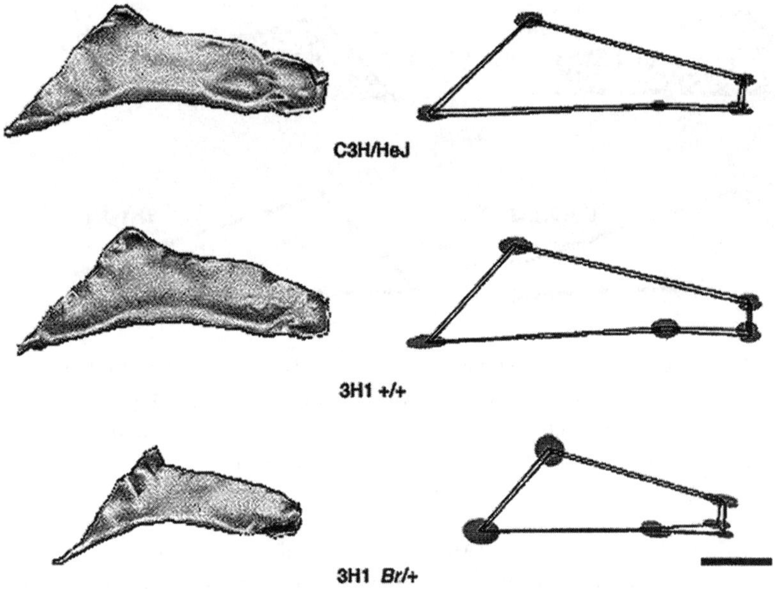

Figure 19.7 Representative anterior cranial bases (left column) of C3H/HeJ (external control) and 3H1 +/+ (internal control) mice compared with 3H1 *Br*/+ mutants. Note that the greatest difference in the morphology appears to be located posteriorly in the presumptive sphenoethmoidal region, which is smaller with an abnormal shape in the *Br* model; the anterior nasal septal region is roughly equivalent between groups. Procrustes models (right column) show that *Br*/+ cranial bases are significantly different from controls. For details, see Lozanoff *et al.* (1994). Bar = 2.0 mm.

region) and the other positioned anteriorly (presumptive nasal septal region).

This result was contradictory to the anticipated result since the nasal septal region was expected to display a larger difference between the *Br* and normal controls whereas the posterior region was expected to differ very little between groups. Instead, results showed that the posterior region of the cranial base, namely the presumptive sphenoethmoidal region, was associated with midfacial retrognathia rather than the anterior region representing the presumptive nasal septum. Therefore, the sphenoethmoidal region of the *Br* mouse would be expected to display cellular growth deficiencies rather than the presumptive nasal septal region when compared to the normal condition. In order to test the hypothesis that the anterior cranial base displays regionalized growth and the presumptive sphenoethmoidal region is affected, rather than the presumptive nasal septal region, a set of in vitro and in vivo experiments were conducted.

19.5 Regionalization of Cellular Growth Activity in the *Br* Cranial Base

Based on the morphometric findings, I hypothesized that cellular proliferation in the cranial base of normal mice should be greater in the sphenoethmoidal region compared to *Br* animals. In order to test this hypothesis, a routine autoradiographic analysis of

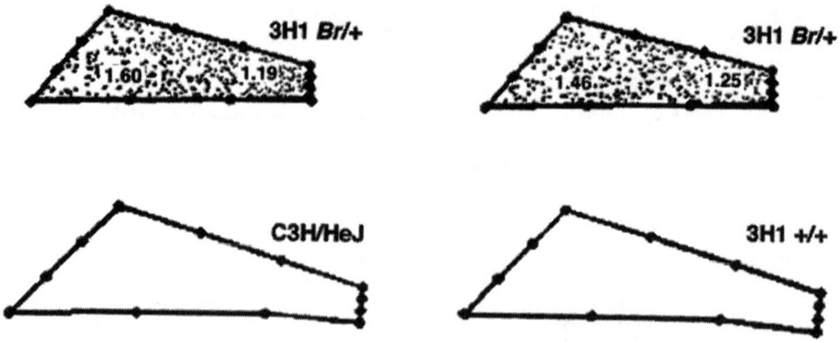

Figure 19.8 FEM maps of size change values when the average *Br* model is compared to 3H1 +/+ and C3H/HeJ models. Size change values (product of three principal extensions at each point) grouped into anterior (presumptive nasal septum) and posterior (presumptive sphenoethmoidal) clusters. For details see Lozanoff *et al.*, (1994).

cranial base growth was undertaken (Ma and Lozanoff, 1999). Pregnant dams were weighed and injected with [3H]-thymidine, collected between 15 and 19 (newborn) days of gestation, and labeling indices were determined. Results showed that chondrocytic-labeling indices were not constant in the cranial base of normal mice. Rather, labeling indices displayed a time- and position-dependent pattern characteristic of transient growth sites. Furthermore, the *Br* cranial base showed a different cellular proliferation pattern compared to normal animals. The *Br* sphenoethmoidal region, but not the presumptive nasal septal region, had significantly smaller labeling indices compared to the normal condition. And transient growth sites present in normal animals were not positioned equivalently in the *Br* cranial base. These transient growth sites are either deficient or absent in the presumptive sphenoethmoidal region of midfacially retrognathic *Br* mice.

Regulation of these transient growth sites is most probably under local control through autocrine and/or paracrine actions of growth factors. Many growth factors that have effects on bone and cartilage are produced locally and in large amounts. During the early period of cartilage formation, epithelial growth factor (EGF) and its receptor have been reported to be expressed in precartilage and growth cartilage. Immunohistochemical staining of the murine cranial base using antibodies to EGF was undertaken to verify that this protein was present in the cranial base of normal and *Br* mice. PAP immunocytochemistry was used and results showed that EGF and its receptor EGF-R were present in the cranial base of *Br* and normal mice. In addition, northern blot analysis was performed using a mouse 1.8 kb EGF precursor cDNA cloned in the vector-pBR 322 and expressed in mouse submandibular gland (gift of Dr Graeme I. Bell, Departments of Biochemistry and Molecular Biology and Medicine, the University of Chicago, Illinois, USA). A mouse transcript EGF precursor mRNA was expressed in the cranial base that corresponded in size to the band derived from the submandibular gland tissue of the same animals. These results indicated that EGF is produced locally within the cranial base (Ma, 1996).

Subsequently, experiments were undertaken to assess the effects of EGF on cranial base cell cultures derived from the presumptive sphenoethmoidal and nasal septal regions of the cartilaginous anterior cranial base. The working hypothesis was that cellular proliferation and extracellular matrix production in the cranial base of normal mice should be greater in the sphenoethmoidal region compared to *Br* animals, whereas growth indices for the nasal septum should not differ significantly between groups. Embryos were collected at 17 days of gestation and the cranial base was dissected free and cleaned of all adherent soft tissues. The crista galli was identified and the cranial base was separated into two parts, anterior and posterior, based on a vertical line intersecting this landmark (Figure 19.9). Results showed that graded concentrations of EGF in serum-free medium stimulated incorporation of [3H]-thymidine in chondrocytes from the presumptive nasal septal and sphenoethmoidal regions of the cranial base. The optimal concentration of EGF that simultaneously induced a large incorporation of [3H]-thymidine and [35S]-sulfate in the cultures of presumptive nasal septal and sphenoethmoidal chondrocytes was 1.0ng/ml medium. So this concentration was used in the culture experiments.

For the chondrocytes from the nasal septal region, EGF stimulated incorporation of [3H]-thymidine and [35S]-sulfate above untreated controls for both normal and *Br* samples (by a factor of 2.5 to 3.0); however, the stimulatory effect was essentially equivalent, with no significant difference occurring between groups. Chondrocytes from the sphenoethmoidal region show less responsiveness (1.7 times and 1.1 times untreated controls for +/+ and *Br* cultures, respectively) compared to the nasal septal region (approximately three times untreated controls for both normal and *Br* cultures). However, the effect was differential with EGF stimulation of [3H]-thymidine incorporation significantly higher than untreated controls for normal cultures compared to the *Br* cultures (Figure 19.10). EGF stimulated the incorporation of [35S]-sulfate above untreated controls for both groups; in the nasal septal by a factor of 2.2 and in the sphenoethmoidal region by a factor $\sim$ 2.5. However, a significant difference between groups was not seen. As a result of these experiments, growth in the cranial base appears regionalized. Chondrocytes from the presumptive sphenoethmoidal region in the midfacially retrognathic *Br* mouse display reduced growth activity compared to mice with normal midfacial prognathism.

19.6 Discussion

Although numerous computerized system exist that provide useful models (Haas and Fischer, 1997; David *et al.*, 1998), the system described here operates on an inexpensive computer platform (Mac or Windows 95), provides accurate three-dimensional models avoiding registration problems, and can be integrated with various morphometric routines (Lozanoff *et al.*, 1994; Weniger *et al.*, 1998; Chapter 20). Therefore, computerized models can be generated, permitting an investigator to localize and quantitatively describe a morphological change and eventually facilitating hypotheses concerning underlying cellular growth mechanisms. One can move to the wet bench in order to test these hypotheses experimentally. Generally, numerical modeling methodologies do not prove or disprove biological hypotheses, but they are crucial in generating those hypotheses. Using a series of numerical modeling techniques, I was able to identify a morphologically deficient sphenoid that led to a series of hypotheses concerning cellular growth deficiencies in the presumptive sphenoethmoidal region; these hypotheses were then tested experimentally.

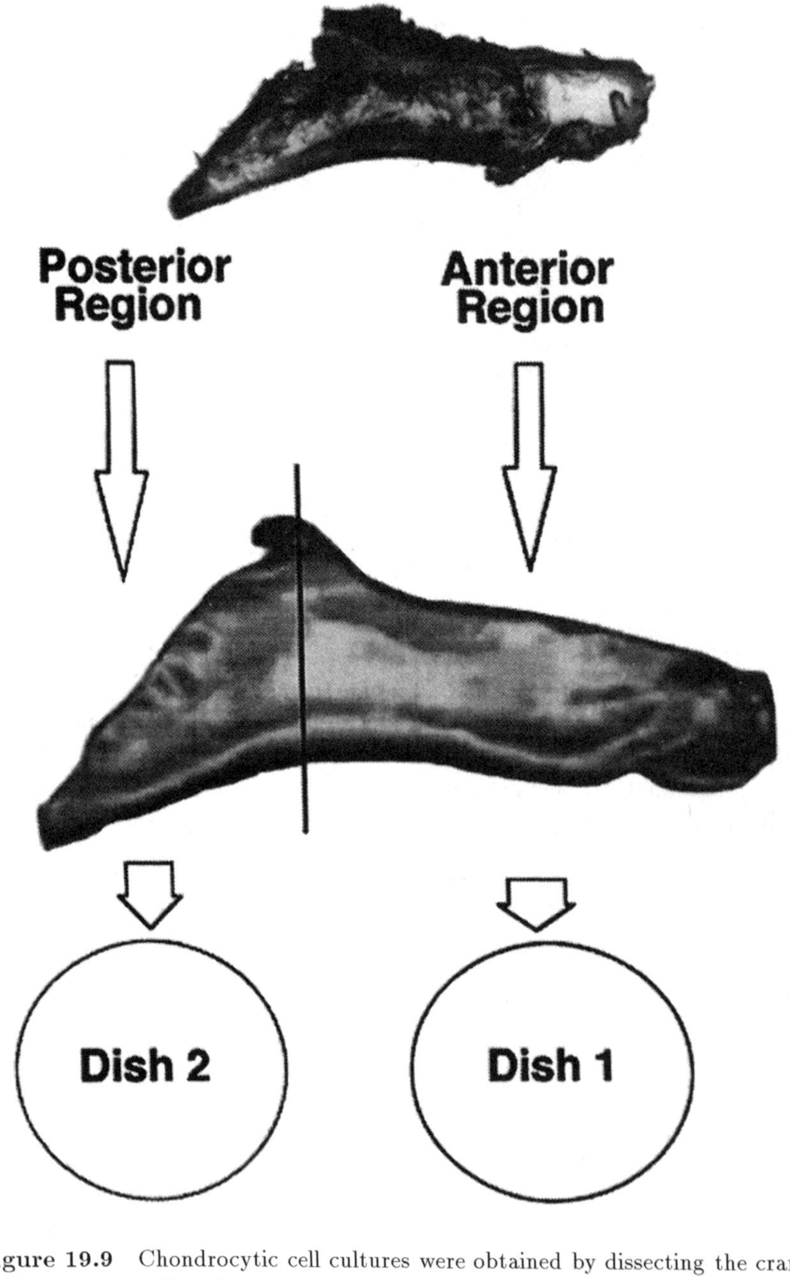

Figure 19.9 Chondrocytic cell cultures were obtained by dissecting the cranial base from surrounding tissue, cutting it along a vertical line through the crista galli and then seeding cells into separate dishes. Normal and *Br* cultures were grown in serum-free medium. One group of *Br* and +/+ cultures was stimulated with EGF and a second corresponding set of cultures remained untreated. Cultures were pulsed with [3H] and [35S] and incorporation was measured and compared to corresponding untreated controls (either +/+ or *Br*/+), providing measures of control.

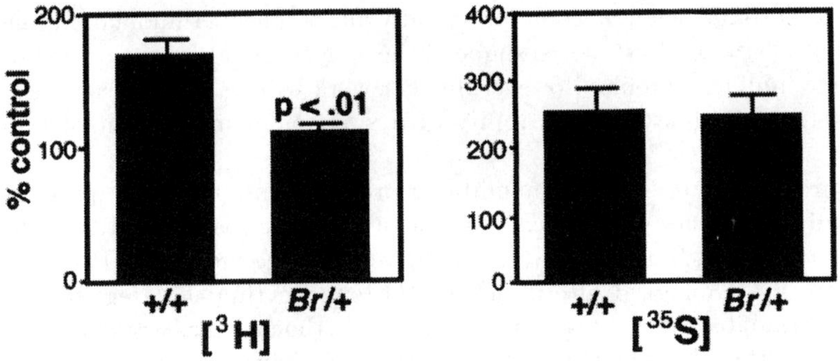

Figure 19.10 Chondrocytic culture incorporation of [3H] and [35S] above untreated controls for the presumptive sphenoethmoidal region. [3H] incorporation for +/+ cultures was significantly greater compared to *Br*/+ cultures whereas [35S] incorporation was greater in normal than *Br* cultures, but the difference was not statistically significant.

Specific craniofacial units responsible for midfacial retrognathia have not been fully delineated (Singh, 1999). However, the cranial base is a likely candidate. Based on the nasal septal theory of midfacial advancement (Scott, 1953; Latham, 1970, Copray, 1986; Mooney *et al.*, 1989), it was expected that the *Br* nasal septum would display growth abnormalities. However, the presumptive sphenoethmoidal region displayed deficiencies. The sphenoethmoidal region appears crucial for the proper positioning of the midface, possibly acting through the coronal ring (Burdi *et al.*, 1988; Rosenberg *et al.*, 1997). Delayed growth of the sphenoethmoidal region could be expected to prevent proper projection of the midface anteriorly by mechanically restricting forward growth through its dural attachments to the frontal and midfacial bones. This may occur independently of nasal septal growth, long considered the primary growth unit providing forward translation of the midface. Later in the prenatal period, the sphenoidal region is critical for normal midfacial advancement. This advancement probably relates to a tissue-separating force in the area of the developing sphenoethmoidal synchondrosis as well as the basisphenoidal and spheno-occipital synchondroses (Hoyte, 1991; Kasai *et al.*, 1995; Sarnat, 1997). This area may act as a propulsive region serving to translate the midface forward. Therefore, if the sphenoidal region fails to develop normally, a short cranial base is expected and midfacial retrognathia is likely to follow.

It has long been recognized that the human midface is unique among mammals since the anterior cranial fossa, which encases the olfactory lobes and nerves, lies directly superior to the maxilla. The premaxilla remains independent in many mammalian groups, and along with nerves of olfaction, this bony complex projects far forward of the anterior cranial fossa. Lieberman (1998) suggests that ontogenetic reduction of the sphenoid may account for the rapid retrogression of the human midface during evolution, achieving a modern cranial head form. Although evolutionary factors causing retrognathic regression of the human maxilla probably involve growth processes more complex than simple neoteny (Dean and Wood, 1984), based on the experiments involving the *Br* mouse it appears that a particular genetic aberration, such as a chromosomal deletion, could profoundly affect sphenoethmoidal morphology. The reduction in size or alteration in shape of the sphenoethmoidal region could cause a retrognathic midfacial profile. However, further work will be required among human populations to determine the plausibility of a sphenoidal mechanism of midfacial evolution.

Maxillary retrognathia, without mandibular prognathism, occurs in approximately 20–35% of adult subjects of European-American ancestry (Sanborn, 1955; Ellis and McNamara, 1984) and Singh *et al.* (1997) showed that young Class III orthodontic patients show morphological alterations in the sphenoethmoidal region, resulting in increased cranial base kyphosis and deficient orthocephalization. Guyer *et al.*, (1986) reported that 50% of Class III 5–7 year-olds have retruded maxillae. This high percentage of midfacial retrognathia in a preadolescent group may have been confounded by mixed dentition exacerbating the appearance of mandibular prognathism. Nonetheless, midfacial retrognathia may be established early in morphogenesis and this region may be directly affected by cranial base growth deficiencies. This is particularly logical, due to dramatic growth activity in the cranial base during the late embryonic and early fetal periods. Midfacial retrognathia may be established very early in development, and many clinicians argue that Class III malocclusions should be treated as early as possible (Guyer *et al.*, 1986; Tollaro *et al.*, 1996).

The identification of genes responsible for midfacial retrognathia would greatly facilitate early diagnosis of Class III malocclusion, thus enabling therapeutic intervention much sooner than currently possible. Work is now focusing on the genetic basis of the *Br* gene and preliminary results suggest a deletion on murine chromosome 17, which may correspond to one genetic aberration responsible for holoprosencephaly in humans (Schell *et al.*, 1996). Of course numerous genes are responsible for facial development and these are being identified in gene knockout and product expression studies (Richman and Mitchell, 1996). However, midfacial retrognathia may exist along a morphological continuum ranging from normal midfacial prognathism to midfacial retrognathia and possibly holoprosencephaly and median midfacial clefting. Future work must be directed at the identification of genes responsible for midfacial development as well as gene expression and function within the facial prominences contributing to proper as well as improper midfacial advancement.

Acknowledgements

The studies contributing to this review benefited by interactions with numerous individuals. Most notably, I wish to acknowledge V.M. Diewert, G.D. Singh, S. Jureczek, J. Deptuch, J. Johnston, and D. Moody for their continued assistance. Special thanks to my student, W. Ma, who performed the cell culture experiments described in this review as part of his PhD dissertation. B.K. Lozanoff provided the illustrations. Information concerning Surfdriver software can be found at `http://www.surfdriver.com`.

REFERENCES

Boissonnat, J.D. (1988) Shape reconstruction from planar cross sections. *Comput. Vis. Graph. Image Proc.* 44: 1–29.

Bookstein, F.L. (1991) *Morphometric Tools for Landmark Data: Geometry and Biology.* Cambridge University Press, Cambridge.

Burdi, A.R. (1969) Cephalometric growth analysis of the human upper face region during the last two trimesters of gestation. *Am. J. Anat.* 125: 133–142.

Burdi, A.R., Lawton, T.J. and Grosslight, J. (1988) Prenatal pattern emergence in early human facial development. *Cleft Palate J.* 25: 8–15.

Copray, J.C.V.M. (1986) Growth of the nasal septal cartilage of the rat in vitro. *J. Anat.* 144: 99–111.

David, M.K., McLachlan, J.C., Aiton, J.E., Whiten, S., Smart, S.D., Thorogood, P.V. and Crockard, H.A. (1998) Cartilaginous development of the human craniovertebral junction as visualized by a new three-dimensional reconstruction technique. *J. Anat.* 196: 269–277.

Dean, M.C. and Wood, B.A. (1984) Phylogeny, neoteny and growth of the cranial base in hominoids. *Folia Primatol.* 43: 157–180.

Dibbets, J.M.H. (1996) Morphological associations between the Angle classes. *Europ. J. Orthod.* 18: 111–118.

Diewert, V.M. (1983) A morphometric analysis of craniofacial growth and changes in spatial relations during secondary palatal development in human embryos and fetuses. *Am. J. Anat.* 167: 495–522.

Diewert, V.M. and Lozanoff, S. (1993a) Growth and morphogenesis of the human embryonic midface during palate formation analyzed in frontal sections. *J. Craniofac. Genet. Dev. Biol.* 13: 161–183.

Diewert, V.M., and Lozanoff, S. (1993b) A morphometric analysis of human embryonic craniofacial growth in the medial plane during primary palate formation. *J. Craniofac. Genet. Dev. Biol.* 13: 147–161.

Diewert, V.M., Lozanoff, S. and Choy, V. (1993) Computer reconstructions of human embryonic craniofacial morphology showing relations between face and brain during primary palate formation. *J. Craniofac. Genet. Dev. Biol.* 13: 193–201.

Ellis, E. and McNamara, J.A. Jr. (1984) Components of adult Class III malocclusion. *J. Oral Maxillofac. Surg.* 42: 295–305.

Gaunt, W.A. and Gaunt P.N. (1978) *Three-Dimensional Reconstruction in Biology.* Pitman Medical Press, Kent, UK.

Guyer, E.C., Ellis, E., McNamara, J.A. Jr. and Behrents, R.G. (1986) Components of Class III malocclusion in juveniles and adolescents. *Angle Orthod.* 56: 7–30.

Haas, A. and Fischer, M.S. (1997) Three-dimensional reconstruction of histological sections, using modern product design software. *Anat. Rec.* 249: 510–516.

Hoyte, D.A.N. (1991) The cranial base in normal and abnormal skull growth. *Neurosurg. Clin., N. Amer.* 2: 515–537.

Kasai, K., Moro, T., Kanazawa, E. and Iwasawa, T. (1995) Relationship between cranial base and maxillofacial morphology. *Eur. J. Orthod.* 17: 403–410.

Latham, R.A. (1970) Maxillary development and growth: The septo-premaxillary ligament. *J. Anat.* 107: 471–478.

Lieberman, D.E. (1998) Sphenoid shortening and the evolution of modern human cranial shape. *Nature* 393: 158–162.

Lozanoff, S. (1992) Accuracy and precision of computerized models of the anterior cranial base in young mice. *Anat. Rec.* 234: 618–624.

Lozanoff, S. (1993) Midfacial retrusion in adult Brachyrrhine mice. *Acta Anat.* 147: 125–132.

Lozanoff , S. and Deptuch, J. (1991) Implementing Boissonnat's approach for generating surface models of craniofacial cartilages. *Anat. Rec.* 229: 556–564.

Lozanoff, S. and Diewert, V.M. (1986) Measuring histological form change with finite element methods: An application using diazo-oxo-norleucine (DON) treated rats. *Am. J. Anat.* 177: 187–201.

Lozanoff, S. and Diewert, V.M. (1989) A computer graphics program for measuring two- and three-dimensional form change in developing craniofacial cartilages using finite-element methods. *Comput. Biomed. Res.* 22: 63–82.

Lozanoff, S., Diewert, V.M., Sinclair, B. and Wang, K.-Y. (1988) Surface modelling of craniofacial form in human embryos with a limited graphics terminal. *J. Craniofac. Genet. Dev. Biol.* 8: 101–109.

Lozanoff, S., Zingeser, M.R. and Diewert, V.M. (1993) Computerized modelling of nasal capsular morphogenesis in prenatal primates. *Clin. Anat.* 6: 37–47.

Lozanoff, S., Jureczek, S., Feng, T. and Padwal, R. (1994) Anterior cranial base morphology in mice with midfacial retrusion. *Cleft Palate-Craniofac. J.* 31: 417–428.

Ma, W. (1996) A sphenoidal mechanism of midfacial retrognathia in the brachyrrhine mouse. PhD dissertation, University of Saskatchewan, Saskatoon, Saskatchewan, Canada.

Ma, W. and Lozanoff, S. (1993) External craniofacial features, body size and renal morphology in prenatal Brachyrrhine mice. *Teratology* 47: 321–332.

Ma, W. and Lozanoff, S. (1996) Morphological deficiency in the prenatal anterior cranial base of midfacially retrognathic mice. *J. Anat.* 188: 547–555.

Ma, W. and Lozanoff, S. (1999) Spatial and temporal distribution of cellular proliferation in the cranial base of normal and midfacially retrusive mice. *Clin. Anat.* (in press).

Marcus, L.F., Corti, M., Loy, A., Naylor, G.J.P. and Slice, D.E. (eds). (1996) *Advances in Morphometrics.* Plenum: New York.

Moody, D. and Lozanoff, S. (1998) SURFdriver: A practical computer program for generating three-dimensional models of anatomical structures using a PowerMac. *Clin. Anat.* 11: 132.

Mooney, M.P., Siegel, M.I., Kimes, K.R. and Todhunter, J.S. (1989) A test of two midfacial growth models using path analysis of normal human fetal material. *Cleft Palate J.* 26: 93–99.

Richman, J. and Mitchell, P.J. (1996) Knockout mice take one on the chin. *Current Biol.* 6: 364–367.

Richstmeier, J.T., Cheverud, J.M. and Lele, S. (1992) Advances in anthropological morphometrics. *Ann. Rev. Anthropol.* 21: 283–305.

Rosenberg, P., Arlis, H.R., Haworth, R.D., Heier, L., Hoffman, L. and LaTrenta, G. (1997) The role of the cranial base in facial growth: experimental craniofacial synostosis in the rabbit. *Plast. Reconstr. Surg.* 99: 1369–1407.

Sanborn, R.T. (1955) Differences between the facial skeletal patterns of Class III malocclusion and normal occlusion. *Angle Orthod.* 25: 208–222.

Sarnat, B.G. (1997) Some methods of assessing postnatal craniofacial growth: A retrospective of personal research. *Cleft Palate Craniofac. J.* 34: 159–172.

Schell, U., Wienberg, J., and Muenke, M. (1996) Molecular characterization of breakpoints in patients with holoprosencephaly and definition of the HPE2 critical region 2p21. *Hum. Mol. Genet.* 5: 223–229.

Scott, J.H. (1953) The cartilage of the nasal septum. *Br. Dent. J.* 95: 37–43.

Singh, G.D. 1999. Morphologic determinants in the etiology of Class III malocclusions: A review. *Clin. Anat.* in press.

Singh, G.D., J.A. McNamara and S. Lozanoff. (1997) Morphometry of the cranial base in subjects with Class III malocclusion. *J. Dent. Res.* 76: 694–703.

Tollaro, I., Baccetti, T. and Franchi, L. (1996) Craniofacial changes induced by early functional treatment of Class III malocclusion. *Am. J. Orthod. Dentofac. Orthoped.* 109: 310–318.

Weniger, W.J., Meng, S., Streicher, J. and Müller, G.B. (1998) A newepiscopic method for rapid 3-D reconstruction: applications in anatomy and embryology. *Anat. and Embryol.* 197: 341–348.

20

Ontogeny and Phylogeny: Some Morphometric Approaches to Skeletal Growth and Evolution

Paul O'Higgins

Evolutionary Anatomy Unit, Department of Anatomy and Developmental Biology, University College London, Rockefeller Building, University Street, London WC1E 6JJ, UK

20.1 Introduction

One aspect of the D'Arcy Thompson's (1917) work, "the theory of transformations, or the comparison of related forms", in particular the "method of coordinates", has particularly captured the imagination of biologists working in diverse fields. Having noted that in the study of organic form (which he terms "morphology") "our essential task lies in the comparison of related forms" he goes on to state that, in comparing forms, "the *deformation* of a complicated figure may be a phenomenon easy of comprehension" although the figure itself may have to be left unanalysed. He therefore points towards an important aspect of the analysis of differences of form. Although full details of the processes underlying morphogenesis may be unknown, when we compare related forms we are considering not the processes producing each aspect of each form, but the differences in process which may be very much simpler than the whole.

In modern terms, even if we do not know all the details of the regulation of morphogenesis, we can make headway in understanding the basis of differences between closely related forms. The commonalities are not necessarily important in making headway in understanding the differences. Imagine two taxa that at the end of embryonic patterning are "identical" but which come to differ after growth

On Growth and Form: Spatio-temporal Pattern Formation in Biology.
Editor Mark A.J. Chaplain, G.D. Singh, J.C. McLachlan

to adulthood. In attempting to understand the differences between closely related adult forms, we need to concentrate on the source of the principal differences, growth itself, and for this we use tools to characterise ("modelling" in a loose, empirical, sense) and compare growth trajectories. Such tools allow the testing of hypotheses of differentiation either through experimental intervention (by varying nutrition, exercise, hormonal environment, etc) or through the study of naturally occurring transformations (due to sex, ecological differences, etc). The application of such tools is the focus of this chapter. Their use will be illustrated with examples from the study of cranial growth and variation in primates. Growth is the issue focused on here because it is an extremely important source of morphological variation between closey related species, yet it is increasingly being sidelined as an issue in modern biology.

In vertebrate developmental biology, for instance, the vast majority of research concerns fundamental mechanisms of morphogenesis at the molecular and genomic levels. Thus, a major endeavour is the attempt to relate patterning and proportioning in the early embryo to the temporal and spatial expression domains of genes. This reductionist approach is, without doubt, extremely powerful; enormous insights into fundamental patterning mechanisms have come about in recent years. These are reviewed by several of the other contributors in this volume. There remains however a large part of biological diversity which is not explained and never will be explained by such approaches.

For instance, among mammals, growth contributes considerably to adult form. As such, a relatively large proportion of the total diversity between related taxa arises not because of variations in tightly specified patterning and proportioning processes but rather because of variations in "downstream" morphogenetic regulation. Such downstream regulation may be locally acting, e.g. localised remodelling response to a force (Oxnard, 1991). Alternatively it may be more generally acting, e.g. general response to activity or hormones (Lieberman 1996). The possible interactions between genome, environment and behaviour are myriad, so it becomes increasingly difficult to tie down specific morphological features and differences to specific morphogenetic processes. Yet the result is that recognisable and (to some degree) distinct adult morphologies arise.

In considering the transformations in form which occur during growth, and of course during evolution, form variation at these post-patterning levels has to be considered holistically and methods come into play that allow the study of form differences and the "modelling" of transformations of form. It is such methods that Thompson first glimpsed with his method of coordinates. This chapter presents examples that indicate how some of the tools of the modern morphometric armoury (geometric morphometrics) might be applied in specific cases. It also gives a non-mathematical exposition for general biologists of the tools and their underlying assumptions. In particular, the methods are illustrated by examples relating to growth of the facial skeleton of monkeys.

The skeleton of the primate face is a good example of a structure which varies considerably between related taxa but whose adult form is likely determined through a web of interactions between genome, tissues and environment. It is formed mainly by ossification within a membrane whose form is largely secondary and adaptive to surrounding tissues and spaces. Ossification centres arise during growth and they meet at sutures as they grow. These centres provide for growth through the deposition of bony material at the sutural boundaries of bones while they are being displaced by soft tissues and expanding functional spaces (functional matrices), e.g. nasal capsule (Moss and Young, 1960). During growth displacement, spatial and functional alignment of skeletal elements is thought to be maintained through remodelling of bony surfaces, known as cortical growth remodelling (Boyde and Jones, 1972). It is therefore hypothesised that the sizes, spatial and temporal relations as well as the activity types (deposition/resorption) and rates of remodelling fields might be useful indicators of growth displacements (Enlow, 1966, 1968, 1975; Bromage, 1986).

In one species of Old World monkey, the mangabey *Cercocebus torquatus*, the facial cortical growth remodelling fields are well described and appear more or less consistent in distribution and nature between development of full deciduous dentition and eruption of the third permanent molar (O'Higgins *et al.*, 1991). It is predicted therefore that the displacements of bones and indeed the general pattern of growth is more or less consistent throughout this period. This has been investigated in detail by O'Higgins and Jones (1998) and it seems that, in general, the prediction is supported. This earlier study is briefly reviewed and new analyses are presented that compare the growth in *Cercocebus* with growth in a distantly related primate, the New World monkey *Cebus apella*. This comparison is an explicit analysis of the relationship between evolutionary divergence and growth divergence. Together these studies serve to illustrate how the methods of geometric morphometrics can lead to highly informative statistical and visual reports of growth and differences in growth.

In attempting to better understand the interaction of remodelling and displacement in growth and how differences in growth generate evolutionary variability, it is not feasible or necessary to undertake detailed description of each process and its interaction. We can, however, apply morphometric methods to the whole in order to compare and contrast different natural and experimental variants. In so doing, we follow D'Arcy Thompson in leaving the common or shared aspects of morphogenesis unanalysed while focusing on the differences which generate variations.

20.2 What Can We Learn of Form Variations Using the Morphometric Toolkit?

Lele and Mardia in Chapters 17 and 18 address certain mathematical aspects of the analysis of form variations. For the general biologist, the mathematical treatments of shape analysis (Bookstein, 1991; Lele, 1993; Dryden and Mardia, 1998) can be difficult to relate to biology and the questions which drive biological research. For this reason it is worth attempting a verbal description of the motivation for the application of morphometric methods and the tools that may be called upon in biological studies of form variation. It is essential to consider the models and assumptions upon which these tools are based. Lele (1993) has already drawn attention to many of them. Others are considered by Dryden and Mardia (1998).

20.2.1 Motivation

In this chapter the focus is form descriptions in terms of landmarks but in other studies we may be interested in curves, areas or volumes (O'Higgins, 1997). It is essential that the chosen description of form is appropriate in terms of the biological hypothesis being addressed, and it needs to be adequate for testing the hypothesis.

In the study of growth we are principally concerned to produce reports of variations within our quantitative descriptions of form. These reports must have certain properties; they should be biologically plausible, they should be suggestive and they should have some statistical authority. Several items are of major interest to the biologist studying of a sample of forms: estimation of mean form, localisation of form differences, description of patterns of form variation and importantly reports of the relationships between patterns of size and shape variation and external factors. Bookstein (1991) puts it elegantly in asking two questions: Is there evidence for covariance between form and some factor? What is the nature of this covariance? Thus, morphometrics is the exploration of the relationship between extrinsic factors (geography, species, sex, etc.) and intrinsic factors (growth, etc.) and patterns of form variation. It legitimately concerns itself with the description of variations in terms of the nature of such form differences (e.g. localisation and character).

The tools employed in any morphometric study should at least be capable of producing insights into biological phenomena. These insights should be informative in terms of the biological issues addressed, and for this reason they should allow the analysis of covariations with form. Much debate in the morphometric literature has focused on the relative merits of "registration free" approaches to morphometrics versus "registration dependent" (e.g. Bookstein, 1978, 1991; Lele, 1993; Richtsmeier *et al.*, 1993a, 1993b; Dryden and Mardia, 1998, p. 286). Lele addresses these points again in this volume. The arguments can seem overly complicated to non-statisticians, with the result that biological workers might be put off carrying out studies using perfectly legitimate and highly appropriate methodology. For this reason it is worth rehearsing the basic models and assumptions underpinning morphometric studies in order to identify those methods which are useful and appropriate for particular biological questions.

Terminology. Form is defined here as the configuration of landmarks, shape as the aspects of form with size removed. Size is a loose term with different meanings under different circumstances but in geometric morphometric studies, the centroid size (see below) is a natural size measure. The issues surrounding the definition of size variables are discussed more fully in O'Higgins (1997) and O'Higgins and Jones (1998).

Models. The basic model underlying most statistical studies of shape variation is simply that landmarks are perturbed about a mean form. Particular variants of this model may assume specific distributions of these perturbations, such as an isotropic independent distribution (iid) at each landmark or dependence between landmarks. Alternatively, general covariance matrices can be used (Dryden and Mardia, 1991).

Other models are possible, such as the log-normal model for interlandmark distances which derives from the work of Mosimann (1970) and which has recently been applied specifically in relation to the analysis of form variations by Rao and Suryawanshi (1996) and Mardia *et al.*. (1996). Inference under this model may be carried out using standard multivariate methods (Rao and Suryawanshi, 1996). In general, however, this model suffers difficulties in relation to the estimation of mean forms (Lele, 1993) and the distance metric between specimens has highly undesirable properties for certain kinds of shape analysis (Rohlf, personal communication). For this reason most analyses of shape variability take some variant of the perturbation model as their basis.

Under the perturbation model alone there are limits to what can be estimated. Thus the "mean reflection size and shape" (Dryden and Mardia, 1998, p. 281) can be estimated using interlandmark distances. Mean shape and mean size and shape can be estimated with the multidimensional scaling approach (Dryden and Mardia, 1998, pp. 281–286). Offset normal distributions in shape space provide efficient estimators for shape or size-shape under Gaussian iid models (Mardia and Dryden,1989). These models can be generalised to non-iid perturbations but the procedure is complicated (Dryden and Mardia, 1991). Alternatively the complex Bingham distribution (Kent, 1994) is often a reasonable model inference.

Mean reflection size and shape can also be estimated through Euclidean distance matrix analysis, EDMA (Lele, 1993). This is expressed as a matrix of all possible interlandmark distances on the mean (the form distance matrix). It is not possible to estimate the relative movements of landmarks since this calls for certain additional assumptions. EDMA does, however, allow some aspects of the variability of landmarks to be estimated but not direction.

EDMA enables the examination of form differences through the calculation of a form difference matrix (FDM) the ratios of equivalent pairs of all possible interlandmark distances from each. Examination of this matrix (which for k landmarks is $k(k-1)/2$ in size; it becomes very large rapidly as the number of landmarks increases) can indicate which landmarks are most changed, but not relative movement (since additional assumptions would be required). EDMA can also be used to assess the significance of differences in reflection mean size and shape between two groups. Lele and Richtsmeier (1991) propose the use of the ratio of the maximum to the minimum elements of the FDM as the basis of such a test. Bootstrap procedures are used to estimate the test statistic. More recently Lele and Cole (1995) have proposed EDMA-II, an alternative test deals with shape differences (after scaling the form distance matrices). Issues concerning this alternative statistic are discussed by Dryden and Mardia (1998, p. 284). If the additional assumption is made that the landmarks are independently and isotropically distributed, it is possible to estimate mean shape consistently, but not mean size-shape, using Procrustes superimposition techniques (Kent and Mardia, 1997). However, the bias in size is small for small variations and the shape part of the estimate is consistent. Procrustes analysis is not consistent under general covariances. But importantly, for small variations, any bias is small and issues of consistency are of mathematical interest only.

In summary, minimal assumptions allow us to estimate mean size and shape using a variety of approaches. These estimates will differ slightly when variations are large (to the biologist such variations would appear immense in most applications and would probably not require statistical analysis in any case). In biology, however, the estimation of mean shape is trivial in most applications; the mean does not impart any information about variability and covariations with form. To proceed in a useful way, we generally need to be able to compare means and describe patterns of form variation. EDMA allows "influential" landmarks to be discerned but it does not allow relative landmark movements to be addressed, so it cannot help with localisation and characterisation of form differences. In order to do so, certain additional assumptions need to be made (entirely reasonable in the vast majority of biological studies).

20.3 Localisation and Characterisation of Form Differences

Further assumptions enable additional and essential analyses of size and shape to be carried out. In particular, if the relative displacements of mean landmarks are to be examined then assumptions need to be made in relation to superimposition of forms. Likewise if the deformation of one form into another is of interest, assumptions must be made about the way in which deformations occur.

For the analysis of relative landmark displacements, it is essential that forms are registered with respect to each other. In biology it is nearly always impossible to carry out such a registration in an unequivocal way, i.e. in a way that exactly mimics the biological truth (Lele, 1991; Chapter 17). The reason for this is simply that many structures do not have a "natural" register with each other (e.g. skulls where, in some sense, all bones move away from each other). Note that even if registration between two diagrams were certainly wrong, it would still be possible to make useful descriptions of shape differences. Consider two skulls, one having a larger face than the other; superimposed first over the occiput and then over the maxilla, we would conclude from both analyses that the face is relatively larger in one. Arrows drawn over landmarks do not indicate the "true" differences in landmarks, but the relative differences given the particular registration.

There are several practical possibilities for registration but three have received considerable attention (Dryden and Mardia, 1998). One might register (i.e. translate, rotate and scale) to a common baseline e.g. Bookstein two-point registration (Bookstein, 1984); register so that most points fit well i.e. robust or resistant fit (Siegel and Benson, 1982) or register by minimising the sum of squared distances between the equivalent landmarks of forms using generalised Procrustes analysis (GPA). GPA of n specimens, each represented by a $k \times m$ matrix of landmark coordinates X_i, $i = 1, ..., n$ results in registered specimens denoted X_i', for which the sum of squared differences d_F^2 between them is minimised, i.e.

$$d_F^2 = \sum_{i=1}^{n} \sum_{j=i+1}^{n} (X_i' - X_j')^2 \ . \tag{20.1}$$

Scaling is according to centroid size (the square root of the sum of squared Euclidean distances from each landmark to the centroid, which is the mean of the landmark coordinates).

20.4 Shape Spaces for Registered Landmarks

Once registration is carried out, each shape (scaled form) can be represented as a point in a "shape space". The shape space resulting from Bookstein two-point registration and Euclidean distance has undesirable statistical properties under the assumption of iid landmark distributions (Dryden and Mardia, 1998, p. 33). The space with a distance resulting from generalised Procrustes analysis, Kendall's shape space (Kendall, 1984), has the desirable property that iid of landmarks results in isotropic distributions of points representing specimens in the shape space (Kent 1994; Dryden and Mardia, 1998, p. 137). So if we assume this superimposition is adequate in terms of biology, and if landmarks vary in location according to an isotropic model we can, expect to find an isotropic distribution of specimens in the shape space. Conversely, deviations from iid landmark variations will lead to a non-isotropic distribution of specimens in the shape space. Such distributions of specimens in the shape space can be of biological interest since the principal directions of variation might be related to interesting biological variables such as age, "size", sex and species.

With registration it is therefore of considerable interest to consider the adequacy of the superimposition method in relation to biology. Different registrations will lead to different distributions of specimens in the resulting shape space (different variance–covariance matrices) and so *possibly* to different biological conclusions.

It turns out, however, that if variations are small (in relation to full Procrustes distance, not in relation to a vague notion of biological variation) then all registrations will lead to approximately similar results (Kent, 1994). Thus, under the assumptions that the registration is reasonable in terms of biology and that variations are small, we can undertake studies of landmark displacement vectors. Dryden and Mardia (1998, p. 287) give a tentative suggestion that "if the data lie within full Procrustes distance of about $d_F = 0.2$ of an average shape then methods give very similar conclusions". Increasing experience of Procrustes methods also indicates that as the number of landmarks increases, the approach is increasingly robust in many circumstances. It seems that no biological data set has yet been published which violates these assumptions to the extent that erroneous biological conclusions are reached.

Statistical inference in Kendall's shape space is not straightforward since it is non-Euclidean (i.e. it is curved). For triangles the space is equivalent to the surface of a sphere of unit diameter (Figure 20.1a) but for more than three landmarks (k landmarks in m dimensions) the space is high-dimensional and more complex. Because of this, great care is needed in carrying out statistical analyses. One approach which is particularly appealing, since it naturally allows the study of allometry, is to carry out principal components analysis (PCA) in the tangent plane to Kendall's shape space (Kent, 1994; Dryden and Mardia, 1993). For triangles we take the scatter of points on the spherical shape space representing variation within our sample and project it into a Euclidean tangent plane in exactly the same way as a cartographer might project a map from a globe onto a flat sheet of paper (Figure 20.1b). The coordinates of the points representing specimens are no longer given in terms of the sphere, but rather as coordinates in the plane. As long as the projection has not resulted in excess distortion (as might occur if the projection encompasses a large proportion of the sphere), we can carry out useful analyses in this plane. For higher dimensions the tangent space to the shape space can be imagined as a space of $km - m - \frac{1}{2}m(m-1) - 1$ dimensions.

Procrustes tangent coordinates can be estimated using the partial Procrustes tangent space projection given by Dryden and Mardia (1993). This projection results in a $(k-1)m$ vector of tangent space shape coordinates with respect to the mean for each specimen. Both of these vectors of tangent space coordinates are of rank $km - m - \frac{1}{2}m(m-1) - 1$. Principal components analysis can be carried out using tangent space coordinates to extract $km - m - \frac{1}{2}m(m-1) - 1$ eigenvectors; these are the principal components of variation of shape (Figure 20.1b).

Whenever growth changes in shape are the major source of variation, it is reasonable to expect that the first few principal components will serve as an adequate representation of allometric shape changes. Note that, since Procrustes analysis eliminates scaling, the variations we examine through PCA are shape rather than form variations. If we wish to examine the relationship between size and shape

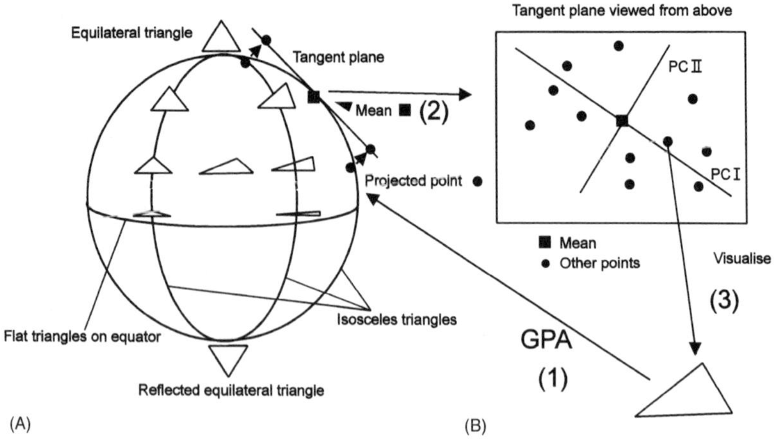

Figure 20.1 (A) An approximate sketch of Kendall's shape space for triangles. Equilateral triangles lie at the poles; the southern hemisphere is a reflection of the northern. The sphere is divided into twelve equal half-lunes (six in each hemisphere); if the apices of the triangles are unlabelled and reflections are ignored, all triangles lie in one half-lune. Isosceles triangles lie along the lines dividing lunes and flat triangles at the equator. (B) A schematic indicating the projection of points representing triangles in Kendall's shape space into a space tangent to the mean triangle (arrows) and the principal components of shape variability (PC I, PC II) in this tangent space.

(allometry), we can do this by examining plots and correlations or regressions of principal component (PC) scores versus centroid size for the significant principal components. It is possible to visualise variations in shape represented by the principal components by simply warping (morphing) the mean shape along each PC of interest. This is achieved by adding to the mean coordinates the product of the eigenvectors (for the PC of interest) and the score on that PC (Kent, 1994). If the PCA is based on tangent space coordinates, it is necessary to carry out a projection of the hypothetical specimen's coordinates into configuration space (the space of the original specimens).

20.5 Shape Space for Interlandmark Distances

Given the assumptions required to carry out PCA of registered landmark coordinates (the registration is reasonable, and there are only small variations) it might be reasonably asked if PCA of the complete set of scaled interlandmark distances (ilds) might be more sensible. This would be an extension of the log-distance approach to the estimation of mean reflection size and shape (Rao and Suryawanshi, 1996; Mardia *et al.*, 1996) in which ilds are scaled to their geometric mean. Coordinate representations of figures could be generated through multidimensional scaling (Mardia *et al.*, 1979).

There are difficulties, however, with such an approach. For triangles, when the coordinates of the vertices are iid perturbed, the distribution of points representing these triangles within the shape space depends on the shape of the original triangle. The consequence is that results of analyses of shape variation (perhaps through the extraction of principal components) depend on the mean shape. The space becomes much more complicated as the number of landmarks increases (F.J. Rohlf, personal communication). And although other assumptions about form perturbations might yield more satisfactory distributions around one particular mean triangle, they will yield different distributions around another. It seems, therefore, that present knowledge of the shape space for interlandmark distances does not allow sensible analyses of patterns of variation among specimens.

20.6 Deformations

A completely different class of morphometric methods attempts to describe form differences in terms of the deformation of the space in the vicinity of a reference specimen into the space in the vicinity of a target specimen. Such approaches include the methods of finite element analysis (FEA) and finite element scaling analysis (FESA); Chapter 19 and Lewis *et al.,* (1980). The forms are divided into finite elements (triangles, tetrahedra, hexahedra, octahedra, etc.) whose interior is considered deformed between reference and target. The deformation can be expressed in terms of the mapping of internal points (such as the nodes of a regular Cartesian grid) between the elements of the reference and those of the target (using mathematical functions often termed homology functions). Alternatively the principal strains which arise in deforming the elements between reference and target can be calculated at equivalent points within the elements.

These methods assume several things (Lele considers them in Chapter 17). For the simplest elements (triangles in 2D and tetrahedra in 3D) only one set of principal strains (representing the principal magnitudes and directions of deformation) need be drawn. The underlying assumption is that the interiors of the finite elements deform uniformly in relation to the landmarks defining the element. For more complex elements ($k > 3$ in 2D and $k > 4$ in 3D) inhomogeneity of deformation is allowed; several principal strains can be drawn at homologised internal points (using the chosen homology function). Here additional assumptions are made that relate to the way in which the chosen homology function matches points between target and reference elements. With all finite element methods a further problem arises in that discontinuities can occur at element boundaries. This means that the principal directions and magnitudes of deformation or the matching of points between target and reference can show a marked change at element boundaries.

Alternative approaches to the study of deformations use homology functions derived from the deformation of the whole configuration of reference landmark coordinates into target coordinates. Such functions can be used to draw Cartesian transformation grids that resemble those first proposed by Thompson (1917). By homologising the space in the vicinity of the reference form to that in the vicinity of the target it is simple to map the nodes of a regular grid over the reference into a those of a deformed grid over the target. One commonly used function is the thin plate spline, TPS (Bookstein, 1989; Marcus *et al.*, 1996; Dryden and Mardia, 1998). This function minimises the bending energy of the deformation and so results in minimal local variation of grid elements with respect to their neighbours.

The resulting grids do not suffer from the problems of element design and discontinuity between elements encountered in FESA since they are derived from a smooth mathematical function applied to the whole landmark configuration. The grids can be interpreted as indicating how the space in the region of a reference shape might be deformed into the space in the region of the target such that landmarks in the reference map exactly into those of the target. The basic assumption in choosing the thin plate spline is that this deformation involves minimum bending. The statistical and graphical models of shape transformations which result from these approaches are readily interpretable and highly visual (Bookstein, 1978, 1989; O'Higgins and Dryden, 1992; Marcus *et al.*, 1996). Care should be taken, however, in balancing their visual appeal against the underlying assumptions in their construction. From a biological perspective it is vital to remember that this mapping is purely mathematical and it is based only on the locations of a few important points (the original landmarks) whose homology or equivalence is known a priori from biological arguments.

20.7 Morphometrics: Models, Assumptions and Biology

The foregoing discussion has considered several commonly applied models, assumptions and methods for the study of form variation. In considering the landmark perturbation model alone, it is possible to estimate mean shape, mean form and variability in shape or form, although not necessarily consistently. But this issue of consistency is principally of mathematical interest rather than practical interest.

To the biologist the estimation of mean shape is fairly trivial. Of much greater importance is the reporting of the nature of form differences and covariances with form. EDMA allows "influential" landmarks to be identified but does not lead to holistic representations or descriptions of the localisation and character of form differences. To achieve such descriptions, further (highly reasonable) assumptions are made with respect to the choice of superimposition or deformation method. Generalised Procrustes superimposition has desirable statistical properties in this regard if variations are small (i.e. within the limits usually encountered in biological studies). If deformations are of interest, alternative assumptions must be made with respect to the biological validity of the chosen homology function. When variations are small, all reasonable methods will lead to similar biological conclusions.

The assumptions underlying any one study must be well understood and justifiable in terms of biology. Increasing experience in the application of the methods outlined above indicates, however, that the methods are robust in terms of the biological conclusions. And, when the number of landmarks is large many of the methods perform robustly (e.g. TPS).

The biologist can take steps to confirm the conclusions reached through the use of diverse approaches with different underlying assumptions (e.g. TPS or FEA and superimposition), different sets of landmarks, different finite elements, etc. Furthermore, where possible, independent testing of assumptions should be considered. Such independent tests might include studies of bone remodelling or vital staining in analyses of growth and evolutionary divergence in growth.

20.8 Example Studies

The rest of this chapter gives some examples of using the morphometric methods considered in the previous sections. The first concerns the growth of the facial skeleton in the Old World monkey *Cercocebus torquatus* and the second examines differences

Number	Definition: based on anatomical orientation of the face
1 & 19	Most lateral point on zygomatico-frontal suture on orbital rim
2 & 20	Most supero-lateral point on supraorbital rim
3 & 21	Uppermost point on orbital aperture
4 & 22	Zygomatico-frontal suture at the lateral aspect of the orbital aperture
5 & 23	Fronto-lacrimal suture at medial orbital margin
6 & 24	Zygomatico-maxillary suture at inferior orbital margin
7 & 25	Superior root of zygomatic arch
8 & 26	Inferior root of zygomatic arch
9 & 27	Zygomatico-maxillary suture at root of zygomatic arch
10 & 28	Most posterior point on maxillary alveolus
11 & 29	Deepest point in maxillary fossa
12 & 30	Maxillary–premaxillary suture at alveolar margin
13 & 31	Nearest point to maxillary–premaxillary suture on nasal aperture
14	Upper margin of supraorbital rim in the midline
15	Naso-frontal suture in the midline
16	Tip of nasal bones in the midline
17	Premaxillary suture at the inferior margin of the nasal aperture in the midline
18	Premaxillary suture at alveolar margin

Table 20.1 Definitions of landmarks

in growth between this species and the New World monkey *Cebus apella*.

Growth series of each species were selected such that specimens representing all stages of growth between completion of deciduous dentition and adulthood were included. Each monkey face is defined in terms of the coordinates of 31 landmarks (Table 20.1). These landmarks are registered through generalised Procrustes analysis (Figures 20.2a and 20.2b), mean shapes (Figures 20.2c and 20.3) are estimated from the registered coordinates and the coordinates are submitted to principal components analysis in the tangent space (O'Higgins and Jones, 1998). Figure 20.1 is a summary of these methods.

20.8.1 Growth of the face in Cercocebus torquatus

An earlier study (O'Higgins and Jones, 1998) is summarised here in order to illustrate how the tools of geometric morphometrics can provide insights into skeletal growth. The concern is to examine the relationship between facial shape and facial size during growth (covariance between shape and size during growth). Earlier remodelling studies (O'Higgins *et al.*, 1991) have indicated that surface remodelling fields are consistent in location and character over a substantial part of the growth period. The findings were

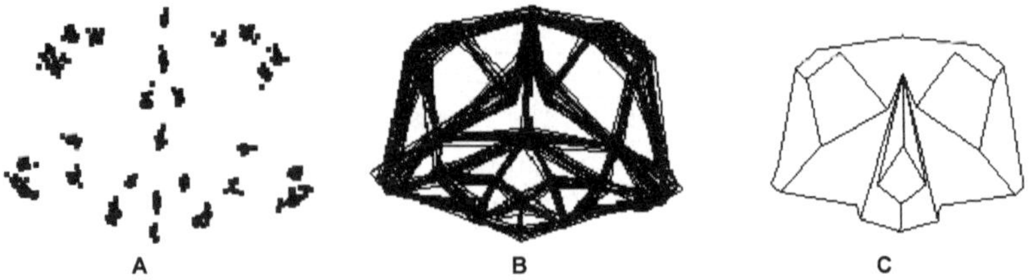

Figure 20.2 Estimation of mean shape through Procrustes analysis. (A) Registered landmarks after GPA; the scatters at each landmark indicate variability. (B) A wire frame representation of these fitted data. (C) The estimated mean shape with wireframe delineating major facial bones.

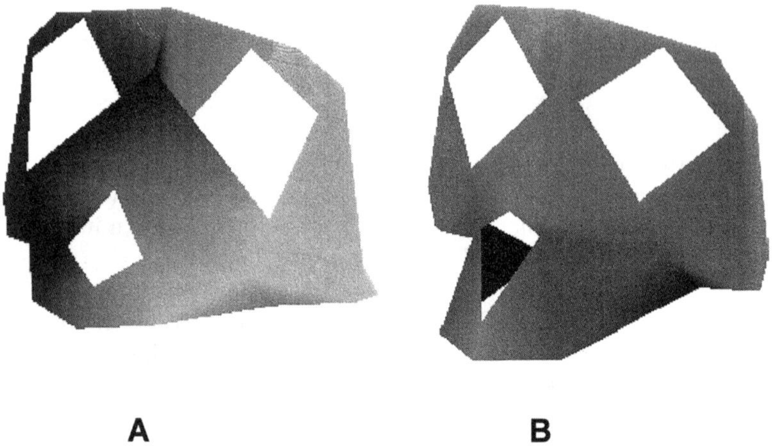

Figure 20.3 Mean rendered faces (not to scale). (A) the mean of the *Cebus apella* growth series, (B) the mean of the *Cercocebus torquatus* growth series. These are smooth rendered representations of the mean shapes of each growth series The polygons used as the basis for these renderings are those shown in Figure 20.2b. The smooth surface effect is achieved through Gouraud shading.

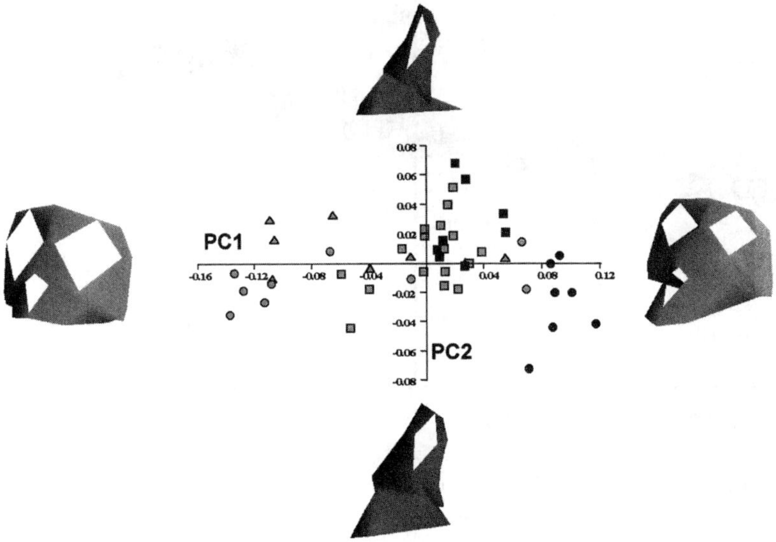

Figure 20.4 Plot of the first two principal components from the analysis of the *Cercocebus torquatus* growth series. Horizontal axis: PC 1, 52% total variance. Vertical axis: PC 2, 8% of total variance. Circles = males, squares = females, black= adult, grey=sub-adult. See text for discussion.

that the external surface of the face is depository everywhere except in the maxillary fossae, which show consistent resorption. It is therefore hypothesised that the general growth vector (i.e. changes in size and shape resulting from bone displacements and surface remodelling) is also consistent over this period. If falsified it would have to be concluded that surface remodelling and bone displacement are not as intertwined as is suggested by Enlow (1966, 1968, 1975) and Bromage, 1986. If not falsified then it is concluded that, at least in this species, the relationship between growth remodelling and growth vectors appears predictable and it would require further testing in other species.

Figure 20.4 presents a plot of the first two principal components (PC 1 vs PC 2) that result from analysis of tangent space coordinates. The shape variability represented by each axis is indicated by the four rendered images at the extremes of the plot. These images were produced by warping the mean shape (score zero on all PCs) to the extremes of the first and then the second principal components. Thus, PC 1 (from left to right, negative to positive extremes) describes differences in shape such that specimens with relatively large orbits and relatively small midfaces sited beneath the orbits are towards the left whereas specimens with relatively small orbits, large and prognathic midfaces are towards the right. PC 2, on the other hand, describes a different pattern of shape variation (as is to be expected since PCs are uncorrelated). Specimens towards the top (positive extreme) relative to those towards the bottom (negative extreme) have a posterior maxilla situated well towards the back (relative to the rest of the face), the maxilla and premaxilla are relatively shortened superoinferiorly and rotated relatively anterosuperiorly. PC 2 therefore captures sample variability in the set of the midface with respect to the upper face.

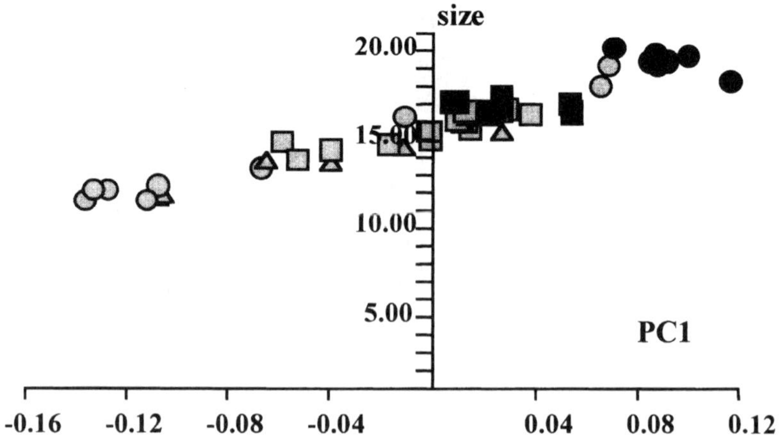

Figure 20.5 Plot of centroid size (vertical axis) versus PC 1 score for the *Cercocebus torquatus* growth series. Circles = males, squares = females, black = adult, grey = sub adult. Correlation between centroid size and PC 1, $r = 0.95, p < 0.001$.

The points plotted in Figure 20.4 are differentiated such that males are shown by circles, females by squares and specimens of unknown sex by triangles. Adults are black and sub-adults grey. PC 1 differentiates adults from sub-adults such that adults have positive scores and sub-adults more negative scores. PC 2, on the other hand, separates adult males (negative scores) from adult females (positive scores). In turn only the adults of each sex appear to be separated by this axis. Thus it appears that adults show differences in morphology which appear only late in growth and these differences relate to the set of the midface with respect to the upper face. A t-test confirms that the adults of each sex are significantly different in these aspects of shape (scores on PC 2; $p < 0.003$).

It has already been noted that PC 1 appears to order specimens according to age. Figure 20.5 extends this finding by demonstrating that scores of specimens on PC 1 are strongly related to centroid size. Thus, large specimens tend to have positive scores and small, negative. The correlation between centroid size and score on PC 1 is $r = 0.95, p < 0.001$. Interestingly, in this aspect of shape variation (PC 1) adult males appear to be scaled versions of adult females (although the additional sexual differences shown on PC 2 should also be considered in relation to the ontogeny of sexual dimorphism). No other PC shows any significant evidence of a relationship with age, size or sex.

There is evidence that in *Cercocebus torquatus* the general vector of growth is consistent throughout a large part of the growth period (males and females diverge late in growth, PC 2). This concords with the relative stability of remodelling during growth and fails to falsify the hypothesis that cortical remodelling can be used to infer general growth pattern.

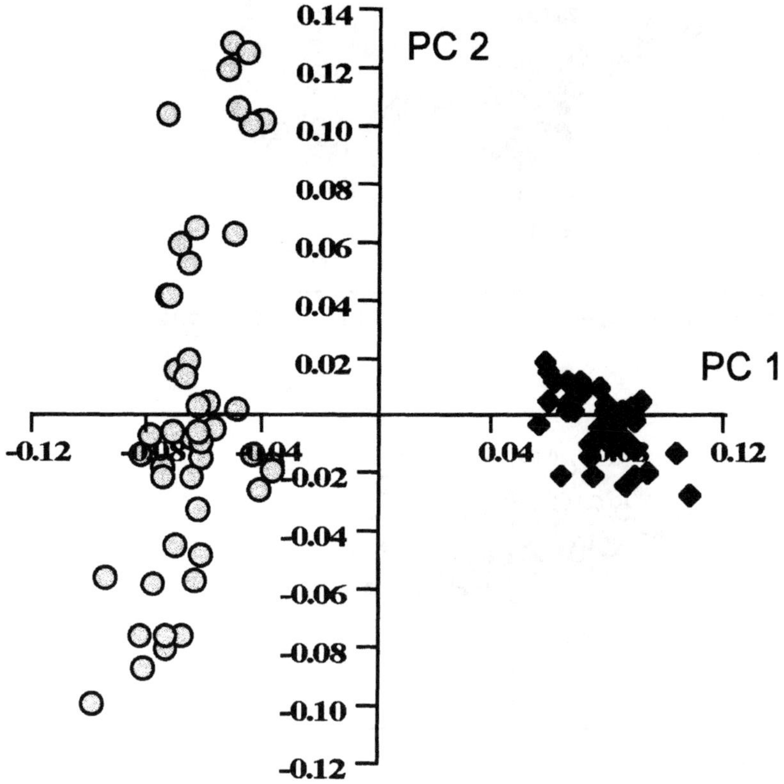

Figure 20.6 Plot of the first two principal components from the combined analysis of the *Cercocebus torquatus* and *Cebus apella* growth series. Horizontal axis: PC 1, 48% total variance. Vertical axis: PC 2, 20% of total variance. Grey circles = *Cercocebus torquatus*; black diamonds = *Cebus apella*.

20.8.2 Comparative growth; Cercocebus torquatus vs Cebus apella

The second example study extends the findings of the first by presenting new data on growth of the face in the New World monkey *Cebus apella* and comparing growth in this species with growth for the mangabey *Cercocebus torquatus*. These two monkeys are widely separated in evolutionary terms (Fleagle, 1988) although there remain strong superficial similarities in facial form.

Recent and as yet unreported scanning electron microscope studies in my laboratory indicate that, unlike *Cercocebus*, the facial bone external surfaces are entirely depository, no resorption is noted in the maxillary fossae. It is predicted therefore that the general vectors of facial growth will differ between these species. This example study is therefore concerned to test the null hypothesis that no differences exist in facial growth patterns. If falsified it aims to report the nature of any such differences. Thus the study considers how shape differences and growth differences are related to evolutionary divergence.

Figure 20.6 shows a plot of the first pair of principal components from the analysis of both species combined (grey circles = *Cercocebus*, black diamonds = *Cebus*). PC

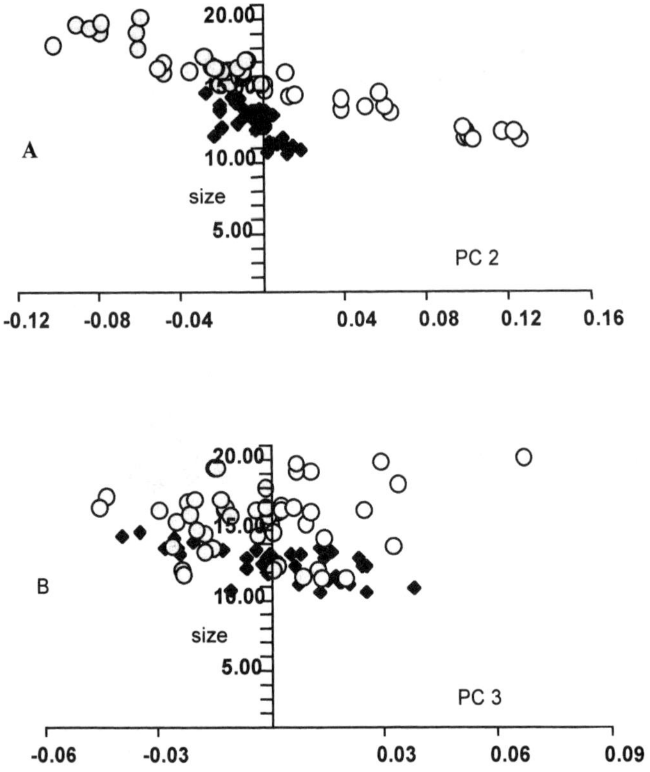

Figure 20.7 (A) Plot of centroid size (vertical axis) versus PC 2 (20% of total variance). (B) Plot of centroid size (vertical axis) versus PC 3 (4% total variance) for the combined growth series. Grey circles = *Cercocebus torquatus*; black diamonds = *Cebus apella*.

1 completely separates *Cercocebus* from *Cebus* and PC 2 shows a wide scatter of *Cercocebus* plus a much more concentrated scatter of *Cebus*. Figure 20.7a plots PC 2 against centroid size and Figure 20.7b plots PC 3 against centroid size. It is clear from these plots that *Cercocebus* shows much greater ontogenetic shape variability than *Cebus* on PC 2, whereas on PCs 1 and 2 the spreads are approximately the same for both species (as indeed they are for all other PCs). Thus there is evidence that *Cercocebus* shows more ontogenetic variation in shape than *Cebus* and that this aspect of ontogenetic variation is principally described by PC 2.

In considering the relationship between centroid size and shape (Figures 20.7a and 20.7b) a further contrast emerges between *Cercocebus* and *Cebus*. In Figure 20.7a, the scores of *Cercocebus* show a strong linear relationship with centroid size. This is reminiscent of the results of the analysis of *Cercocebus* alone (Figure 20.5). However, the scores of *Cebus* on PC 2 show evidence of a different relationship with centroid size. The situation is different with respect to PC 3; the scores of *Cebus* show evidence of a strong linear relationship with size whereas those of *Cercocebus* do not. The conclusion is that although both species show evidence of a size-related shape variation (allometry), the nature of this variation is different in each, in terms of the aspects of

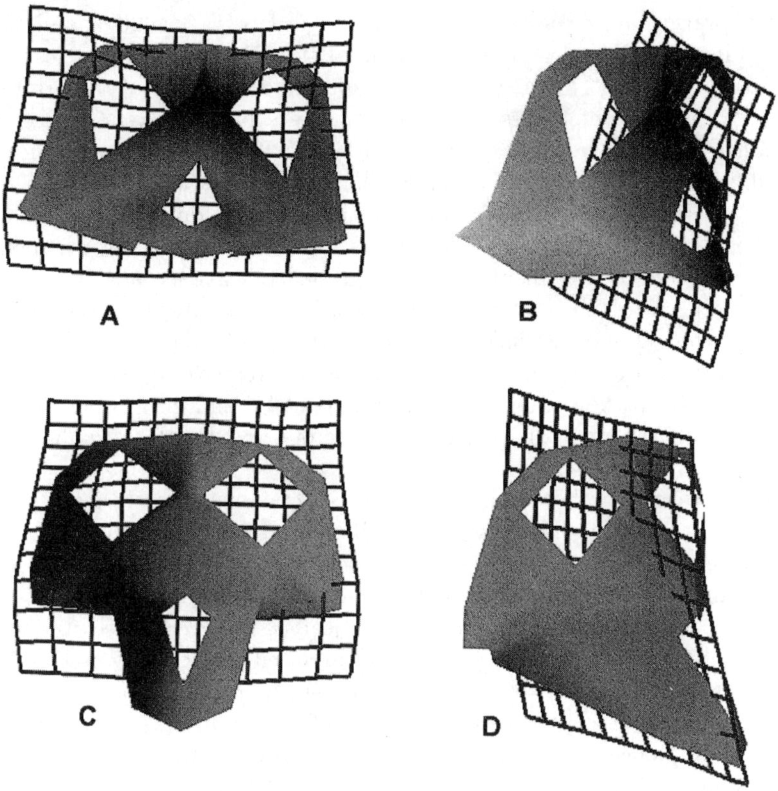

Figure 20.8 Cartesian transformation grids drawn using thin plate splines. Grids are calculated between the extremes of PC 1 representing small (reference) and large (target) specimens from PCA of age series of single species. The grids are drawn over the target shapes. (A, B) *Cebus apella*, (C , D) *Cercocebus torquatus.* (A , C) Frontal view, grid sited approximately through zygomatic root and orbital rim. (B , D) view of right face, grid drawn approximately in midline.

shape that covary with size.

This difference is confirmed by a permutation test which indicates that, from analyses of *Cebus* alone and *Cercocebus* alone, the angle between the first PC (which represents size-related shape variation in each analysis) is 78° which is highly significant, $p < 0.0009$. The analysis of *Cebus apella* alone results in scores on PC 1 that, as in the analysis of *Cercocebus* alone (Figure 20.5), show a significant correlation with size ($r = 0.76$, $p < 0.001$). No other PC from this analysis shows any evidence of size-related shape variation. Those aspects of growth-related shape variation represented by PC 1 in the analysis of *Cebus* alone are illustrated in Figure 20.8. Figures 20.8a and b show transformation grids drawn over a warped mean representing the adult extreme of PC 1. The transformation grids are calculated between the (infant and adult) extremes of this PC. Likewise, Figure 20.8c and d show the differences between the (infant and adult) extremes of PC1 from the analysis of *Cercocebus torquatus.* Figure 20.8 therefore allows growth in *Cebus* (Figure 20.8a and b) to be appraised in relation to growth in *Cercocebus* (Figure 20.8c and d). The frontal views (Figure 20.8a and c) show some similarities. Thus, in both species, growth results in relative widening of the zygomatic region and narrowing of the orbital and interorbital regions. The differences between species are principally in degree rather than in the nature of shape variation with growth. Figure 20.8b and d show much more pronounced differences between the species. Thus in *Cercocebus* (Figure 20.8d) the deformation of the transformation grid indicates that the development of prognathism during growth is a pronounced feature. The grid is smoothly deformed from superior to inferior, indicating this prognathism involves a graded relative expansion of midface and nose with respect to the upper face. This contrasts starkly with the grid from the analysis of *Cebus* (Figure 20.8b) which shows that the subnasal region becomes relatively prognathic, and the nasal region retracted relative to the interorbital. Thus the inference from the combined PCA (Figures 20.6 and 20.7) that *Cebus* and *Cercocebus* show different growth trajectories is supported by the visualisations of Figure 20.8.

The hypothesis that growth vectors do not differ between these species is falsified. There is abundant evidence that they grow differently and it is tempting to take the morphometric and remodelling studies to indicate that cortical remodelling is a useful indicator of general facial growth (which arises from bone displacements and remodelling). A word of caution should be injected, however; these studies have examined remodelling and growth in two distantly related species only. Facial growth remodelling is probably a much more subtle process than can be represented on a map of depository and resorptive areas alone. Differences in rate and timing can also have major influence on the pattern of facial growth. Many more studies of cortical remodelling and geometry are called for (especially among species with similar remodelling maps, e.g. *Macaca* and *Cercocebus*) before any solid conclusions can be drawn. The studies presented here do, however, indicate possible approaches to better understanding of the integration and evolution of facial growth.

20.9 Concluding Remarks

The review and studies presented in this chapter are intended to illustrate important aspects of the methods of geometric morphometrics and their potential for application in the study of growth. This work can be traced back to the original ideas of D'Arcy Thompson and in particular to his "method of coordinates". The studies of facial growth in monkeys serve to underline Thompson's original notion that although we may not understand the processes giving rise to form itself, we can make progress through focusing on deformations or transformations and by comparing such transformations between related species. Ongoing studies are using the tools of the geometric morphometric toolkit to unravel the relative contributions of growth differences and patterning differences to skeletal differentiation between species. This is a long-term endeavour that will eventually lead to deeper understanding of evolutionary adaptation of skeletal form.

This chapter has tried to consider the models and assumptions underlying the current armoury of morphometric tools. From this review and from the studies presented it should be clear that morphometrics is essentially concerned with the study of form variation and of covariances with form. In order to make progress and produce reports of form variation which have biological utility, it is essential that certain assumptions are made with respect to registration or homology functions. All the evidence from published studies indicates that these assumptions are robust in the face of a wide variety of data sets. This is because when variations are small, all sensible registration methods and homology functions will lead to very similar statistical and identical biological conclusions. Without these assumptions we are able to estimate means and some aspects of variability (excluding relative landmark movements) but these are in themselves of limited utility in biology.

Thus, Thompson's contribution to the analysis of form variations continues to inspire and motivate generations of biologists and mathematicians. The development of the geometric morphometric toolkit is likely to have a significant impact in the future. In particular, the applications of these methods to developmental biology, diagnostic imaging, and growth prediction are in their early stages at present but seem set to transform the possibilities for understanding biological form and morphogenesis.

Acknowledgements

I should like to thank Mark Chaplain for having the insight, energy and ambition to organise the meeting that has led to the production of this volume. I could not have begun this work without the enormous support of my statistical colleagues Kanti Mardia, Ian Dryden and John Kent of the Department of Statistics, University of Leeds. Other, much more statistically knowledgeable workers than myself have contributed valuable advice. These include Subhash Lele, who spent many hours arguing every point in this chapter, and Fred Bookstein, who gave thoughtful advice and insightful comments. Ian Dryden also made helpful and much appreciated comments on an earlier draft.

The programming that underpins the studies presented in this chapter relied heavily on Nicholas Jones, who made it all work together and reliably. The *Cebus* data were gathered with Pippa Chadfield. I thank the Natural History Museum, London, especially Paula Jenkins, for granting access to the collection of primate material.

Software availability

Tools for geometric morphometric analysis are widely available on the internet. Useful addresses include the SUNY Stony Brook morphometrics site, `http://life.bio.sunysb.edu/morph/`, the University of Leeds, Department of Statistics site `http://www.amsta.leeds.ac.uk/Statistics/software` and `http://evolution.anat.ucl.ac.uk/morph/helphtmls/morph.html`, from which can be obtained the program used in the examples given in this chapter.

REFERENCES

Bookstein, F.L. (1978) *The Measurement of Biological Shape and Shape Change.* Springer: New York.

Bookstein, F.L. (1984) A statistical method for biological shape comparisons. *J. Theor. Biol.* 107: 475–520.

Bookstein, F.L. (1989) Principal warps: thin-plate splines and the decomposition of deformations. *IEEE Trans. in Pattern Analysis and Machine Intelligence.* 11: 567–585.

Bookstein, F.L. (1991) *Morphometric Tools for Landmark Data: Geometry and Biology.* Cambridge University Press: Cambridge.

Boyde, A. and Jones, S.J. (1972) Scanning electron microscopic studies of the formation of mineralised tissues. In: *Developmental Aspects of Oral Biology.* Slavkin, H.C. and Baveta, L.A., pp. 243–274. Academic Press: New York.

Bromage, T.G. (1986) A comparative scanning electron microscope study of facial growth remodelling in early hominids. PhD thesis, University of Toronto.

Dryden, I.L. and Mardia, K.V. (1991) General shape distributions in a plane. *Adv. Appl. Prob.* 23: 259–276.

Dryden, I.L. and Mardia, K.V. (1993) Multivariate shape analysis. *Sankhya* 55(A): 460–480.

Dryden, I.L. and Mardia, K.V. (1998) *Statistical Shape Analysis.* John Wiley: Chichester.

Enlow, D.H. (1966) A comparative study of facial growth in *Homo* and *Macaca. Am. J. Phys. Anthropol.* 24: 293–308.

Enlow, D.H. (1968) *The Human Face: An Account of the Postnatal Growth and Development of the Craniofacial Skeleton.* Harper and Row: New York.

Enlow, D.H. (1975) *Handbook of Facial Growth.* W.B. Saunders: Toronto.

Fleagle, J.G. (1988) *Primate Adaptation and Evolution.* Academic Press: New York.

Kendall, D.G. (1984) Shape manifolds, Procrustean metrics and complex projective spaces. *Bull. Lond. Math. Soc.* 16: 81–121.

Kent, J.T. (1994) The complex Bingham distribution and shape analysis. *J. Roy. Stat. Soc. Ser. B.* 56: 285–299.

Kent, J.T. and Mardia, K.V. (1997) Consistency of Procrustes estimators. *J. Roy. Stat. Soc. Ser. B.* 59: 281–290

Lele, S. (1991) Some comments on co-ordinate free and scale invariant methods in morphometrics. *Am. J. Phys. Anthropol.* 85: 407–418.

Lele, S. (1993) Euclidean distance matrix analysis: estimation of mean form and form difference. *Math. Geol.* 25: 573–602.

Lele, S. and Cole, T.M. (1995) Euclidean distance matrix analysis: a statistical review. In: *Current issues in Statistical Shape Analysis.* Mardia, K.V. and Gill, C.A., eds. pp. 49–53. Leeds University Press: Leeds.

Lele, S. and Richtsmeier, J.T. (1991) Euclidean distance matrix analysis: a coordinate free approach for comparing biological shapes using landmark data. *Am. J. Phys. Anthropol.* 86: 415–427.

Lewis, J.L., Lew, W. and Zimmerman, J. (1980) A non-homogenous anthropometric scaling method based on finite element principles. *J. Biomech.* 13: 815–24.

Lieberman, D.E. (1996) How and why humans grow thin skulls: Experimental evidence for systemic cortical robusticity. *Am. J. Phys. Anthropol.* 101: 217–236.

Marcus, L.F., Corti, M., Loy, A., Naylor, G.J.P. and Slice, D. (eds.) (1996) *Advances in Morphometrics.* Plenum Press: New York.

Mardia, K.V. and Dryden, I.L. (1989) The statistical analysis of shape data. *Biometrika* 76: 71–282.

Mardia, K.V., Kent, J.T. and Bibby, J.M. (1979) *Multivariate Analysis.* Academic Press: London.

Mardia, K.V., Coombes, A., Kirkbride, J., Linney, A. and Bowie, J.L. (1996) On statistical problems with face identification from photographs. *J. Appl. Stat.* 23: 655–675.

Mosimann, J.E. (1970) Size allometry: Size and shape variables with characterisations of the log-normal and generalised gamma distributions. *J. Am. Stat. Assoc.* 65: 930–948.

Moss, M.L. and Young, R.W. (1960) A functional approach to craniology. *Am. J. Phys. Anthropol.* 18: 281–292.

O'Higgins, P. (1997) Methodological issues in the description of forms. In: *Fourier descriptors and their applications in biology.* Lestrel, P.E. (ed.), pp. 74–105. Cambridge University Press: Cambridge.

O'Higgins, P. and Dryden, I.L. (1992) Studies of craniofacial growth and development. *Perspectives in Human Biology 2, Archaeology in Oceania* 27: 5–104.

O'Higgins, P. and Jones, N. (1998) Facial growth in *Cercocebus torquatus*: An application of three dimensional geometric morphometric techniques to the study of morphological variation. *J. Anat.* 193: 251–272.

O'Higgins, P., Bromage, T.G., Johnson, D.R., Moore, W.J. and McPhie, P. (1991) A study of facial growth in the sooty mangabey *Cercocebus torquatus*. *Folia Primatologica*. 56: 86–94.

Oxnard, C.E. (1991) Mechanical stress and strain at a point: Implications for biomorphometric and biomechanical studies of bone form and architecture. In: *Human Biology, an Integrative Science: Proceedings of the Australasian Society for Human Biology*, no 4. O'Higgins, P. (ed.), Centre for Human Biology, University of Western Australia.

Rao, C.R. and Suryawanshi, S. (1996) Statistical analysis of shape of objects based on landmark data. *Proc. Natl. Acad. Sci. USA* 93: 12132–12136.

Richtsmeier, J.T., Cheverud, J.M., Dahaney, S.E., Corner, B.D. and Lele, S. (1993a) The onotogeny of sexual dimorphism in the crab-eating macaque (*Macaca fascicularis*). *J. Human Evol.* 25: 1–30.

Richtsmeier, J.T., Corner, B.D., Grausz, H.M., Cheverud, J.M. and Danahey, S.E. (1993b) The role of post natal growth in the production of facial morphology. *Sys. Biol.* 42: 307–330.

Siegel, A.F. and Benson, R.H. (1982) A robust comparison of biological shapes. *Biometrics* 38: 341–50.

Thompson, D.W. (1917) *On Growth and Form.* Cambridge University Press: Cambridge.

21

D'Arcy Thompson and the Problem of Biological Form

Brian C. Goodwin

Schumacher College, The Old Postern, Dartington, Totnes, Devon TQ9 6EA, UK

21.1 Introduction

D'Arcy Thompson's contribution to biology remains enigmatic. Whereas there is virtually universal achnowledgement that his insights into the principles that underlie organismic forms and their transformations are of unique value, it is not clear how to connect these insights with those that have emerged from the steady focus that this century, and particularly the last 20 years, has brought to bear on the role of genes in the emergence of organismic form during development. There seem to be two disjunctions here: (1) gene action during development is described in terms of information expressed in a genetic program whereas D'Arcy Thompson's (1917) treatment of form is entirely in terms of energetics and global transformations; (2) gene action is dynamic and transient during development, while the principles that D'Arcy used to describe biological forms are based on static energy minimisation, defining equilibrium states. These disjunctions are by no means irreconcilable, but neither is their resolution a trivial problem. They represent two quite different ways of describing the world and they have resulted in a profound separation between biology and physics, to the detriment of each. However, the dialectic of science results in periodic movements to reconcile differences and seek unification, and there are signs that we may be moving towards one of these at the moment. So this seems an auspicious time to explore the issues involved.

21.2 Historical Narratives and Scientific Explanations

D'Arcy Thompson, like his contemporary William Bateson, reacted strongly against Darwin's claim that his evolutionary theory of descent with modification provided an explanation of biological forms and their transformations during evolution. Darwin (1859) based this view on his claim that "the chief part of the organisation of every being is simply due to inheritance". The direct descendent of this claim is

On Growth and Form: Spatio-temporal Pattern Formation in Biology.
Editor Mark A.J. Chaplain, G.D. Singh, J.C. McLachlan

the contemporary belief of many biologists that "DNA provides the programme which controls the development of an embryo and brings about epigenesis" (Wolpert, 1991). There are two aspects of such claims that need to be distinguished. The first relates to material reductionism. In Darwin's perspective the organism is reduced to a collection of independently variable characters such as limbs, eyes, heart, etc., which are derived historically by descent from a common ancestor in which the character first appeared, such as the first pentadactyl (five-digit) limb in tetrapods. This is assumed to be historically given, and evolutionary explanations then address the question of modifications of the ancestral form by hereditary variation and selection of fitter forms, leading to wings in birds, flippers in seals, hands in humans, etc. Bateson's criticism was that, even if there are common ancestral forms (which in general there are not), they need to be accounted for; that is, their initial emergence in evolution needs to be explained in terms of what forms organisms can generate, or equivalently, how these forms are possible as expressions of the living state. D'Arcy Thompson's comment on Darwin's type of explanation was: "So long and so far as 'fortuitous variation' and the 'survival of the fittest' remain engrained as fundamental and satisfactory hypotheses in the philosophy of biology, so long will these 'satisfactory and specious causes' tend to stay 'severe and diligent inquiry ... to the arrest and prejudice of future discovery' ". The last two internal quotations here are from Francis Bacon's comments on the unfortunate overuse of explanations in terms of teleology and final causes that he saw in scholasticism and which Thompson saw being overused in Darwinism.

In the contemporary perspective of molecular genetics, material reductionism extends beyond the characters of an organism to the molecules it is made of, molecules which are produced by the genes. It is often assumed that a complete knowledge of the sequence of molecular compositions that constitute a developing organism at every stage of its development is sufficient to define its form, via a kind of self-assembly postulate. The scientific inadequacies of this position have been pointed out on numerous occasions (Goodwin, 1982, 1994; Webster and Goodwin, 1982, 1996). The argument is simply that molecular composition is not sufficient, in general, to define form in any science (physics, chemistry, biology). What is required in addition is an understanding of the organisational principles that define space-time order in the system under study, together with initial and boundary conditions. These are known as field theories. In biology these principles are described as the properties of morphogenetic fields, which D'Arcy Thompson described in terms of energetic interactions.

In addition to material reductionism, the second aspect of Darwin's conceptual approach to biology and its contemporary descendents is explanation by historical narratives in which the role of heredity is to generate the diversity (variation) on which natural selection acts, while also ensuring that offspring tend to resemble their parents. Modern genetics provides detailed molecular explanations for these properties (up to a point) while retaining explanations of evolutionary process by historical narratives. This introduces an irreconcilable difference of explanatory principle from that employed by D'Arcy Thompson, and it is this that causes the present hiatus. The difficulty is as follows.

Evolutionary theory à la Darwin sees species taxa as historical individuals, hence as accidents of their evolutionary histories (Hull, 1976, 1984). The only necessity they satisfy is survival; they have no necessary intrinsic properties, or natures. Hence their properties, including their forms, cannot be explained by scientific principles, which are based upon concepts of type or essence. Purely accidental and contingent properties do not have scientific explanations; they are unique events that just happen. So if species taxa are historical individuals (unique collections of accidental characters that satisfy the requirement of survival in some habitat) then it is impossible to produce explanations of their forms of the type sought by D'Arcy Thompson. Darwinian biology is not a science in the sense of physics and chemistry. It is a purely historical study. Hence the clash between Darwin and D'Arcy Thompson is absolute, irreconcilable. There can be no rational science of form in Darwinian biology.

However, there is no reason to believe that Darwin's version of biology is definitive. It may be more reasonable to view current evolutionary theory as a limited aspect of a more comprehensive biology that is now struggling to emerge from the interesting interactions that are occurring in the hinterland between biology, mathematics and physics, where physics is becoming more historical and biology is becoming more rational. (The term "rational" here is simply to contrast it with historical or contingent; it does not imply that Darwinian biology is illogical, since historical narratives can have their own logic.) There is no intrinsic reason why a science cannot have both historical and rational ingredients; and indeed, this is already the case for physics. The deeper conceptual shifts required if biology is to have a science of form of the type sought by Bateson and D'Arcy Thompson lie in the structure of biology, not of physics.

Physics has always been grounded in dynamics, the principles of motion. This involves both historical contingencies and laws. The former enter dynamics as initial conditions, which describe the arbitrary starting values of variables whose subsequent behaviour is dictated by the dynamical laws of the system under study. These can be the initial position and velocity of a planet or other celestial body under the action of gravitational forces, or the initial values and boundary conditions of relativistic field equations describing early cosmic evolution, involving the production of galaxies or black holes or other structures in relativity theory. The move made by Darwin was to entirely eliminate any law-like (typological) concepts from biology, defining it as a purely historical science in which initial conditions (for organismic life-cycles, say) and the particularities of species development are defined by inheritance and environmental conditions, which together exhaustively describe the evolution of organisms. These initial conditions and particularities vary randomly, and natural selection is a stability condition: those organisms (strictly speaking, those genomes) persist which are fit enough to leave sufficient offspring to perpetuate the species which they define. The tautological nature of natural selection is evident: persistence is defined in term of fitness, which is defined in terms of persistence. This is characteristic of all stability concepts: stability is defined in terms of invariance (something that does not change, such as persistence) while invariance is the definition of stability. This does not mean that such concepts are of no value. On the contrary, it is often very instructive to determine what features of the dynamics result in stability, and in biology it is interesting to enquire into the properties of species that result in their perpetuation in particular habitats. However, it is important to understand that natural selection cannot explain the forms of organisms. The most it can do is to suggest how particular organismic morphologies may contribute to survival in particular habitats. Natural selection says nothing explanatory about the origins of species, understood as organisms of specific form.

21.3 A Generative Biology: Morphogenetic Fields

To get a biological theory that has the possibility of saying something informative about organisms and their morphologies, it is necessary to change the conceptual structure of current biology in a fundamental manner. First, the notion that species taxa are historical individuals defined by purely accidental properties must be extended to some concept of organisms as entities with necessary natures which define their intrinsic properties, allowing the construction of a rational taxonomy of biological forms as well as a historical genealogy. But if this is to be a truly dynamic theory, then this taxonomy must deal not with a description of adult forms in terms of static equilibrium states, as D'Arcy Thompson does in terms of variational principles (e.g. energy minimisation) or coordinate transformations of species morphologies. What is required is a generative taxonomy: a theory about the causal processes intrinsic to organisms that generate a set of forms, starting from eggs or buds, and proceeding through development to the adult. Genes are fundamentally involved in producing and stabilising the diversity of morphologies that has emerged during evolution, but they are not the causes of the generative processes that give rise to organisms of specific form.

The reason for this is that genes are responsible for producing the proteins and other macromolecules that determine the molecular specificity of organisms at any stage in their development. However, as stated earlier, molecular composition is not sufficient to determine three-dimensional form in any science. This requires a theory describing the relational order that is distinctive to the system under study. In physics these field theories include gravitational, magnetic, hydrodynamic, and quantum electrodynamic fields, which define the causal principles that generate different types of dynamic order.

In biology the relevant field concept is the morphogenetic field. Different types of morphogenetic field have different causal powers: limb fields generate limbs; eye fields generate eyes; floral meristems are fields that generate flowers. Particular genes are active within these fields, but their activities are not in themselves sufficient to account for the structures generated. This requires the definition of relational order within the morphogenetic fields in mathematical or equivalent logical terms, defining the forces that act in the production of particular biological structures. To explain the development of a specific adult form from an egg, say, it is necessary to show how the initial unitary morphogenetic field of the developing embryo gives rise to a sequence of sub-fields, such as those generating limbs, eyes, heart, etc.; or leaves and flower organs in a plant. Such theories require properties of the type described, for example, by excitable media that can undergo cascades of symmetry-breaking events whereby an initial whole undergoes bifurcation to a set of organised parts or sub-fields. These sub-fields themselves undergo further bifurcations to smaller fields so there emerges a hierarchically nested set of elements, organised spatially relative to one another (Goodwin, 1994; Webster and Goodwin, 1996). Genes contribute to the excitable dynamics of these fields, as is evident from, for example, the pseudo-waves of oscillation in avian hairy gene product during the generation of somites in the chick (Palmeirim *et al.*, 1997).

The result is a dynamic field theory that uses energetic principles of the type described by D'Arcy Thompson, whose example of the oil droplet undergoing progressive bifurcations in a liquid medium to produce a medusa-like structure is the prototype. All that is added is that the oil droplet itself acquires the energy to generate the sequence of bifurcations to organised complexity, as described in the morphogenesis of an algal cell to whorled form (Goodwin and Brière, 1992; Goodwin, 1994). This just makes explicit what was already implicit in D'Arcy Thompson's initial insight, but it required the elaboration of an armoury of tools and techniques in non-linear dynamical analysis and dynamical systems theory that has taken half a century to develop. This is not an easy domain to work in. It is much simpler to talk about information in the DNA and genetic programs that generate morphology. But in the end, science requires explicit demonstrations of processes via models that reliably simulate the phenomena one seeks to understand, indicating the nature of the causal mechanisms involved. No one has ever generated three-dimensional form of the type observed in organisms by a genetic program, which simply generates molecules. The realisation that this kind of genetic reductionism is scientifically erroneous is now beginning to be recognised by the biological community addressing these problems, as indicated by a landmark paper by Gilbert *et al.* (1996). To quote from their article: " Just as the cell (and not the genome) functions as the unit of organic structure and function, so the morphogenetic field (and not the genes or the cells) is seen as a major unit of ontogeny, whose changes bring about changes in evolution."

21.4 Genetic Networks, Morphogenesis and Evolution

The role of gene activities in morphogenesis needs now to be defined. Essentially, genes determine the particularities of different morphogenetic fields, the properties that distinguish the power of a limb field to generate a limb from that of an eye field; or the power of a rose floral meristem to produce a rose as distinct from a geranium. However, roses and geraniums, limbs and eyes have properties in common as well as differences, and it is these invariant or unchanging qualities that allow us to speak of morphogenetic fields as the embodiments of the capacity of living organisms to generate three-dimensional form. Genes produce specific types of protein which influence the details of the form generated, but they can do this only within the context of a system that is dynamically organised according to certain general principles. It is instructive now to consider a particular model that attempts to capture basic features of this organisation in mathematical form, and to then see how gene activities may be understood to have both the capacity to stabilise particular forms (of different species) and to explore the production of new forms, as occurs in evolution. At the moment this will be restricted to animals.

The model that I shall use for this purpose has been developed by Cummings (1994). He defines a general class of non-linear field equations that generate spatial patterns of gene products within an initially defined spatial domain. These gene products are called "morphogens" by Cummings because they are assumed to have an influence on cell shape via such processes as cell–cell adhesion, differential contraction of cells, cell growth and division, etc. This influence is described specifically in terms of curvature. Cummings describes metazoan form in terms of the curvatures of sheets of cells which can change shape in particular ways in response to local concentrations of morphogens. His cells are elements of structure defining the deformable sheets; they need not correspond precisely to biological cells. He imposes realistic constraints on how these "cells" can change their shapes, from regular cuboids to elements contracted on one side and expanded on the other, including expansion and contraction along different axes so that saddle-like deformations can occur. From these allowed local transformations he describes the changes of curvature that can result in the whole surface. He shows that the constraints allow the sheets of cells to transform in all the characteristic ways that occur during metazoan development: gastrulation, neural tube (cylinder) formation, buckling out or in of the surface, pinching off structures, periodicities of structure as in segmentation, and so on. The particular transformations that the system undergoes depend upon local morphogen values which determine cell shape and hence local curvature, involving a single parameter. This parameter is assumed to depend upon local morphogen values. These morphogens (only two in Cummings' model, which he suggests may be cell adhesion molecules) have values determined by morphogenetic field equations. These equations also have parameters which determine their detailed behaviour, and they are assumed to depend on gene activity.

Here is a model that goes from gene activity via a morphogenetic field to the precise specification of three-dimensional form, the morphology of the developing organism. As gene activity changes, morphogenesis occurs and the organism changes its form. However, a series of morphogenetic transformations will occur even if the only activity of the genes is to allow growth to occur, or some process equivalent to growth in terms of the model (e.g. change of cell size or modification of the effective communication distance between cells). That is to say, genes don't have to do very much to produce morphogenesis in this model because of the coupling between morphogens and curvature on the one hand, and reciprocal effect of curvature (shape) and "size" on the morphogenetic field solutions. The model captures the intrinsic dynamic organisation of living organisms to generate form, and the role of the genes in any morphological species is to stabilise the choice of developmental pathway that leads to a particular form.

Genes are also involved in the process whereby the morphospace of possible forms can be explored by evolving organisms. This requires genetic networks within the "cells" of the organism that can modify parameters and thus alter morphogenetic sequences. It also requires that the states of the genetic networks are affected by the local values of the morphogenetic field. One way of achieving this is to have gene activity affected by local curvature. The implication is that gene activity is influenced by variables that depend upon mechanical strain in cells, such as substances whose intracellular levels depend upon strain-dependent membrane pumps and channels, or by the state of the cytoskeleton and the nuclear skeleton influencing gene activities via modification of translation and transcription rates. However, what is important in the model is not so much the precise details of the coupling as the logical structure of the whole system. To generate a space of possible forms (a morphospace) the system must be autonomous or logically closed in the following sense (assuming a constant environment for the moment): the states of the genetic network control the parameters that specify the morphogen patterns and hence the curvature, and the local state of the "cells" influences the states of genes in the network. This type of system has the property that nothing is in overall control because the cycle of influences is dynamically closed.

The genetic networks require two basic properties to assist in the exploration of morphospace. There must be a source of variation of the "genes" and their interactions, which can be either random or directed. And the networks must have the capacity to stabilise dynamic patterns which result in parameter changes that produce life cycle closure. By this I mean that the "organism" starts with a particular shape (say a hollow sphere of cells), proceeds through a sequence of morphogenetic transformations, and generates from itself more of the initial reproductive structures (spheres of cells) which then repeat the cycle. Such sequences represent asexual reproduction by a budding process, as in zooid formation in coelenterates (cnidarians and ctenophores), bryozoa, tunicates, and other phyla. Morphogenetic sequences of this type can be generated in Cummings' model as the morphogenetic field passes through a bifurcation sequence and local morphogen values induce shape changes via their effects on local curvature.

This pattern of life cycle closure can be taken to be the simplest criterion of natural selection or dynamic stability of "successful" morphological reproduction. For this to occur, it is necessary for the spherical "bud" that is generated by the adult form in the model to consist of cells whose genetic networks are in the same (or a sufficiently similar) state to those in the initiating sphere of cells that gave rise to the adult form, so that the life cycle will repeat. The genetic networks must learn to stabilise such closed life cycles. Since these are the only developmental sequences that repeat, the condition is virtually tautological. That is to say, the only dynamically stable states (attractors) in the space of genetic networks are limit cycles, quasi-periodic cycles or strange attractors. This provides a conceptual framework to explore evolution as a process that involves form and transformation in organisms as real biological entities that are not reducible to genes, though genes are clearly part of the dynamical process in which organisms are engaged. This model is now under research study. Such an expanded framework allows for a synthesis of the extremely effective theory of genes and their action that has been the major triumph of twentieth-century biology, with the fundamental principles required for a theory of biological form and the evolution of organisms. It extends D'Arcy Thompson's vision in ways that are compatible with his explanatory principles of growth and form, while accommodating the theory of inheritance.

REFERENCES

Cummings, F.W. (1994) Aspects of growth and form. *Physica D* 79: 149–163.

Darwin, C. (1859) *On the Origin of Species by Natural Selection.* Reprinted 1968 by Penguin Books.

Gilbert, S.F., Opitz, J.M., and Raff, R.A. (1996) Resynthesizing evolutionary and developmental biology. *Dev. Biol.* 173: 357–372.

Goodwin, B.C. (1982) Development and evolution. *J. Theor. Biol.* 97: 43–55.

Goodwin, B.C. (1994) *How The Leopard Changed Its Spots: The Evolution of Complexity.* London: Weidenfeld and Nicolson.

Goodwin, B.C., and Brière, C. (1992) A mathematical model of cytoskeletal dynamics and morphogenesis in Acetabularia. In *The Cytoskeleton of the Algae*, ed. D. Menzel, pp. 219–238. Boca Raton FL: CRC Press.

Hull, D.L. (1976) Are species really individuals? *Systematic Zool.* 25: 174–191.

Hull, D.L. (1984) Historical entities and historical narratives. In *Minds, Machines and Evolution*, ed. C. Hookway. Cambridge: Cambridge University Press.

Palmeirin, I., Henrique, D., Ish-Horowicz, D., and Pourquie, O. (1997) Avian hairy gene expression identifies a molecular clock linked to vertebrate segmentation and somitogenesis. *Cell* 91: 639–648.

Thompson, D.W. (1917) *On Growth and Form.* p. 5 of abridged edition, 1966 (ed. J.T. Bonner). Cambridge: Cambridge University Press.

Webster, G.C. and Goodwin, B.C. (1982) The origin of species: a structuralist approach. *J. Sociological and Biological Structures* 5: 15–47.

Webster, G.C., and Goodwin, B.C. (1996) *Form and Transformation: Generative and Relational Principles in Biology.* Cambridge: Cambridge University Press.

Wolpert, L. (1991) *The Triumph of the Embryo.* Oxford: Oxford University Press.

Index